MAGNETISM

Volume IIA

MAGNETISM

A Treatise on Modern Theory and Materials

Volume I:

Magnetic Ions in Insulators. Their Interactions, Resonances, and Optical Properties. 1963

Volume II (In two parts):

Statistical Models, Magnetic Symmetry, Hyperfine Interactions, and Metals. Part A. 1965

Part B in preparation

Volume III:

Spin Arrangements and Crystal Structure, Domains, and Micromagnetics. 1963

MAGNETISM

EDITED BY **George T. Rado**

Magnetism Branch
U.S. Naval Research Laboratory
Washington, D.C.

Harry Suhl

Department of Physics
University of California, San Diego
La Jolla, California

Volume II
Part A

Statistical Models,
Magnetic Symmetry,
Hyperfine
Interactions,
and Metals

1965

ACADEMIC PRESS
New York and London

Copyright © 1965, by Academic Press Inc.
ALL RIGHTS RESERVED.
NO PART OF THIS BOOK MAY BE REPRODUCED IN ANY FORM,
BY PHOTOSTAT, MICROFILM, OR ANY OTHER MEANS, WITHOUT
WRITTEN PERMISSION FROM THE PUBLISHERS.

ACADEMIC PRESS INC.
111 Fifth Avenue, New York, New York 10003

United Kingdom Edition published by
ACADEMIC PRESS INC. (LONDON) LTD.
Berkeley Square House, London W.1

LIBRARY OF CONGRESS CATALOG CARD NUMBER: 63-16972

PRINTED IN THE UNITED STATES OF AMERICA

Contributors to Volume IIA

Numbers in parentheses refer to the page on which the author's contribution begins.

Robert Brout (43), *Faculté des Sciences, Université Libre de Bruxelles, Brussels, Belgium*

C. Domb (1), *Wheatstone Laboratory, King's College, London, England*

R. J. Elliott (385), *Clarendon Laboratory, Oxford, England*

Arthur J. Freeman (167), *National Magnet Laboratory, Massachusetts Institute of Technology, Cambridge, Massachusetts*

Rosalia Guccione* (105), *Department of Physics, University of British Columbia, Vancouver, Canada*

V. Jaccarino (307), *Bell Telephone Laboratories, Murray Hill, New Jersey*

R. H. Lindquist (357), *California Research Corporation, Richmond, California*

W. Opechowski** (105), *Department of Physics, University of British Columbia, Vancouver, Canada*

A. M. Portis (357), *University of California, Berkeley, California*

Richard E. Watson (167), *Bell Telephone Laboratories, Murray Hill, New Jersey and Quantum Chemistry Group, University of Uppsala, Uppsala, Sweden*

* Present address: Department of Physics, University of Toronto, Toronto, Canada
** Present address: Lorentz Institute, University of Leiden, Leiden, Holland

Preface

This treatise attempts to provide an up-to-date and reasonably concise summary of our understanding of magnetically ordered materials. Thus it deals almost exclusively with ferromagnetism, ferrimagnetism, and antiferromagnetism, i.e., with cooperative phenomena characterized by ordered arrangements of magnetic moments subject to strong mutual interactions.

Although research in magnetism during the past fifteen years has experienced a tremendous expansion, the existing books cover only a few areas of present knowledge. Many of the available review articles are addressed to small circles of specialists, and the periodical literature is voluminous and highly dispersed. The need for a consolidation of almost all theoretical and experimental aspects of magnetically ordered materials is the motivation for the present work. It is hoped that students with physics or chemistry backgrounds as well as professionals will find this treatise useful for study and reference.

As shown by the Table of Contents, the unusually broad scope of this work includes the most diverse aspects of ferromagnetism, ferrimagnetism, and antiferromagnetism in insulators as well as in metals. The chapters range from discussions of abstract quantum mechanical and statistical models to the analysis of actual magnetic structures, from the theory of spin interactions in solids to the phenomenology of ferromagnets, and from electronic and nuclear resonance effects to neutron diffraction and optical phenomena in magnetically ordered materials. An effort was made to represent both theoretical and experimental points of view, to discuss each topic selectively rather than encyclopedically, and to incorporate in most chapters a discussion of the fundamentals. Since the most recent theories and materials are covered, several chapters deal with subjects, controversial and otherwise, which did not even exist a few years ago. While some aspects of the current technological applications of magnetism are also treated, the emphasis is on their physical basis, potentialities, and limitations.

In order to emphasize the recent developments and to cover the whole field of magnetically ordered materials, various recognized and active specialists were invited to write the chapters. Efforts were made to establish a reasonable amount of coherence among the chapters and to minimize unnecessary duplication. For practical reasons, on the other hand, no attempt was made to establish a uniform notation throughout the exposition or to achieve unity of approach and style.

The three volumes of the work bear the following partially descriptive subtitles:

Volume I: Magnetic Ions in Insulators, Their Interactions, Resonances, and Optical Properties.

Volume II: Statistical Models, Magnetic Symmetry, Hyperfine Interactions, and Metals.

Volume III: Spin Arrangements and Crystal Structure, Domains and Micromagnetics.

The publication schedule calls for Volume III to appear first, followed by Volumes I and II.

The editors wish to express their deep appreciation to the authors who prepared the chapters, in most cases even while pursuing active research programs. Thanks are also due to V. J. Folen, D. R. Fredkin, and N. R. Werthamer for editorial assistance, and to Academic Press for friendly cooperation.

G. T. Rado
H. Suhl

Contents

Contributors to Volume IIA v
Preface vii
Contents of Volume I xiii
Contents of Volume IIB (Tentative) xiv
Contents of Volume III xv

1. Statistical Mechanics of Critical Behavior in Magnetic Systems
C. Domb

I. Introduction	2
II. High-Temperature Expansions	7
III. Susceptibility of a Ferromagnet above the Curie Point . .	15
IV. Critical Behavior of Other Thermodynamic Functions . .	24
V. Antiferromagnetism	29
VI. Further Theoretical Developments Required	34
VII. Experimental Results	35
Appendix A. Zero-Field Energy and Susceptibility Coefficients for the Ising Model (Lattice)	39
Appendix B. Zero-Field Energy and Susceptibility Coefficients for the Heisenberg Model (fcc Lattice) . .	40
References	41

2. Statistical Mechanics of Ferromagnetism
Robert Brout

I. Introduction	43
II. Ising Model	47
III. Heisenberg Model	67
IV. Band Theory of Ferromagnetism	83
V. Random Spin Systems	91
Appendix A	97
Appendix B	100
References	102

3. Magnetic Symmetry
W. Opechowski and Rosalia Guccione

I. Introduction	105
II. Magnetic Groups; Definition and Method of Construction	106

    III. Construction of the Magnetic Groups 114
    IV. Invariant Spin Arrangements 133
        Appendix A.   List of Some of the Symbols Used . . . 163
        References 164

4. Hyperfine Interactions in Magnetic Materials
Arthur J. Freeman and Richard E. Watson

      I. Introduction 168
     II. Experimental Methods 182
    III. The Hartree-Fock Method in Its Conventional and
         "Unrestricted" Forms. 223
     IV. Interpretation of Measured Hyperfine Interactions . . . 259
         Appendix A.   Comments on the Solution of Conventional
         Hartree-Fock Radial Equations for Atoms 287
         Appendix B.   Hartree-Fock (r^n) Values for $3d$, $4d$, and $4f$
         Ions 290
         References 292

5. Nuclear Resonance in Antiferromagnets
V. Jaccarino

      I. Introduction 307
     II. Some General Considerations on Nuclear Resonance . . 308
    III. Nuclear Hamiltonian: Origins and Magnitudes of the
         Internal Fields 312
     IV. Temperature Dependence of the Sublattice Magnetization
         from the Time-Averaged Field 319
      V. Relaxation and Line Widths 333
     VI. Summary 347
         Appendix A.   Spin Waves on a Two-Sublattice Antiferro-
         magnet 348
         Appendix B.   The Isotropic Hyperfine Interaction Ex-
         pressed in Terms of the Spin Wave Operators 351
         Appendix C.   Importance of Spin Wave-Phonon Inter-
         actions 352
         References 353

6. Nuclear Resonance in Ferromagnetic Materials
A. M. Portis and R. H. Lindquist

      I. Introduction 357
     II. Radio-Frequency Excitation 358
    III. Field Contributions. 363
     IV. Nuclear Relaxation 376
         References 380

7. Theory of Magnetism in the Rare Earth Metals
R. J. Elliott

- I. Introduction 385
- II. Phenomenological Theory 397
- III. The Crystal Field 403
- IV. Theory of Exchange Interactions 406
- V. Conclusion 421
- References 421

Author Index 425
Subject Index 440

Contents of Volume I

Spin Hamiltonians
  K. W. H. Stevens

Exchange in Insulators: Superexchange, Direct Exchange, and Double Exchange
  P. W. Anderson

Weak Ferromagnetism
  Tôru Moriya

Anisotropy and Magnetostriction of Ferromagnetic and Antiferromagnetic Materials
  Junjiro Kanamori

Magnetic Annealing
  John C. Slonczewski

Optical Spectra in Magnetically Ordered Materials
  Satoru Sugano and Yukito Tanabe

Optical and Infrared Properties of Magnetic Materials
  Kenneth A. Wickersheim

Spin Waves and Other Magnetic Modes
  L. R. Walker

Antiferromagnetic and Ferrimagnetic Resonance
  Simon Foner

Ferromagnetic Relaxation and Resonance Line Widths
  C. Warren Haas and Herbert B. Callen

Ferromagnetic Resonance at High Power
  Richard W. Damon

Microwave Devices
  Kenneth J. Button and Thomas S. Hartwick

Author Index—Subject Index

Contents of Volume IIB (Tentative)

Exchange Interactions in Metals
    C. Herring

On s-d and s-f Interactions
    T. Kasuya

Antiferromagnetism in Metals and Alloys
    A. Arrott

Magnetism and Superconductivity
    H. Suhl

Author Index

Subject Index

Contents of Volume III

Magnetism and Crystal Structure in Nonmetals
    John B. Goodenough

Evaluation of Exchange Interactions from Experimental Data
    J. Samuel Smart

Theory of Neutron Scattering by Magnetic Crystals
    P. G. de Gennes

Spin Configuration of Ionic Structures: Theory and Practice
    E. F. Bertaut

Spin Arrangements in Metals
    R. Nathans and S. J. Pickart

Fine Particles, Thin Films and Exchange Anisotropy (Effects of Finite Dimensions and Interfaces on the Basic Properties of Ferromagnets)
    I. S. Jacobs and C. P. Bean

Permanent Magnet Materials
    E. P. Wohlfarth

Micromagnetics
    S. Shtrikman and D. Treves

Domains and Domain Walls
    J. F. Dillon, Jr.

The Structure and Switching of Permalloy Films
    Donald O. Smith

Magnetization Reversal in Nonmetallic Ferromagnets
    E. M. Gyorgy

Preparation and Crystal Synthesis of Magnetic Oxides
    C. J. Kriessman and N. Goldberg

Author Index—Subject Index

MAGNETISM

Volume IIA

1. Statistical Mechanics of Critical Behavior in Magnetic Systems

C. Domb

*Wheatstone Laboratory, King's College,
London, England*

I. Introduction	2
1. Critical Behavior in Magnetic Systems	2
2. Exact Solutions	3
3. Series Expansions	5
4. Closed-Form Approximations	6
II. High-Temperature Expansions	7
1. General Method of Deriving Expansions	7
2. Ising Model	8
3. Heisenberg Model	15
III. Susceptibility of a Ferromagnet above the Curie Point	15
1. Ising Model. Two-Dimensional Lattices	15
2. Ising Model. Three-Dimensional Lattices	18
3. Heisenberg Model	22
4. Effect of Lattice Structure	24
IV. Critical Behavior of Other Thermodynamic Functions	24
1. Critical Values of Energy and Entropy	24
2. Specific Heat	27
3. Spontaneous Magnetization and Low-Temperature Properties	29
V. Antiferromagnetism	29
1. Introduction	29
2. Susceptibility of the Ising Model ($s = \frac{1}{2}$)	30
3. Superexchange Model	32
4. Perpendicular Susceptibility	33
VI. Further Theoretical Developments Required	34
1. Ferromagnetism	34
2. Antiferromagnetism	34
VII. Experimental Results	35
1. Susceptibility of a Ferromagnet	35
2. Specific Heat and Thermodynamic Properties	37
3. Susceptibility of an Antiferromagnet	37

Appendix A. Zero-Field Energy and Susceptibility Coefficients for the Ising Model (fcc Lattice) 39
Appendix B. Zero-Field Energy and Susceptibility Coefficients for the Heisenberg Model (fcc Lattice) 40
References 41

I. Introduction

1. Critical Behavior in Magnetic Systems

Many magnetic systems exhibit characteristic singularities in their thermodynamic behavior. The best known example of such a singularity is the Curie point of a ferromagnet, which is associated with a specific heat anomaly marking the disappearance of ferromagnetism. Below the Curie temperature, T_c, the system possesses a spontaneous magnetization, and above the Curie temperature the system is paramagnetic, the magnetic susceptibility becoming infinite as the temperature approaches T_c from above. An antiferromagnet has analogous singularities, the Curie point now corresponding to the disappearance of long-range antiferromagnetic ordering. In fact, even if a system is paramagnetic over a wide range of temperature, the third law of thermodynamics tells us that at sufficiently low temperatures the entropy arising from randomness in the orientation of the spins must disappear and order will set in, and there are strong theoretical indications that this will be accompanied by thermodynamic singularities.

It has long been recognized that the source of the singularities is the mutual interaction of the magnetic carriers, and if J is the magnitude of such an interaction energy, an estimate of order of magnitude of the Curie temperature can be obtained from the relation $kT_c \sim J$. The mechanism of production of singularities was also understood as being "cooperative" in nature. For example, a ferromagnet at absolute zero consists of aligned spins; as the temperature is raised some spins overturn and break the alignment, and each overturned spin makes it easier for others to overturn; thus a "cascade" process is generated which leads to the disappearance of ferromagnetism at the Curie temperature. Early theoretical treatments, such as the Weiss "mean-field" approximation, tried to give mathematical expression of these qualitative ideas, and were able to provide a satisfactory explanation of the existence of characteristic singularities. However, the detailed predictions of the critical behavior of thermodynamic quantities near T_c did not agree with experiment.

The exact statistical mechanical solution of a particular two-dimensional model by Onsager in 1944 [1] showed clearly that such approximate treatments could furnish no reliable information regarding critical behavior, and revealed the essential complexity of the mathematical problem involved. In the next section we shall summarize the critical properties of this two-dimensional Ising model. No such exact solutions are yet available for three-dimensional models, but a great deal of effort has gone into refining approximate methods in order to make a more reliable assessment of critical behavior. Certain aspects can now be treated with confidence, whereas others require further theoretical development. It is the purpose of this article to survey the current position, and summarize the relevant conclusions in regard to critical behavior.

We shall concern ourselves solely with the Ising and Heisenberg models. For some substances these represent a satisfactory approximation, but for others they are inadequate and a collective electron treatment based on a band model is more satisfactory. However, theoretical treatments using the latter approach have not developed beyond elementary statistical approximations which cannot provide detailed information of behavior near the Curie point; hence we shall not refer to them. We shall also not be concerned with the "spherical model" [2], for although it provides exact solutions, the interaction which it uses does not correspond to physical observation, and direct comparison with experiment does not seem possible.

2. Exact Solutions

Onsager's exact solution applies to the Ising model of the rectangular lattice (i.e., interactions can be different in the two principal lattice directions) in zero external magnetic field. Many improved methods have been proposed for deriving this exact solution; they are adequately described elsewhere [3, 4], and we shall briefly summarize the relevant conclusions for the quadratic case in which the interactions in different directions are equal.

The specific heat manifests a striking singularity at the Curie point, and is reproduced in Fig. 1. It is logarithmically infinite on both sides of the Curie temperature, and possesses a large high-temperature "tail." This can be seen more clearly from the critical value of the entropy, which shows that the entropy change in the region of temperature above T_c ($= 0.387$) is greater than that in the region below T_c ($= 0.306$).

The spontaneous magnetization differs only slightly from its maximum value except in the immediate neighborhood of the Curie point; it

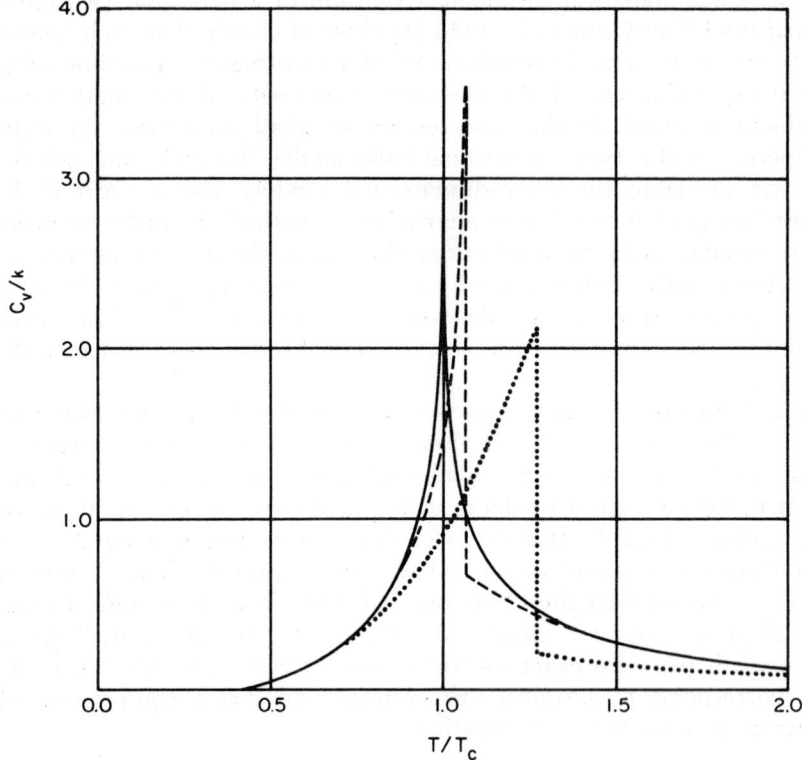

FIG. 1. Specific heat of quadratic Ising lattice (spin $\frac{1}{2}$) (after Onsager). Solid line, exact solution; dashed line, approximation of Kramers and Wannier (= Kikuchi); dotted line, approximation of Bethe. (From Domb [3], p. 291.)

decreases very rapidly to zero in the latter region, being of the order $(1 - T/T_c)^{1/8}$.

Another feature of significance provided by the solution is the behavior of the correlation between two spins as a function of distance at the critical point; it is found that at large distances k, this correlation falls off as $k^{-1/4}$.

The solution provides exact expressions for the partition function and its first derivative with respect to the external magnetic field, H, (for $H = 0$) but not for the second and higher derivatives; hence it does not tell us the critical behavior of the initial magnetic susceptibility, χ_0. However, Fisher [5] has used the correlations to show that in the neighborhood of T_c, $\chi_0 \sim [(T/Tc) - 1]^{-7/4}$ ($T > T_c$), and we shall discuss this further in Section III, 1c.

Corresponding exact solutions have been derived for a variety of other two-dimensional lattices [3]. The critical behavior is very similar to that of the Onsager solution, and critical properties are very insensitive to lattice structure. Thus, for example, the entropy at $T = T_c$ changes by less than 1% in going from the quadratic to the triangular lattice.

In the presence of a nonzero magnetic field some exact information arises from a special model of an antiferromagnet due to Fisher [6], and we shall discuss this in Section V.

3. Series Expansions

In the absence of an exact solution in closed form it is possible to derive series expansions at high and low temperatures which are exact as far as they go. The derivation of higher-order terms of such series expansions is a complex technical problem whose difficulty increases rapidly with the order of the term. However, special methods have been devised in particular cases which have enabled substantial numbers of terms to be calculated; in other cases the calculation of additional terms is very desirable, and the matter will be considered further in Section II.

Series expansions are most useful when all the terms are consistent in sign, so that the series converges right up to the singularity [6a]. The asymptotic behavior of the coefficients then provides immediate information regarding critical properties but we must have enough terms for initial irregularities to have been smoothed out. Since we have a good idea on physical grounds of the general pattern to be expected, our problem is not really one of extrapolation but of optimum estimation of parameters from the number of terms available.

The high-temperature expansions for the magnetic susceptibility in zero field of the Ising and Heisenberg models are series of positive terms, and belong to this category.* For the two-dimensional triangular lattice, for example, this susceptibility, χ_0, (Ising model $s = \frac{1}{2}$), is given by [7]

$$\frac{kT\chi_0}{Nm^2} = 1 + 6w + 30w^2 + 138w^3 + 606w^4 + 2586w^5$$
$$+ 10818w^6 + 44574w^7 + 181542w^8 + 732678w^9$$
$$+ 2935218w^{10} + 11687202w^{11} + 46296210w^{12}$$
$$+ \cdots, \tag{1.1}$$

---

* Recent work by the author and his colleagues indicates that for the Heisenberg model this is only valid for close-packed lattices [33].

where $w = \tanh(J/kT)$ (see Section II for notation). The radius of convergence of this series, w_c, corresponds to the Curie point, and is known exactly. Writing the nth term, a_n, in the form $a_n = \phi(n)/w_c{}^n$, it is possible to predict with confidence from the above data that $\phi(n) \sim An^{3/4}$, and hence that $\chi_0 \sim (1 - T_c/T)^{-7/4}$ for $T \sim T_c$. We shall deal with this further in Section III.

When the terms of a series expansion are not consistent in sign the series does not usually converge near the Curie point. Some transformation or regrouping must be sought, and it is more difficult to draw direct conclusions regarding critical behavior. However, a great step forward recently in this connection was the introduction by G. A. Baker of the Padé approximant [6b]. This can provide an analytical continuation of the function beyond the spurious singularities which have no physical significance; and it has furnished valuable information regarding critical behavior. We shall refer in more detail to the Padé approximant in Sections III and IV.

4. Closed-Form Approximations

The alternative to series expansions is the use of closed-form approximations which ignore some particular aspects of the statistical problem to be solved, thereby enabling the partition function to be calculated. The simplest approximations of this kind, the mean-field approximation, and its improvement by Bethe, while accounting qualitatively for critical behavior, provide little reliable information near the Curie point. Many ingenious methods have been devised for improving these approximations (see Chapter 4 of reference [3] for further details). However, even improved single approximations do not tell us very much, and it is only from a steady series of successive approximations that we can learn of the true thermodynamic properties of a given model.

It seems to us that the approach which lends itself most readily to such a series of approximations is that of Yvon [8], which makes use of a cluster integral development. Each cluster integral can be represented by a diagram of points and bonds on the lattice, and by restricting the total number of points or bonds, we obtain a natural series of closed-form approximations. The method was applied by Fournet to a number of problems [9a], but the approximations were not pursued sufficiently for the conclusions regarding critical behavior to be very reliable. Recently Domb and Hiley [9b] have extended the method, and we shall reproduce some of their results regarding the susceptibility of the Ising model in Section III, 1.

Domb and Hiley's work demonstrates clearly that when series expansions of positive terms are available they are better to use than series of closed-form approximations. But when the terms in the series expansion are not consistent in sign, closed-form approximations can be useful.

II. High-Temperature Expansions

1. General Method of Deriving Expansions

For the Ising model of spin s the Hamiltonian may be put in the form

$$\mathcal{H}^I = -\frac{J}{s^2} \sum_{i,j} s_{zi} s_{zj} - \frac{mH}{s} \sum_i s_{zi}. \tag{2.1}$$

Here s_{zi}, the z-components of spin, can take integral values $-s$, $-(s-1)$, ..., $(s-1)$, s, and $\pm J$ are the maximum and minimum energies of a pair of interacting spins; the maximum magnetic moment of a spin is m; the sum i, j is taken over all nearest-neighbor pairs in the lattice, and the sum i is taken over all spins in the lattice. The atoms in the lattice are connected cyclically in the usual manner to avoid boundary effects. This choice of interaction parameters ensures that for given J and m the maximum internal energy and saturation magnetization remain constant as s varies, and hence a direct comparison is possible of the properties of the model for different values of s. For the Heisenberg model $s_{zi}s_{zj}$ must be replaced by $\mathbf{s}_i \cdot \mathbf{s}_j$ ($= s_{xi}s_{xj} + s_{yi}s_{yj} + s_{zi}s_{zj}$), where $\mathbf{s}$ denotes the vector spin operator.

The partition function is given by

$$Z_N(\beta, H) = \langle e^{-\beta\mathcal{H}} \rangle = 1 - \beta \langle \mathcal{H} \rangle + \frac{\beta^2 \langle \mathcal{H}^2 \rangle}{2!} + \cdots \frac{(-\beta)^r}{r!} \langle \mathcal{H}^r \rangle + \cdots \tag{2.2}$$

($\langle \ \rangle$ denotes $(2s+1)^{-N}$ trace, $\beta = 1/kT$), where the rth term in the expansion will be a polynomial of order r in N. This is the analog of a moment expansion in statistics. We should expect on physical grounds that for sufficiently large N,

$$Z_N(\beta, H) \sim [Z(\beta, H)]^N, \tag{2.3}$$

where $Z(\beta, H)$ does not depend on N. If formally we take the logarithm of Eq. (2.2) we obtain the analog of a cumulant expansion

$$\ln \langle e^{-\beta\mathcal{H}} \rangle = -\beta \langle \mathcal{H} \rangle + \frac{\beta^2}{2!} [\langle \mathcal{H}^2 \rangle - \langle \mathcal{H} \rangle^2]$$
$$- \frac{\beta^3}{3!} [\langle \mathcal{H}^3 \rangle - 3 \langle \mathcal{H}^2 \rangle \langle \mathcal{H} \rangle + 2 \langle \mathcal{H} \rangle^3]$$
$$+ \cdots \tag{2.4}$$

and each term is now of order N. Thus, by comparison with Eq. (2.3), we are furnished with an expansion in powers of β of $\ln Z(\beta, H)$; an alternative formal method [9] of obtaining the rth term in Eq. (2.4) is to pick out the coefficient of N in the corresponding term in Eq. (2.2).

A justification of this formal procedure would seem to lie in selecting the maximum term in the expansion (2.2), and following an argument similar to that in the Mayer cluster integral theory. There is little doubt that the results to which the method leads are correct; it was first introduced by Opechowski [9] who calculated four terms of the expansion for the Heisenberg model of spin $\tfrac{1}{2}$ (a serious error was corrected by Zehler [10]). A number of authors have contributed to the subsequent development, and to the application to other models (see reference [3] for more detailed references).

The contributions to $\langle \mathscr{H}^r \rangle$ can be divided into various groups corresponding to different types of linear graph. Two basic calculations are then required.

(a) The number of independent graphs of a particular type which can be formed from r nearest-neighbor links of the lattice.

(b) The mean value of the spin operators in Eq. (2.1) for the particular graph.

These calculations are the limiting factor in the evaluation of higher-order terms in the expansions (2.2) and (2.4). For certain models simplifying features arise in (a) and (b) and the expansions can be pushed quite a long way. For others the complexity increases rapidly and fewer terms are available. We now proceed to discuss the matter in more detail.

2. Ising Model

a. *Ising model* ($s = \tfrac{1}{2}$). The greatest simplification arises for the simple model ($s = \tfrac{1}{2}$) for which the Hamiltonian (2.1) can be written in the form

$$\mathscr{H}^I = -J \sum_{i,j} \sigma_i \sigma_j - mH \sum_i \sigma_i, \tag{2.5}$$

where the σ_i are variables taking on values ± 1 ($s_{zi} = \tfrac{1}{2} \sigma_i$). The partition function for any given lattice can now be written in the form

$$Z_N^I(\beta, H) = \sum_{\sigma_i, \sigma_j = \pm 1} \exp \left\{ \beta \left[J \sum_{i,j} \sigma_i \sigma_j + mH \sum_i \sigma_i \right] \right\}, \tag{2.6}$$

where the outside sum must be taken over 2^N possible values of the σ_i for all the N lattice points, and i, j are nearest-neighbors; we can thus consider the first term in Eq. (2.6) as a sum over all nearest-neighbor links in the lattice. Clearly the σ_i variables commute, and Eq. (2.6) can be written as a product:

$$Z_N{}^I(\beta, H) = \sum_{\sigma_i, \sigma_j = \pm 1} \prod_{i,j} \exp(K\sigma_i\sigma_j) \prod_i \exp(\beta m H \sigma_i) \qquad (K = \beta J = J/kT). \tag{2.7}$$

The $\sigma_i\sigma_j$ satisfy the relation

$$(\sigma_i\sigma_j)^2 = (\sigma_i\sigma_j)^4 = \cdots = 1, \qquad (\sigma_i\sigma_j) = (\sigma_i\sigma_j)^3 = (\sigma_i\sigma_j)^5 = \cdots,$$

and hence we can write

$$\exp(K\sigma_i\sigma_j) = \cosh K + \sigma_i\sigma_j \sinh K. \tag{2.8}$$

We can expand the first product in Eq. (2.7) as follows [11]:

$$\prod_{i,j}(\cosh K + \sigma_i\sigma_j \sinh K) = \prod_{i,j}[1 + \sigma_i\sigma_j \tanh K]$$

$$= \cosh^{\mathcal{N}} K \sum_{ij\ldots kl} [1 + (\tanh K)(\sigma_i\sigma_j)$$

$$+ \tanh^2 K(\sigma_i\sigma_j)(\sigma_k\sigma_l) + \cdots], \tag{2.9}$$

where $\mathcal{N}$ is the number of nearest-neighbor links in the lattice. This can now be interpreted topologically. The first term $\sigma_i\sigma_j$ corresponds to any one of $\mathcal{N}$ nearest-neighbor links in the lattice; the second term corresponds to any nonidentical pairs of nearest-neighbor links, and there will be $\mathcal{N}(\mathcal{N}-1)/2!$ such pairs; similarly for higher terms (see Fig. 2).

The simplification in this model springs from the relation (2.8), which ensures that the graphs to be counted contain only single bonds; thus the complexity is substantially reduced. The mean value calculation for a given graph is also very straightforward. Each lattice point, i, will give rise to a factor $\sigma_i{}^a$ where a is the number of bonds meeting at i (including zero for unoccupied lattice sites). If we now sum over the second factor in Eq. (2.7) we obtain $(\mu^{-1/2} + \mu^{1/2})$ when a is even and $(\mu^{-1/2} - \mu^{1/2})$ when a is odd [$\mu = \exp(-2\beta mH)$]. Hence, taking out a factor $(\mu^{-1/2} + \mu^{1/2})^N$, we may write

$$Z_N{}^I(\beta, H) = (\mu^{-1/2} + \mu^{1/2})^N (\cosh K)^{\mathcal{N}}[1 + \tanh K\, \Phi_1(\mathcal{N}, \tau)$$

$$+ \tanh^2 K \Phi_2(\mathcal{N}, \tau) + \cdots + \tanh^r K \Phi_r(\mathcal{N}, \tau) + \cdots], \tag{2.10}$$

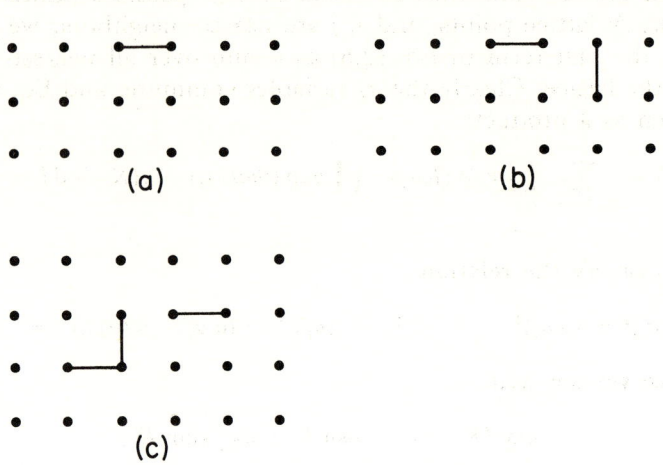

FIG. 2. Typical contributions to the high temperature expansion. (a) First term $\mathcal{N}$ contributions; (b) second term $\mathcal{N}(\mathcal{N}-1)/2!$ contributions; (c) third term $\mathcal{N}(\mathcal{N}-1)(\mathcal{N}-2)/3!$ contributions. (From Domb [3], p. 176.)

where $\tau = (1-\mu)/(1+\mu)$. Any given graph contributes a factor τ^b, where b is the number of odd vertices in the graph, and unoccupied sites can be ignored.

When $H = 0$, $\tau = 0$ and the only nonzero contributions are from graphs all of whose vertices are even (zero-field configurations, Fig. 3a). Hence we can write

$$Z_N{}^I(\beta, 0) = 2^N (\cosh K)^{\mathcal{N}} \left[1 + \sum_{l=1}^{\mathcal{N}} p(N, l) \tanh^l K \right], \qquad (2.11)$$

where $p(N, l)$ is the number of closed graphs all of whose vertices are even which can be constructed with l bonds of the lattice.

The susceptibility in zero field, χ_0, is obtained from

$$(\partial^2/\partial H^2) [\ln Z_N{}^I]_{H=0},$$

and is hence derived from the coefficient of τ^2, or from graphs with two odd vertices (susceptibility configurations, Fig. 3b). It is easy to show that

$$\chi_0 = \frac{Nm^2}{kT} \left[1 + 2 \sum_{l=1}^{\infty} b_l \tanh^l K \right], \qquad (2.12)$$

where b_l is the coefficient of N in the sum of susceptibility configurations of l lines.

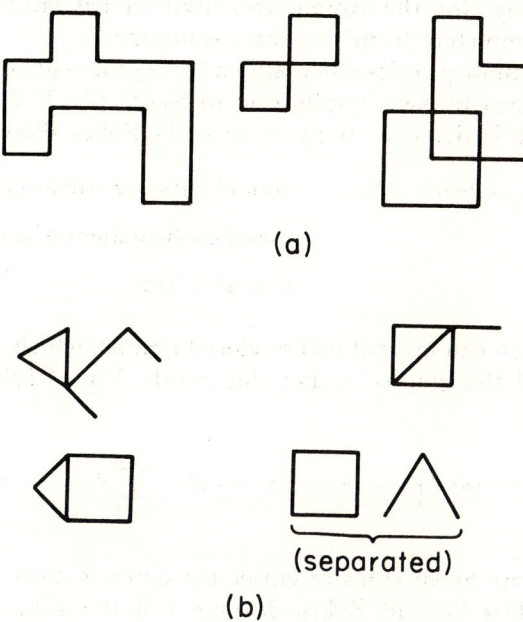

Fig. 3. Typical zero-field and susceptibility configurations (Ising model spin $\frac{1}{2}$). (a) Zero-field configurations have all even vertices. (b) Susceptibility configurations have two odd vertices. (From Domb [3], pp. 177, 323.)

Systematic methods of calculating the number of graphs of various types for simple lattices are outlined in Chapter 5 of reference [3]. They can all be expressed algebraically as functions of certain "lattice constants," which are the numbers corresponding to closed graphs. As examples of zero-field series we quote the terms which have been derived for the sc [12] and fcc lattices [13]:

sc:
$$Z^I(\beta, 0) = 2(\cosh K)^3[1 + 3w^4 + 22w^6 + 192w^8 + 2046w^{10} \\ + 24853w^{12} + \cdots] \quad (2.13)$$

fcc:
$$Z^I(\beta, 0) = 2(\cosh K)^6[1 + 8w^3 + 33w^4 + 168w^5 + 962w^6 + 5928w^7 \\ + 38907w^8 + 268056w^9 + \cdots].$$

Here $Z^I(\beta, 0)$ is the partition function per lattice site, i.e.,
$$\lim_{N \to \infty} [Z_N^I(\beta, 0)]^{1/N}.$$

Needless to say, for the simple two-dimensional lattices these series are known completely from the exact solutions.

The susceptibility series calculations (2.12) for regular* lattices were greatly advanced by two results due to Sykes [7]. If the coordination number of the lattice is q, and $\sigma = (q-1)$, Sykes showed, first that

$$b_{l+1} - 2\sigma b_l + \sigma^2 b_{l-1} = \text{Sum of lattice constants of closed configurations of order } (l+1) \text{ or less.} \quad (2.14)$$

Hence attention can be confined to closed graphs, which are only a small fraction of all the graphs. Using this result, Eq. (2.12) can be put in the form

$$\frac{kT\chi_0}{Nm^2} = (1 - \sigma w)^{-2} \left[1 - (\sigma - 1)w - \sigma w^2 + \sum_{l=3}^{\infty} d_l w^l \right] \quad (w = \tanh K), \quad (2.15)$$

where the d_l are linear sums of lattice constants of closed configurations of order l or less. Second, Sykes decomposed the series $\sum d_l w^l$ into the following:

$$\sum_{3}^{\infty} d_l w^l = \frac{2wE(w)}{J} + (\sigma + 1)w^2 + 8(1+w)^2 \sum_{l=5}^{\infty} g_l w^l, \quad (2.16)$$

where $E(w)$ represents the internal energy, and the g_l involve a much sparser set of lattice constants. For simple two-dimensional lattices $E(w)$ is known exactly, and the relation (2.16) has enabled the susceptibility series to be substantially extended; 15 terms have been evaluated for the quadratic lattice, 24 for the honeycomb lattice, and 12 for the triangular lattice [quoted in Eq. (1.1)]. These terms are adequate in practice for the calculation of the susceptibility right up to the Curie point.

Fewer terms are available for the three-dimensional lattices [14] but they show steady behavior and enable useful conclusions to be drawn. We quote as an example the series for the face-centered cubic lattice:

$$\frac{kT\chi_0}{Nm^2} = 1 + 12w + 132w^2 + 1404w^3 + 14652w^4 + 151116w^5$$
$$+ 1546332w^6 + 15734460w^7 + 159425580w^8 + \cdots . \quad (2.17)$$

---

* A regular lattice is one in which all sites and bonds are equivalent.

b. Ising model (general s). The elimination of multiple bonds is possible only for $s = \frac{1}{2}$, and for general s more complex configurations such as those shown in Fig. 4 must be taken into account. The number of configurations increases with great rapidity and soon limits the practicability of the calculation. For the particular case $s = 1$ a relation

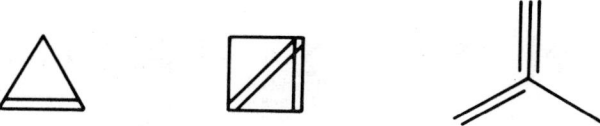

FIG. 4. Typical configurations with multiple bonds. (From Domb [3], p. 325.)

similar to Eq. (2.8) can be established, and many more terms could be calculated; this does not seem to have been properly exploited so far.

The partition function of the Ising model with general s can be written

$$Z_N{}^I(\beta, H) = \sum_{s_{zi}=-s}^{s} \prod_{i,j} \exp(4K s_{zi} s_{zj}) \prod_{i} \exp(2L s_{zi}). \qquad (2.18)$$

Here $K = \beta J/4s^2$, $L = \beta m H/2s$, so as to maintain consistency with our previous notation for $s = \frac{1}{2}$. We now expand each term of the first product as follows:

$$\exp(4K s_{zi} s_{zj}) = 1 + 4K s_{zi} s_{zj} + \frac{(4K)^2}{2!}(s_{zi} s_{zj})^2 + \cdots$$
$$+ \frac{(4K)^r}{r!}(s_{zi} s_{zj})^r + \cdots. \qquad (2.19)$$

When we substitute Eq. (2.19) into (2.18) and pick out the term in $(4K)^l$, we shall obtain a contribution from each possible configuration of l lines; every such configuration is represented by an appropriate product of $s_{zi}, s_{zj} \ldots s_{zk}$, there being one $s_{zi} s_{zj}$ for each bond. We must now multiply these products by $\prod_i \exp(2L s_{zi})$ and sum over values of s_{zi} equal to $-s, -(s-1), \ldots, (s-1), s$.

Let us define

$$t_0 = \sum_{s_{zi}=-s}^{s} \exp(2L s_{zi}) = \mu^{-s} + \mu^{-(s-1)} + \cdots + \mu^{(s-1)} + \mu^{s}$$

$$t_a = \sum_{s_{zi}=-s}^{s} s_{zi}^a \exp(2L s_{zi}) = s^a \mu^{-s} + (s-1)^a \mu^{-(s-1)} + \cdots$$
$$+ (-s+1)^a \mu^{(s-1)} + (-s)^a \mu^s$$
$$\mu = \exp(-2L) = \exp(-\beta m H/s). \qquad (2.20)$$

Then in any configuration each free vertex gives rise to a factor t_0, and each vertex which is the meeting point of a lines gives rise to a factor t_a. Let C_l be the number of occurrences of a given configuration on the lattice, and suppose that the bonds of this configuration have multiplicities $\alpha, \beta, \gamma, \ldots$, and that the vertices have multiplicities $a, b, c, \ldots$ $[\alpha + \beta + \gamma + \cdots = \frac{1}{2}(a + b + c + \cdots) = l]$. We can then write

$$Z_N^I(\beta, H) = t_0^N \left[1 + \sum_{l=1}^{\infty} (4K)^l \sum_{\substack{\text{configuration} \\ \text{with } l \text{ lines}}} \frac{C_l \tau_a \tau_b \tau_c \cdots}{\alpha! \beta! \gamma!} \right] \quad \left(\tau_a = \frac{t_a}{t_0} \right). \tag{2.21}$$

In the absence of a magnetic field τ_a are zero for odd a, and only configurations all of whose vertices are even contribute. For the zero-field susceptibility, configurations with not more than two odd vertices contribute. To evaluate this susceptibility we must expand each τ_a as far as the second term in H, and pick out the coefficient of H^2 in Eq. (2.21). The coefficients in these expansions are given by successive derivatives of $t_0 = [\mu^{-s} - \mu^{(s+1)}/(1 - \mu)]$ at $\mu = 1$, and can readily be obtained from the Taylor expansion of t_0 about $\mu = 1$. For example,

$$\tau_1 = -\frac{2X}{3} L + \cdots$$

$$\tau_2 = \frac{X}{3} \left[1 + \frac{8X - 6}{15} L^2 + \cdots \right] \qquad X = s(s+1) \tag{2.22}$$

$$\tau_3 = -\frac{2X}{3} \left[\frac{3X - 1}{5} L + \cdots \right]$$

The zero-field partition function can be conveniently written in the form

$$\ln Z_N^I(\beta, 0) = N \ln (2s + 1) + N \frac{q}{2} \sum_{r=2}^{\infty} c_r(s) K^r / r! s^{2r}, \tag{2.23}$$

and the zero-field susceptibility, χ_0, is given by

$$\frac{3s^2 \chi_0}{X \beta N m^2} = \sum_{r=0}^{\infty} h_r(s) K^r / s^{2r}. \tag{2.24}$$

The calculations of Domb and Sykes [15, 18a] determine c_r up to c_8 and h_r up to h_6 for a variety of lattices. We shall confine our attention to the fcc lattice for which convergence is most rapid; the coefficients for this lattice are tabulated in Appendix A as functions of s. It will be noted that the coefficient of $(J/kT)^r$ remains finite as $s \to \infty$.

3. Heisenberg Model

For the Heisenberg model $s_{zi}s_{zj}$ in Eqs. (2.18) and (2.19) must be replaced by $\mathbf{s}_i \cdot \mathbf{s}_j$. The configurational problem is identical with that of the previous section, but the mean value calculation is much more difficult since it involves evaluating the traces of products of non-commuting spin operators. For each configuration all possible permutations of the factors must be considered, and, despite some simplification due to symmetry, the work is very lengthy.

Some simplification arises when $s = \tfrac{1}{2}$, but the number of terms available so far is no greater than for general s.[*] The most comprehensive calculations are due to Rushbrooke and Wood [16] (see also Section 5.3.3 in reference [3]); since they are described in detail we shall not reproduce them here. However, we again quote in Appendix B the zero-field and susceptibility coefficients c_r, h_r for the fcc lattice, the form of the expansions remaining as in Eqs. (2.23) and (2.24).

III. Susceptibility of a Ferromagnet above the Curie Point

1. Ising Model. Two-Dimensional Lattices

a. *Estimate of critical behavior from series expensions.* In Eq. (1.1) we quoted the first 12 terms of the susceptibility expansion for the triangular lattice ($s = \tfrac{1}{2}$), and pointed out that the radius of convergence of this series, w_c, was known exactly $(2 - \sqrt{3})$. Thus if we write $a_n [= 2b_n$, Eq. (2.12)$] \sim \varphi(n)/w_c^n$ we know that

$$\lim_{n \to \infty} [\varphi(n)]^{1/n} = 1. \tag{3.1}$$

It is natural to try to fit $\varphi(n)$ by an expression of the form

$$\varphi(n) \sim An^g. \tag{3.2}$$

To assess whether this estimation is in reasonable accord with the numerical data, it is convenient to plot a_n/a_{n-1} as a function of $1/n$. If expression (3.2) is valid,

$$a_n/a_{n-1} \simeq (1/w_c)(1 + g/n), \tag{3.3}$$

and hence we should approach a straight line whose slope determines g. In Fig. 5 we have plotted a_n/qa_{n-1} a function of $1/n$,[†] and it will be seen

---

[*] Recently a new method has been introduced which has already added two new terms for loose packed lattices [24a, 33].

[†] q is a scaling factor useful in the comparison of different lattices.

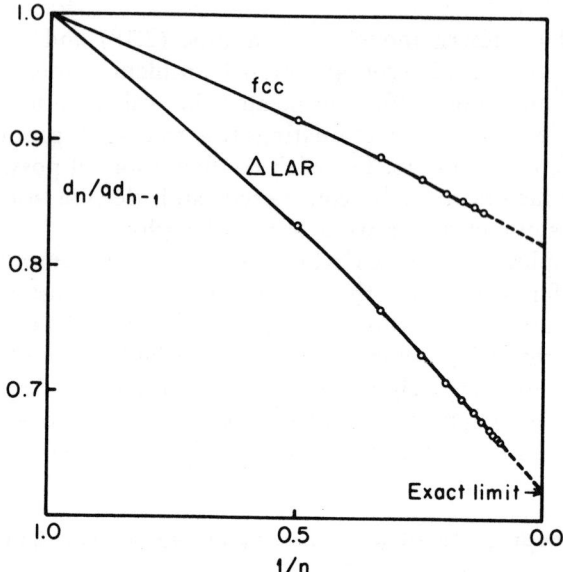

FIG. 5. Ising model. Successive ratios in the susceptibility expansions of the triangular and fcc lattices as functions of $1/n$. (From Domb and Sykes [14].)

that, after initial variation, the slope settles down to a steady value; numerical analysis gives strong indication that this value is $\frac{3}{4}$ [14]. If Eq. (3.2) is satisfied, the asymptotic form of the series near w_c is $(1 - w/w_c)^{-g-1}$; in fact, this binomial expansion satisfies Eq. (3.3) exactly, and the close fit, even for relatively small n, indicated by Fig. 5 shows that such a function provides a good approximation for the susceptibility.

For the simple quadratic (sq) lattice, although more terms are available, there is a marked oscillation between "odd" and "even" ratios. Nevertheless, it is possible to smooth this effect [14], and for this lattice one can with equal confidence conclude that near the Curie temperature $\chi_0 \simeq C(1 - T_c/T)^{-7/4} \, (T > T_c)$.

b. Exact argument based on the correlation functions. The magnetic susceptibility, χ, is given by the thermodynamic formula

$$\chi = \frac{1}{\beta} \frac{\partial^2}{\partial H^2} [\ln Z_N(\beta, H)] = \frac{1}{\beta} \left[\frac{\partial^2 Z_N/\partial H^2}{Z_N} - \left(\frac{\partial Z_N/\partial H}{Z_N} \right)^2 \right]. \qquad (3.4)$$

If we apply this formula at $H = 0$ to a ferromagnet above the Curie temperature, the second term vanishes since there is no spontaneous magnetization. Substituting in Eq. (2.6), we find that

$$\frac{\chi_0}{\beta m^2} = \frac{\sum_p \sigma_p \sum_q \sigma_q \exp(\beta J \sum_{i,j} \sigma_i \sigma_j)}{\exp(\beta J \sum_{i,j} \sigma_i \sigma_j)} = \sum_{i,j} \langle \sigma_i \sigma_j \rangle, \qquad (3.5)$$

$\langle \sigma_i \sigma_j \rangle$ representing the correlation between the spins on the ith and jth lattice sites in the absence of a magnetic field.

The work of Kaufman and Onsager [17] (Section 3.5.2 of [3]) showed how to calculate these correlations but did not provide complete information over all ranges of temperature and distance which are required for the calculation of χ_0 from Eq. (3.5). However, Fisher [5] pointed out that sufficient information was available for the critical behavior of χ_0 to be determined.

Kaufman and Onsager had shown that at the Curie point the correlation along a row between two spins b sites apart was asymptotically of the form $A/b^{1/4}$; there was further evidence that the dependence on distance was similar in other directions in the lattice, and that we might write more generally for two sites distance $\mathbf{k}$ apart

$$\langle \sigma_0 \sigma_\mathbf{k} \rangle \sim A(\theta)/k^{1/4}, \qquad (3.6)$$

where θ is an angle made with a reference direction on the lattice. To generalize this to temperatures above the Curie temperature we may write

$$\langle \sigma_0 \sigma_\mathbf{k} \rangle \sim A(\theta) k^{-1/4} \exp[-f_\mathbf{k}(T)] \qquad (T \geqslant T_c) \qquad (3.7)$$

where $f_\mathbf{k}(T_c) = 0$ and $f_\mathbf{k}(T) > 0$ for $T > T_c$. By considering the propagation of short-range order above T_c from the matrix viewpoint it can be shown that this is determined by the ratio of the second largest to the largest eigenvalue (Section 3.3.1 of [3]); and that approximately

$$f_\mathbf{k}(T) \simeq k g(T) \qquad (k \to \infty), \qquad (3.8)$$

where near the Curie temperature

$$g(T) \simeq d(1 - T/T_c). \qquad (3.9)$$

Substituting Eqs. (3.7)–(3.9) into (3.5), and replacing the sums by integrals, we find

$$\frac{\chi_0}{\beta m^2} = N \int_0^\infty dk \int_0^{2\pi} d\theta \, A k^{3/4} \exp[-kd(1 - T_c/T)]$$
$$= C(1 - T_c/T)^{-7/4}. \qquad (3.10)$$

Some refinement of the argument is needed to take account of the remaining spectrum of eigenvalues, but the result (3.10) remains essentially unchanged.

2. Ising Model. Three-Dimensional Lattices

a. *Series expansion* ($s = \frac{1}{2}$). The method of analysis in Section III, 1a can equally well be applied to three-dimensional lattices, but now w_c is not known exactly and must also be estimated numerically from the data. Although fewer terms are available for three-dimensional lattices the convergence is more rapid, as will be seen from the fcc data in Fig. 5. Domb and Sykes [14] concluded from numerical analysis that the appropriate g value for three-dimensional lattices is $\frac{1}{4}$, and hence that

$$\chi_0 \sim C'(1 - T_c/T)^{-5/4}. \qquad (3.11)$$

Having estimated g, they suggested that a better estimate of w_c could be obtained by plotting

$$\beta_n = na_n/qa_{n-1}(n+g) \qquad (3.12)$$

against $1/n$. By virtue of Eq. (3.3), β_n tends to $1/qw_c$ as $n \to \infty$; the additional factor $(n+g)$ has the effect of straightening out the limit horizontally. Even if the estimate of g is in error, the limit of β_n is still $1/qw_c$, but the approach to this limit is not quite horizontal.

In Fig. 6A the successive β_n are plotted against $1/n$ for the simple quadratic lattice with $g = \frac{3}{4}$. The values $\frac{1}{2}\%$ above and below the exact limit (0.60355 denoted by β_a) have been marked, and the last seven values are all between these limits.

In Fig. 6B and Fig. 6C the corresponding β_n for the simple cubic and body-centered cubic lattices are plotted taking $g = \frac{1}{4}$. It will be seen that the behavior is very regular, and remarkably similar for the two lattices, suggesting that the value $g = \frac{1}{4}$ is exact. Domb and Sykes considered that the critical values for these two lattices, denoted by β_b and β_c, respectively. probably lay between the pairs of curves in Fig. 6B and 6C. Their final estimates for three-dimensional lattices were as follows:

$$1/qw_c = \begin{cases} 0.7640 \pm 0.0010 & \text{sc} \\ 0.8004 \pm 0.0010 & \text{bcc} \\ 0.8192 \pm 0.0010 & \text{fcc.} \end{cases} \qquad (3.13)$$

The results of the last two sections are confirmed in a remarkable manner by the use of the Padé approximant [6b]. The $[N, M]$ approxi-

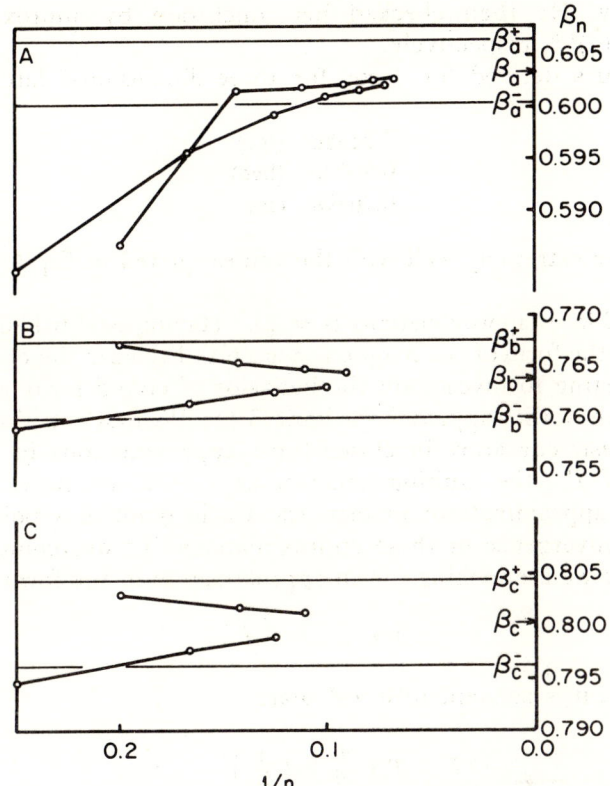

Fig. 6. Ising model. Estimation of critical temperatures by extrapolation of β_n. $\beta_n = na_n/ga_{n-1}(n+g)$. A. Simple quadratic lattice; $g = 3/4$. Exact limit $\beta_a = 0.6036$. $\beta_a^+ = 1.005\beta_a$, $\beta_a^- = 0.995\beta_a$. B. Simple cubic lattice; $g = 1/4$. Extrapolated limit $\beta_b = 0.7640$. $\beta_b^+ = 1.005\beta_b$, $\beta_b^- = 0.995\beta_b$. C. Body-centered cubic lattice; $g = 1/4$. Extrapolated limit $\beta_c = 0.8004$. $\beta_c^+ = 1.005\beta_c$, $\beta_c^- = 0.995\beta_c$. (From Domb and Sykes [14].)

mant to a function seeks an approximation of the form $P(z)/Q(z)$ where $P(z)$ and $Q(z)$ are polynomials of degrees M, N, respectively. A particularly useful example is the $[N, N]$ approximant which is invariant under the group of homographic transformations $z = Aw/(1 + Bw)$.

The Padé approximant can be particularly useful in representing integral functions whose only singularities are poles, and these can be located as zeros of $Q(z)$. The form $(1 - w/w_c)^{-h}$ for the susceptibility χ suggests that $(d/dw) \ln \chi$ should have a simple pole whose residue is equal to h. Baker [6b] found that values of h derived on this basis were extremely close to 7/4 and 5/4 for two- and three-dimensional lattices, respectively. He then checked his conclusion by approximating to $(\chi)^{4/7}$ and $(\chi)^{4/5}$, respectively.

The values derived for $1/qw_c$ for three-dimensional lattices are as follows:

$$\begin{aligned} &0.81886 \quad \text{(fcc)} \\ &0.80036 \quad \text{(bcc)} \\ &0.76398 \quad \text{(sc)}. \end{aligned} \tag{3.13a}$$

These agree extremely well with the values quoted in Eq. (3.13).

b. Closed-form approximations ($s = \frac{1}{2}$). Having established with some confidence the form of the magnetic susceptibility near the critical point, it is interesting to investigate the behavior of closed-form approximations to see how they approach the limit (3.11). We first note that although $1/\chi_0$ manifests curvature in closed-form approximations in the neighborhood of T_c, the limiting behavior as $T \to T_c$ is always linear, for any given approximation; in fact, the Curie point is a point of nonuniform convergence of these approximations. To overcome this difficulty, since we are looking for an approximation of the form

$$1/\chi_0 \sim (T - T_c)^\gamma, \tag{3.14}$$

as $T \to T_c$, it is convenient to calculate

$$(T - T_c) \frac{d}{dT} \ln \left(\frac{1}{\chi_0} \right) = \gamma(T) \tag{3.15}$$

in the neighborhood of $T = T_c$.

The curves of $\gamma(T)$ against $T^*(= T/T_c)$ reproduced in Fig. 7 are due to Domb and Hiley [9b], and represent successive closed-form approximations in the method of Yvon for the fcc lattice. In any particular approximation $\gamma(T)$ rises to a maximum as T^* decreases and then drops to 1 when T^* becomes unity; the approximation can only be regarded as reasonable above the maximum. As the order of the approximation increases the maximum should approach the limiting value of $\gamma(T)$, i.e., 1.25. The curve E is the expected behavior based on series expansions.

It will be seen that any one individual approximation provides rather sparse information. But extrapolation based on a steady series of approximations can provide an estimate which is not too far removed from the true value. It is clear that in this case series expansions lead more readily to reliable conclusions, although it is encouraging to find that the behavior of closed-form approximations is consistent with these conclusions.

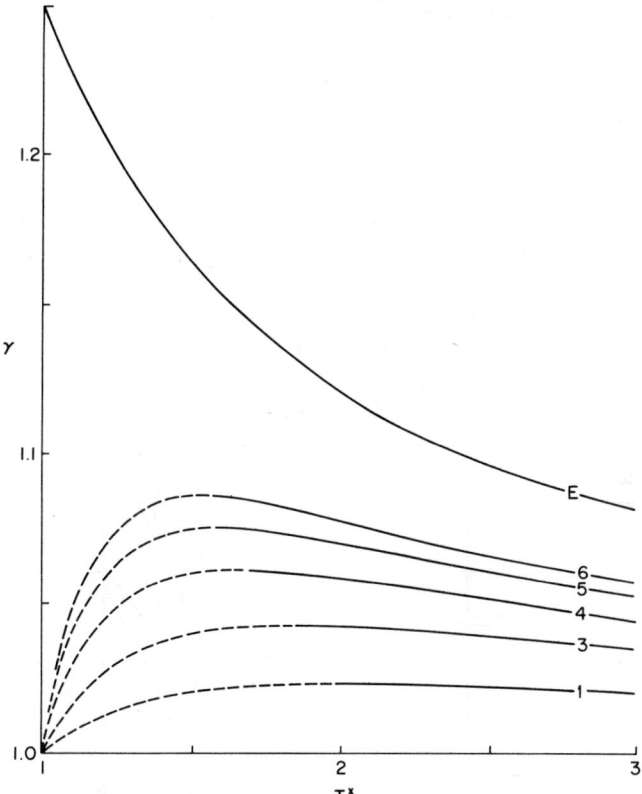

FIG. 7. Use of successive closed-form approximations to estimate critical behavior. Curve E is an estimate based on series expansions and should be very close to the exact value.

c. General s. The series expansion (2.24) for the fcc lattice was used by Domb and Sykes [15] to investigate the effect of change of s on the form of the susceptibility near the Curie point. If only two terms of the expansion are taken into account the approximation to the Curie temperature is given by

$$kT_c = qJ(s+1)/3s. \qquad (3.16)$$

This naturally suggests using the quantity $u_n = 3a_n s/(s+1)qa_{n-1}$, and seeing how this varies with s. Domb and Sykes tabulated the ratios

$$R_n(1) = u_n(s=1)/u_n(s=\tfrac{1}{2})$$
$$R_n(\infty) = u_n(s=\infty)/u_n(s=\tfrac{1}{2}) \qquad (3.17)$$

and found that they converged rapidly to limits

$$R_n(1) \to 1.0424$$
$$R_n(\infty) \to 1.0709. \tag{3.18}$$

Hence we can write

$$T_c(s)/T_c(\tfrac{1}{2}) = e(s)(s+1)/3s \tag{3.19}$$

where $e(\tfrac{1}{2}) = 1$, $e(1) = 1.0424$, and $e(\infty) = 1.0709$, so that $e(s)$ is a slowly varying function. The convergence of the ratio $R_n(s)$ to a limit as a geometrical progression shows that the asymptotic form (3.11) established for $s = \tfrac{1}{2}$ remains valid for all s. The rapidity of convergence leads us to expect that the detailed shape of the susceptibility curve will be insensitive to changes in s. Figure 8 shows the magnitude of the change to be expected as s varies from $\tfrac{1}{2}$ to ∞.

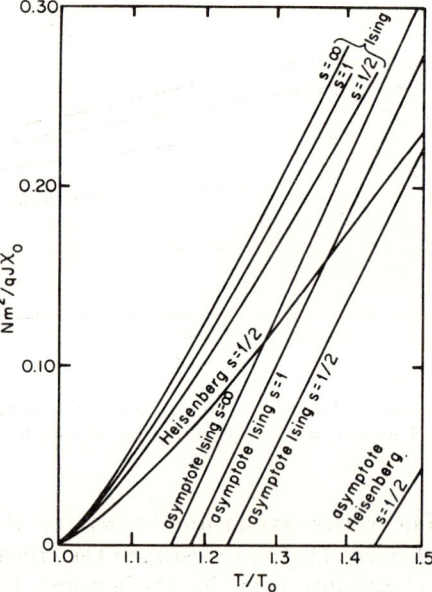

FIG. 8. Reciprocal susceptibility of face-centered cubic for different interactions. (From Domb and Sykes [15].)

3. Heisenberg Model

The terms in the high-temperature susceptibility expansion for the Heisenberg model remain positive (see [33]) but they are less regular in behavior than the Ising model. From spin wave theory valid at low temperatures [4] we should expect the Curie point to be at $T = 0$ for all two-dimensional lattices. The first few terms of the series expansions show no clear evidence of such behavior, and it is evident that higher-order terms play a dominant part in determining critical properties. Physically this means that at higher temperatures the behavior is quite similar to the Ising model, and it is only when the inverse susceptibility becomes small that it curves sharply to cut the axis at $T = 0$.

The three-dimensional lattices improve in regularity as the coordination increases from the sc through the bcc to the fcc for which the terms are quite smooth. All the Curie temperatures for the Heisenberg model are lower than the corresponding temperatures for the Ising model; for example, the estimates of Domb and Sykes [15] are

$$kT_c/qJ = \begin{cases} \text{fcc} & \text{Ising} \quad \text{Heisenberg} \\ \text{fcc} & 0.816 \quad 0.695 \\ \text{bcc} & 0.794 \quad 0.659 \\ \text{sc} & 0.752 \quad 0.61. \end{cases} \quad (3.20)$$

An analysis of the behavior of successive terms for the Heisenberg model (fcc, $s = \frac{1}{2}$) indicates that the curvature of the inverse susceptibility does not differ greatly from the Ising model, and an asymptotic behavior of the form

$$\chi_0 = B(1 - T_c/T)^{-1.33} \quad (3.21)$$

seemed consistent with the numerical data. The difference between the reciprocal susceptibility on the Ising and Heisenberg models is illustrated in Fig. 8.

The behavior of the Heisenberg model for general s is discussed in detail by Rushbrooke and Wood [16]. They give a useful formula providing an approximation to the Curie point for a variety of lattices and spins as follows:

$$kT_c/J = 5(q - 1)(11X - 1)/192s^2. \quad (3.22)$$

Again they find that change of s makes little difference to the form of the susceptibility (Fig. 12).

Domb and Sykes [18a] have recently re-examined the data for the fcc lattice in which the terms in the expansion are fairly regular, and the procedure of Section III, 2c can be used to assess the effect of changing s. The ratios $R_n(1)$ and $R_n(\infty)$ again converge rapidly, the limiting values for $s = 1$, $s = \infty$ (1.104, 1.175) being somewhat larger than those of Eq. (3.18). The regularity improves with increasing s, the coefficients for $s = \infty$ being sufficiently well behaved for the methods of Section III, 2a to be applied. As a result they suggest that the asymptotic behavior of the susceptibility of the three-dimensional Heisenberg model is

$$\chi_0 \simeq B(1 - T_c/T)^{-4/3}, \quad (3.23)$$

and that the limit kT_c/qJ for $s \to \infty$ is 0.2660; this value provides a more reliable basis than that given in Eq. (3.20) for $s = \frac{1}{2}$, and the latter estimate should be modified [using the limiting value of $R_n(\infty)$] to $kT_c/qJ = 0.679$. The form (3.23) has also been confirmed by Padé approximant studies [18b].

4. Effect of Lattice Structure

Dimensionality seems to be the dominant factor in determining the form of the magnetic susceptibility, details of the lattice structure playing only a secondary role; in particular the coordination number has no primary significance as might be assumed from simple closed-form approximations. For example, both the two-dimensional triangular lattice and the three-dimensional sc lattice have coordination number 6; the susceptibility of the former is similar to that of the quadratic lattice, but very different from the latter, which is similar to the fcc lattice. Some idea of the variation to be expected from change of lattice structure can be derived from Fig. 8 for the Ising model ($s = \frac{1}{2}$) [15].

IV. Critical Behavior of Other Thermodynamic Functions

1. Critical Values of Energy and Entropy

The energy and entropy represent integrals of C_v, the specific heat, and C_v/T, respectively. Hence they provide information about the area under the specific heat curve, and, in particular by studying their critical values, we can learn of the magnitude of the "tail" of the specific heat anomaly above the Curie temperature. For more detailed knowledge of the shape of the specific heat curve near the Curie point we must study the mathematical behavior of the specific heat itself.

In order to compare the specific heat curves for the various models and lattices, it is convenient to take T_c as a unit of temperature, and consider C_v as a function of $t(= T/T_c)$. Then

$$S_c = \int_0^1 \frac{C_v}{t} dt$$

$$S_\infty - S_c = \int_1^\infty \frac{C_v}{t} dt \qquad (4.1)$$

so that the values of S_c and $S_\infty - S_c$ enable us to compare the magnitude of the specific heat curves below and above the Curie point. The sum the two terms, S_∞, is usually known for any model and tells us the total magnitude of the specific heat anomaly. Critical values of the entropy are particularly useful for comparison with experiment since they are independent of the interaction energy.

Similarly, for the internal energy it is useful to study

$$(E_c - E_0)/kT_c = \frac{1}{k} \int_0^1 C_v \, dt,$$

$$(E_\infty - E_c)/kT_c = \frac{1}{k} \int_1^\infty C_v \, dt. \tag{4.2}$$

These values represent directly the areas under the specific heat curve below and above the Curie temperature, and are also independent of the interaction energy.

a. Ising model ($s = \frac{1}{2}$). We have pointed out in Section I, 2 that a striking characteristic of the Onsager specific heat curve is its large "tail"; it is important to know whether this arises from the two-dimensional character of the model. Table I (taken from reference [3]) gives the values of Eqs. (4.1) and (4.2) for a number of lattices. Values for the two-dimensional lattices are exact, and those for the three-dimensional lattices are estimated from series expansions; errors should not be larger than a few percent.

The difference in passing from a two to a three-dimensional lattice is quite striking. For all the two-dimensional lattices more than 50% of the entropy change occurs above the Curie temperature, but for the three-dimensional lattices this is much reduced. For a given dimension the effect of lattice structure is only small, the "tail" decreasing steadily as the coordination increases.

TABLE I[a]

CRITICAL VALUES FOR THE ISING MODEL ($s = \frac{1}{2}$)

Lattice structure	q	$kT_c/qJ(=t_c)$	S_c/k	$(S-S_c)/k$	$(E_c-E_0)/kT_c$	$(E_\infty-E_c)/kT_c$
Linear chain	2	0	0	0.693	0	∞
Honeycomb	3	0.506	0.265	0.428	0.227	0.761
Simple quadratic	4	0.567	0.306	0.387	0.258	0.623
Triangular	6	0.607	0.330	0.363	0.275	0.549
Diamond	4	0.676	0.511	0.182	0.418	0.322
Simple cubic	6	0.752	0.560	0.153	0.447	0.218
Body-centered cubic	8	0.794	0.586	0.107	0.460	0.169
Face-centered cubic	12	0.816	0.597	0.102	0.463	0.150

[a] From reference [27a].

b. Ising model (general s).* The total entropy change from $T = 0$ to $T = \infty$ for a model of given s is $\ln(2s + 1)$. Hence the area under the curve $C_v(t)/t$ increases steadily but slowly as s increases; as $s \to \infty$ this area becomes logarithmically infinite. The corresponding area under the $C_v(t)$ curve is determined by $-E_0/kT_c$, and from Eq. (3.19) this increases approximately as $3s/(s + 1)$, i.e., by a factor of 3 as s goes from $\frac{1}{2}$ to ∞. We may thus expect that for large s the major change in C_v occurs near $t = 0$.

To compare the critical values of E_c/kT_c for different s we can evaluate the coefficients c_r from Appendix A, and use the kT_c estimates of Eq. (3.19) to derive the following series for the high-temperature expansion $-E/kT_c$ (fcc lattice):

$$-E/kT_c = \begin{cases} 0.06257t'(1 + 0.4085t' + 0.2260t'^2 + 0.1420t'^3 \\ \quad + 0.09639t'^4 + 0.06985t'^5 + 0.05324t'^6 \\ \quad + 0.04207t'^7 + \cdots) \quad (s = \tfrac{1}{2}) \\ 0.05756t'(1 + 0.3919t' + 0.2627t'^2 + 0.1763t'^3 \\ \quad + 0.1254t'^4 + 0.09375t'^5 + 0.07272t'^6 + \cdots) \quad (s = 1) \\ 0.05456t\,(1 + 0.3815t' + 0.2805t'^2 + 0.1915t'^3 \\ \quad + 0.1411t'^4 + 0.1081t'^5 + 0.08550t'^6 + \cdots) \quad (s = \infty) \end{cases} \quad (4.3)$$

($t' = 1/t = T_c/T$). The term outside the bracket, which dominates at high temperatures, decreases slowly as s increases; however, when t' approaches unity, the series within the brackets increases as s increases, the net result being a slight increase in $-E_c/kT_c$ as s increases from $\frac{1}{2}$ to ∞. Estimates of the critical values for $s = \frac{1}{2}, 1, \infty$ are 0.150, 0.160, and 0.167, respectively. Hence only a small fraction of the increase in the area under the specific heat curve occurs in the region above the Curie temperature, and most of it must be in the region below the Curie temperature.

c. Heisenberg model. We can use the estimates of the Curie temperature of the Heisenberg model given at the end of Section III, 3 to put the series for this model in the same form as Eq. (4.3) (fcc lattice):

$$-E/kT_c = \begin{cases} 0.2712t'\,(1 + 0.3682t' + 0.02511t'^2 - 0.02157t'^3 \\ \quad + 0.01702t'^4 + 0.03131t'^5 + \cdots) \quad (s = \tfrac{1}{2}) \\ 0.2005t'\,(1 + 0.4156t' + 0.2240t'^2 + 0.1371t'^3 + 0.09001t'^4 \\ \quad + 0.06291t'^5 + \cdots) \quad (s = 1) \\ 0.1964t'(1 + 0.4178t' + 0.2335t'^2 + 0.1459t'^3 + 0.09733t'^4 \\ \quad + 0.06883t'^5 + \cdots) \quad (s = \infty). \end{cases} \quad (4.4)$$

* The remainder of this chapter draws largely on reference [18a].

The terms are irregular for $s = \frac{1}{2}$ and have not settled down to steady behavior. However, the regularity improves rapidly with increasing s, and it seems that the one negative term in the $s = \frac{1}{2}$ series may represent an isolated small number effect. It will be seen by comparison with Eq. (4.3) that the term outside the parentheses is three or four times as large for the Heisenberg model. Estimates of critical values based on Eq. (4.4) are not quite as reliable as those for the Ising model, but should still be sufficiently accurate to delineate the major differences in critical behavior between the two models. We find that for the Heisenberg model

s	$\frac{1}{2}$	1	∞
$-E_c/kT_c =$	0.439	0.459	0.474
	(0.150)	(0.160)	(0.167)
$(S_\infty - S_c)/k =$	0.265	0.306	0.322
	(0.102)	(0.114)	(0.123).

(4.5)

(Figures in parentheses correspond to the Ising model.)

Hence for this model the "tail" of the specific heat curve is appreciably larger.

2. Specific Heat

If we wish to analyze the detailed behavior of the high-temperature specific heat curve, and its variation with change of dimension, lattice structure, spin, or model, it is useful to examine and compare reduced high-temperature expansions for this quantity. Let us first investigate the effect of change of dimension on the Ising model ($s = \frac{1}{2}$). It is convenient to compare the triangular lattice, which is close-packed two-dimensional, with the fcc, as follows [18]:

$$C_v/k = \begin{cases} 0.2263t'^2 (1 + 1.0986t' + 0.8298t'^2 + 0.5525t'^3 \\ \quad + 0.4021t'^4 + 0.3384t'^5 + 0.2971t'^6 \\ \quad + 0.2614t'^7 + \cdots) \quad \text{(triangular)} \\ 0.06257t'^2 (1 + 0.8170t' + 0.6779t'^2 + 0.5680t'^3 \\ \quad + 0.4819t'^4 + 0.4191t'^5 + 0.3727t'^6 \\ \quad + 0.3365t'^7 + \cdots) \quad \text{(fcc)}. \end{cases} \quad (4.6)$$

In comparing these two expansions we may say roughly that the term outside the parentheses represents the magnitude of the "tail" of the specific heat curve, and the term inside the parentheses represents the shape of the curve, particularly near the Curie point. Hence the magnitude of the tail is much smaller for the fcc lattice as was evidenced by critical values. But it will be seen that the coefficients decrease less rapidly for the fcc lattice and the specific heat curve is therefore sharper. The extent of this sharpness can be assessed by the methods of Section III, 1a; although the series are not as well behaved as those for the susceptibility, and conclusions are less reliable, there are indications that the asymptotic behavior is approximately represented by

$$C_v/k \sim A(1 - T_c/T)^{-1/5}. \tag{4.7}$$

It can similarly be shown that change of lattice structure in a given dimension produces only a small effect, but in the same direction, i.e., a decrease in magnitude and increase in sharpness as the coordination number gets larger.

To demonstrate the effect of change of spin we now give corresponding series expansions for the lattice with $s = 1, \infty$ as follows:

$$C_v/k = \begin{cases} 0.05756 t'^2 (1 + 0.7838 t' + 0.7882 t'^2 + 0.7052 t'^3 \\ \qquad + 0.6268 t'^4 + 0.5625 t'^5 + 0.5091 t'^6 + \cdots) \quad (s = 1) \\ 0.05456 t'^2 (1 + 0.7629 t' + 0.8414 t'^2 + 0.7660 t'^3 \\ \qquad + 0.7055 t'^4 + 0.6487 t'^5 + 0.5985 t'^6 + \cdots) \quad (s = \infty). \end{cases} \tag{4.8}$$

Comparing Eqs. (4.8 and (4.6), we see that the effect of increase of spin is similar qualitatively to that of increase of coordination number, but the increase in sharpness is quite appreciable; it may even be sufficient to change the index in Eq. (4.7) to $\frac{1}{4}$ for $s = 1$ and $\frac{1}{3}$ for $s = \infty$.

Finally we give corresponding series expansions for fcc lattice on the Heisenberg model:

$$C_v/k = \begin{cases} 0.2712 t'^2 (1 + 0.7364 t' + 0.07533 t'^2 - 0.08629 t'^3 \\ \qquad + 0.08511 t'^4 + 0.1879 t'^5 + \cdots) \quad (s = \tfrac{1}{2}) \\ 0.2005 t'^2 (1 + 0.8312 t' + 0.6719 t'^2 + 0.5485 t'^3 \\ \qquad + 0.4500 t'^4 + 0.3775 t'^5 + \cdots) \quad (s = 1) \\ 0.1964 t'^2 (1 + 0.8357 t' + 0.7006 t'^2 + 0.5836 t'^3 \\ \qquad + 4866 t'^4 + 0.4130 t'^5 + \cdots) \quad (s = \infty). \end{cases} \tag{4.9}$$

The tail is much larger than for the Ising model, but the sharpness is decreased. The series in parentheses for $s = \infty$ is comparable with the corresponding Ising series, Eq. (4.6), for $s = \frac{1}{2}$. For the Heisenberg model of spin $\frac{1}{2}$ the terms are not yet smooth enough for analysis, but from the general trend established, the specific heat will be less steep than Eq. (4.7) and may be logarithmically infinite, or even finite.

3. Spontaneous Magnetization and Low-Temperature Properties

Low-temperature series expansions for three-dimensional lattices on the Ising model are not usually consistent in sign; hence evaluation of critical behavior is more difficult, and the information available until very recently was rather scant.

The earliest systematic investigation of the behavior of the spontaneous magnetization near the Curie point is due to Burley [19], who used the Kikuchi closed-form approximation, and a series regrouping introduced by Domb and Sykes [20]. He showed clearly that the spontaneous magnetization, L_0, for three-dimensional lattices approaches zero less steeply than $(1 - T/T_c)^{1/8}$, the two-dimensional exact value, and suggested that if $I_0 \sim A'(1 - T/T_c)^b$, then b lies between $\frac{1}{2}$ and $\frac{1}{4}$. Recently more precise information has become available from two independent sources, the Padé approximant [6b, 19a], and the statistics of the diamond lattice [19b] for which the terms of the low-temperature series are all positive in sign. The first results of Padé approximant investigations suggested that b was approximately 3/10 [6b]. Subsequently a more refined treatment indicated a value of 5/16 [19a], and this was also the conclusion reached from a study of the diamond lattice [19b].

Low-temperature expansions for the Heisenberg model are discussed elsewhere in this volume [4]; they are much more difficult to derive, and have not developed sufficiently to provide any information in the critical region.

V. Antiferromagnetism

1. Introduction

The Ising model of an antiferromagnet corresponds to the Hamiltonian (2.1) with the sign of J changed (likewise the Heisenberg model). Critical properties can again be manifested but they differ in character from those of a ferromagnet. For even lattices (i.e., those containing only closed polygons with even numbers of sides) the zero-field solution is unchanged, and hence the discussion of specific heat and other thermodynamic properties in Section IV remains valid. For odd lattices, however, critical properties usually disappear. The behavior in a nonzero magnetic field is completely different from a ferromagnet. The zero-field magnetic susceptibility no longer becomes infinite but passes through a maximum, and has a lower-order singularity at the critical point (Néel temperature). Series expansions for the susceptibility now alternate in sign, and can be used to determine critical behavior only after suitable transformation.

A ferromagnet does not retain any thermodynamic discontinuity in the presence of a nonzero magnetic field, but an antiferromagnet still has a critical point, and detailed knowledge of the thermodynamic behavior in its neighborhood would be of great significance for a general understanding of cooperative behavior. Again, series expansions are not consistent in sign and provide little immediate information on critical properties.

A reliable assessment of the critical behavior of the zero-field susceptibility for the Ising model of spin $\frac{1}{2}$ has resulted largely from the work of Sykes and Fisher [5, 6, 7, 21, 22, 22a], which we shall now discuss in more detail. For the Heisenberg model theoretical work has not developed beyond elementary approximations.

2. Susceptibility of the Ising Model ($s = \frac{1}{2}$)

a. Series expansions. The high-temperature expansion of the zero-field susceptibility for an antiferromagnet (sq lattice) is given by [7, 22]

$$\frac{kT\chi_0}{Nm^2} = 1 - 4w' + 12w'^2 - 36w'^3 + 100w'^4 - 276w'^5 + 740w'^6$$
$$- 1972w'^7 + 5172w'^8 - 13492w'^9 + 34876w'^{10} - 89764w'^{11}$$
$$+ 229628w'^{12} - 585508w'^{13} + 1486308w'^{14} - 3763460w'^{15}$$
$$+ 9497380w'^{16} + \cdots \quad (w' = \tanh |K| = -w). \tag{5.1}$$

We know from the zero-field partition function that the critical point corresponds to $w' = (\sqrt{2} - 1)$, and hence that the series (5.1) has a singularity at this point. Because of the alternation of sign of terms in the series, it is not possible to obtain direct information regarding the nature of the singularity.

We have referred [Section II, 2a, Eqs. (16), (17)] to the decomposition of the susceptibility series, which for an antiferromagnet assumes the form

$$(1 + \sigma w')^{-2} \left(1 + (\sigma - 1)w' + w'^2 + 2w' \frac{E(w')}{|J|} + 8(1 - w')^2 \sum_{l=5}^{\infty} (-1)^l g_l w'^l \right). \tag{5.2}$$

The last term in Eq. (5.2) is denoted by $G(w')$; for the quadratic lattice,

$$G(w') = 8(1 - w')^2(-2w'^7 + 2w'^8 - 20w'^9 + 38w'^{10} - 146w'^{11}$$
$$+ 368w'^{12} - 1070w'^{13} + 2824w'^{14} - 7680w'^{15} + 19996w'^{16} \cdots). \tag{5.3}$$

Sykes and Fisher call the approximation to the susceptibility obtained by neglecting $G(w')$ the "energetic susceptibility," $\chi_E(w')$. Numerical evaluation of $G(w')$ indicates that it contributes not more than a few percent to χ_0 at temperatures in the region of the critical point and above. Because of this, the behavior of the susceptibility in the critical region is effectively determined by the energy alone. For the quadratic lattice this is known exactly [1], and the susceptibility in the critical region is given by

$$a + b(T - T_c) \ln (T - T_c). \tag{5.4}$$

For the sc and bcc lattices the energy is not known exactly in the critical region; but there are clear indications that the specific heat is infinite, its high-temperature behavior being approximately determined by Eq. (4.7). Hence the susceptibility should again have a vertical tangent at temperature T_c.

From these observations Sykes and Fisher conclude that the maximum in the susceptibility vs. temperature curve occurs above the critical temperature, the latter being characterized by a vertical tangent (e.g., Fig. 10). Numerical calculation indicates that the difference between the temperature of the maximum and T_c is about 40% for two-dimensional lattices and 8% for three-dimensional lattices.

The argument can be substantiated for the sq lattice [22] for which the critical temperature and ferromagnetic critical behavior are known exactly. This suggests removing a factor $(w - w_c)^{-7/4}$ in Eq. 5.1, and, following a method introduced by Park [23], Fisher and Sykes write

$$\ln \chi_0(w') = (-7/4) \ln (1 + 2w' - w'^2) + \ln \psi(w'), \tag{5.5}$$

where

$$\begin{aligned}
\ln \psi(w') = {} & 0.500000 w' + 1.250000 w'^2 + 1.66667 w'^3 \\
& + 2.875000 w'^4 + 4.100000 w'^5 + 8.416667 w'^6 \\
& + 17.07143 w'^7 + 32.43750 w'^8 + 61.38889 w'^9 \\
& + 120.2500 w'^{10} + 241.0455 w'^{11} + 496.9583 w'^{12} \\
& + 1036.038 w'^{13} + 2192.179 w'^{14} \\
& + 4633.567 w'^{15} + 9900.218 w'^{16} + \cdots.
\end{aligned}$$

The series (5.5) has all positive terms and its radius of convergence is equal to w_c'; hence the methods of Section III, 1a can be used to determine its critical behavior.

b. *Exact argument.* Fisher also adapted the argument of Section III, 1b to apply to an antiferromagnet. The approximations used previously, represented by Eqs. (3.7) and (3.8), remain valid, but the sum (3.5) contains alternating signs, and needs special treatment. Writing

$$\langle \sigma_0 \sigma_k \rangle = \omega_{lm}(T), \quad \mathbf{k} = (l, m) \tag{5.6}$$

and regrouping the terms in Eq. (3.5), we have

$$\frac{\chi_0}{N_m \beta m^2} = 1 - \omega_{01}(T) - \frac{1}{4} \sum_{l,m} \nabla^2 \omega_{lm}(T), \tag{5.7}$$

where the sum now runs only over those lattice points for which $l + m$ is even and excludes the origin. The operator ∇^2 is the two-dimensional finite-difference Laplace operator defined by

$$\Delta^2 f_{i,j} = f_{i,j+1} + f_{i,j-1} + f_{i+1,j} + f_{i-1,j} - 4f_{ij}.$$

Using the approximations (3.7) and (3.8) one can see that the terms $\nabla^2 \omega_{lm}$ are all positive and essentially proportional to ω_{lm}. The term $\omega_{01}(T)$ corresponds to the internal energy, and gives rise to the vertical tangent at the critical point and all the other $\omega_{lm}(T)$ have the same type of singularity at this point; hence the singularity remains in the expression for χ_0. The sum in Eq. (5.7) may be evaluated by summing the first few terms explicitly, and estimating the remainder with the aid of the divergence theorem. We obtain a physical interpretation of the decomposition (5.2), the energetic susceptibility corresponding to near-neighbor correlations and $G(w')$ to more distant correlations. Numerical results at the critical point derived by evaluating Eq. (5.7) are in good agreement with those obtained from series expansions.

3. Superexchange Model

Using the decoration transformation (Section 3.4.4 of reference [3]) Fisher [6] was able to derive from the Onsager solution a closed form solution of a special model of an antiferromagnet in a magnetic field. The model is illustrated in Fig. 9; nonmagnetic spins (i.e., Ising spins with zero magnetic moment, denoted by open circles) provide the coupling between magnetic spins, the latter being arranged as a normal antiferromagnet in an sq array. Although the model differs from a standard Ising model in a number of respects, exact information on the nature of the singularities is extremely useful. It is found that the form of the zero-field susceptibility near the critical point is precisely as suggested in Section V, 2a (Fig. 10). However, in the presence of a nonzero

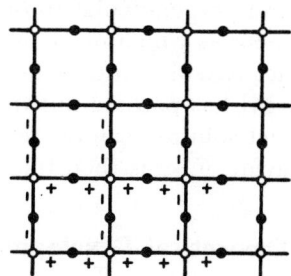

Fig. 9. A superexchange antiferromagnetic lattice. The black circles denote magnetic spins and the open circles nonmagnetic spins. The vertical bonds are "ferromagnetic," while the horizontal bonds are "antiferromagnetic" in character. (From Fisher, *Proc. Roy. Soc.* **254**, 69.)

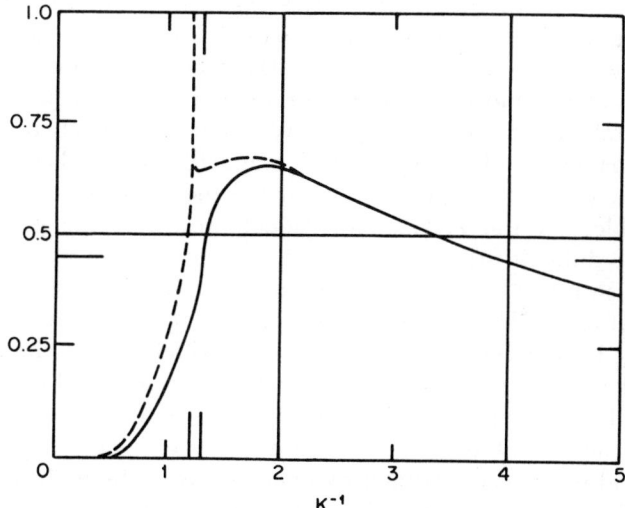

Fig. 10. The temperature dependence of the zero-field susceptibility (solid curve) and the susceptibility in a magnetic field (dashed curve). The coordinates are marked on the axes. (From Fisher, *Proc. Roy. Soc.* **254**, 82.)

field the character of the singularity changes and a logarithmic infinity is added. The nature of the specific heat anomaly remains unchanged as a logarithmic infinity in the presence of a magnetic field.

4. Perpendicular Susceptibility

One of the characteristic anisotropic features of an antiferromagnet is the difference between the susceptibilities $\chi_\parallel$ and $\chi_\perp$ corresponding to magnetic fields parallel and perpendicular to the direction of orientation of the spins. The usual theories which make use of closed-form approximations do not provide reliable information regarding the critical behavior of $\chi_\perp$. Fisher [24] has pointed out that for the particular case of the Ising model an exact solution can be derived for two-dimensional lattices; the critical behavior of $\chi_\perp$ is then the same as that of $\chi_\parallel$.

VI. Further Theoretical Developments Required

We shall briefly summarize in this section further investigations which we consider of primary importance in the development of the theory.

1. Ferromagnetism

Solutions (exact or substantial series expansions) for the two-dimensional Ising model for spins other than $\frac{1}{2}$ would help to determine whether the specific heat singularity varies with spin. For three-dimensional lattices the most urgent requirement is for information on the low-temperature side of the singularity which would tell us about the critical behavior of the spontaneous magnetization and specific heat. For the Ising model of spin $\frac{1}{2}$ some of the methods used above should be adaptable; for the Heisenberg model, however, the problems are much more difficult [4], and new methods would seem to be necessary. There is also a need for more terms in the zero-field high-temperature expansion for the Heisenberg model of spin $\frac{1}{2}$, since steady behavior has not yet been reached. A new approach which avoids the evaluation of averages of noncommuting operators has been initiated recently [24a, 33]. Instead, partition functions corresponding to finite clusters are used, and these can be obtained from computations on electronic computers. The method has already yielded two new terms in the high-temperature series, and seems to be capable of further development. Even where predictions have been possible because of steady behavior, more terms in the series expansions would still be useful to substantiate the predictions.

2. Antiferromagnetism

For the Ising model little precise information is available regarding critical behavior in the presence of a non-zero field [24b]; closed-form approximations based on a cluster integral development or Padé approximant studies may here provide a suitable approach. For the Heisenberg model information is required on all aspects of critical behavior, particularly the zero-field susceptibility and specific heat [24c].

VII. Experimental Results

1. Susceptibility of a Ferromagnet

The first comparison between theory and experiment was naturally sought in the ferromagnetic metals, and the most useful experimental results for this purpose seem to be those of Weiss and Forrer for nickel [25] and Sucksmith and Pearce [26] for nickel and iron. The curvature of the reciprocal susceptibility curve near the Curie temperature was in disagreement with early closed-form approximations which assumed short-range interactions. Néel [27] showed that if such approximations were to be used interactions between each atom and some 750 neighbors were needed to account for the experimental results. Domb and Sykes [15] showed that this surprising result could be attributed completely to the inadequacy of simple closed-form approximations near the Curie point. Estimates of the susceptibility based on series expansions and using the methods of Section III, 1a could account for the experimental results if one assumed only short-range interactions (Fig. 11). Further

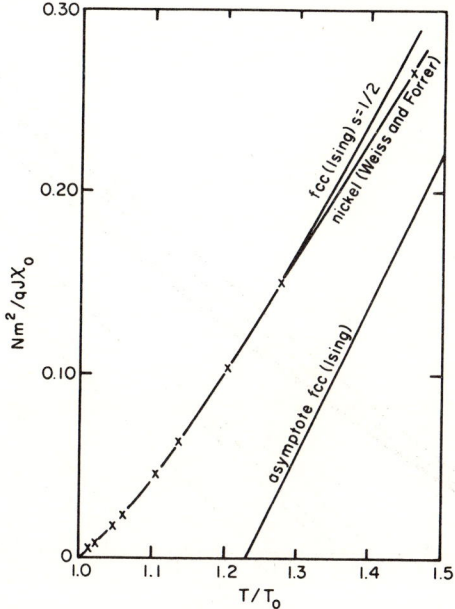

Fig. 11. Comparison of the reciprocal susceptibility of the Ising model (face-centered cubic, $s = \frac{1}{2}$) with that of nickel (Néel, after Weiss and Forrer). (From Domb and Sykes [15].)

away from the Curie point the experimental and theoretical curves began to deviate, but this might well be due to the shortcomings of a localized interaction model for nickel. (The number of effective magnetic electrons, as determined from the saturation magnetization, is only 0.6 per atom).

Rushbrooke and Wood [16] attempted an absolute comparison with experiment replacing the magnetic moment m by $g\beta'$ (β' devoting the Bohr magneton), and taking $g = 2$. For nickel fair agreement with experiment could only be secured by a reducing factor 0.6 as suggested above. For iron, however, the results were more gratifying (Fig. 12); the Heisenberg model could be regarded as a better approximation for this substance, although the fact that the experimental curves are closer to the theoretical calculations for $s = 3$ than for $s = 1$ needs further investigation.

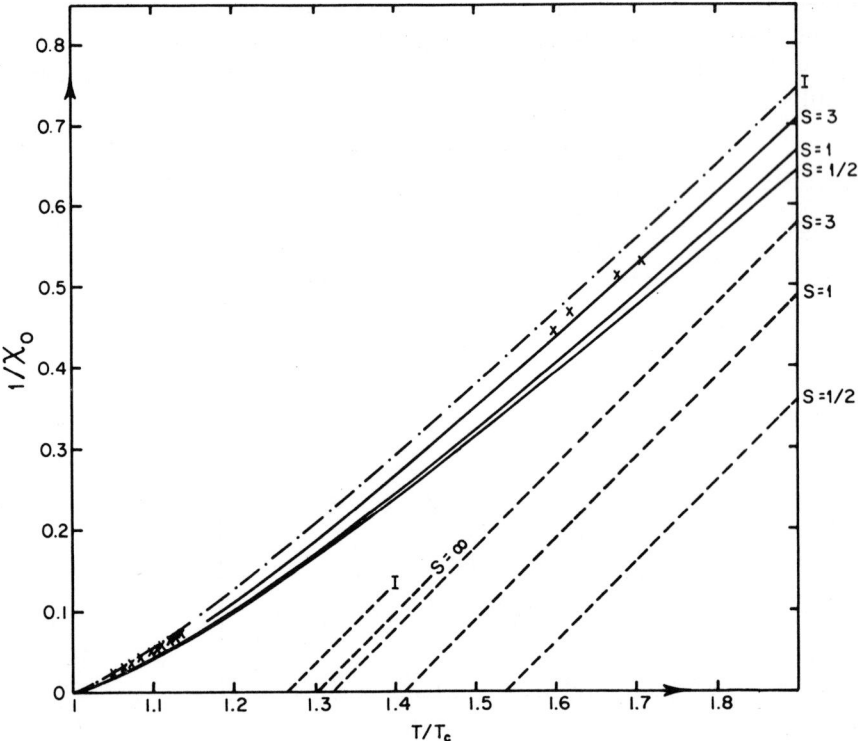

FIG. 12. Comparison of theoretical curves for reciprocal susceptibility (bcc lattice) with experimental results for Fe (Sucksmith and Pearce). (From Rushbrooke and Wood [16].)

Recently a number of ferromagnetic insulators have been discovered, and they seem to offer better prospects of identification with the Ising and Heisenberg models. A careful analysis of experimental data available so far has been undertaken recently [27a]. Two copper salts, $CuK_2Cl_4 \cdot 2H_2O$ and $Cu(NH_4)_2Cl_4 \cdot 2H_2O$, seem to provide a good representation of the Heisenberg model, and the zero-field susceptibility near the Curie point is proportional to $(1 - T_c/T)^{-n}$ with $n = 1.36$, 1.37, respectively; this result is in good agreement with the prediction of Section III, 3. The $\frac{4}{3}$ power law has also been investigated recently for iron [34] and nickel [35]. The agreement with the experiment is surprisingly good.

2. Specific Heat and Thermodynamic Properties

An early measurement of a specific heat of nickel made by Néel [28] indicates a finite maximum and discontinuity of slope at the Curie point. Recent measurements of the specific heats of $NiCl_2 \cdot 6H_2O$ and $CoCl_2 \cdot 6H_2O$ (which are antiferromagnetic) by Robinson and Friedberg [29] show logarithmic infinities near the Curie point. It might therefore be worth repeating Néel's measurements with more modern equipment. In any case it is clear that the experimental results differ appreciably from theoretical calculations for the three-dimensional Ising model with spin $\frac{1}{2}$ drawn in Fig. 13, and estimates of the entropy at the Curie point indicate that agreement would be better with the Heisenberg model. A more detailed comparison with the latter model must await further theoretical calculations.

It is perhaps worth emphasizing that the specific heat curve is quite a sensitive test to the type of interaction, and differs appreciably in this respect from the susceptibility.

For the two copper salts mentioned in Section VII, 1 (of spin $\frac{1}{2}$) the experimental value of $(S_\infty - S_c)/k$ is 0.22; this compares favorably with the theoretical estimate for the Heisenberg model [Eq. (4.5)]. Critical parameters for a number of antiferromagnetic salts have also been examined, and the best agreement with the Ising model is obtained from cobalt tutton salts [27a].

3. Susceptibility of an Antiferromagnet

Early measurements did not differentiate between the susceptibility maximum and the Néel point, and assumed that they were coincident [30]. Recent measurements in Leiden [31] and Oxford [32] make a clear distinction between them, giving rise to a susceptibility behavior near

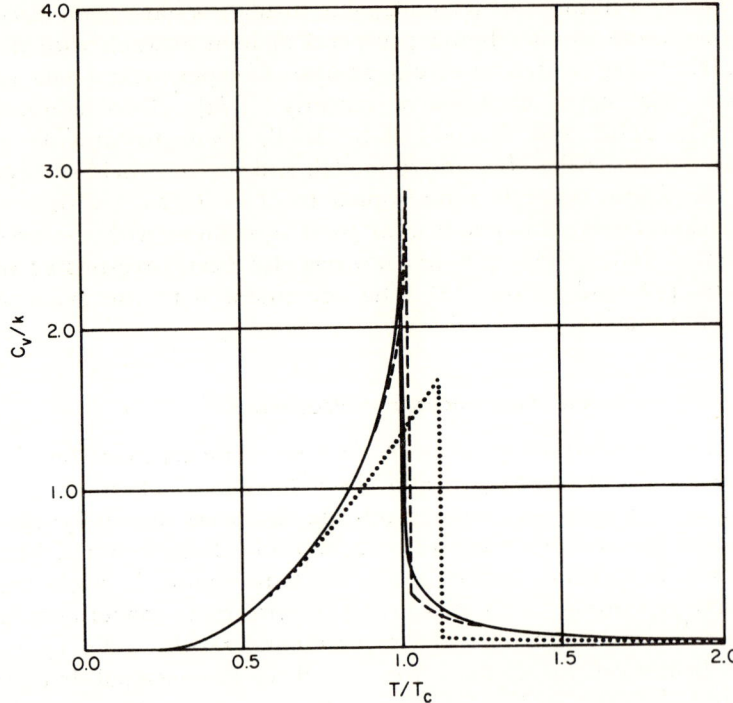

FIG. 13. Specific heat of fcc Ising lattice (spin $\frac{1}{2}$). Solid line, estimated from series expansions; dashed line, Kikuchi approximation; dotted line, first-order Bethe approximation. (From Domb [3], p. 290.)

the critical point in substantial agreement with the ideas of Sykes and Fisher [21, 22] (see Section V, 2).

Acknowledgments

The writer is indebted to his colleagues Dr. M. E. Fisher, Dr. M. F. Sykes, and Dr. B. J. Hiley for helpful discussion; to Professor L. Néel for advice on experimental results; and to The Royal Society, Messrs. Taylor & Francis Ltd., and The Journal of Mathematical Physics for permission to reproduce diagrams.

Appendix A. Zero-Field Energy and Susceptibility Coefficients for the Ising Model (Lattice)

$$c_2(s) = \frac{X^2}{9}$$

$$c_3(s) = \frac{8X^3}{27}$$

$$c_4(s) = \frac{X^2}{225}[514X^2 - 116X + 1]$$

$$c_5(s) = \frac{2X^3}{405}[184X^2 - 56X + 1]$$

$$c_6(s) = \frac{X^2}{297675}[83599648X^4 - 36144288X^3 + 4664376X^2 - 118584X + 675]$$

$$c_7(s) = \frac{8X^3}{14175}[7996592X^4 - 4275072X^3 + 817524X^2 - 35076X + 435]$$

$$c_8(s) = \frac{X^2}{212625}[18568249616X^6 - 11735319488X^5 \\ + 3100557664X^4 - 343347552X^3 \\ + 14868306X^2 - 246780X + 945]$$

$$h_0(s) = 1$$

$$h_1(s) = 4X$$

$$h_2(s) = \frac{2X}{5}(38X - 1)$$

$$h_3(s) = \frac{2X}{75}(2124X^2 - 136X + 1)$$

$$h_4(s) = \frac{X}{3150}(656648X^3 - 70772X^2 + 2331X - 15)$$

$$h_5(s) = \frac{X}{330750}(251682608X^4 - 39096208X^3 \\ + 2440236X^2 - 49104X + 225)$$

$$h_6(s) = \frac{X}{1984500}(5480403392X^5 - 1125263472X^4 \\ + 105206144X^3 - 4607196X^2 \\ + 79290X - 315)$$

$$\ln Z(\beta, 0) = \ln(2s+1) + 6\sum_{r=2}^{\infty} c_r(s) K^r/s^{2r} r!$$

$$E(\beta, 0) = -6J\sum_{r=2}^{\infty} c_r(s) K^{r-1}/s^{2r}(r-1)! \qquad (K = J/kT)$$

$$\frac{C_v}{k} = 6\sum_{r=2}^{\infty} c_r(s) K^r/s^{2r}(r-2)!$$

$$\chi_0 = \frac{\beta m^2 X}{3s^2} \sum_{r=0}^{\infty} h_r(s) K^r/s^{2r}$$

Appendix B. Zero-Field Energy and Susceptibility Coefficients for the Heisenberg Model (fcc Lattice)

$$c_2(s) = \frac{X^2}{3}$$

$$c_3(s) = \frac{X^2}{18}(16X - 3)$$

$$c_4(s) = \frac{2X^2}{45}(107X^2 - 78X + 3)$$

$$c_5(s) = \frac{X^2}{54}(2048X^3 - 2220X^2 + 420X - 9)$$

$$c_6(s) = \frac{X^2}{5670}(2288704X^4 - 3209664X^3 + 1383465X^2 - 111420X + 1728)$$

$$c_7(s) = \frac{X^2}{1620}(8854016X^5 - 14814272X^4 + 9484000X^3 - 1886331X^2 + 98928X - 1242)$$

$$h_0(s) = 1$$
$$h_1(s) = 4X$$
$$h_2(s) = \frac{X}{3}(44X - 3)$$
$$h_3(s) = \frac{X}{45}(2328X^2 - 382X + 12)$$
$$h_4(s) = \frac{X}{540}(96368X^3 - 26432X^2 + 2352X - 45)$$

$$h_5(s) = \frac{X}{14175}(8600616X^4 - 3377996X^3 + 526440X$$
$$-30060X + 432)$$
$$h_6(s) = \frac{X}{340200}(694722560X^5 - 358715504X^4 + 81267018X^3$$
$$-8691207X^2 + 367290X - 4347)$$

Thermodynamic formulas are as for Ising Model.

References

1. L. Onsager, *Phys. Rev.* **65**, 117 (1944).
2. T. H. Berlin and M. Kac, *Phys. Rev.* **86**, 821 (1952).
3. C. Domb, *Advan. Phys.* **9**, 149, 245 (1960).
4. R. Brout, Chapter 2 in this volume, p. 43.
5. M. E. Fisher, Physica **25**, 521 (1959).
6. M. E. Fisher, *Proc. Roy. Soc.* **254**, 66; **256**, 502 (1960).
6a. P. Dienes, "The Taylor Series," Chapt. 14. Oxford Univ. Press, London and New York, 1931.
6b. G. A. Baker, *Phys. Rev.* **124**, 768 (1961).
7. M. F. Sykes, *J. Math. Phys.* **2**, 52 (1961).
8. J. Yvon, *Cahiers Phys.* **28** (1945); **31**, 32 (1948).
9. W. Opechowski, *Physica* **4**, 181 (1937).
9a. G. Fournet, *J. Phys. Radium* **13**, 14 (1952); D. Taupin and G. Fournet, *ibid.* **20**, 477 (1959).
9b. C. Domb and B. J. Hiley, *Proc. Roy. Soc.* **268**, 506 (1962).
10. V. Zehler, *Z. Naturforsch.* **5a**, 344 (1950).
11. B. L. Van der Waerden, *Z. Physik* **118**, 473 (1941).
12. A. J. Wakefield, *Proc. Camb. Phil. Soc.* **47**, 419 (1951).
13. C. Domb and M. F. Sykes, *Phil. Mag.* [8] **2**, 733 (1957).
14. C. Domb and M. F. Sykes, *J. Math. Phys.* **2**, 63 (1961).
15. C. Domb and M. F. Sykes, *Proc. Roy. Soc.* **A240**, 214 (1957).
16. G. S. Rushbrooke and P. J. Wood, *Mol. Phys.* **1**, 257 (1958).
17. B. Kaufman and L. Onsager, *Phys. Rev.* **76**, 1244 (1949).
18. C. Domb and M. F. Sykes, *Phys. Rev.* **108**, 1415 (1957).
18a. C. Domb and M. F. Sykes, *Phys. Rev.* **128**, 168 (1962).
18b. W. Marshall, J. Gammel, and L. Morgan, *Proc. Roy. Soc.* **275**, 257 (1963).
19. D. M. Burley, *Phil. Mag.* [8] **5**, 909 (1960).
19a. J. W. Essam and M. E. Fisher, *J. Chem. Phys.* **38**, 802 (1963).
19b. J. W. Essam and M. F. Sykes, *Physica* **28**, 378 (1963).
20. C. Domb and M. F. Sykes, *Proc. Roy. Soc.* **A235**, 247 (1956).
21. M. F. Sykes and M. E. Fisher, *Phys. Rev. Letters* **1**, 321 (1958).
22. M. F. Sykes and M. E. Fisher, *Physica* **28**, 919 (1962).
22a. M. E. Fisher and M. F. Sykes, *Physica* **28**, 939 (1962).
23. D. Park, *Physica* **22**, 932 (1956).
24. M. E. Fisher, *Physica* **26**, 618 (1960); *J. Math. Phys.* **4**, 124 (1963).
24a. C. Domb and D. W. Wood, *Phys. Letters* **8**, 21 (1964).

24b. D. M. Burley, *Physica* **27**, 768 (1961).
24c. M. E. Fisher, *Phil. Mag.* **7**, 1731 (1962).
25. P. Weiss and R. Forrer, *Ann. Physik* [4] **5**, 153 (1926); [5] **12**, 279 (1929).
26. W. Sucksmith and R. R. Pearce, *Proc. Roy. Soc.* **A167**, 189 (1938).
27. L. Néel, *J. Phys. Raduim* **5**, 104 (1934).
27a. C. Domb and A. R. Miedema, "Progress in Low Temperature Physics," Vol. VI North-Holland, Amsterdam, 1964.
28. L. Néel, *Compt. Rend.* **207**, 1384 (1938).
29. W. K. Robinson and S. A. Friedberg, *Phys. Rev.* **117**, 402 (1960).
30. A. B. Lidiard, *Rep. Progr. Phys.* **17**, 201 (1954).
31. M. A. Lasheen, J. van den Broek, and C. J. Gorter, *Physica* **24**, 1061, 1076 (1958).
32. A. H. Cooke, R. Lazenby, F. R. McKim, J. Owen, and W. P. Wolf, *Proc. Roy. Roy. Soc.* **A250**, 97 (1959).
33. C. Domb, N. W. Dalton, G. S. Joyce, and D. W. Wood, *Proc. Intern. Conf. on Magnetism, Nottingham, 1964*.
34. J. E. Noakes and A. Arrot, *J. Appl. Phys.* **35**, 931 (1964).
35. M. E. Fisher, *Proc. Intern. Conf. on Magnetism, Nottingham, 1964*.

2. Statistical Mechanics of Ferromagnetism[*]

Robert Brout

*Faculté des Sciences, Université Libre de Bruxelles,
Brussels, Belgium*

I. Introduction	43
II. Ising Model	47
1. The Weiss Field	47
2. Cluster Methods and the Ring Approximation	51
3. Refinements on the Ring Approximation	62
III. Heisenberg Model	67
1. Spin Waves	68
2. Graphical Analysis of the Heisenberg Model	78
IV. Band Theory of Ferromagnetism	83
V. Random Spin Systems	91
Appendix A	97
Appendix B	100
References	102

I. Introduction

The statistical mechanics of ferromagnetism is divided in its interests between ferromagnetism per se and the general theory of phase transitions. The former point of view naturally stresses various exchange mechanisms in different physical situations and tends to solve the statistical mechanics of realistic situations by "physical" reasoning rather than rigorous methods. Notable among these arguments are those of the Weiss field and spin waves. The point of view of the theory of phase transitions is that ferromagnetism, more particularly the Ising

---

[*] This research was supported in part by the U.S. Air Force under Grant No. AF6 EOAR 6351 through the European Office of Aerospace Research.

model, is a very useful abstraction from which one may extract by both qualitative and quantitative analysis certain results of general validity in many or all phase transitions. Questions of interest are:

(1) Is it possible to isolate in a particular mathematical approach that part of the problem which manifests the singular behavior characteristic of the critical phenomena?

(2) What is the nature of the singularity and what are the critical parameters (in the case of ferromagnetism, the critical temperature)?

(3) How does the system accommodate itself to the singularity below the critical temperature?

In this chapter, we try to compromise between the two points of view. Our principal aim is to extract out of the formalism of statistical mechanics the main qualitative aspects of ferromagnetism as well as semiquantitative estimates of quantities of interest. No discussion will be made of exchange mechanisms, but a few models will be considered, which presumably scan the principal features of the phenomenon of ferromagnetism.

The models to be considered are as follows:

(1) *Ising model.*

$$\mathcal{H} = -\frac{1}{2}\sum_{i \neq j}^{N} v_{ij}\mu_i\mu_j - H\sum_{i=1}^{N} \mu_i \tag{1.1}$$

Here μ_i is a number taking on the values ± 1 (thought of as the z-component of spin at site i). The Ising model is then a classical model since all operators of the problem commute with each other. It is a convenient model to use as a guinea pig for more realistic problems. The function v_{ij} for the moment is an arbitrary function defined on lattice sites and H is the magnetic field (the magnetic moment per spin is taken to be unity).

(2) *Heisenberg model.*

$$\mathcal{H} = -\frac{1}{2}\sum v_{ij}\mathbf{S}_i \cdot \mathbf{S}_j - H\sum S_i^z \tag{1.2}$$

Here the full $\mathbf{S}_i$ is used in an isotropic exchange coupling. Otherwise the notation is the same as for the Ising model. The Heisenberg model is thought to be suitable for many antiferromagnets and ferrimagnets, and probably highly unsuitable for metallic ferromagnets, apart from the rare earths which have a highly specialized form of v_{ij} due to indirect coupling of the f electrons through the conduction band. We shall show that at high T the Heisenberg and Ising models will be similar in their behavior, whereas at low T, the Heisenberg model, having spin waves, is much richer in theoretical content.

(3) *Band model.* In this case the electrons are placed in bands and an exchange coupling is assumed between them. Such a model is applicable to the transition elements. A limiting case is the Heisenberg model where the Bloch functions taken in a Wannier representation become highly localized.

We discuss these three models in turn. Of the three, the Ising model has received the most extensive theoretical treatment. There are roughly speaking three classes of theories: (1) the exact solution of the two-dimensional near-neighbor case; (2) the very precise moment calculations; and (3) theories which are based on the Weiss model as zeroth approximation.

In this chapter only a particular category of the third class is discussed, for the following reasons. The two-dimensional near-neighbor solution of Onsager [1], a true *tour de force* in mathematical physics, has not led to further generalization. It remains a monumental achievement in the theory of the many-body problem that unfortunately stands loftily aloof. It has been remarked by Kac [2] that there are reasons to suspect that this particular model is solvable by a unique freak. In any case all other efforts in ferromagnetism as well as other many-body problems are based on approximations whose origins are through either physical insight or mathematical convenience or both. Since it is our principal aim to develop techniques applicable to physical models, it would be somewhat amiss to present this aspect of the theory of ferromagnetism in the present chapter. Moreover, there already exists the excellent review article of Montroll and Newell [3], to which the interested reader is referred.

As for the second class of theory, the precise numerical calculations from moment expansions are reviewed elsewhere in this volume [4]. We regard these calculations as the probable solution of the problem of the three-dimensional Ising model with near-neighbor interactions in much the same light as one regards a definitive experiment. It is the job of the analytical theory to duplicate these results, at least approximately.

The third class of theory is based on the Weiss field. In this article we discuss in great detail the theory of the Weiss field. In the opinion of the author, the essential phenomenon of ferromagnetism (and indeed of all phase transitions) is contained therein. The difficulty of the problem is in the proper treatment of the short-range correlations, which are ignored in the Weiss approximation. For a number of years the most popular approach to this problem was the combinational method,* begun by Bethe [6] in 1935. The disadvantage of this method is that there is no singular precursor behavior in the free energy for temperatures greater than the critical temperature. More recently, the author [7]. using clusters series methods, has summed infinite sets of graphs which do indeed present the necessary critical fluctuations giving rise to precursor singularities. These methods have been further developed by Horwitz and Callen [8] and Englert [9] and will be reviewed extensively in this article. The problem has by no means been solved by this method, but certain important qualitative features have turned up which seem to characterize very aptly the approach to the phase transition from both the high- and low-temperature regions. In particular, the singularity is obtained, as well as the mechanism (Weiss field) through which the system adapts itself to the singularity. The difficulty is in finding the nature of the singularity. The simplest form of the theory makes direct contact with the Ornstein–Zernike (O-Z) macroscopic theory of critical fluctuations [10]. This should be gratifying but unfortunately the O-Z result is incorrect. Presently, efforts are being made to improve the cluster theory in order to bring it into accord with the moment expansion calculations.

Before leaving the Ising model, we mention one last technique, that of functional integration. This is reviewed in an article by Helfand [11]. Unfortunately, the only real success of this method is in a special case of one dimension. Here a phase transition does not occur except for infinite-range forces where the Weiss method is applicable in any case. However, Siegert [12] has duplicated many of the simple graphical results by this method and one must await further theoretical developments to see if this technique is capable of producing the refinements necessary to an adequate theory.

We now turn to the Heisenberg model. It will be shown that for the Ising model the Weiss theory is exact at low temperatures. In this theory, an energy gap to create an excitation develops in the ferromagnetic phase. For this reason, the Ising model magnetization at low temperatures falls off exponentially in the temperature from its saturation value at zero temperature. In the Heisenberg model, this gap becomes filled with a continuum of excitations, the spin waves. These are introduced and discussed in Section III. The most interesting problem in the Heisenberg model is the interpolation of the low-temperature spin wave region with the high-temperature critical region.

---

* For a review of these methods, see reference [5].

This will be discussed in detail in Section III along with the low-temperature deviations of Dyson [13]. This section is based on the work of Stinchcombe *et al.* [14].

In Section IV, the band model is discussed, the Stoner model and Bloch spin waves are reviewed. The more recent developments of Englert [15], which are the beginnings of a realistic band theory, are discussed. The long-wavelength spin waves are more like Frenckel excitons in character whereas the shorter-wavelength spin waves are like the spin waves of the Heisenberg model. The band theory of ferromagnetism, though the most important from the point of view of conventional metallic ferromagnetism, is still at a most rudimentary stage. It will probably remain so for some time to come because it is so intimately tied up with the complicated band structure of the transition elements.

Up to this point we have been discussing ferromagnetism in pure materials. An equally interesting problem is that of ferromagnetism of exchange-coupled spins dissolved in a nonmagnetic substratum. Here there are several new questions to ask, one of the more enticing of which is: If the impurity spins are coupled by near-neighbor interactions, what is the critical concentration necessary to sustain ferromagnetism? In Section V, the problem of dilute ferromagnets is formulated and some detailed questions are considered.

Though the basic motivations are simple, the development of the statistical mechanics of ferromagnetism is often formal and tedious. In order to facilitate reading, we have decided to relegate to appendices all formal developments of cluster theory. At critical junctures, the text will present the theorems necessary to characterize the expansion in question and proceed from there.

II. Ising Model

1. The Weiss Field

The Weiss field is obtained as follows: a given spin i is surrounded by spins j of spin μ_j. Each μ_j contributes, according to Eq. (1.1), a contribution to the exchange field at i, the value $-v_{ij}\mu_j$. Thus the total exchange field at i is $H_i = -\sum_j' v_{ij}\mu_j$, where the prime means $j \neq i$. The average field is $\langle H_i \rangle = -\sum_j' v_{ij}\langle \mu_j \rangle$. Since our lattice of spins has complete translational symmetry $\langle \mu_j \rangle$ is independent of the site j where the spin sits. Thus we relieve it of its index. Call it R_W. The subscript stands for the long-range order in the Weiss approximation. In general, we have the definition

$$R = \frac{\text{(Number of up spins)} - \text{(Number of down spins)}}{N} \qquad (2.1)$$

Finally we have

$$\langle H \rangle_W = -v(0) R_W; \qquad v(0) \equiv {\sum_j}' v_{ij} \qquad (2.2)$$

The Weiss energy of the spin is $\mu_i \langle H \rangle_W$. Hence the average Weiss energy is $R_W \langle H \rangle_W$. The total Weiss energy is the sum on all spins. Remember, however, to multiply by $\frac{1}{2}$ so as not to count the same pair twice. Thus,

$$E_{\text{Weiss}} = -\frac{N}{2} v(0) R^2 \qquad (2.3)$$

We see that this result is obtainable from Eq. (1.1) through the approximation $\langle \mu_i \mu_j \rangle \simeq \langle \mu_i \rangle \langle \mu_j \rangle$. That is, the Weiss approximation neglects correlations among spins. Clearly this is in error, for if a given spin is up, its neighbor will have a stronger than average predilection for following. Given Eq. (2.3), it is an easy matter to compute R as a function of T (we use $\beta = 1/kT$). We do this in two ways. The first method is through the local field $\langle H \rangle = -v(0) R_W$. Hence we may compute the local partition function of a single spin subject to this field

$$Z_{\text{Single}} = \sum_{\mu_i = \pm 1} \exp(-\beta \langle H \rangle_W \mu_i) = 2 \cosh \beta \langle H \rangle_W$$

$$\langle \mu_i \rangle_W = \partial \log Z / \partial \beta \langle H \rangle_W = \tanh \beta \langle H \rangle_W \qquad (2.4)$$

This supplies an algebraic equation for R_W

$$R_W = \tanh \beta v(0) R_W \qquad (2.5)$$

If $\beta v(0) > 1$, there will be a solution $R_W \neq 0$ as well as $R_W = 0$. For $\beta v(0) < 1$, we have the unique solution $R_W = 0$. For $\beta v(0) > 1$, $R_W \neq 0$ is clearly the solution that gives lower free energy, since at $T = 0$ ($\beta = \infty$), all spins are surely parallel. In the presence of an external field H, Eq. (2.5) becomes

$$R_W = \tanh[\beta(v(0) R_W + H)] \qquad (2.6)$$

The point $\beta v(0) = 1$ or $kT_c = v(0)$ gives the temperature of a transition to spontaneous magnetization and hence identifies a Curie point. The solutions of Eqs. (2.5) and (2.6) are sketched in Figs. 1 and 2.

Figure 2 shows the behavior of R_W vs. H for $T > T_c$ and $T < T_c$. In the latter case the solution tends to the nonvanishing values of spon-

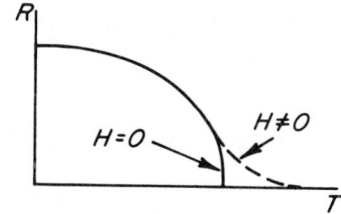

Fig. 1. Magnetization curve at fixed field in the Weiss approximation.

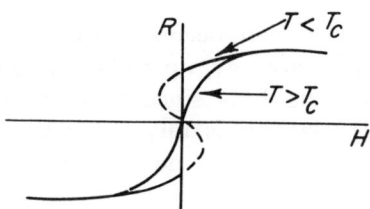

Fig. 2. Magnetization curve at fixed temperature in the Weiss approximation.

taneous magnetization at $H = 0$. The dotted lines are easily shown to be spurious. This follows from the argument below [Eq. (2.9)], where it is evident that the minimum in the free energy follows the solid lines of Fig. 2. For $T < T_c$, R_W is discontinuous in H at $H = 0$. The dotted lines in Fig. 2 may, however, correspond to the existence of a metastable phase in the case of hysteresis. According to Eq. (2.3), at low T we have $R_W \cong 1$. The energy it takes to make a wrong spin is then $2v(0)$. Hence $\lim_{T \to 0}(1 - R) = \exp[-2\beta v(0)]$, which is what (2.6) gives. Thus the low-temperature magnetization sticks to full saturation in the Ising model in an exponential way and the Weiss theory is exact in this limit.

The second way of obtaining Eq. (2.6) is to calculate the free energy of the system as a whole [16].

$$\log Z = -\beta F = -\beta \langle E \rangle + S/k \tag{2.7}$$

As E_{Weiss} is a function of R_W only and we expect R to fluctuate only by $O(1/\sqrt{N})$ at a given T, we may calculate S/k by merely enumerating the number of ways of arranging the spins to get R, i.e., $(S/k)_{\text{Weiss}} = \log W(R_W)$ where $W(R_W)$ is the degeneracy of states with fixed R_W. According to Eq. (2.1) we have

$$W(R_W) = \binom{N}{\frac{N(1 + R_W)}{2}} \tag{2.8}$$

i.e., the binomial coefficient counts the number to have $N(1 - R_W)/2$ spins down and $N(1 + R_W)/2$ spins up.

$$-\beta F_W = \frac{N}{2}(\beta v(0))R_W^2 + N\beta H R_W + \log\left(\frac{N}{\frac{N(1+R_W)}{2}}\right) \quad (2.9)$$

To find R_W, one minimizes F_W with respect to R_W. E_W is minimal at $R = 1$ and $-TS_W$ is minimal at $R = 0$, so clearly there will be a range in β where an extremum is possible. The extremal equation is (2.6).

It is clear that the Weiss field becomes exact at very low T ($\beta v(0) \gg 1$), for then almost all spins are, say, up and turning one spin down cannot very much affect its neighbors since they are held fixed in turn by their neighbors. This argument is only true in the presence of long-range order and hence fails for one-dimensional models with finite-range potential.

At high T, the Weiss approximation is no longer valid. In fact for $T > T_c$ the Weiss theory puts $\langle E \rangle = 0$, which is quite wrong because of the effects of "local order" mentioned above. Let us inquire, however, if there is a limiting mathematical case where the Weiss theory becomes exact. Consider a potential $v(\mathbf{R}_i - \mathbf{R}_j)$ which has very long range, covering z particles in its range, and always positive. Further, let it have square shape of strength $J = v(0)/z$ out to a radius of $z^{1/3}$. Now go to the limit $z \to \infty$, $v(0)$ fixed (by $z \to \infty$ we mean $z \to N \to \infty$). In this case all spins are coupled to each other by the same force and v_{ij} becomes independent of its indices. Thus

$$E \underset{z \to N}{\to} \frac{v(0)}{2N} \sum_{i \neq j} \mu_i \mu_j = \frac{v(0)}{2} NR^2 + O(1) \quad (2.10)$$

where we have used Eq. (2.1). Equation (2.10) establishes the Weiss theory as the correct limit for infinitesimally weak, infinitely long-range forces. In a crystal, with at worst near-neighbor interactions ($z \simeq 10$), $1/z$ is in fact a true parameter of smallness, thereby permitting one to adopt the Weiss field as a zeroth-order approximation.

The following simple argument shows that corrections of $O(1/z)$ do in fact come in. The Curie point is given by $kT_c \simeq zJ$, where J characterizes the interaction strength and $z^{1/3}$ its range in units of the lattice distance. Consider a temperature slightly above T_c and a particular spin which is in the up direction. The spins in its neighborhood will be polarized up with respect to this given spin by a factor $\sim [\exp(J/kT_c) - 1] \simeq J/kT_c = O(1/z)$. Since there are z spins in the range this gives $E(T \cong T_c) \cong J$ or, in terms of dimensionless variables, $\beta_c E = O(1/z)$. In these terms, we also see that the Curie point predicted from the Weiss field is necessarily an upper limit since in Weiss theory for $T > T_c$ we have $E = 0$, i.e., a temperature is reached where *all* the exchange energy is randomized. Clearly the correct T_c will be less than this.

2. Cluster Methods and the Ring Approximation

There are two distinct cluster expansions available for work on the Ising model, that of the author [7] and that of Horwitz and Callen [8] and Englert [9]. One is a rearrangement of the other. The difference in the expansions is that in the expansion of Brout once a spin is used in a graph it is not permitted to appear again in the same graph. The rearrangement of Horwitz and Callen eliminates this restriction. The formal result is mathematically simpler than that of Brout, but the sense of a mathematical cluster associated with a physical cluster of spins is lost. Because of mathematical facility we will present the Horwitz and Callen form of the expansion, following the procedure of Englert [9].

Let O be any spin operator. Particular O's of importance are μ_i and $\frac{1}{2} \sum v_{ij} \mu_i \mu_j$, in which cases we are dealing with the average magnetization per particle or the total energy, respectively. We wish to develop in a power series in β the ensemble average

$$\langle O \rangle = \text{tr } e^{-\beta \mathcal{H}} O / \text{tr } e^{-\beta \mathcal{H}} \tag{2.11}$$

where

$$\mathcal{H} = \mathcal{H}_0 + V$$
$$\mathcal{H}_0 = -H \sum \mu_i; \quad V = -\frac{1}{2} \sum v_{ij} \mu_i \mu_j \tag{2.12}$$

We have split the Hamiltonian into an unperturbed part $\mathcal{H}_0$ and a perturbed part V. The unperturbed density matrix as defined by $\rho_0 \equiv \exp(-\beta \mathcal{H}_0)$ has the advantage that it is factorizable in each of the spins, i.e., $\rho_0 = \prod \exp \beta H \mu_i$. Averages taken with ρ_0 are denoted $\langle \rangle_0$ so that

$$\langle O \rangle = \frac{\text{tr exp}(-\beta \mathcal{H}_0) e^{-\beta V} O}{\text{tr exp}(\beta \mathcal{H}_0) e^{-\beta V}} = \frac{\langle e^{-\beta V} O \rangle_0}{\langle e^{-\beta V} \rangle_0} \tag{2.13}$$

This is now developed in a power series in V to yield a linked cluster expansion. We write this as follows:

$$\langle O \rangle = \sum_{n=0}^{\infty} \beta^n B_n(O) \tag{2.14}$$

To calculate $B_n(O)$ one is essentially presented with a set of graphical rules. It is clear from Eq. (2.13) that in lowest order in V one has $\langle O \rangle = \langle O \rangle_0$. The terms that follow contain intermediate spins coming from the expansion of $e^{-\beta V}$. In nth order one has n powers of the interactions v_{ij} appearing in $B_n(O)$. Each v_{ij} is represented by a bond between points i and j. Let O consist of ν given spins. A graph is said to be linked if the components of the graph connecting spins to each other through v_{ij} bonds are connected by a path of bonds to some of the fixed spins in O. Note, for example, that if $O = \mu_i \mu_j$ then both Figs. 3(a) and 3(b) are linked.

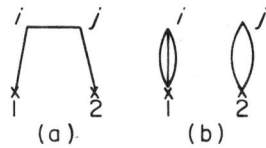

FIG. 3. Typical linked graphs contributing to $\langle \mu_1 \mu_2 \rangle$. (a) $v_{1i} v_{ij} v_{j2} (M_2^0)^4$; (b) $[(1/2!)(v_{1i})^2 M_3 M_2][(1/3!)(v_{2j})^3 M_4 M_3]$ (note $g = 2! \times 3!$).

With these definitions, the rules for calculating $B_n(O)$ are the following:

(1) Draw all possible linked graphs containing n bonds ending at the ν fixed spins of O.

(2) With each bond associate a factor v_{ij}.

(3) With each vertex associate a semi-invariant factor M_s^0 (s is the number of lines joined at a vertex; M_s^0 is defined after rule 5). A fixed point in O is counted as a line in case the vertex is at such a fixed point.

(4) Divide each graph by g where g is the order of the symmetry group of the graph, i.e., g is the number of symmetry operations whereby a graph is transformed into itself, considering the points in O distinct.

(5) Sum over all spin indices except the ν, which are fixed.

We now define M_s^0, called the semi-invariant of order s.

$$M_s^0 \equiv \partial^s \log Z_0 / \partial (\beta H)^s \qquad (2.15)$$

where Z_0 is the unperturbed partition function of a single spin.

$$Z_0 \equiv \log \cosh (\beta H) \qquad (2.16)$$

The proof of these rules is simple and is found in Appendix A. The manipulation of them is less simple and it is suggested that the reader try a few examples in addition to the two presented in Fig. 3.

We now use the cluster expansion to derive some results of interest. As has been shown, the Weiss field becomes valid in the limit $z \to \infty$. Since physical values of z are $O(10)$ or greater, we will adopt $1/z$ as a parameter of smallness. We then may classify graphs according to powers in $1/z$. It will suffice to do this for temperatures, T, in the region of the Curie point, T_c, where $kT_c \cong zJ$. J measures some characteristic strength of the potential. Since each bond carries a factor βJ, we have the result that in the critical region a bond is of order of magnitude $(1/z)$. Now consider the magnetization $R \equiv \langle \mu_i \rangle$. The first few graphs which arise are depicted in Fig. 4. These we classify in powers of $1/z$.

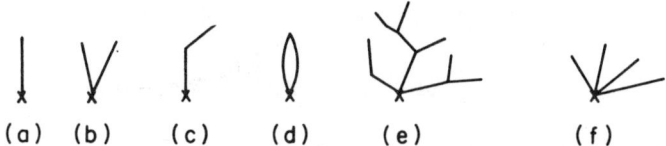

FIG. 4. Graphs used to classify orders in $(1/z)$. Figure 4(e) is a typical Cayley tree.

Figure 4(a) has one bond and the end point is summed on z sites. Hence it is of order of magnitude $(1/z) \times z = 1$. Figure 4(b) and 4(c) contain two bonds and a double summation over z^2 sites and hence is $O(1)$. Figure 4(d) containing two bonds and a single summation is $O(1/z)$. It is then clear that graphs which are of $O(1)$ are the graphs which do not close on themselves. This maintains the maximal number of free summations. These are the Cayley trees, a general example of which is Fig. 4(e).

We will now show that the summation of Cayley trees leads to the Weiss molecular field theory. This is expected since they are the terms of $O(1)$. This derivation of the Weiss theory due to Horwitz and Callen [8] is probably the deepest because it gives insight into the way the interactions reinforce an external field by the creation of an internal field through multiple spin cooperation.

For orientation we first evaluate Figs. 4(a) and 4(b).

$$\text{Fig. 4(a)} = \sum v_{1j} M_1^0 M_2^0$$
$$= [v(0)R_0] M_2^0 \tag{2.17a}$$

$$\text{Fig. 4(b)} = \frac{1}{2}! \sum_j v_{1j} \sum_k v_{1k} M_1^0 M_1^0 M_3^0$$
$$= \frac{1}{2!} [v(0)R_0]^2 M_3^0 \tag{2.17b}$$

Here we introduced the notation $v(\mathbf{q})$ for the qth Fourier transform of the potential, a quantity which will enter into our considerations often:

$$v(\mathbf{q}) = \sum_j v_{ij} \exp\left[i\mathbf{q} \cdot (\mathbf{R}_i - \mathbf{R}_j)\right] \tag{2.18}$$

Since v_{ij} is a function of $(\mathbf{R}_i - \mathbf{R}_j)$, the right-hand side is independent of i. Note also the appearance of the g factor $\frac{1}{2!}$ in Eq. (2.17b) corresponding to the exchange of the two bonds in Fig. 4(b). Also we have introduced $R_0 = \tanh(\beta H)$.

It is clear that the generalization of Eq. (2.17) to the graph with n bonds emanating from μ_1 [Fig. 4(f)], each of which is terminated at its end, is

$$(1/n!)[v(0)R_0]^n M_{n+1}^0 \tag{2.17c}$$

The sum of all of these graphs is then

$$\sum_{n=0}^{\infty} (1/n!)[\beta v(0)R_0]^n M_{n+1}^0 = \sum (1/n!)[\beta v(0)R_0]^n \, \partial^n/\partial(\beta H)^n R_0$$
$$= \exp[\beta v(0)R_0(\partial/\partial\beta H)] \tanh(\beta H) \tag{2.19}$$
$$= \tanh[\beta(H + v(0)R_0)]$$

where we have used the generating function (2.15) according to

$$M_{n+1}^0 = [\partial^{n+1}/\partial(\beta H)^{n+1}] \log Z_0 = [\partial^n/\partial(\beta H^n)] \tanh(\beta H) \tag{2.20}$$

Equation (2.19) is not yet the self-consistent Weiss theory. However, it is immediately evident that each of the ends of Fig. 4(f) can itself be the origin of a new tree [see, for example, Fig. 4(e)]; hence each end which is R_0 in Eq. (2.19) itself becomes converted to R_W (here R_W is that of value of $\langle\mu_i\rangle$ which is determined from the Cayley trees) Thus the infinite summation of trees is equivalent to the solution of the equation

$$R_W = \tanh[\beta(v(0)R_W + H)] \tag{2.21}$$

In this way we see how a given spin sits in the field of the other spins which displaces the external field to the molecular field.

One of the more interesting results of this summation is that in the limit $H \to 0$ a solution of Eq. (2.21) corresponding to a nonvanishing R is possible. It is easily verified that in any finite order of perturbation theory this is not possible; only $R = 0$ is a possible solution. This is a consequence of the symmetry of the problem. If $H = 0$, then only

$R = 0$ is a possible solution. However, in infinite order of perturbation theory, the intrinsic symmetry of the problem is broken and $R \neq 0$ is a possible solution. However, if R is a solution for $H = 0$, so is $-R$. The physical state of the system is then a function of its preparation. If the system is magnetized in a positive field and the field turned off, then $+R$ is the appropriate solution. On the other hand, if no direction was ever established then $\langle \mu_i \rangle$ can be taken to be that linear combination of the two solutions which gives zero, i.e., equal amounts of $+R$ and $-R$. These considerations will play a very important role when we discuss the much richer symmetry group of the Heisenberg model.

Finally, we also remark that, when one calculates the energy, to $O(1)$ one considers in the graphs contributing to $\langle \mu_1 \mu_2 \rangle$ only the double trees of Fig. 5. Thus $\langle \mu_1 \mu_2 \rangle_{\text{trees}} = R_W^2$ and $E_W = -(N/2)v(0)R_W^2$,

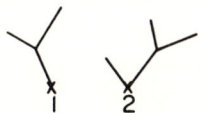

FIG. 5. Energy graphs in the Weiss approximation.

which is the Weiss result. In this way it is seen that terms to $O(1)$ do indeed neglect short-range order effects. We now turn to these neglected effects, hoping that they can be evaluated precisely to $O(1/z)$. Unfortunately, due to divergence difficulties, this turns out not to be the case, and at the present time a systematic development in $(1/z)$ at the Curie point appears impossible. Nevertheless much is to be learned by the present approach.

We will evaluate the energy to $O(1/z)$, first in the absence of long-range order so that the graphs of Fig. 5 vanish (this requires $H = 0$). Remembering that each bond in the critical region carries order of magnitude $(1/z)$, we wish to maximize the number of summations over spins. In the evaluation of $\langle \mu_1 \mu_2 \rangle$ this leads to the selection of the set of cycle graphs of Fig. 6. The nth member of this series has n bonds and

FIG. 6. Ring graph contribution to $\langle \mu_1 \mu_2 \rangle$.

$(n - 1)$ summations; hence is of order $(1/z)^n (z)^{n-1} = O(1/z)$. All other graphs so classified are higher order in $(1/z)$.

The sum on cycles is easily affected

$$\langle \mu_1 \mu_2 \rangle = \langle \mu_1 \mu_2 \rangle_0 + \beta v_{12}(M_2^0)^2 \qquad (2.22)$$
$$+ \sum_{n=1}^{\infty} \beta^n \sum_{i_1 \cdots i_n} [v_{1i_1} v_{i_1 i_2} \cdots v_{i_n 2}][M_2^0]^{n+2}$$

Consider v_{ij} as a matrix. Then the nth term in Eq. (2.22) is the matrix element $[v^n]_{12}$, which is easily calculated if the eigenvalues and eigenfunctions of v_{ij} are known. From translational symmetry the eigenvalues are $v(\mathbf{q})$ and the eigenfunctions are $\exp(i\mathbf{q} \cdot \mathbf{R}_i)$, i.e., the solutions of the equation

$$\sum_j v_{ij} a_j = \lambda a_i \qquad (2.23)$$

are $\lambda = v(\mathbf{q})$ and $a_i = \exp(i\mathbf{q} \cdot \mathbf{R}_i)$ where $\mathbf{q}$ is taken to be in the first Brillouin zone. Because $v_{ii} = 0$ we have

$$\sum_\mathbf{q} v(\mathbf{q}) = 0 \qquad (2.24)$$

This result is then used to evaluate Eq. (2.22). In the absence of H, $\langle \mu_1 \mu_2 \rangle_0 = 0$. Taking the Fourier transform of Eq. (2.22) gives

$$\frac{1}{N} \sum_{i \neq j} \langle \mu_i \mu_j \rangle \exp[i\mathbf{q} \cdot (\mathbf{R}_i - \mathbf{R}_j)] = \sum_{n=1}^{\infty} M_2^0 [\beta v(\mathbf{q}) M_2^0]^n$$
$$= M_2^0 \frac{[\beta v(\mathbf{q}) M_2^0]}{1 - \beta v(\mathbf{q}) M_2^0} \qquad (2.25)$$

In the absence of H, $M_2^0 = 1 - \langle \mu_i \rangle^2 = 1$. We specialize to this case in what follows.

Adding unity to Eq. (2.25) in the form $(1/N) \sum \mu_i^2$ gives in the ring approximation

$$\frac{1}{N} \sum_{i,j} \langle \mu_i \mu_j \rangle \exp[i\mathbf{q} \cdot (\mathbf{R}_i - \mathbf{R}_j)] \equiv \langle |\mu_\mathbf{q}|^2 \rangle = \frac{1}{1 - \beta v(\mathbf{q})} \qquad (2.26)$$

where we have introduced the Fourier transform of the spin fluctuation

$$\mu_\mathbf{q} \equiv \frac{1}{\sqrt{N}} \sum \mu_i \exp(i\mathbf{q} \cdot \mathbf{R}_i) \qquad (2.27)$$

The inverse Fourier transform of Eq. (2.26) gives the two-particle distribution function

$$\langle \mu_i \mu_j \rangle = \frac{1}{N} \sum_\mathbf{q} \frac{\exp[i\mathbf{q} \cdot (\mathbf{R}_i - \mathbf{R}_j)]}{1 - \beta v(\mathbf{q})} \qquad (\mathbf{R}_i \neq \mathbf{R}_j) \qquad (2.28)$$

2. STATISTICAL MECHANICS OF FERROMAGNETISM

We analyze this expression below after first presenting some further properties in this approximation.

(1) *Energy.* The average value of the Hamiltonian is expressed directly in terms of the two-particle distribution function

$$E = -\frac{1}{2}\sum v_{ij}\langle\mu_i\mu_j\rangle = -\frac{1}{2}\sum v(\mathbf{q})\langle|\mu_\mathbf{q}|^2\rangle \tag{2.29}$$

where we have used Eq. (2.24). In the ring approximation we then have (2.26)

$$E_{\text{Ring}} = -\frac{1}{2}\sum_q \frac{v(\mathbf{q})}{1-\beta v(\mathbf{q})} \tag{2.30}$$

(2) *Specific Heat.* Differentiation of Eq. (2.30) yields

$$C_V/k = \frac{1}{2}\sum_q \frac{[\beta v(\mathbf{q})]^2}{[1-\beta v(\mathbf{q})]^2}$$

(3) *Susceptibility.* Though all the present considerations have been made in the absence of H, it is still possible to obtain an expression for the zero field susceptibility. The partition function is

$$Z = \text{tr}\exp[\beta(H\sum\mu_i - V)] \tag{2.31}$$

whence

$$\frac{\partial R}{\partial \beta H} \equiv kT\chi = \frac{1}{N}\frac{\partial^2 \log Z}{\partial(\beta H)^2} = \frac{1}{N}\left[\left\langle\left(\sum\mu_i\right)^2\right\rangle - \left\langle\sum\mu_i\right\rangle^2\right] \tag{2.32}$$

In particular, at $H = 0$

$$\chi = \lim_{q\to 0}\beta\langle|\mu_q|^2\rangle \xrightarrow[\text{Rings}]{} \frac{\beta}{1-\beta v(0)} \tag{2.33}$$

Equation (2.33) is the same as that obtained from the Weiss expression (2.7) for $T > T_c$. Indeed, if we expand the hyperbolic tangent in Eq. (2.7) (valid for infinitesimal H and $T > T_c$), then

$$\chi_{\text{Weiss}} = \frac{\beta}{1-\beta v(0)} \tag{2.34}$$

(4) *Free Energy.* We use the thermodynamic relation $E = \partial\beta F/\partial\beta$ or

$$\beta F = \int_0^\beta E\,d\beta' - N\log 2 \tag{2.35}$$

where the integration constant is simply given by the entropy at $T = \infty$ where all the energy is randomized. Thus

$$\beta F_{\text{Rings}} = \frac{1}{2} \sum_{\mathbf{q}} \log \left[1 - \beta v(\mathbf{q})\right] - N \log 2 \qquad (2.36)$$

We are now prepared to discuss the physical nature of the approximation we have obtained for $T > T_c$ and $H = 0$. For near-neighbor interactions a plot of $v(\mathbf{q})$ against $\mathbf{q}$ along a crystal axis direction is the cosine curve of Fig. 7. For long-range forces the curve is modified as

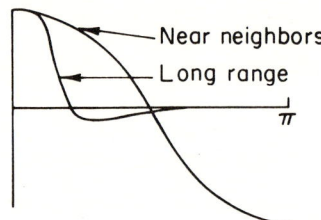

FIG. 7. $v(\mathbf{q})$ vs. $(\mathbf{q})$ for near neighbors along a crystal direction and for long-range force.

indicated. In particular, $v(\mathbf{q})$ approximates to zero for $\mathbf{q} \gtrsim z^{1/3}$ and $v(\mathbf{q})$ is maximal for $\mathbf{q} = 0$. Thus the radius of convergence of the series (2.25) is $\beta v(0) = 1$, which is precisely at the temperature where the Weiss theory yields long-range order. In other words, the theory developed here is that of the precursor phenomenon of short-range order for $T > T_c$ developing into the long-range order for $T < T_c$ given by the Weiss theory. We will see shortly how the long-range order rapidly damps the wild fluctuations represented in the divergence of $\langle |\mu_{\mathbf{q}}|^2 \rangle$ for small $\mathbf{q}$.

To study the singular character of the onset of long-range order we develop $v(\mathbf{q})$ as a power series in $\mathbf{q}$ for small $\mathbf{q}$, where the divergence takes place.

$$v(\mathbf{q}) = v(0)[1 - \alpha q^2]$$
$$\alpha = O(z^{2/3}) \qquad (2.37)$$

Then in the region above but near the Curie point given by $\beta_c v(0) = 1$ we have for small $\mathbf{q}$

$$\langle |\mu_{\mathbf{q}}|^2 \rangle_{\text{Rings}} = \frac{1}{\beta_c v(0) - \beta v(q)} = \frac{1}{(1 - T_c/T) + \alpha q^2} \qquad (2.38)$$

Taking the Fourier transform of Eq. (2.38) gives

$$\langle \mu(0)\mu(\mathbf{R})\rangle = \frac{1}{4\pi\alpha}\frac{e^{-KR}}{KR} \qquad (2.39)$$

where

$$K = \frac{1}{\sqrt{\alpha}}\sqrt{\frac{T-T_c}{T_c}} \qquad (2.40)$$

This shows how, at $T = T_c$, the short-range order develops an infinite correlation length according to $(1/\sqrt{T-T_c})$. This correlation length can be measured by neutron scattering [7].

The other important prediction of Eq. (2.39) is that at $T = T_c$ one has $\langle \mu(0)\mu(R)\rangle \sim 1/R$. The moment expansion [4] gives $\langle \mu(0)\mu(R)\rangle \sim 1/R^{7/4}$.* Hence the present simple theory on the approach to the singularity, though qualitatively satisfying, is incorrect in detail. It also duplicates the macroscopic theory of Ornstein and Zernike [10], which is therefore also incorrect. We discuss some of these features subsequent to our presentation of the more refined versions of these ideas.

Completing our discussion of the thermodynamic behavior for $T > T_c$ at the Curie point, we have

(1) $\qquad \chi \sim \dfrac{1}{T-T_c} \qquad$ rings

$\qquad \chi \sim \dfrac{1}{(T-T_c)^{5/4}} \qquad$ moment expansion

(2) $\qquad C_V \sim \dfrac{1}{\sqrt{T-T_c}} \qquad$ rings

$\qquad C_V \sim -\log|T-T_c| \qquad$ moment expansion
or a weak power
law like $|T-T_c|^{0.2}$

(3) Both energy and free energy are continuous and at $T = T_c$ the order of magnitude of the energy is $Nv(0)/z$ in accord with our qualitative arguments. In Fig. 8 we present schematic plots of C_V and χ for $T > T_c$.

The values of T_c obtained from the moment expansion for various crystal structures and near-neighbor interactions (i.e., $v(0) = zJ$) are as follows: simple cubic, $kT_c = 0.75v(0)$; body centered cubic, $kT_c = 0.79v(0)$; face-centered cubic, $kT_c = 0.82v(0)$.

It is seen that for $z = 12$, the Weiss Curie point is not too bad. The

---

* Note added in proof: This number is now revised on the basis of less rigid initial assumptions. The present behavior is $e^{-KR}/R^{1.06}$; $K \sim (T-T_c)^{0.65}$.

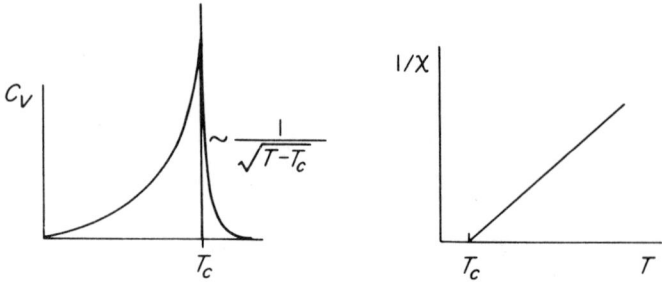

Fig. 8. C_V and $(1/\chi)$ in ring approximation.

conclusion is that the approach of the ring approximation to the Curie point is qualitatively sensible. It gives a Curie point which is as expected $O(1/z)$ too high. However, the quantitative details of the singularity are incorrectly given.

At high $T[(T/T_c) > (1 + O(1/z))]$, our approximation is very good ($\sim 1\%$). The error in our approximation is for powers of interaction greater than z and these terms only contribute within $1/z$ of T_c. What we have done is to isolate the most divergent set of graphs, and apparently the true singularity is concerned with such graphs combined in various ways in very high order. This precludes the possibility of an expansion in $(1/z)$ in the critical region. We return to this point after first showing how the singularity is accommodated in the condensed phase.

We have remarked that the radius of convergence of our series at $\beta v(0) = 1$ coincides with a temperature below which the Weiss theory gives nonvanishing values of R in the absence of H. We then must sum our series under conditions such that each spin sits in a molecular field. The result that we need is essentially Eq. (2.25), but in which M_2^0 (the semi-invariant $1 - R_0^2 = 1 - \tanh^2 \beta H$) is replaced by M_2^W where

$$M_2^W = 1 - \tanh^2 |[\beta(v(0)R_W + H)] \equiv 1 - R_W^2 \qquad (2.41)$$

Having established that the Weiss approximation is the summation of trees, it is evident that the graphs needed for $\langle \mu_1 \mu_2 \rangle$ in addition to those of Fig. 5 (pure Weiss approximation) and Fig. 6 are the rings of Fig. 9 (i.e., a tree added at each vertex). The result is then obtained by simply summing the trees articulated to the points on the ring in the same way as one summed to Eqs. (2.19) and (2.21). Each vertex is then

$$\sum_{n=0}^{\infty} \frac{[\beta v(0)R_W]^n}{n!} [\partial/\partial \beta H]^n [1 - \tanh^2 \beta H] = M_2^W \qquad (2.42)$$

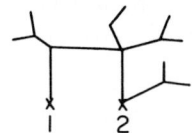

FIG. 9. Cycles in the ordered phase as corrected in Weiss approximation.

We then have in ring approximation

$$\frac{1}{N}\sum_{i\neq j}\langle\mu_i\mu_j\rangle \exp[i\mathbf{q}\cdot(\mathbf{R}_i-\mathbf{R}_j)] = \frac{\beta v(\mathbf{q})M_2^W}{1-\beta v(\mathbf{q})M_2^W} \quad (2.43)$$

Adding $(1/N)\sum[\langle\mu_i^2\rangle - \langle\mu_i\rangle^2] = M_2^W$ gives for $\mathbf{q}\neq 0$

$$\frac{1}{N}\sum_{i,j}[\langle\mu_i\mu_j\rangle - \langle\mu_i\rangle\langle\mu_j\rangle]\exp[i\mathbf{q}\cdot(\mathbf{R}_i-\mathbf{R}_j)] = \frac{M_2^W}{1-\beta v(\mathbf{q})M_2^W} \quad (2.44)$$

whence

$$\langle|\mu_\mathbf{q}|^2\rangle = \frac{1-R_W^2}{1-\beta v(\mathbf{q})(1-R_W^2)}; \quad q\neq 0 \quad (2.45a)$$

$$= NR_W^2 + O(1); \quad q=0 \quad (2.45b)$$

We have used $\sum_i \exp(i\mathbf{q}\cdot\mathbf{R}_i) = N\delta_{0\mathbf{q}}$. The energy is then given by

$$E = -\frac{N}{2}v(0)R_W^2 + \frac{1}{2}\sum_{q\neq 0}\frac{v(\mathbf{q})(1-R_W^2)}{1-\beta v(\mathbf{q})(1-R_W^2)} \quad (2.46)$$

If we remember the rapid rise of R_W with decreasing T (Fig. 1), the denominator of (2.45a) becomes positive nonvanishing for $T < T_c$. This is easily established since the initial rise of R_W with T has infinite slope. At $T=0$, $R_W=1$ and the fluctuation $\langle|\mu_\mathbf{q}\cdot|^2\rangle$ vanishes completely.* For small T, $1-R_W\sim e^{-2\beta v(0)}$ so that the fluctuation vanishes exponentially. In this way the system, by establishing its own internal field, succeeds in accommodating to the singularity, by simply eliminating the fluctuations which give rise to the singularity. At low T, the simple Weiss theory takes over. This occurs for $(T_c-T)/T_c \gtrsim 1/z$.

In short, at high T $[(T-T_c)/T_c > 1/z]$ the first few terms of a power series in the interaction describe the situation precisely and for $(T-T_c)/T_c < -1/z$ the Weiss field is accurate. Within $(1/z)$ of T_c

---

* In fact, it is easily established that the $q=0$ term, which is the most singular, is proportional to the Weiss value of the susceptibility and hence is positive.

on either side a simple approximation is available, but is wrong in its analytical detail. We now turn to attempts at refinement.

3. Refinements on the Ring Approximation

The first thing that comes to mind in improving the above is to note that we have used both trees and rings to calculate the energy but only trees to calculate the magnetization. The initial attempts at refinement were to rectify this. On the one hand the author [7], using his expansion, summed a set of graphs to yield what is called the spherical model [18], and on the other hand Horwitz and Callen [8], using the expansion presented in this review, summed a similar set of graphs to obtain a very similar approximation. As we will need both results in the sequel we present the spherical model by a nongraphical argument and then the Horwitz—Callen result graphically. For the graphical derivation of the spherical model we refer the reader to reference [7].

The simplest way to find the spherical model is to note that the sum rule

$$\sum_i \mu_i^2 = \sum |\mu_\mathbf{q}|^2 = N \tag{2.47}$$

is violated. One can restore this by simply introducing an appropriate factor into $\langle |\mu_\mathbf{q}|^2 \rangle$. This cannot be done by a simple multiplicative factor since $\lim_{\mathbf{q}\to 0} |\mu_\mathbf{q}|^2 \to \infty$ at $T = T_c$. We therefore write (taking first the case $R = 0$) $\langle |\mu_\mathbf{q}|^2 \rangle = [\lambda - \beta v(\mathbf{q})]^{-1}$ or in more convenient form

$$\langle |\mu_\mathbf{q}|^2 \rangle = \frac{1}{1 - \beta(v(\mathbf{q}) - \mu)} \tag{2.48a}$$

where μ is determined by

$$\sum \langle |\mu_\mathbf{q}|^2 \rangle = N \tag{2.48b}$$

We now identify $-\mu/2$ with the energy per particle

$$\begin{aligned}
E &= -\frac{1}{2} \sum_\mathbf{q} v(\mathbf{q}) \langle |\mu_\mathbf{q}|^2 \rangle = -\frac{1}{2} \sum_\mathbf{q} \frac{v(\mathbf{q})}{1 - \beta(v(\mathbf{q}) - \mu)} \\
&= -\frac{\mu}{2} \sum \frac{1}{1 - \beta(v(\mathbf{q}) - \mu)} = \frac{1}{2} \sum \frac{(v(\mathbf{q}) - \mu)}{1 - \beta(v(\mathbf{q}) - \mu)} \\
&= -(N/2)\mu
\end{aligned} \tag{2.49}$$

The Curie point is displaced to

$$kT_c^{\text{Sph}} = v(0) - \mu \tag{2.50}$$

and the specific heat is clearly no longer divergent because of Eq. (2.48b) rather $C_V/N(T = T_c^+) = 2k$. The susceptibility at T_c is obtained from $\lim_{q \to 0} \langle | \mu_q |^2 \rangle = [1 - \beta(v(0) - \mu(T))]^{-1} \sim [T - T_c]^{-2}$. The behavior of C_V and χ is sketched in Fig. 10. It appears that the fluctuations are

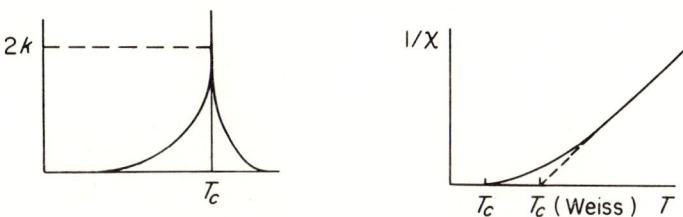

FIG. 10. C_V and $(1/\chi)$ in spherical model approximation.

oversuppressed by the spherical model and the behavior indicated by the moment expansion seems to lie between the spherical and ring approximations. The Curie point is too low. We note that μ at the Curie point is $O(1/z)$ and vanishes at both high and low temperatures.

In the presence of long-range order, one must set

$$\langle | \mu_q |^2 \rangle = \frac{1 - R^2}{1 - \beta(v(\mathbf{q}) - \mu)(1 - R^2)} \tag{2.51}$$

$$\sum_{q \neq 0} \langle | \mu_q |^2 \rangle = N(1 - R^2) \tag{2.52}$$

$$E = -(N/2)v(0)R^2 - \sum_{q \neq 0} v(\mathbf{q})\langle | \mu_q |^2 \rangle - NHR \tag{2.53}$$

To obtain R as a function of T and H one must find the free energy and minimize it with respect to R. This is found by integration with respect to β along a line of constant R. One must also remember that μ is a function of β through Eq. (2.52); this can be handled variationally (for details see reference [7]). The result is [compare Eq. (2.9)]

$$-\beta F = \log W(R) + N\beta H R + (N/2)\beta v(0)R^2$$
$$- \sum_{q \neq 0} \log [1 - \beta(v(\mathbf{q}) - \mu)(1 - R^2)] + \frac{N\beta\mu}{2}(1 - R^2) \tag{2.54}$$

The first term is the constant of integration at $T = \infty$, for fixed R. The next three terms are obtained by direct integration and the last term takes care of the dependence of μ on T. Observe that $\partial F/\partial \mu = 0$ regenerates Eq. (2.52).

Minimizing F with respect to R then gives

$$R = \tanh[\beta(v(0) - \mu)R + H] \qquad (2.55)$$

Notice that $\mu = |2E|/N$ constitutes a small reduction of the molecular field. It is shown in reference [7] that Eq. (2.55) is the value of magnetization which corrects the Weiss value due to rings. We discuss the weaknesses of this theory after first presenting the Horwitz–Callen result.

In addition to articulating all trees to a point to calculate R, we now articulate all rings (Fig. 11). A single ring articulated to a point has the

FIG. 11. Graphs of the Horwitz–Callen approximation to $\langle \mu_1 \mu_2 \rangle$ (only one decorated vertex is shown).

value

$$\text{Ring attached to a point} = \sum_{n=1}^{\infty} \beta^n \sum_{i_1 \cdots i_n} v_{1 i_1} \cdots v_{i_n 1}[M_2^0]^n$$

$$= \frac{1}{N} \sum_{\mathbf{q}} \frac{\beta v(\mathbf{q})}{[1 - \beta v(\mathbf{q}) M_2^0]} \equiv G_2^0 \qquad (2.56)$$

We further define

$$\text{Single bond attached to a point} \equiv \beta v(0) R_0 \equiv G_1^0 \qquad (2.57)$$

where we have *not* included in G_i^0 the semi-invariant factor that accompanies the articulation. Following the same procedure as that leading to Eq. (2.19) we then have

$$R_{\text{Trees+rings}} = \sum_{n,m=0}^{\infty} \frac{1}{n!m!} [G_1^0]^n [G_2^0]^m [\partial/\partial \beta H]^{n+2m} R^0$$

$$= \exp[G_1^0(\partial/\partial \beta H) + G_2^0(\partial/\partial \beta H)^2] \tanh \beta H \qquad (2.58)$$

To arrive at Eq. (2.58) we have introduced the factor $n!m!$ because there are n identical single bonds and m identical rings. Attaching a ring series raises the index of the semi-invariant by 2, attaching a single bond, by 1. In Eq. (2.58), the differential operators $\partial/\partial H$ and $\partial^2/\partial H^2$ do not operate on $G_1{}^0$ and $G_2{}^0$.

We now proceed to the next natural step and articulate full trees and rings on to each of the vertices in question. Thus at each vertex in a ring the M_2 must be replaced by the infinite summation given in Fig. 11. This process is called vertex renormalization (in the present case, by trees and rings). Each vertex in a ring is then

$$M_2 \equiv \sum_{n,m=0}^{\infty} \frac{1}{n!m!} [G_1]^n [G_2]^m [\partial/\partial \beta H]^{n+2m} M_2{}^0$$

$$= \exp\left[G_1(\partial/\partial H) + G_2(\partial/\partial \beta H)^2\right][1 - \tanh^2 \beta H] \quad (2.59)$$

and each single point (i.e., the magnetization) is replaced by

$$M_1 = R = \exp\left[G_1(\partial/\partial \beta H) + G_2(\partial/\partial \beta H)^2\right] \tanh(\beta H) \quad (2.60)$$

where

$$G_1 \equiv v(0)R \quad (2.61a)$$

$$G_2 \equiv \frac{1}{N} \sum_q \frac{\beta v(\mathbf{q})}{1 - \beta v(\mathbf{q}) M_2} \quad (2.61b)$$

In the same approximation $\langle \mu_1 \mu_2 \rangle$ is given by the graphs of Fig. 11:

$$\langle \mu_1 \mu_2 \rangle = R^2 + \frac{1}{N} \sum_q \frac{\beta v(\mathbf{q}) M_2{}^2 \exp[i\mathbf{q} \cdot (\mathbf{R}_1 - \mathbf{R}_2)]}{1 - \beta v(\mathbf{q}) M_2} \quad (2.62)$$

and the energy therefore is given by

$$E = -(N/2)v(0)R^2 - \frac{1}{2} \sum_q \frac{v(\mathbf{q}) M_2}{[1 - \beta v(\mathbf{q}) M_2]} \quad (2.63)$$

The free energy is obtained by integration with respect to β. This integration is affected in a subtle way which involves the proof of some important variational principles. We present the result first and prove the result subsequently. One defines M_0 analogous to Eqs. (2.59) and (2.60).

$$M_0 = \exp\left[G_1(\partial/\partial \beta H) + G_2(\partial/\partial \beta H)^2\right] \log \cosh \beta H \quad (2.64)$$

Then (we mix notation, recalling $M_1 = R$ and $G_1 = v(0)R$ to bring out the symmetry and meaning of the formula),

$$-\frac{\beta F}{N} = M_0 - M_1 G_1 - M_2 G_2 + \tfrac{1}{2} \beta v(0) R^2$$

$$- \frac{1}{2N} \sum_q \log[1 - \beta v(\mathbf{q}) M_2] \quad (2.65)$$

The latter two terms are found by direct integration of Eq. (2.63) whereas the first three terms are constants of integration arising from the β dependence of M_1 and M_2 in Eq. (2.63). Note the variational properties of F:

$$\partial F/\partial G_1 = \partial F/\partial G_2 = 0 \qquad (2.66)$$

$$\partial F/\partial M_1 = \partial F/\partial M_2 = 0 \qquad (2.67)$$

Equation (2.66) recovers the definitions of M_1 and M_2 and (2.67) gives back Eqs. (2.61) for G_1 and G_2.

Using Eqs. (2.66) and (2.67) one finds the energy (2.63) from (2.65) by differentiation with respect to β. Further, when $\beta = 0$, our result gives $-\beta F = \log \cosh (\beta H)$, which is the correct infinite temperature limit, thereby establishing the validity of Eq. (2.65).

These results [either (2.65) or (2.54)] both are consistent theories in that they correctly give both magnetization and energy due to rings and trees, which at first sight is consistent to $O(1/z)$. In particular, one can show that R as determined from

$$R = \partial F/\partial \beta H \qquad (2.68)$$

is the same as the R obtained from the sum of trees and rings articulated to a point. This is left as an exercise for the reader.

Equation (2.68) is a very natural requirement of consistency. One can show that the expressions (2.54) and (2.65) are essentially equivalent and differ graphically only in a minor set of graphs.

Because of the similar nature of the two theories we analyze the faults of the spherical model only. Unfortunately, they are duplicated in detail by the Horwitz–Callen model. When one seeks for the solution of Eq. (2.55) for $T > T_c$ one finds perfectly reasonable solutions of the type of Fig. 2 for small H, but for larger H when one is within $(1/z)$ of T_c, the solution appears as in Fig. 12. According to a theorem of Lee and Yang [19], F is an analytical function of H and T at $H \neq 0$ and hence no such behavior is permitted. Such behavior persists for

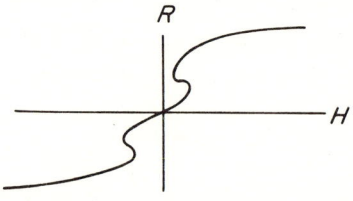

FIG. 12. Loop phenomenon in spherical model.

$T < T_c$ again within $1/z$ of T_c. This phenomenon is traced to the fact that in the present approximation one violates flagrantly the equality

$$kT\chi = \partial R/\partial \beta H = \frac{1}{N}\sum \left[\left\langle\left(\sum \mu_i\right)^2\right\rangle - \left\langle\sum \mu_i\right\rangle^2\right] \quad (2.69)$$

where $\partial R/\partial H$ is calculated by taking the second derivative of the free energy in the present approximation and the right-hand side of Eq. (2.69) is calculated directly graphically. In fact the extra loop in Eq. (2.69) is due to $\chi < 0$ in a certain region.

Englert [9] has analyzed the conditions under which Eq. (2.69) is true and the reader is referred to his article for the proof that one can maintain the validity of this relationship only if one takes all terms in the perturbation series. At the present time it is not known to what extent the violation of Eq. (2.69) in any given approximation is serious.

After the above difficulties turned up, both Horwitz and Englert [20] pointed out the possibility of eliminating the undesirable loops in Fig. 12 by replacing the bonds themselves by rings. This program has been partially carried out by Englert [9]. At the time of writing, it appears not impossible that the approximation of summing all graphs generated from rings gives a logarithmic specific heat.

III. Heisenberg Model

For simplicity, we shall discuss the Heisenberg model for spin $\frac{1}{2}$ using Pauli matrices σ_i instead of $\mathbf{S}_i$. We define in the usual way

$$\sigma_i^\pm = \sigma_i^x \pm i\sigma_i^y \quad (3.1)$$

with the commutation rules

$$[\sigma_i^+, \sigma_j^-] = 4\delta_{ij}\sigma_i^z$$
$$[\sigma_i^z, \sigma_j^\pm] = \pm 2\delta_{ij}\sigma_i^\pm \quad (3.2)$$

The Heisenberg Hamiltonian then becomes

$$\mathscr{H} = \frac{1}{2}\sum_{i\neq j} v_{ij}[\sigma_i^z\sigma_j^z + \tfrac{1}{2}(\sigma_i^+\sigma_j^- + \sigma_i^-\sigma_j^+)]$$

$$= \frac{1}{2}\sum v_{ij}[\sigma_i^z\sigma_j^z + \sigma_i^-\sigma_j^+] \quad (3.3)$$

or in terms of the Fourier transform operators

$$\mathcal{H} = \frac{1}{2} \sum_{\mathbf{q}} v(\mathbf{q})[|\sigma_{\mathbf{q}}^z|^2 + \sigma_{\mathbf{q}}^- \sigma_{-\mathbf{q}}^+] \tag{3.4}$$

where

$$\sigma_{\mathbf{q}} \equiv \frac{1}{\sqrt{N}} \sum_i \sigma_i \exp[i\mathbf{q} \cdot \mathbf{R}_i] \tag{3.5}$$

1. Spin Waves

We shall first introduce the reader to the idea of spin waves so that, in the full statistical theory, we have a simple point of reference on which to hinge the theory.

In the Ising model, at low temperatures a gap of $2v(0)$ develops. That this gap should disappear in the Heisenberg model can be reasoned on grounds of symmetry. As was discussed for the Ising model, up-down symmetry implies solutions $\pm R$ for the magnetization in the absence of H. In the Heisenberg model, the symmetry of isotropy implies that points on a sphere of radius R are possible solutions. Therefore the ground state is $(N+1)$-fold degenerate, corresponding to the $N+1$ points on a unit sphere. There therefore exist motions of the spins at $T = 0$ corresponding to rotation of the system as a whole which require zero energy to excite. These motions correspond to the $\mathbf{q} = 0$ spin wave introduced below. Thus the spectrum of excitation energies must have zero frequency at $\mathbf{q} = 0$. At higher values of $\mathbf{q}$, large spin units rotate against each other so that if the interaction is not of infinite range the frequency must pass continuously to zero as the wavelength of the excitation increases. In this way the richer symmetry of the Heisenberg model implies the filling of the Weiss gap. We now make this quantitative.

We shall consider an infinitesimal magnetic field to be present in the z-direction so that our ground state has all spins up (called $|0\rangle$) with energy

$$E_0 = -\tfrac{1}{2} \sum v_{ij} \langle 0 | \boldsymbol{\sigma}_i \cdot \boldsymbol{\sigma}_j | 0 \rangle = -(N/2)v(0) \tag{3.6}$$

To calculate the excited states we note that since H is scalar it commutes with $\sum \sigma_i^z$ and hence there are no matrix elements connecting states of different S_z. Further, the eigenfunctions of H are also eigenfunctions of S^2, but this is less useful. We expect the first excited states, then, to have one down ("wrong") spin. The states under consideration are $\sigma_i^-|0\rangle$, $i = 1 \ldots N$, and hence the eigenfunctions of one wrong spin will be of the form

$$\psi = \sum a_i \sigma_i^- |0\rangle \tag{3.7}$$

We may derive the secular equation by considering

$$H\sigma_i^-|0\rangle = E_0 \sigma_i^-|0\rangle + [H, \sigma_i^-]|0\rangle \tag{3.8}$$

Using Eq. (3.2), we have

$$[H, \sigma_i^-] = 2\sum_j v_{ij}[-\sigma_j^- \sigma_i^z + \sigma_j^z \sigma_i^-] \tag{3.9}$$

Substituting Eqs. (3.9) and (3.8) into (3.7) gives (with $\sigma_i^z|0\rangle = |0\rangle$)

$$(E - E_0)\sum a_i \sigma_i^-|0\rangle = 2 \sum_{i,j} a_i v_{ij}[\sigma_i^- - \sigma_j^-]|0\rangle \tag{3.10}$$

The term in $-\sigma_j^-|0\rangle$ on the right-hand side shows the admixture of other states into $\sigma_i^-|0\rangle$ owing to spin flips of i against j because of the exchange. Without this term, $\sigma_i^-|0\rangle$ would be an eigenstate of energy $2v(0)$. Writing $\omega = (E - E_0)$ (we shall use $\hbar = 1$), we have

$$[\omega - 2v(0)]\sum a_i \sigma_i^-|0\rangle = -2\sum a_i v_{ij} \sigma_j^-|0\rangle \tag{3.11}$$

Since v_{ij} has translational symmetry, $v_{ij} = v(\mathbf{R}_i - \mathbf{R}_j)$, we expect $a_i \sim \exp(i\mathbf{q} \cdot \mathbf{R}_i)$ as in ordinary band theory. In fact, writing v_{ij} in terms of its Fourier transform, we see immediately that (3.11) is satisfied by $a_i = \exp(i\mathbf{q} \cdot \mathbf{R}_i)$ for $\omega = 2[v(0) - v(\mathbf{q})]$. The conclusion is that the first excited states are given by

$$\begin{gathered}\psi_\mathbf{q} = \sigma_\mathbf{q}^-|0\rangle; \quad \mathbf{q} \text{ in first zone} \\ \omega(\mathbf{q}) = 2[v(0) - v(\mathbf{q})]\end{gathered} \tag{3.12}$$

These excitations are the spin waves. Their spectrum is plotted in Fig. 13 for near-neighbor interaction and for a long-range force. The Weiss energy is thus spread out into a band over a range $q = O(z^{-1/3})$. In the presence of an external field H in the z-direction the commutator (3.8) has the term $2H$ added to it so that

$$\omega(\mathbf{q}; H) = 2[v(0) - v(\mathbf{q})] + 2H \tag{3.13}$$

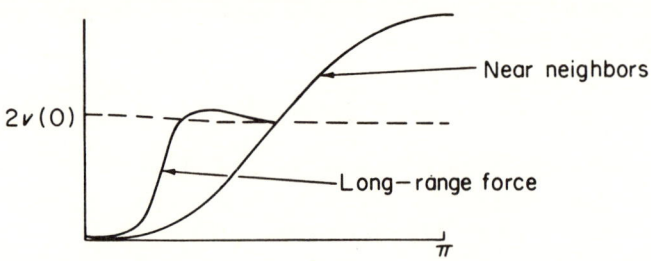

FIG. 13. Spin wave spectrum for near-neighbor forces (q in a crystal axis direction) and long-range forces.

Though Eq. (3.12) is a true eigenstate, it is no longer so that a two-spin-wave state $\sigma_q^- \sigma_{q'}^- |0\rangle$ is an eigenstate since such a function has terms for which the same spin is turned down twice. This function, however, is almost an eigenstate and only misses by $O(1/N)$. However, at finite temperature there is a finite fraction of wrong spins and the description in terms of spin wave states becomes an approximation in some power of $(kT/v(0))$ rather than $1/N$. Since it will turn out that the number of spin waves at low T goes like $T^{3/2}$ and one needs at least two spin waves to give an error in the spin wave description, the deviation from the crude spin wave theory must go at least like T^3. Because of certain cancellations, to be discussed below, the error goes like T^4. Thus, at low T, the states $\prod_q (\sigma_q^-)^{n_q}/n_q! |0\rangle$ comprise an approximate set of normalized eigenstates to be described in terms of Bose statistics (since n_q is perfectly arbitrary). Thus we have (remark that $[\sigma_q^+, \sigma_{-q}^-] = 4$ so that the boson operator is $\sigma_q^-/2 = S_q^-$)

$$\langle n_q \rangle = \langle S_q^- S_q^+ \rangle \equiv \frac{\operatorname{tr} S_q^- S_q^+ e^{-\beta \mathcal{H}}}{\operatorname{tr} e^{-\beta \mathcal{H}}} = [e^{\beta \omega(q)} - 1]^{-1} \qquad (3.14)$$

We now observe by the unitarity of the Fourier transformation that

$$\sum_q S_q^- S_q^+ = \sum_{i=1}^N S_i^- S_i^+ = \frac{\text{Number of}}{\text{wrong spins}} = \frac{N(1-R)}{2} \qquad (3.15)$$

which gives the formula for $R(T)$ in spin wave theory:

$$\frac{(1 - R(T))}{2} = \frac{1}{N} \sum_q [e^{\beta \omega(q)} - 1]^{-1} \qquad (3.16)$$

For a power series in temperature, we observe that for large β, small q contribute predominantly, so that one may expand $\omega(q)$ and extend the integration to all space:

$$\omega(q) = 2v(0)[\alpha q^2 + \gamma q^4 + \cdots] \qquad (3.17)$$

2. STATISTICAL MECHANICS OF FERROMAGNETISM

where $\alpha = O(z^{2/3})$, whereupon

$$[1 - R(T)]/2 = A(kT/v(0))^{3/2} + B(kT/v(0))^{5/2} + \cdots$$

$$A = \frac{1}{8\pi^2} \alpha^{-3/2} \int_0^\infty \frac{x^2\, dx}{\exp(x^2) - 1}$$

$$B = \frac{\gamma \alpha^{-7/2}}{8\pi^2} \int_0^\infty \frac{x^6\, dx \exp(x^2)}{[\exp(x^2) - 1]^2}$$

(3.18)

We now proceed to the study of corrections of spin wave theory at low T. We proceed through the Fourier transform of the equations of motion (3.9):

$$i\dot\sigma_\mathbf{q}^- = [\mathcal{H}, \sigma_\mathbf{q}^-] = \frac{2}{\sqrt{N}} \sum_{\mathbf{q}'} (v(\mathbf{q}') - v(\mathbf{q} - \mathbf{q}'))\sigma_{\mathbf{q}'}^z \sigma_{\mathbf{q}-\mathbf{q}'}^-, \qquad (3.19)$$

At low T, almost all spins are up and hence $\lim_{T\to 0} \sigma_{\mathbf{q}'}^z = \delta_{0\mathbf{q}'}\sqrt{N}$, which upon substitution into Eq. (3.19) gives as required

$$i\dot\sigma_\mathbf{q}^- = \omega(\mathbf{q})\sigma_\mathbf{q}^- \qquad (3.20)$$

One possible approach to generalization to finite T is to use the idea of the random phase approximation (RPA), treating each Fourier component as independent.* This approximation replaces the operator $\sigma_{\mathbf{q}'}^z$ in Eq. (3.19) by its expectation value at finite T: $\langle \sigma_{\mathbf{q}'}^z \rangle = \sqrt{N} R \delta_{0\mathbf{q}'}$, in which case one would obtain

$$i\dot\sigma_\mathbf{q}^- = \omega(\mathbf{q}) R \sigma_\mathbf{q}^- \qquad (3.21)$$

Thus RPA gives a set of bosons whose frequency is a function of T through $\omega(\mathbf{q}; T) = \omega(\mathbf{q})R$. In order to obtain the consequences of this approximation, we observe that we could have obtained these same equations of motion from the original Hamiltonian, by mutilation of the original commutation rules (3.2). These commutation rules in Fourier space read

$$[\sigma_\mathbf{q}^+, \sigma_{\mathbf{q}'}^-] = (4/\sqrt{N})\sigma_{\mathbf{q}+\mathbf{q}'}^z \qquad (3.22a)$$

$$[\sigma_\mathbf{q}^z, \sigma_{\mathbf{q}'}^\pm] = \pm (2/\sqrt{N})\sigma_{\mathbf{q}+\mathbf{q}'}^\pm \qquad (3.22b)$$

The mutilation required is obtained by taking the expectation value of the right-hand side of Eq. (3.22a) and considering (3.22b) to give a

---

* This idea was introduced independently by S. V. Tyablikov in Russia [21] and R. Brout and H. Haken in the United States [22].

nonvanishing commutator only for $\mathbf{q} = 0$ (i.e., only σ_0^z is allowed in the theory; no fluctuation effects due to $\sigma_\mathbf{q}^z$ are allowed to affect the transverse equations of motion). Thus, we have

$$[\sigma_\mathbf{q}^+, \sigma_{\mathbf{q}'}^-]_{\text{RPA}} = 4R\delta_{\mathbf{q},-\mathbf{q}'} \tag{3.23a}$$

$$[\sigma^z, \sigma_{\mathbf{q}'}^\pm]_{\text{RPA}} = \pm (2/\sqrt{N})\delta_{0\mathbf{q}}\sigma_\mathbf{q}^\pm \tag{3.23b}$$

Observe that RPA becomes exact at low T. Our previous considerations lead us to expect that it becomes a good approximation in $O(1/z)$ near the Curie point since ring graphs do not mix Fourier components. Hence, it is possible to use it as an interpolation tool which takes us from one region to the other.

Equations (3.23) imply that the boson creation operator is now to be taken as $\sigma_\mathbf{q}^-/2\sqrt{R}$; hence

$$\langle S_\mathbf{q}^- S_{-\mathbf{q}}^+ \rangle_{\text{RPA}} = \frac{R}{e^{\beta\omega(\mathbf{q})R} - 1} \tag{3.24}$$

so that the equation of $R(T)$ is given by (we include a magnetic field H in the result)

$$\frac{1-R}{2} = \frac{1}{N} \sum_\mathbf{q} \frac{R}{\exp[\beta(\omega(\mathbf{q})R + H)] - 1} \tag{3.25}$$

This is a result derived by Englert [23] on the basis of the fluctuation dissipation theorem, and independently due to Tyablikov [21] through a Green's function approach.

Before proceeding further, we shall now inquire into the validity of RPA at low T. From Eq. (3.25), it is readily established that the first deviations from spin wave theory set in with T^3. This conclusion turns out to be incorrect owing to an important cancellation effect mentioned above. The error lies in the fact that in Eq. (3.19) the expectation value of $\sigma_{\mathbf{q}'}^z$ was taken too soon. In fact, $\sigma_{\mathbf{q}'}^z$ is strongly correlated to $\sigma_{-\mathbf{q}'}^-$, a feature which is completely ignored in our calculations. To show this we proceed as follows. Since we are at low T, we shall consider that the spin wave states are a proper description of the system. To be sure, they are redundant in our spin space, but since there are few spin waves around, this redundancy is unimportant (in fact, Dyson [13] demonstrates conclusively that this redundancy gives an effect that vanishes like $\exp[-2\beta v(0)]$ at low T). We then define a set of pseudo-spin bose operators $\hat{\sigma}_\mathbf{q}^-$ which have the property that

$$[\hat{\sigma}_\mathbf{q}^+, \hat{\sigma}_{-\mathbf{q}}^-] = 4\delta_{\mathbf{q}\mathbf{q}'} \tag{3.26}$$

2. STATISTICAL MECHANICS OF FERROMAGNETISM

The factor 4 is introduced in analogy to Eq. (3.22a). The operator σ_q^z is then constructed from the operators $\hat{\sigma}_q^\pm$ in such a way that Eq. (3.22b) is satisfied. Consequently, we replace σ_q^z by

$$\sigma_q^z \to \frac{-1}{2\sqrt{N}} \sum_{q'} \hat{\sigma}_{q-q'}^- \hat{\sigma}_{q'}^+ + \delta_{0q} \sqrt{N} \tag{3.27}$$

The second term is added in order to make the result come out correctly for $q = 0$ where the ordering of the operators is ambiguous (we require $\sigma_0^z = \sqrt{N}R$). Substituting Eq. (3.27) into Eq. (3.19) and replacing $\sigma_q^\pm$ by $\hat{\sigma}_q^\pm$ gives

$$i\dot{\hat{\sigma}}_q^- = \omega(\mathbf{q})\hat{\sigma}_q^- - \frac{1}{N} \sum_{q'q''} [v(\mathbf{q}') - v(\mathbf{q} - \mathbf{q}')]\hat{\sigma}_{q-q'}^- \hat{\sigma}_{q'-q''}^- \hat{\sigma}_{q''}^+ \tag{3.28}$$

We now observe that Eq. (3.28) could have been equally derived from a pseudo-Hamiltonian in terms of the operators $\hat{\sigma}_q$:

$$H_{\text{int}} = \frac{-1}{4N} \sum_{qq'q''} [v(\mathbf{q}'') + v(\mathbf{q}'' - \mathbf{q}' + \mathbf{q}) - v(\mathbf{q} - \mathbf{q}'') - v(\mathbf{q}' + \mathbf{q}'')]$$

$$\times [\hat{\sigma}_{q-q''}^- \hat{\sigma}_{q'+q''}^- \hat{\sigma}_{-q'}^+ \hat{\sigma}_{-q}^+] \tag{3.29}$$

This Hamiltonian was derived by Dyson in a different way by considering matrix elements between the various spin wave functions. If the eigenvalues of H could be found, then barring the redundancy question, the problem could be solved.* This is, of course, not possible. However, since Eq. (3.29) gives an interaction between two spin waves, it is clear that a perturbation series in H^{int} can be arranged to give a power series in T. The lowest correction is the sum of all diagrams in which only two spins interact and hence will be of the form

$$\langle H_{\text{int}} \rangle_{\text{low}T} = \frac{1}{2} \left(\frac{4}{N}\right) \sum_{q,q'} t(\mathbf{q}, \mathbf{q}') \langle n(\mathbf{q}) \rangle \langle n(\mathbf{q}') \rangle \tag{3.30}$$

where t is the forward scattering amplitude for the process $(\mathbf{q}, \mathbf{q}') \to (\mathbf{q}, \mathbf{q}')$, which may be evaluated at $T = 0$. The approximation (3.30) is familiar in the theory of Fermi liquids, and in fact, serves as the basis of Landau's theory of quasiparticles. We use these ideas to interpret the meaning of Eq. (3.30) in terms of spin waves. Then we have

$$(E - E_0)_{\text{low}T} = \sum_q \omega(\mathbf{q})\langle n_q \rangle + \left(\frac{4}{2N}\right) \sum_q t(\mathbf{q}, \mathbf{q}')\langle n(\mathbf{q}) \rangle \langle n(\mathbf{q}') \rangle \tag{3.31}$$

---
* Insofar as the redundance problem is unimportant, Dyson has demonstrated that one can use the many-boson state $\Pi_q(\hat{\sigma}_q^-)^{n_q}/n_q! \mid 0\rangle$ to determine the thermodynamic properties, provided the interaction of Eq. (3.29) is taken into account.

Observe that if a spin wave **q** is removed this requires an energy

$$\omega(\mathbf{q}, T) = \omega(\mathbf{q}, 0) + \left(\frac{4}{N}\right) \sum t(\mathbf{q}, \mathbf{q}') \langle n(\mathbf{q}') \rangle \qquad (3.32)$$

Hence, Eq. (3.31) implies a shift in frequency of the spin wave at finite temperature to the indicated value.

In first approximation $t(\mathbf{q}, \mathbf{q}')$ is given by Born approximation. This is obtained from Eq. (3.29) by setting $q'' = 0$ to give

$$t_{\text{Born}}(\mathbf{q}, \mathbf{q}') = [v(0) + v(\mathbf{q} - \mathbf{q}') - v(\mathbf{q}) - v(\mathbf{q}')] \qquad (3.33)$$

$$= \omega(\mathbf{q}, T = 0)/2 + [v(\mathbf{q}') - v(\mathbf{q} - \mathbf{q}')] \qquad (3.33\text{a})$$

In terms of Eq. (3.33a) one may write (3.32) (with $(1/N) \sum \langle n(\mathbf{q}) \rangle = (1 - R)/2$):

$$\omega(\mathbf{q}, T) = \omega(\mathbf{q}; T = 0)R + \frac{2}{N} \sum [v(\mathbf{q}') - v(\mathbf{q} - \mathbf{q}')] \langle n(\mathbf{q}') \rangle \qquad (3.34)$$

The first term is the RPA result (3.21). The second term of Eq. (3.34) is a consequence of the pairing of indices which arises from the fact that $\sigma_\mathbf{q}^z$ is affected by the presence of spin waves. It corresponds to an exchange correction. We now show how the T^3 effect disappears because of cancellation in the two terms (3.33a). From Eq. (3.31), we see that $t(\mathbf{q}, \mathbf{q}')$ need only be evaluated for small q, q', whereupon, expanding Eq. (3.33), we have

$$\begin{aligned} t_{\text{Born}} = {} & \alpha[(\mathbf{q} - \mathbf{q}')^2 - \mathbf{q}^2 - \mathbf{q}'^2] \\ & + \gamma[(\mathbf{q} - \mathbf{q}')^4 - \mathbf{q}^4 - \mathbf{q}'^4] \end{aligned} \qquad (3.35)$$

If this is inserted into Eq. (3.32) and we use inversion symmetry $\langle n(\mathbf{q}) \rangle = \langle n(-\mathbf{q}) \rangle$ we obtain

$$\omega(\mathbf{q}; T) = \omega(\mathbf{q}; T = 0) - 12 \frac{\gamma}{N} \sum_{\mathbf{q}'} (\mathbf{q}' \mathbf{q}')^2 \langle n(\mathbf{q}') \rangle$$

$$= \omega(\mathbf{q}; T = 0) - \frac{12\gamma}{3N} \mathbf{q}^2 \sum_{\mathbf{q}'} \mathbf{q}'^2 \langle n(\mathbf{q}') \rangle$$

The second term is proportional both to $\omega(\mathbf{q})$ and $(E(T) - E(0)) = \sum \omega(\mathbf{q}) n(\mathbf{q})$ so that we obtain in Born approximation

$$[\omega(\mathbf{q}; T) - \omega(\mathbf{q}; T = 0)] = -CT^{5/2} \omega(\mathbf{q}; T = 0) \qquad (3.36)$$

This result was first obtained by Keffer and Loudon [24] as a physical interpretation of Dyson's theory. It is seen that the original shift of

$O(T^{3/2})$ is cancelled out. For near-neighbor interactions the result can be sharpened a bit since for this case

$$\sum v(\mathbf{q} - \mathbf{q}')\langle n(\mathbf{q}')\rangle = J \sum_{\text{near neighbors}} \cos[(\mathbf{q} - \mathbf{q}') \cdot \mathbf{a}]\langle n(\mathbf{q}')\rangle$$

$$= J \cos(\mathbf{q} \cdot \mathbf{a}) \sum \cos(\mathbf{q}' \cdot \mathbf{a})\langle n(\mathbf{q}')\rangle$$

where a term odd in $\mathbf{q}'$ has been discarded. Substituting into Eqs. (3.33) and (3.32), we have

$$\left. \frac{\omega(\mathbf{q}; T) - \omega(\mathbf{q}; T=0)}{\omega(\mathbf{q}; T=0)} \right|_{\text{near neighbor}} = \frac{\sum \omega(\mathbf{q}'; T=0)\langle n(\mathbf{q}')\rangle}{v(0)} = \frac{E(T) - E(0)}{E(0)} \quad (3.37)$$

As far as higher-order contributions to t are concerned, it is an easy matter to show that they are proportional to $(\mathbf{q} \cdot \mathbf{q}')^2$ for small q, q', and hence Eq. (3.36) is merely multiplied by a constant. This has been calculated by Dyson for near-neighbor interactions. At the worst, the deviation from Born approximation is $\sim 50\%$. The deviation goes like $1/S$ for arbitrary spin S so that spin $1/2$ is the least favorable.

The results presented here, which were first established by Dyson, have been obtained by Oguchi [25] using the method of Holstein and Primakoff [26]. This method arranges the power series according to both $1/S$ and $(kT/v(0))$ and suffers from poor convergence. Nevertheless, keeping terms to a given power in $1/S$ in a consistent fashion leads to the right answer. For spin $1/2$, the method is of doubtful value, but for large spin it can be of help. Both the above method and that of Holstein and Primakoff are immediately taken over for antiferromagnetism, but we shall not do this here.

It is then seen that the T^3 dependence in R as obtained from Eq. (3.25) is spurious. However, it is easy to show that this deviation from the exact result is very minor from the point of view of the R vs. T curve, but nevertheless important in a measurement of the temperature dependence of the spin wave spectrum. We estimate below the importance of the error and show that at most it can amount to about 1% in the magnetization at temperatures where the Dyson approximation still makes sense. To do this we use *RPA*.

For $H = 0$, we have

$$\langle \sigma_\mathbf{q}^- \sigma_\mathbf{q}^+ \rangle = \frac{4R}{\exp(2\beta(v(0) - v(\mathbf{q}))R] - 1} \quad (3.38)$$

The first thing to notice is that for $z \to \infty$ (i.e., $v(\mathbf{q}) \to \delta_{0q} v(0)$) this theory goes over correctly to the Weiss theory, as is seen by rewriting Eq. (3.25).*

$$R = 1 - \frac{1}{N} \sum_q \frac{2R}{\exp[2\beta v(0)R] - 1} = 1 - \frac{2R}{\exp[2\beta v(0)R] - 1}$$

or

$$R = \tanh[\beta v(0)R] \tag{3.39}$$

Now for finite z, $v(\mathbf{q})$ will cut off at $q \sim z^{-1/3}$. Hence, in the sum of $\mathbf{q}$ in Eq. (3.28) there will be $0(1/z)$ terms of the "spin wave" type and $0[1 - (1/z)]$ of the molecular-field type. At low T only the first type gives a contribution; the remaining terms will be exponentially small. Hence, at low T one will have a pure spin wave type of behavior. To see this more clearly, let us approximate the potential by

$$v(0) - v(q) \sim \begin{cases} v(0) z^{2/3} q^2 & q < z^{-1/3} \\ v(0) & q > z^{-1/3} \end{cases}$$

The magnetization is then

$$R = \left[1 - \frac{2R}{2\pi^2 z} \int_0^1 \frac{d^3\hat{q}}{\{\exp[2\beta v(0) R \hat{q}^2] - 1\}} - \frac{2R\left(1 - \frac{1}{2\pi^2 z}\right)}{\{\exp[2\beta v(0) R] - 1\}} \right] \tag{3.40}$$

In the second term of the bracket we have changed variables to $\hat{q} = q z^{1/3}$. This term is the spin wave contribution and the third term is the molecular-field term. These two becomes equal at a temperature which we denote by T_e, given approximately by

$$\frac{1}{2\pi^2 z} \left(\frac{kT_e}{2v(0)} \right)^{3/2} = \exp\left[-\frac{2v(0)}{kT_e} \right] \left[1 - \frac{1}{2\pi^2 z} \right] \tag{3.41}$$

For $T < T_e$, the spin wave contributions dominate, and Dyson's corrections will have a role in this region. For $T > T_e$, the molecular-field terms dominate and the principal effects are a result of what Dyson calls the kinematic interaction, i.e., Weiss field effects.

In the limit of infinite z, Eq. (3.47) gives $T_e = 0$. For finite z, representative solutions of the transcendental equation are: for $z = 10$, $T_e/T_c \simeq 0.2$; for $z = 100$, $T_e/T_c \simeq 0.15$. (We have put $kT_c = v(0)$), Thus, we find that spin wave effects dominate the behavior of the model

---

* By $z \to \infty$, we mean that the range goes to ∞.

up to about 20% of the Curie point. In this range, Dyson's correction, which is $O((T/T_c)^{5/2})$ of the usual spin wave theory, is quite negligible $O(1\%)$. Therefore, for the calculation of equilibrium properties it is probably no great error to forsake spin wave corrections in order to use Eq. (3.28) as an interpolation formula to the Curie point.

This interpolation is quite remarkable in that formula (3.25) extrapolates to the spherical model at and above T_c in the presence of small field. This is seen as follows. At low T, a solution of Eq. (3.28) for $H = 0$ gives a possible $R \neq 0$. Increasing T decreases R. The Curie point is then obtained by expanding Eq. (3.28) about $R = 0$ and $H = 0$, with the result that

$$1 = \frac{1}{N} \sum_{\mathbf{q}} \frac{1}{\beta_c[v(0) - v(\mathbf{q})]} \tag{3.42}$$

This is evidently the equation for the spherical-model Curie point, as is easily seen by substituting into the sum rule

$$\frac{1}{N} \sum_{\mathbf{q}} \frac{1}{1 + \beta\mu - \beta v(\mathbf{q})} = 1 \tag{3.43}$$

the value of β at the Curie point given by $\beta_c(v(0) - \mu) = 1$. For $T > T_c$, then, one must pass to the limit $R \to 0$, $H \to 0$ simultaneously such that $(R/H) = \chi$. Then Eq. (3.28) gives

$$1 = \frac{1}{N} \sum_{\mathbf{q}} \frac{1}{\beta[v(0) - v(\mathbf{q}) + (1/\chi)]} \tag{3.44}$$

This supplies an equation for χ which is the same as the spherical model value as shown by the usual spherical value of the susceptibility

$$\chi = \frac{\beta}{1 - \beta(v(0) - \mu)}; \quad 1 + \beta\mu = \beta v(0) + (\beta/\chi) \tag{3.45}$$

and substituting into the sum rule (3.43).

Thus Eq. (3.25) supplies an interpolation formula which links the low-temperature spin wave region to the critical region where it gives the qualitative singular behavior characteristic of the phase transition. It is then of interest to inquire to what extent this approximation can be duplicated by a systematic treatment such as has been developed for the Ising model. This analysis has been carried out graphically [14] and is reported below.

2. Graphical Analysis of the Heisenberg Model

We first present the linked cluster expansion of the Heisenberg model. This is most easily obtained by the same techniques as Eq. (2.11), (2.12), and (2.13), where now

$$V = -\frac{1}{2}\sum v_{ij}\, \boldsymbol{\sigma}_i \cdot \boldsymbol{\sigma}_j \qquad (3.46)$$

Since V and $\mathcal{H}_0$ do not commute, it is necessary to introduce the time-ordered expansion of Dyson:

$$\exp[-\beta(H_0 + V)] = \exp(-\beta H_0) \sum_{n=0}^{\infty} (1/n!) T \int_0^\beta \cdots \int_0^\beta d\beta_1 \cdots d\beta_n V(\beta_1) \cdots V(\beta_n) \qquad (3.47)$$

where

$$V(\beta') = \exp(\beta' \mathcal{H}_0)\, V \exp(-\beta' \mathcal{H}_0) \qquad (3.48)$$

and T is Dyson's time-ordering operator.

The linked cluster expansion for any operator O containing ν indices is then of the form

$$\langle O \rangle = \sum_{n=0}^{\infty} \xi^n B_n(O) \qquad (3.49)$$

where for the sake of classification we have introduced a coupling constant ξ which may be set equal to unity at the end of the calculation. The coefficient $B_n(O)$ then contains n powers of the interaction.

Before giving the rules to calculate $B_n(O)$ we discuss the modifications required because of (a) noncommutativity and (b) the presence of the two types of interaction in Eq. (3.4): longitudinal, $v_{ij}\sigma_i^z\sigma_j^z$, and transverse, $v_{ij}\sigma_i^-\sigma_j^+$. To handle the noncommutativity problem one introduces the time-ordered semi-invariants of interaction representation operators which are defined in the usual way by

$$\sigma_i^\pm(\beta') = \exp(\beta' \mathcal{H}_0)\sigma_i^\pm \exp(-\beta' \mathcal{H}_0) = \exp(\pm 2\beta' H)\, \sigma_i^\pm$$
$$\sigma_i^z(\beta') = \sigma_i^z \qquad (3.50)$$

Here β' is the "time" at which the interaction involving the spin occurs. The time-ordered semi-invariants are semi-invariants with a time-ordering operator in front of the expectation value of all products, e.g.,

$$TM_2(\sigma_i^+(\beta_i)\sigma_j^-(\beta_j)) \equiv T\langle \sigma_i^+(\beta_i)\sigma_j^-(\beta_j)\rangle_0$$
$$- T\langle \sigma_i^+(\beta_i)\rangle_0 \langle \sigma_j^-(\beta_j)\rangle_0 \qquad (3.51)$$

To distinguish between longitudinal and transverse bonds, we first note that $\langle \sigma_i^{\pm}\rangle_0 = 0$, so that a transverse bond must be joined by other transverse bonds at a vertex. We designate a transverse bond by an arrow directed from the index carrying σ^+ to the index carrying σ^-. Transverse graphs then are all closed and are composed of oriented cycles. Longitudinal bonds do not have an arrow. A vertex "i" containing operators σ_i^+, σ_i^-, σ_i^z either from powers of the interaction V or from O represents a time-ordered semi-invariant the order of which is the number of bonds joined at the vertex plus the number of spins from O. The spins from O are explicitly labeled $+$, $-$, or z so that the value of the vertex can be read unambiguously from the graph. The number of σ^+'s (σ^-'s) in the semi-invariant is given by the number of bonds carrying arrows out of (into) the vertex plus the number from O; the number of σ^z's is the number of undirected bonds joined at the vertex plus those from O. For example, the vertex of Fig. 14 corresponds to the

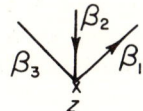

FIG. 14. A vertex in a Heisenberg model graph corresponding to the semi-invariant $M_4(T\sigma^+(\beta_1)\sigma^-(\beta_2)\sigma^z(\beta_3)\sigma^z(0))$.

semi-invariant $M_4(T\sigma^+(\beta_1)\sigma^-(\beta_2)\sigma^z(\beta_3)\sigma^z(0))$. Note that the spins from O carry argument $\beta' = 0$.

The rules for calculating $B_n(O)$ are then the same as those of the Ising model [Eqs. (2.14), (2.15), and (2.16)] with the following supplements. Rule 1 must be supplemented by stating that one must include both oriented and nonoriented bonds; the oriented bonds never terminate. Rule 2 is supplemented by labeling each bond with the "time" "β_i" at which the interaction occurs. Rule 3 is supplemented by the use of the time-ordered semi-invariants as discussed above. In Rule 4 one must also include the factor $\frac{1}{2}^{n_t}$ where n_t is the number of transverse bonds, owing to the fact that these bonds are oriented. In Rule 5, include also integration over $\beta_1 \ldots \beta_n$ from 0 to β.

The proof of this expansion is given in Appendix B.

The most important result to recover graphically is that of spin waves which we know to be rigorous at low temperatures. We must then see how such graphs behave at high temperatures. We will show below that spin waves are contained in the transverse ring graphs which together with the longitudinal rings at high temperatures are the correct set to take in an expansion in $(1/z)$. Consequently one has an immediate framework on which to build a theory which can span the whole temperature region. As usual, divergence troubles in the critical region frustrate simplicity.

Before summing on rings we first point out that the sum on trees in the Heisenberg model is the same as that in the Ising model, hence leading to the Weiss theory. This is because a tree always has free ends at the end of its branches. Therefore it cannot contain transverse operators. Since it can contain only longitudinal operators, and therefore only operators which commute, the time-ordering operator drops out of the calculation and one recovers the previous results (2.17) to (2.21).

We will show below that the sum on transverse rings for $T > T_c$ gives the classical result (2.26) for $\langle \sigma_q^+ \sigma_q^- \rangle$. For $T < T_c$, an ordered phase is imposed and a consistent theory is possible only when the vertices in the rings have trees addended to them in the manner of the Ising model, Eq. (2.41). For the transverse rings, rather than finding (2.41) one recovers spin waves. For the longitudinal rings, Eq. (2.41) still applies and hence longitudinal rings vanish exponentially at low T.

We sum first the simple rings of Fig. 15 to obtain the simplest terms in

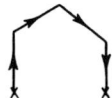

FIG. 15. Transverse ring graph contribution to $\langle \sigma_1^+(\beta_1)\sigma_2^-(\beta_2)\rangle$.

the ring approximation to the transverse propagator $\equiv \langle T\sigma_i^+(\beta_i)\sigma_j^-(\beta_j)\rangle$, following closely reference [14].

The unperturbed propagator is

$$\langle T\sigma_i^+(\beta_1)\sigma_j^-(\beta_2)\rangle_0 = 4\delta_{ij} \frac{\exp[-2(\beta_1 - \beta_2)H]}{e^{\beta H} + e^{-\beta H}} \begin{matrix} |e^{\beta H} & \beta_1 > \beta_2 \\ |e^{-\beta H} & \beta_1 < \beta_2 \end{matrix} \quad (3.52)$$

As a function of $(\beta_1 - \beta_2)$, this function is periodic with period β and hence may be Fourier analyzed:

$$\langle T\sigma_i^+(\beta_i)\sigma_j^-(\beta_j)\rangle_0 = 4\delta_{ij} \sum_{k=-\infty}^{\infty} \frac{\tanh(\beta H)}{(2\beta H - i\lambda_k)} \exp[-i\lambda_k(\beta_i - \beta_j)] \quad (3.53)$$

where $\lambda_k = 2\pi k/\beta$. The unperturbed propagator (3.54) occurs as follows in the sum over rings:

$$\langle T\sigma_i^+(\beta_i)\sigma_j^-(\beta_j)\rangle_{\text{Rings}} = \sum_{n=0}^{\infty} [(\tfrac{1}{2})^n v_{ii_1} v_{i_1 i_2} \cdots v_{i_n j}$$

$$\times \int_0^\beta \cdots \int_0^\beta d\beta_1 \cdots d\beta_n \langle T\sigma_i^+(\beta_i)\sigma_{i_1}^-(\beta_1)\rangle_0$$

$$\times \langle T\sigma_{i_1}^+(\beta_1)\sigma_{i_2}^-(\beta_2)\rangle \cdots \langle T\sigma_j^+(\beta_n)\sigma_j^-(\beta_j)\rangle_0 \quad (3.54)$$

Fourier decomposition of both v_{ij} and the propagator (3.53) then leads to

$$\langle T\sigma_i^+(\beta_i)\sigma_j^-(\beta_j)\rangle_{\text{Rings}} = 4\sum_{\mathbf{q}} \sum_{n=0}^{\infty} \sum_{k=-\infty}^{+\infty} \exp[i\mathbf{q}\cdot(\mathbf{r}_i - \mathbf{r}_j)] \exp[-i\lambda_k(\beta_i - \beta_j)]$$

$$\times [2\beta v(\mathbf{q})]^n [\tanh(\beta H)/(2\beta H - i\lambda_k)]^{n+1} \quad (3.55)$$

$$= \sum_{\mathbf{q}} \sum_k \exp\{i[q\cdot(\mathbf{r}_i - \mathbf{r}_j) - \lambda_k(\beta_i - \beta_j)]\}$$

$$\times \frac{\tanh \beta H}{\beta[2H - i\lambda_k - 2v(\mathbf{q})\tanh \beta H]}$$

The sum over all integers k may be replaced by a contour integral with respect to z by using the function $\pm \beta[\exp(\pm \beta z) - 1]^{-1}$ which has poles with unit residue at the points $z = (2\pi i/\beta)j$. The function which makes the integrand converge for all large z is chosen. The result is

$$\langle T\sigma_i^+(\beta_i)\sigma_j^-(\beta_j)\rangle_{\text{Rings}} = 4\sum_{\mathbf{q}} [\exp[i\mathbf{q}\cdot(\mathbf{R}_i - \mathbf{R}_j)] \tanh(\beta H)$$

$$\times \exp[-(\beta_i - \beta_j)(2H - 2v(\mathbf{q})\tanh \beta H)] \times \begin{cases} g^-(\mathbf{q}) & \beta_i > \beta_j \\ g^+(\mathbf{q}) & \beta_i < \beta_j \end{cases} \quad (3.56)$$

where

$$g^\pm(\mathbf{q}) = \pm\{\exp[\pm\beta(2H - 2v(\mathbf{q})\tanh(\beta H))] - 1\}^{-1} \quad (3.57)$$

It is proved in reference [14] that the effect of addending trees to each vertex in the ring is to displace H to $[H + \beta v(0)R_W]$ where R_W is the Weiss magnetization (2.6).

We then insert this renormalization correction into Eqs. (3.56) and (3.57). Each of the factors in (3.57) which contains H arises from an unperturbed semi-invariant. Thus the renormalization with longitudinal trees causes everywhere a translation of H to the value $H + v(0)R_W$. Our result is then

$$\langle T\sigma_i^+(\beta_i)\sigma_j^-(\beta_j)\rangle_{\text{Rings with tree renormalization}}$$
$$= 4\sum_q \exp[i\mathbf{q}\cdot(\mathbf{R}_i - \mathbf{R}_j)]\tanh[\beta(H + v(0)R_W)]$$
$$\times \exp\{-(\beta_i - \beta_j)[2H + 2v(0)R_W - 2v(q)\tanh(\beta(H + v(0)R_W))]\}$$
$$\times \begin{cases} \tilde{g}^-(\mathbf{q}) & \beta_i > \beta_j \\ \tilde{g}^+(\mathbf{q}) & \beta_i < \beta_j \end{cases} \quad (3.58)$$

where

$$\tilde{g}^\pm(\mathbf{q}) = \pm\{\exp[\pm\beta(2H + 2v(0)R_W - 2v(\mathbf{q})\tanh(\beta(H + v(0)R_W)))] - 1\}^{-1} \quad (3.59)$$

This expression is simplified for $T < T_c$ to

$$\langle T\sigma_i^+(\beta_i)\sigma_j^-(\beta_j)\rangle_{\text{Rings}} = \sum_q \exp[i\mathbf{q}\cdot(R_i - R_j)]\langle T\sigma_\mathbf{q}^+(\beta_i)\sigma_{-\mathbf{q}}^-(\beta_j)\rangle_{\text{Rings}}$$

where

$$\langle T\sigma_\mathbf{q}^+(\beta_i)\sigma_{-\mathbf{q}}^-(\beta_j)\rangle_{\text{Rings}} = \exp[-2(\beta_i - \beta_j)[R_W(v(0) - v(\mathbf{q})) + H]\}$$
$$\times 4R_W \begin{cases} \tilde{g}^- & \beta_i > \beta_j \\ \tilde{g}^+ & \beta_i < \beta_j \end{cases} \quad (3.60)$$

where

$$g^\pm = \pm[\exp\{\pm 2\beta[(v(0) - v(\mathbf{q}))R_W + H]\} - 1]^{-1} \quad (3.61)$$

At low temperatures $R_W \to 1$, so that we are led to the correct spin wave frequency

$$\omega(\mathbf{q}) = 2[v(0) - v(\mathbf{q}) + H] \quad (3.62)$$

and we have

$$\langle T\sigma_\mathbf{q}^+(\beta_i)\sigma_{-\mathbf{q}}^-(\beta_j)\rangle = 4\exp[-(\beta_i - \beta_j)\omega(\mathbf{q})] \times \begin{cases} \hat{g}^- & \beta_i > \beta_j \\ \hat{g}^+ & \beta_i < \beta_j \end{cases} \quad (3.63)$$

where

$$\hat{g}^\pm = \pm[\exp(\pm\beta\omega(\mathbf{q})) - 1]^{-1} \quad (3.64)$$

which is in fact the Fourier transform of the spin wave propagator. Equation (3.63) then gives for the spin wave populations [obtained by setting $\beta_i = \beta_j - 0$ in Eq. (3.63)]:

$$\langle n(\mathbf{q})\rangle = \tfrac{1}{4}\langle\sigma_\mathbf{q}^-\sigma_{-\mathbf{q}}^+\rangle = [\exp(\beta\omega(\mathbf{q})) - 1]^{-1} \quad (3.65)$$

To examine Eq. (3.43) for $kT > v(0)$, we take the Fourier transform of Eq. (3.58) for $\beta_i = \beta_j$ to give

$$\langle \sigma_{-q}^- \sigma_q^+ \rangle = \frac{4 \tanh[\beta(H + v(0)R_W)]}{\exp\{2\beta[H + v(0)R_W - v(q)\tanh(\beta(v(0)R_W + H))]\} - 1} \quad (3.66)$$

Taking the limit as $R_W = H = 0$ for $kT > v(0)$ gives

$$\langle \sigma_{-q}^- \sigma_q^+ \rangle = \frac{2}{1 - \beta v(\mathbf{q})} \quad (3.67)$$

which is the same as the classical Ising model result (2.26). It is shown in reference [7] that this result arises because in the nonferromagnetic phase the noncommutativity of spin operators is of no consequence in the ring approximation.

It is shown in reference [14] how to addend cycles to a ring as in the spherical model or Horwitz–Callen approximation to the Ising model. The result of this correction is so close in its consequences to the RPA theory of Eq. (3.28) that we will not derive it here. The essential point is that a graphical derivation of the essential RPA results is possible and hence relieves the theory of some of its arbitrariness. As in the Ising model, the theory will probably give difficulty at finite H for T in the critical region. Presumably these difficulties can be handled in the same way as for the Ising model, but no attempt has been made in this direction.

Of the graphs which are needed in the above refinement to the ring approximation, some contribute to the Dyson correction at low temperatures [Eq. (3.34)]. However, one of the essential graphs is missing (cf. reference [14]) so that a low-temperature T^3 correction still exists. It is therefore impossible to develop a simple approximation which gives both the low-temperature Dyson corrections correctly as well as a respectable qualitative approach to the Curie point.

IV. Band Theory of Ferromagnetism

The ideas which have been sketched above find their counterparts very readily in a band theory. In the extreme case of a free-electron gas, Bloch [27] showed a number of years ago that in Hartree–Fock theory the Coulomb exchange repulsion induces ferromagnetism for a free-electron gas at sufficiently low density. The theory of the ferromagnetic phase can then be treated in Weiss approximation which for a band theory is due to Stoner [28]. This self-consistent theory gives rise to a gap in the same way as the Weiss theory. The gap is then filled with spin waves. The theory of how to do this is contained in recent work of

Englert and Antonoff [15] as well as Kubo *et al.* [29]. The former work shows how to pass continuously from the band theory to the Heisenberg model as the electrons become increasingly localized. Very recent work of Gutswiller [30] has put this whole approach in doubt. Very likely what will occur in future theory is that the structure of the ground state will conform to Gutswiller, but the excitations will remain as presented below.

The theory of the precursor to the phase transition is easily developed and follows the same pattern as for the Ising and Heisenberg models. This work is due to Stern [31] and is the finite temperature generalization of work of Wolff [33] which is related to the quasiparticle formulation of Landau [34].

We first review the elementary Stoner theory. The Hartree–Fock energy of an electron of wave number $\mathbf{k}$, spin σ is

$$E_{\mathbf{k}\sigma} = \epsilon_{\mathbf{k}} + \sum_{\mathbf{k}',\sigma'} [v(0) - \delta_{\sigma\sigma'} v(\mathbf{k} - \mathbf{k}')] n(\mathbf{k}', \sigma') \qquad (4.1)$$

Here $\epsilon_{\mathbf{k}}$ is the unperturbed Bloch energy which is spin independent, and $v(\mathbf{q})$ is the Fourier transform of the interaction potential which in our work is the screened Coulomb interaction. For simplicity we will take $v(\mathbf{q}) = \text{constant} \equiv V/N$. The term in $v(0)$ in Eq. (4.1) is the direct term and the term in $v(\mathbf{k} - \mathbf{k}')$ in the exchange which couples only electrons of the same spin. Since the direct term is constant, it may be taken up in a shift of the origin of energy and we will drop it from this point on. Since $v(\mathbf{q})$ is constant we may then write

$$E_{\mathbf{k}\uparrow} = \epsilon_k - V(N\uparrow/N) \qquad (4.2a)$$

$$E_{\mathbf{k}\downarrow} = \epsilon_k + V(N\downarrow/N) \qquad (4.2b)$$

$$E_{\mathbf{k}\uparrow} - E_{\mathbf{k}\downarrow} = -V\left(\frac{N\uparrow - N\downarrow}{N}\right) = -VR \qquad (4.2c)$$

where R is the magnetization per electron in dimensionless units.

Adding an irrelevant constant to Eq. (4.2) gives

$$\begin{aligned} E_{\mathbf{k}\uparrow} &= \epsilon_{\mathbf{k}} - V(R/2) \\ E_{\mathbf{k}\downarrow} &= \epsilon_{\mathbf{k}} + V(R/2) \end{aligned} \qquad (4.3)$$

The self-consistent field equations are then obtained by remarking that the electrons in Hartree–Fock approximation become quasiparticles whose populations are then given by

$$n_{\mathbf{k}\sigma} = \frac{1}{\exp[\beta(E_{\mathbf{k}\sigma} - \mu)] + 1} \qquad (4.4)$$

where μ is determined by the condition $\Sigma_{k,\sigma} n_{k\sigma} = N$. Substituting Eq. (4.3) into (4.4) and summing over k then gives

$$\frac{1}{N} \sum_k (n_{k\uparrow} - n_{k\downarrow}) = R = \frac{1}{N} \sum_k \qquad (4.5)$$

$$\times \left\{ \frac{1}{\exp[\beta(\epsilon_k - V(R/2) - \mu)] + 1} - \frac{1}{\exp[\beta(\epsilon_k + V(R/2) - \mu)] + 1} \right\}$$

which is the Stoner equation for R vs. (T). Calculational details are to be found in a review article by Wohlfarth [35]. The essential points are that the deviation from perfect magnetization at small T is proportional to T^2. At the Curie point R vs. T has infinite slope. The Curie point is found by setting $R = 0$ in Eq. (4.5) to give

$$1 = -V \sum_k (\partial n_k / \partial \epsilon_k) \qquad (4.6)$$

No ferromagnetism occurs if at $T = 0$ the right-hand side of Eq. (4.6) is < 1; i.e., if $g(\epsilon_F)$ is the density of states at the Fermi surface then the density below which ferromagnetism occurs is given by

$$1 = g(\epsilon_F) V \qquad (4.7)$$

where $\epsilon_F = k_F^2 / 2\tilde{m} \rho^{2/3}$. Equation (4.7) is the Bloch criterion.

The reader who is interested in the graphical theoretical derivation of the above results is referred to the paper of Horwitz et al. [36], where it is shown how to derive Hartree-Fock theory from the graphs of the free energy. It will be noted that the infinite Cayley trees of exchange graphs which are summed in reference [36] are equivalent to the Cayley trees of Horwitz and Callen. Therefore the philosophy of the Stoner-Weiss field is the same as discussed in Section II, 1.

Before going into the spin wave modifications of the Stoner theory we first study briefly the precursor phenomena associated with the Curie point (4.6).* We present Stern's graphical method [31, 32] which is a synthesis of the ideas of Wolff [33] at zero temperature and Englert's

---

* The following paragraphs constitute a chapter in what is typically called the "many-fermion" problem. The techniques are similar to those of the spin problem, but are complicated by kinetic energy considerations of the fermions. To develop the necessary techniques would take us far astray from our task of exploring the statistical methods of ferromagnetism. For this reason what follows below is quite cursory. To the initiated, the formulas should be clear. To those unfamiliar with the many-fermion problem, the results (4.10) and the dispersion relation (4.13) are the central features. An introduction to the many-electron problem will be found in Brout and Carruthers [36a].

[23] treatment of the Heisenberg model at finite temperature. An equally simple method is that of linearized equations of motion of Wolff [33] generalized to finite temperature. As Wolff has shown, the susceptibility is approximately calculated by the elementary set of graphs of Fig. 16. The incoming dotted line is a magnetic field of wave number **q**

FIG. 16. The set of graphs leading to Eq. (4.8).

and the symbol x marks the measurement of the induced spin. This set of graphs is summed directly (for $v(\mathbf{q}) = V/N = $ constant) to

$$kT\chi_\mathbf{q} = \frac{\langle \sigma_\mathbf{q} \rangle}{\beta H} = \frac{\chi_q{}^0(0)}{1 + \frac{V}{N}\sum_\mathbf{k} \frac{n(\mathbf{k}+\mathbf{q}\uparrow) - n(\mathbf{k}\downarrow)}{E(\mathbf{k}+\mathbf{q}\uparrow) - E(\mathbf{k}\downarrow)}} \quad (4.8)$$

where $\chi_q{}^0(\omega)$ is the susceptibility of the free-electron gas given by

$$\chi_q{}^0(\omega) = \frac{1}{N}\sum_k \frac{n(\mathbf{k}) - n(\mathbf{k}+\mathbf{q})}{E(\mathbf{k}+\mathbf{q}) - E(\mathbf{k}) - (\omega + i\epsilon)}$$

(The sum on k does not include a sum on spin, and "up" means in the direction of H.) If H has frequency ω, the generalization of Eq. (4.8) is

$$kT\chi_\mathbf{q}(\omega) = \frac{\chi_q{}^0(\omega)}{1 + \frac{V}{N}\sum_\mathbf{k} \frac{n(\mathbf{k}+\mathbf{q}\uparrow) - n(\mathbf{k}\downarrow)}{E(\mathbf{k}+\mathbf{q}\uparrow) - E(\mathbf{k}\downarrow) - (\omega + i\epsilon)}} \quad (4.9)$$

where we have as usual included a small imaginary part $i\epsilon$ in ω.

In the limit $q \to 0$ then, Eq. (4.8) goes to

$$kT\chi = \frac{(1/N)\sum_k \partial n_k/\partial \epsilon}{1 + (V/N)\sum (\partial n_k/\partial \epsilon)} \quad (4.10)$$

One sees that χ blows up at the same critical point as the Stoner theory [Eq. (4.6)]. It is easily established from Eq. (4.10) that in the vicinity of the Curie point $\chi \sim 1/(T - T_c)$.

The theory of the specific heat is developed by summing the free energy graphs corresponding to the susceptibility graph of Fig. (16).

These are the graphs (Fig. 17). The two are related by the fluctuation dissipation theorem (Englert [23], Stern [32]) and one proves that the expectation value of the interaction due to the graphs of Fig. 17

FIG. 17. Graphs contributing to $\langle V \rangle$ leading to Eq. (4.11).

is given by

$$\langle V \rangle = \sum_{\mathbf{q}} \frac{1}{\pi} \int_{-\infty}^{+\infty} d\omega \, \frac{1}{e^{\beta\omega} - 1} \, \text{Im} \, [kT\chi_{\mathbf{q}}(\omega)] \qquad (4.11)$$

where $\chi_{\mathbf{q}}(\omega)$ is given by Eq. (4.9). The specific heat is found directly from Eq. (4.11) and leads to the not surprising dependence:

$$C_V \sim \frac{1}{\sqrt{T - T_c}} \qquad (4.12)$$

To continue the free energy for $T < T_c$ requires the onset of long-range order since Eq. (4.11) will diverge unless $R \neq 0$. In terms of graphs it means that the Hartree–Fock renormalization graphs of $\langle n_{\mathbf{k}\sigma} \rangle$ lead to Eqs. (4.4) and (4.3) and hence to $R \neq 0$. This is analogous to the Ising and Heisenberg model procedure of addending all tree graphs to the cycles of Section II, 2. In the band theory, the trees are the Hartree–Fock bubbles addended to the electron propagators.

The result is then to replace the $\langle n_{\mathbf{k}\sigma} \rangle$ and $E_{\mathbf{k}\sigma}$ in Eq. (4.9) by Eqs. (4.4) and (4.3), respectively. It is an easy matter to prove that the free energy is convergent. However, much more is gained. We will show below that Eq. (4.9) has poles, corresponding to the spin waves. These poles mean that if the system is exposed to an oscillating field of frequency ω, there will be a resonant response of true excitations of the system. In fact, we know that there must be a mode at $q = 0$, following the symmetry arguments of Section III. The poles of Eq. (4.9) then give the spin wave dispersion relation. The free energy gets a spin wave contribution from the graphs of Fig. 17. This is seen directly from Eq. (4.11) since $\text{Im} \, \chi \sim \delta(\omega - \omega_q)$ where ω_q is the spin wave frequency (see Englert [23], Englert and Brout [37], and Stern [32] for details of how this kind of calculation works).

The spin wave dispersion relation as obtained from the pole of Eq. (4.9) is

$$1 = -\frac{V}{N}\sum_{\mathbf{k}}\frac{n_{\mathbf{k+q}\uparrow} - n_{\mathbf{k}\downarrow}}{E_{\mathbf{k+q}\uparrow} - E_{\mathbf{k}\downarrow} - \omega} \qquad (4.13)$$

$$= -\frac{V}{N}\sum_{\mathbf{k}}\frac{n_{\mathbf{k+q}\uparrow} - n_{\mathbf{k}\downarrow}}{\epsilon_{\mathbf{k+q}} - \epsilon_{\mathbf{k}} + VR - \omega} \qquad (4.14)$$

The roots of ω have been carefully investigated by Englert and Antonoff [15] and Kubo *et al.* [29] where $n_{\mathbf{k}\sigma}$ is given by Eq. (4.4). For small $\mathbf{q}$ (by small $\mathbf{q}$ we mean that both $v_F q \ll VR$ and $\omega \ll VR$ are satisfied) then one finds

$$\omega_q \sim \frac{q^2}{2mR} \qquad (4.15)$$

The mode in question is similar to a Wannier-type exciton where the particle in the down state forms a bound state with the hole in the up state. For $q = 0$, the binding energy is VR, which just eliminates the gap. The set of collective excitations is good so long as ω_q does not exceed the energy needed to make single-particle excitations, for then a branch line arises in Eq. (4.14) corresponding to damping of the excitation. An estimate of the cutoff yields

$$q_c \simeq k_F \left(\frac{VR}{\epsilon_F}\right) \qquad (4.16)$$

In other words the collective modes fill the gap with effective mass proportional to R and cutoff wave number given by (4.16).

At this point we have seen how a free-electron model can develop a ferromagnetic state which is closely analogous in its mathematical structure to the Heisenberg model of localized spins on fixed sites. The big difference is in the nature of the spin waves. The band theory spin wave of frequency given by Eq. (4.15) is analogous to a Wannier-type exciton whereas the Heisenberg model spin wave is analogous to a Frenkel exciton in which the wave propagates by successive flips. Englert [15] has shown how to pass continuously from one picture to the other as the states become increasingly localized.

We will not review the details of Englert's work, but only the essential notions and results. Details are found in his paper. One works in a Bloch representation and for simplicity we will confine ourselves to a single nondegenerate band. (Important physical effects are therefore left out.)

The matrix element for scattering from $\mathbf{k}_1$, $\mathbf{k}_2$ to $\mathbf{k}_3$, $\mathbf{k}_4$ is called $V_{\mathbf{k}_2\mathbf{k}_1}^{\mathbf{k}_3\mathbf{k}_4}$, i.e.,

$$V_{\mathbf{k}_2\mathbf{k}_1}^{\mathbf{k}_3\mathbf{k}_4} = \int\int d\mathbf{r}_2\, d\mathbf{r}_1\, \psi_{\mathbf{k}_4}^* \psi_{\mathbf{k}_3}^* v(\mathbf{r}_2 - \mathbf{r}_1)\psi_{\mathbf{k}_2}\psi_{\mathbf{k}_1} \quad (4.17)$$

where $\psi_{\mathbf{k}}$ are the Bloch functions. If we use a single-zone scheme we can let $\mathbf{k}_4 = \mathbf{k}_2 + \mathbf{q}$, $\mathbf{k}_3 = \mathbf{k}_1 - \mathbf{q}$ with the usual reciprocal lattice translation prescription in the case of Umklapp processes. The essential point is the analysis of Eq. (4.17) in terms of the Wannier functions φ which are related to the Bloch functions by

$$\varphi(\mathbf{r} - \mathbf{R}_m) = \frac{1}{\sqrt{N}} \sum_{\mathbf{k}} \exp(-i\mathbf{k}\cdot \mathbf{R}_m)\psi_{\mathbf{k}}(\mathbf{r}) \quad (4.18)$$

where $\mathbf{R}_m$ is a lattice vector. Then Eq. (4.17) becomes

$$\begin{aligned}V_{\mathbf{k}_2\mathbf{k}_1}^{\mathbf{k}_1+\mathbf{q},\mathbf{k}_2-\mathbf{q}} = \frac{1}{N^2} \sum_{lmnp} \int\int d\mathbf{r}_2\, d\mathbf{r}_1 \varphi^*(\mathbf{r}_1 - \mathbf{R}_l)\varphi^*(\mathbf{r}_2 - \mathbf{R}_m) \\
\times v(\mathbf{r}_1 - \mathbf{r}_2)\varphi(\mathbf{r}_2 - \mathbf{R}_n)\varphi(\mathbf{r}_1 - \mathbf{R}_p) \exp[i\mathbf{k}_1\cdot(\mathbf{R}_p - \mathbf{R}_l)] \\
\times \exp[i\mathbf{k}_2\cdot(\mathbf{R}_n - \mathbf{R}_m)] \exp[i\mathbf{q}\cdot(\mathbf{R}_m - \mathbf{R}_l)]\end{aligned} \quad (4.19)$$

In the tight binding limit, $\varphi(\mathbf{r} - \mathbf{R}_l)$ becomes strongly localized about $\mathbf{R}_l$ and it becomes a good approximation to neglect three- and four-center integrals in Eq. (4.19). This approximation is also valid in the free electron limit since then φ tends to a δ function. Thus if we are interested in an interpolation scheme between the Heisenberg and free-electron models, it is appropriate to neglect the three- and four-center integrals. Then Eq. (4.19) becomes

$$V_{\mathbf{k}_2\mathbf{k}_1}^{\mathbf{k}_1+\mathbf{q},\mathbf{k}_2-\mathbf{q}} = (1/N)[v(\mathbf{q}) + J(\mathbf{k}_1 - \mathbf{k}_2 - \mathbf{q}) + K(\mathbf{k}_1 + \mathbf{k}_2)] \quad (4.20)$$

$$v(\mathbf{q}) = \sum_p \exp(i\mathbf{q}\cdot \mathbf{R}_p) \int\int |\varphi(\mathbf{r}_2 - \mathbf{R}_p)|^2 |\varphi(\mathbf{r}_1)|^2 v(\mathbf{r}_1 - \mathbf{r}_2)\, d\mathbf{r}_1\, d\mathbf{r}_2 \quad (4.20a)$$

$$\begin{aligned}J(\mathbf{k}_1 + \mathbf{k}_2 - \mathbf{q}) = \sum_p \exp[i(\mathbf{k}_1 - \mathbf{k}_2 + \mathbf{q})\cdot \mathbf{R}_p] \int\int d\mathbf{r}_1\, d\mathbf{r}_2 \\
\times \varphi^*(\mathbf{r}_2 - \mathbf{R}_p)\varphi^*(\mathbf{r}_1)v(\mathbf{r}_1 - \mathbf{r}_2)\varphi(\mathbf{r}_1 - \mathbf{R}_p)\varphi(\mathbf{r}_2 - \mathbf{R}_p)\end{aligned} \quad (4.20b)$$

$$\begin{aligned}K(\mathbf{k}_1 + \mathbf{k}_2) = \sum_p \exp[i(\mathbf{k}_1 + \mathbf{k}_2)\cdot \mathbf{R}_p] \int\int d\mathbf{r}_1\, d\mathbf{r}_2 \\
\times \varphi^*(\mathbf{r}_2)\varphi^*(\mathbf{r}_1)v(\mathbf{r}_1 - \mathbf{r}_2)\varphi(\mathbf{r}_1 - \mathbf{R}_p)\varphi(\mathbf{r}_2 - \mathbf{R}_p)\end{aligned} \quad (4.20c)$$

In the free-electron limit $J = K = 0$ and $v(\mathbf{q})$ is the V which appears in all our previous equations. The presence of J and K in Eq. (4.20) gives the effect of localization.

One can make further qualitative progress by drastic simplification and replace $v(\mathbf{q})$ and $K(\mathbf{k}_1 + \mathbf{k}_2)$ by a single average on the Fermi surface. However, one should keep the q dependence of J since Eq. (4.20b) tells us that J is indeed the Fourier transform of the exchange integral used in the Heisenberg model. The simplified V is then taken to be

$$V^{\mathbf{k}_1+\mathbf{q},\mathbf{k}_2-\mathbf{q}}_{\mathbf{k}_2 \mathbf{k}_1} = V + J(\mathbf{q}) \tag{4.21}$$

and the dispersion relation (4.13) is converted to

$$1 = -\frac{V + J(\mathbf{q})}{N} \sum_{\mathbf{k}} \frac{n_{\mathbf{k}+\mathbf{q}\downarrow} - n_{\mathbf{k}\downarrow}}{\epsilon(\mathbf{k}+\mathbf{q}) - \epsilon(\mathbf{k}) + (V + J(0))R - \omega} \tag{4.22}$$

In the free-electron limit $J = 0$, and one recovers the previous results of this section. In the extreme localized limit, one may neglect $\epsilon(\mathbf{k}+\mathbf{q}) - \epsilon(\mathbf{k})$ since the effective mass of the Bloch electrons is then infinite. In that case Eq. (4.22) is converted to

$$1 = \frac{[V + J(\mathbf{q})]R}{[V + J(0)]R - \omega} \tag{4.23}$$

$$\omega = [J(0) - J(\mathbf{q})]R \tag{4.24}$$

which is our random phase approximation result for the Heisenberg ferromagnet. Between the extremes, one can develop a formula for ω at small q in the effective mass approximation for $\epsilon(\mathbf{k})$. Under these conditions, one finds

$$\omega \sim \frac{q^2}{2m^*R} + [J(0) - J(\mathbf{q})]R \tag{4.25}$$

If the second term in Eq. (4.25) dominates, the excitation is primarily spin-flip-type propagation (Frenkel exciton), and if the first term dominates, the excitation is more like a bound particle–hole pair (Wannier exciton).

The reader is referred to Englert's paper [15] for a discussion of how this formalism can give rise to antiferromagnetism as well as states whose local magnetic moments vary incommensurately with the lattice. This reference also contains the theory needed to get the magnetization and thermodynamic functions.

V. Random Spin Systems

We envision a crystal in which spins are dissolved at high temperatures so that they go to random sites to which they are subsequently frozen upon cooling. Thus, the spin positions are completely uncorrelated with the value of the spin in question. This situation is to be distinguished from a gas of particles interacting through spin forces alone, for in the latter case spatial configurations of the spins are very much correlated with the spin configuration. To find the mathematical expression of this situation, consider a portion i of the crystal. Let this portion have free energy F_i, so that by the extensive property of F

$$F_{\text{Total}} = \sum_i F_i \qquad (5.1)$$

Equation (5.1) indicates that it is appropriate to compute the free energy on an ensemble basis with (surface effects neglected)

$$\langle F \rangle = \langle \log Z \rangle \qquad (5.2)$$

computed on an ensemble of crystals of xN spins each, distributed on N sites. The ensemble has all distributions of the spins with equal *a priori* probability. Equation (5.2) is to be sharply contrasted to the gas case. In this case one has thermodynamic equilibrium, all the F_i are equal, and $F = \log Z$ where Z sums the spin partition function on sites as well as spins. The difference between these two problems is extremely important to bear in mind. For example, for forces with a range there will be threshold concentration for ferromagnetism for the frozen-in case; whereas for the gas case this will not occur, since the configurations where the spins will all be close and parallel will be heavily weighted in Z.

We now proceed with a qualitative discussion of some of the implications of Eq. (5.2). If the range of interaction covers z sites, then clearly when $x < O(1/z)$, spins will not be connected by a continuous chain of sites, and hence, no cooperative phenomenon can exist. Thus, there exists a threshold concentration. For near-neighbor interactions in cubic lattices this occurs at $x_t \simeq 2/z$, i.e., the concentration necessary for a given spin to have on the average two spins in its exchange field, thereby allowing propagation of a chain. One may guess that this is a general result and independent of the near-neighbor assumption.

For large x, it is possible to have $xz \gg 1$. For such x, a Weiss field approach is approximately valid, in which case $kT_c \simeq xv(0)$. A sketch of T_c vs. x is given in Fig. 18. For xz large, one has a linear region and then a sharp drop off at $x \simeq 2/z$ as sketched in Fig. 18.

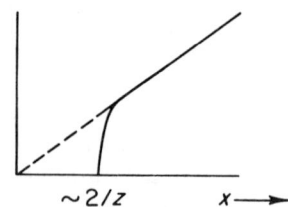

Fig. 18. Schematic plot of T_c vs. x.

Let us now try to construct the magnetization curve. For $xz \gg 1$, the curve will once more be of Weiss character except at low T, where spin waves come in. (The Heisenberg random ferromagnet has not been handled theoretically at all apart from estimating x_l). An interesting feature comes in as x approaches x_l. Consider the near-neighbor case: x_l is the concentration needed to maintain chains of spins connected through the sample. Clearly, many spins will be left out of these chains, giving a spontaneous saturation magnetization $R(T = 0) < 1$. In fact, as we must have $\lim_{x \to x_l} R, (T = 0) = 0$. Elliott [38] has given arguments to show that for near-neighbor forces $R(T = 0, x) \sim \exp[-1/(x - x_l)]$ for x very near x_l. Suppose now that we include the effect of second-neighbor forces. Let these be z_2 in number and of strength J_2 and let there be z_1 first neighbors of strength J_1 with $z_1 J_1 \gg z_2 J_2$. Consider $x = x_l + \epsilon$, where x_l is reckoned with first neighbors alone and $\epsilon \ll x_l$. Then for $T < T_c$, R levels off at the value typical of first-neigbor interactions of $e^{-(1/\epsilon)}$. The second neighbors at this high T are not yet forced parallel to the chains of first neighbors. Now as T decreases to $O(z_2 J_2)$, we will start to line up the second neighbors and R will climb to a much higher value. This is sketched schematically in Fig. 19. This phenomenon is expected only for $\epsilon \ll x_l$, for once x increases to any substantial degree above x_l the second neighbors are brought in much faster, yielding a standard magnetization curve. For long-range forces, this phenomenon is not expected. Further, forces which oscillate

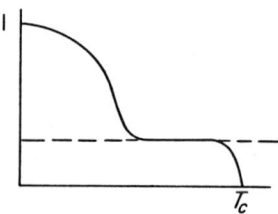

Fig. 19. Effect of weak second-neighbor force for $x = x_l + \epsilon$.

in sign as a function of distance must be treated on an entirely different basis.

To put these ideas on a quantitative basis, we will use the Ising model, since we are interested primarily in phenomena in the Curie-point region. At low T, these results will then be modified by spin waves.

For the Ising model, the cluster development of Section II may be taken over easily with the following modification. In order to compute $\langle F \rangle$, the most convenient formulation is to consider μ_i as a random variable taking on the values $+1$, 0, -1 with probabilities $x(1 + R)/2$, $(1 - x)$, $x(1 - R)/2$, respectively. In the spin algebra we still maintain the rule $\mu_i^{2n} = \mu_i^2$, $\mu_i^{2n+1} = \mu_i$, n integer, so that

$$\langle \mu_i^{2n} \rangle = x \tag{5.3a}$$

$$\langle \mu_i^{2n+1} \rangle = xR \tag{5.3b}$$

For $xz \gg 1$, we may again sum on cycles. The semi-invariant structure is again a product of M_2's where according to Eq. (5.3)

$$M_2 = x - x^2 R^2$$

so that

$$\langle |\mu_q|^2 \rangle_{\text{Rings}} = \frac{x - x^2 R^2}{1 - \beta[v(q)][x - x^2 R^2]} \tag{5.4}$$

The Curie point is then at $kT_c = xv(0)$.

Clearly for $xz = O(1)$, this theory is highly inadequate and does not predict a threshhold concentration. The clue to the theory of the dilute case is supplied by the remark that the power of x is the number of vertices of a graph. Further, any classification in the dilute region according to powers of $\beta_c J$ is clearly spurious since our qualitative discussion indicated that, near x_t, kT_c is a violently varying function of x. Unfortunately, no rigorous theory is presently available even for limiting cases so that here we will only construct a crude theory which carries the qualitative implications discussed above. The method is based on the cluster expansion of Brout [7], specifically designed for the random spin problem.

We will not present the general expansion, but only state the essential point; namely, a graph with ν vertices carries a power of $x^{\nu-1}$. Therefore the set of graphs of two vertices, sketched in Fig. 20 all carry the power x. For any given T, this set will give the correct behavior for small x. However, it is not a suitable set for determination of kT_c since $T_c = T_c(x)$ Nevertheless, we shall see that, for $T < T_c$, this set of graphs does con-

FIG. 20. Graphs contributing to coefficient of x in the free energy.

tain all the necessary qualitative features. This summation is easily performed [7], to give the contribution to $\langle F \rangle$ from the graphs as

$$(1/N)\langle -\beta F \rangle = \log W(R) + \frac{x}{2} {\sum_{j}}' [\cosh \beta v_{ij} + R^2 \sinh \beta v_{ij}] \quad (5.5)$$

The magnetization curve is then obtained from $\partial F/\partial R = 0$ to give

$$R = \tanh \left[xR {\sum_{j}}' \frac{\tanh \beta v_{ij}}{1 - R^2 \tanh \beta v_{ij}} \right] \quad (5.6)$$

and the Curie point from $\partial^2 F/\partial R^2 = 0$:

$$1 = x {\sum_{j}}' \tanh (\beta v_{ij}) \quad (5.7)$$

Equations (5.6) and (5.7) do indeed give the qualitative behavior sketched above. Since the quantitative aspect is unreliable, we will not calculate on detailed models here. Note that Eq. (5.7) does indeed give a threshhold concentration. Indeed, for a potential covering z spins, Eq. (5.7) gives

$$x_t = (1/z) \quad (5.8)$$

The calculation of x_t can be considerably refined without undue effort. The object is to calculate $\partial^2 F/\partial R^2$ as a power series in x, the contribution to F in x^ν coming from all graphs with $(\nu + 1)$ vertices. Setting $kT_c = 0$ in this result gives an equation for x_t. The result of this calculation for the simple cubic lattice gives [39]*

$$1 - xz + 6x^2 + 30x^3 - 114x^4 + 186x^5 + \cdots = 0$$

Plotting successive roots indicates $x_t \cong 0.28$ with an indicated precision of a few percent.

An alternative method due to Elliott *et al.* [40] calculates the divergence of the susceptibility χ directly (i.e., radius of convergence of the

---

* The x_t quoted in reference [39] are in error because of a mistake in the last coefficient. Recalculation gives agreement with reference [40].

series for χ) whereas the above calculation calculates the zero in $1/\chi$. Comparison of the methods is seen as follows. For low H

$$F(T, H) - F(T, 0) = \tfrac{1}{2} \langle R \rangle H = \tfrac{1}{2} R^2/\chi \qquad (5.9)$$

$$= \tfrac{1}{2} (\partial^2 F/\partial R^2)_{R=0} R^2 \qquad (5.9a)$$

$$= \tfrac{1}{2} \chi H^2 \qquad (5.9b)$$

$$= \tfrac{1}{2} (\partial^2 F/\partial H^2)_{H=0} H^2 \qquad (5.9c)$$

Thus $\partial^2 F/\partial R^2$ gives $1/\chi$ whereas $\partial^2 F/\partial H^2$ gives χ. Equation (5.9c) gives a simple method for calculating χ through

$$\chi = \lim_{q \to 0} \langle |\mu_q|^2 \rangle_{H=0} \qquad (5.10)$$

Equation (5.10) can then be calculated by methods of the linked cluster expansion with the coefficient of x^ν again being given by diagrams with $\nu + 1$ vertices. The results of the two methods gibe to the order of approximation. For other lattices, Elliott et al. [40] find $x_t = 0.18$, 0.22 for fcc and bcc, respectively. In summary, for near neighbor lattices, $x_t \simeq 2/z$.

An important remark by Elliott et al. [40] is that the calculation of x_t is the same for Heisenberg and Ising models and arbitrary spin S, as seems intuitively evident. This result, however, cannot be seen from the first term in χ alone. Rather, it follows from the fact that all the other terms in χ as a powers series in x are independent of the spin S. The argument runs as follows.

Using Eq. (5.10) or its quantum mechanical equivalent, we have

$$\chi = \langle |S^z|^2 \rangle$$
$$S^z = \text{total } z \text{ component of spin} \qquad (5.11)$$

Now note that Eq. (5.11) can be evaluated for a given concentration by counting how many spins are in clusters of 1, 2, ..., n, ..., etc. For each such cluster of n we have at zero temperature all spins in the cluster parallel (or in the Heisenberg case, the maximum value of the spin, we take S per single spin).

Thus in a cluster of n spins, we have

$$\langle |S^z|^2 \rangle_n = n^2 S^2 \qquad \text{(Ising)} \qquad (5.12a)$$

$$= \tfrac{1}{3} nS(nS + 1) \qquad \text{(Heisenberg)} \qquad (5.12b)$$

The average number of spins in a cluster of n is readily evaluated (for a particular lattice) in terms of the concentration. The divergence of the series for χ comes from large n so that the behavior in n is entirely the same in (5.12a) and (5.12b) and independent of S. Thus the critical concentration is a function of the lattice alone.

Further, we expect that this same x_l will appear in the theory of the random antiferromagnet. Since the Heisenberg model is more appropriate for this case, the calculation of x_l is of considerable value in interpretation of experiment for the random antiferromagnet. This has been carried out by Elliott and Heap [41] in a very nice corroboration of the theory against experimental data on MnF_2 dissolved in a magnetically neutral fluoride.

For ferromagnets, in order to interpret experiment in terms of the theory presented here one must consider a paramagnetic ion dissolved in a metal such that the band structure of the host is not much distorted. This excludes, for the moment, the transition elements, which certainly require a band theory. A well-known case is Ni in Cu, which is understandable in terms of the Cu electrons filling the 3d hole of Ni. A good case where the host should not be distorted is the rare earths. Matthias et al. have measured T_c vs. x for Gd in La [42] and find a linear plot from 2% to 8%. Below 2%, superconducting effects enter, but the spin seems to enter an ordered phase with T_c still linear in x down to the smallest measured concentrations of 0.5%. Above 8%, nonlinear effects come in, pushing the Curie point higher. It is very likely then, that there are two mechanisms of exchange, a short-range one which drops out of the picture at $x \sim 1/12$, and a very long-range mechanism. An indirect exchange mechanism through a spherical band of conduction electrons cannot produce such a ferromagnetic exchange, since it produces oscillations with period $2k_F$ where $k_F =$ Fermi momentum of the band [43]. It is very likely then, that there is a strong indirect exchange with the 5d electron in La. This will be a narrow band and no theory of indirect exchange has been worked out for this case.

Other cases of long-range ferromagnetism exist, such as Co in Pd. A long-range kind of diffuse antiferromagnetism exists for Mn in Cu. The theory of these solutions is most interesting, but will not be presented here. See, however, Overhauser [44], Marshall [45], and Klein and Brout [46], for the interesting theoretical conflict which has developed about these systems.

Appendix A

We follow closely the proof of Englert which is the most clear and concise in the literature. Expanding the right-hand side of Eq. (2.13) in a power series in β gives

$$\langle O \rangle = \frac{\sum_{n=0}^{\infty} ((-\beta)^n/n!) \langle V^n O \rangle_0}{\sum_{n=0}^{\infty} ((-\beta)^n/n!) \langle V^n \rangle_0} \tag{A-1}$$

where $V = -\frac{1}{2} \sum v_{ij} \mu_i \mu_j$. Let the spins in O be designated $\mu_1 \ldots \mu_\nu$. Then a typical product entering into the numerator of Eq. (A-1) contains a spin product of the form $\langle \mu_i \mu_j \mu_j \mu_k \ldots \mu_1 \ldots \mu_\nu \rangle$. Let μ_1 appear l_1 times ..., μ_ν appear l_ν times and each other μ_i appear l_i times in this product. Then owing to the factorizability of $\exp(-\beta H_0)$ the product in question may be factorized to

$$\prod_{i \notin \nu} \langle \mu_i^{l_i} \rangle [\langle \mu_1^{l_1} \rangle \cdots \langle \mu_\nu^{l_\nu} \rangle] \tag{A-2}$$

The first part of the product involves spins other than the fixed spins $1 \ldots \nu$. Clearly it is the appearance of the spins $\mu_1 \ldots \mu_\nu$ in V^n which is of interest for if these spins did not occur in V^n the average $\langle O \rangle_0$ would factor out of the numerator leaving a factor to cancel against the denominator. The reason that such a cancellation does not occur in general is that $\langle \mu_i^n \rangle \neq \langle \mu_i \rangle^n$ for $i = 1, \ldots, \nu$. It is therefore propitious to arrange the expansion in such a way as to affect the maximal amount of cancellation. To do this one introduces the semi-invariants M_ν^0 to which we now turn.

One writes the product

$$\langle \mu_k^n \rangle = \sum_{\substack{\text{All splits} \\ \text{such that} \\ \Sigma \nu_i = n}} \prod_{i=1}^{n} M_{\nu_i}^0 \tag{A-3}$$

The sum is over all possible splits of the n factors μ_k

$$\begin{aligned} \langle \mu_k \rangle_0 &= M_1^0 \\ \langle \mu_k^2 \rangle_0 &= M_1^0 M_1^0 + M_2^0 \\ \langle \mu_k^3 \rangle_0 &= M_1^0 M_1^0 M_1^0 + 3 M_1^0 M_2^0 + M_3^0 \quad \text{etc.} \end{aligned} \tag{A-4}$$

These are inverted according to

$$\begin{aligned} M_1^0 &= \langle \mu_k \rangle_0 \equiv R_0 \\ M_2^0 &= \langle \mu_k^2 \rangle_0 - \langle \mu_k \rangle^2 = 1 - R_0^2 \\ M_3^0 &= \langle \mu_k^3 \rangle_0 - 3 \langle \mu_k^2 \rangle_0 \langle \mu_k \rangle_0 + 2 \langle \mu_k \rangle_0^3 = -2 R_0 (1 - R_0^2) \end{aligned} \tag{A-5}$$

The general inversion formula and the proof are due to Kahn [47]. This is not needed in the sequel. In the last set of equalities we have introduced the unperturbed magnetization

$$R_0 = \langle \mu_k \rangle_0 = \frac{\partial \log \cosh \beta H}{\partial \beta H} = \tanh \beta H \tag{A-6}$$

together with the properties $\langle \mu_k^{2n} \rangle_0 = 1$, $\langle \mu_k^{2n+1} \rangle_0 = R_0$, $n =$ integer. It is a very easy matter to prove that the generating function of the M_ν^0 is the log of the unperturbed single particle partition function $Z_0 = \text{tr} \exp(\beta \mu_k H) = 2 \cosh \beta H$, i.e.,

$$M_\nu^0 = (\partial/\partial \beta H)^\nu \log Z_0 \tag{A-7}$$

proof is found in the appendix of reference [48].

To return to the original problem, a general term in the numerator of (A-1) is

$$(1/n!)(\tfrac{1}{2})^n (\beta v_{ij})(\beta v_{kl}) \cdots (\beta v_{mn}) \times \langle \mu_i \mu_j \mu_k \mu_l \cdots \mu_m \mu_n O \rangle_0 \tag{A-8}$$

where some indices may be equal. In the case of equal indices, one expands the product $\langle \mu_i^{l_i} \rangle$ into semi-invariants in the manner of Eq. (A-3) so that (A-8) is then a sum of terms. One graphs each member of the sum as follows: Associate a factor $\tfrac{1}{2} v_{ij}$ with each bond, and with each vertex labeled i with n_i lines joined at it associate a factor $M_{n_i}^0$. In Fig. 21 we graph a typical term coming from $O = \mu_1 \mu_2$. The special

FIG. 21. Graphical representation of the term
$(\tfrac{1}{2})^4 \{[(v_{1i})^2 (v_{2i})] M_3^0(1) M_3^0(i) M_2^0(2)\} \{v_{1j} M_1^0(1) M_1^0(j)\}$.

indices 1,2 are designated with an X. In the product written in the figure legend it will be noticed that the term in question has been written as a product of two terms, the first of which involves terms linked to the spins 1, 2 and the second, terms which are unlinked to these spins. (For the precise definition of "linked" see below). It is seen that the semi-invariant expansion accomplishes the desired result of splitting off as a factor terms which are not associated with the given set. It is the next step to show how to cancel off the set of unlinked terms.

The total contribution of a given graph (i.e., of a given topological structure) to the numerator of Eq. (A-1) is then obtained by multiplying all the $\frac{1}{2}v_{ij}$ factors by all the $M_{n_i}^0(i)$, dividing by $(1/n!)$, summing over all spin indices, and multiplying the graph by an appropriate combinatorial factor. This factor is $2n!/g$ where g is the number of symmetry operations which transform the graph into itself. We now turn to the proof of this last point.

Consider a term arising from the nth-order term in the expansion of the numerator of Eq. (A-1). We write this term symbolically in the following way

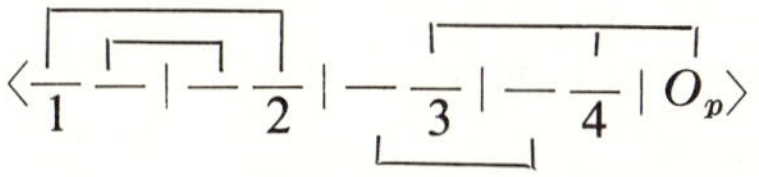

$$\langle \overline{\underset{1}{\sqcap} - | - \underset{2}{\sqcap}} | - \underset{3}{\sqcap} | - \underset{4}{\sqcap} | O_p \rangle \tag{A-9}$$

Here, each vertical line separates two interactions $v_{ij}\mu_i\mu_j$. The μ_i factors are represented by a horizontal line and an mth-order semi-invariant by a contraction sign $\sqcap$. O is any operator and in example (A-9) has only one spin index contracted with interaction factors. The graph corresponding to the above term is the fourth-order graph represented in Fig. 22.

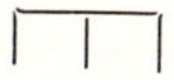

FIG. 22. The graphical representation of Eq. (A-9).

All the terms of the expansion are obtained if one draws all possible sets of contraction signs in any order. Many of these, however, give rise to an identical graph representation. These are the following:

(1) All the terms differing from a given one by a permutation of the n interactions. This contributes a factor $n!$ to the combinatorial coefficient.

(2) All the terms differing by an interchange of the two contractions ending in the same interaction. For instance, the two terms

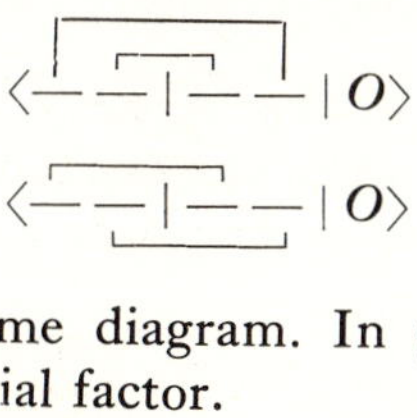

$$\langle \overline{\sqcap - | - \sqcap} | O \rangle$$
$$\langle \overline{\sqcap - | - \sqcap} | O \rangle$$

are represented by the same diagram. In general, this contributes a factor 2^n to the combinatorial factor.

(3) In counting all the diagrams with a factor $2^n n!$, we have overcounted all the terms which are transformed into themselves under the operations performed in (1) and (2). For instance, a permutation of (1) and (2) of (3) and (4) in the term $(A - q)$ does not lead to a new term of the expansion. Thus we must divide the combinatorial factor by 4. In general we clearly have to divide the factor $2^n n!$ by the number g of symmetry operations that transform the graph into itself.

With these preparations, we may now derive the linked cluster expansion. A linked graph is defined as a graph that does not contain a part which is not linked by some path of bonds to the fixed indices 1.... For example, Fig. 21 has an unlinked part. It is clear that if a graph has unlinked parts g is a product of symmetry factors, one for each unlinked part. This is evident from the demonstration presented in the previous paragraph. Also the factor of $(1/n!)$ cancels. It then follows that an unlinked graph is the product of the part linked to the given indices together with all the unlinked parts. Thus if one considers a given linked part, the total of all unlinked parts which multiplies this linked part is the total of all possible graphs with no spins fixed. This is precisely the expansion of the denominator of Eq. (A-1). The result after cancellation is that only the linked parts survive together with the rules announced in the text.

Appendix B

We follow Stinchcombe *et al.* [14] and write

$$\langle O \rangle = \frac{\sum_{n=0}^{\infty} (\xi^n/n!) T \int_0^\beta \cdots \int_0^\beta d\beta_n \cdots d\beta_1 \langle V(\beta_n) \cdots V(\beta_1) O \rangle_0}{\sum_{n=0}^{\infty} (\xi^n/n!) T \int_0^\beta \cdots \int_0^\beta d\beta_n \cdots d\beta_1 \langle V(\beta_n) \cdots V(\beta_1) \rangle_0} \quad \text{(B-1)}$$

Then following the reasoning of Appendix A one is led to a general term in the numerator

$$\int \cdots \int d\beta_1 \cdots d\beta_n (1/n!)(1/2^n) v_{ij} v_{kl} \cdots v_{mn}$$
$$\times T \langle \sigma_i(\beta_1) \cdot \sigma_j(\beta_1) \cdots \sigma_m(\beta_n) \cdot \sigma_n(\beta_n) O(0) \rangle_0 \quad \text{(B-2)}$$

where each $\sigma_\mu \cdot \sigma_\nu$ must be written $\sigma_\mu^z \sigma_\nu^z + \sigma_\mu^+ \sigma_\nu^-$. Again when a given spin i appears a number of times one expands the time-ordered product as in Eq. (A3), but this time in time-ordered semi-invariants such as

$$\langle T(\sigma_k^+(\beta_1) \sigma_k^-(\beta_2) \sigma_k^+(\beta_3) \cdots) \rangle_0 = \sum_{\text{All Splits}} T \prod_{i=1}^n M_{\nu_i}^0(\{\beta_i\}) \quad \text{(B-3)}$$

2. STATISTICAL MECHANICS OF FERROMAGNETISM

We give the first few members of Eq. (B-3), remembering that $\langle \sigma_i^+ \rangle_0 = \langle \sigma_i^z \sigma_i^+ \rangle_0 = 0$ and $\langle \sigma_i^z \rangle_0$ is independent of β_i:

$$\langle \sigma_i^z(\beta_i) \sigma_i^z(\beta_j) \rangle_0 = M_2^0 + M_1^0 M_1^0 \tag{B-4}$$

$$\langle T \sigma_i^+(\beta_1) \sigma_i^-(\beta_2) \rangle_0 = M_2(T(\sigma_i^+(\beta_1) \sigma_i^-(\beta_2)))$$

$$= \exp\left[-(\beta_1 - \beta_2)H\right] \begin{cases} \langle \sigma_i^+ \sigma_i^- \rangle_0 & \beta_1 > \beta_2 \\ \langle \sigma_i^- \sigma_i^+ \rangle_0 & \beta_1 < \beta_2 \end{cases} \tag{B-5}$$

$$\langle T \sigma^z(\beta_k) \sigma^+(\beta_j) \sigma^-(\beta_i) \rangle_0 = M_3 \langle T \sigma^z(\beta_k) \sigma^+(\beta_j) \sigma^-(\beta_i) \rangle$$

$$+ M_1(\sigma^z(\beta_k)) M_2(T \sigma^+(\beta_j) \sigma^-(\beta_i)) \tag{B-6}$$

We now use the semi-invariant development of each of the products that appear in the numerator and denominator of Eq. (B-1).

We represent by graphs each term which arises from expressing the numerator of Eq. (B-1) in terms of the semi-invariants. Each interaction is represented by a bond carrying the temperature label and the indices of the spins which interact. In order to distinguish between the longitudinal and the transverse parts of the interaction, the bonds representing the transverse part $-\frac{1}{2} \sum_j v_{ij} \sigma_i^+ \sigma_j^-$ also carry an arrow directed from the index corresponding to the σ^+ to that corresponding to the σ^-. The semi-invariants are made up of the averages of the spin operators. The spin operators appear either from H_1 or from O. If we represent a vertex containing operators σ^+, σ^-, σ^z from O with the appropriate symbols ($+$, $-$, or z) written inside, the semi-invariants are then given completely by the structure at the vertices: The order of the semi-invariant associated with a given vertex is the number of bonds joined at that vertex plus the number of operators from O associated with the vertex, and the numbers of σ^+'s and σ^-'s in the semi-invariant are given by the number of bonds carrying arrows directed out from and into the vertex, respectively, together with the transverse operators from O. The building blocks from which the graphs are constructed are the operators from O and the individual interaction terms $\frac{1}{2} v_{ij} \boldsymbol{\sigma}_i \cdot \boldsymbol{\sigma}_j$ from V. When more than one interaction involves a given spin index and/or that spin index occurs in O the average $\langle \cdots \rangle_0$ is on a product of spin operators of a given index. One then uses Eq. (B-3) to write this in terms of the semi-invariants. The graphs so constructed are all those that can be made by starting with the isolated bonds, and operators from O, and joining the indices labeled with the same letter together in all possible ways. Thus, the expansion of the numerator of Eq. (B-1) into the semi-invariant diagrams produces all possible diagrams that can be drawn from the bonds and vertices O. The next step is to sum all indices and integrate over all temperatures. Then the same graph appears $n!2^{n_l}/g$ times, where n_l is the number of longitudinal bonds and g is the number of symmetry operations which transform the graph into itself. (The proof is the same as that of Appendix A.)

We again define an unlinked graph as one which separates into one or more parts which are not connected by any bonds to the spins of O. The expansion of numerator and denominator differs only in that the operators of O provide extra vertices to which we can join bonds. The numerator expansion is therefore equivalent to all graphs linked to the operators of O, multiplied by all graphs unlinked to these graphs. This second factor is just the expansion of the denominator. Hence, we arrive at the linked cluster expansion: O is given by summing all graphs linked to one or more of the operators of O according to the rules given in the text.

References

1. L. Onsager, *Phys. Rev.* **65**, 117 (1944).
2. M. Kac, private communication, 1960.
3. G. Newell and E. Montroll, *Rev. Mod. Phys.* **25**, 352 (1953).
4. C. Domb, this volume, Chapt. 1.
5. C. Domb, *Advan. Phys.* **9**, 149 (1960).
6. H. A. Bethe, *Proc. Roy. Soc.* **A150**, 552 (1935).
7. R. Brout, *Phys. Rev.* **115**, 824 (1959); **118**, 1009 (1960); **122**, 469 (1961).
8. G. Horwitz and H. B. Callen, *Phys. Rev.* **124**, 1757 (1961).
9. F. Englert, *Phys. Rev.* **129**, 567 (1963).
10. M. M. Ornstein and F. Zernike, *Proc. Acad. Sci. Amsterdam* **17**, 793 (1914).
11. E. Helfand, *Ann. Rev. Phys. Chem.* to be published, 1963.
12. A. Siegert, to be published.
13. F. J. Dyson, *Phys. Rev.* **102**, 1217, 1230 (1956).
14. R. L. Stinchcombe, G. Horwitz, F. Englert, and R. Brout, *Phys. Rev.* **130**, 155 (1963).
15. F. Englert and M. Antonoff, *Bull. Am. Phys. Soc.* (1961); *Physica* **30**, 429 (1964).
16. W. Bragg and E. Williams, *Proc. Roy. Soc.* **A145**, 609 (1934).
17. R. J. Elliott and W. Marshall, *Rev. Mod. Phys.* **30**, 75 (1958).
18. T. Berlin and M. Kac, *Phys. Rev.* **86**, 821 (1952).
19. T. D. Lee and C. N. Yang, *Phys. Rev.* **87**, 410 (1952).
20. G. Horwitz and F. Englert, private communications, 1961.
21. S. V. Tyablikov, *Soviet Phys. JETP* (*English Transl.*) **11**, 287 (1959).
22. R. Brout and H. Haken, *Bull. Am. Phys. Soc.* [2] **5**, 118 (1960).
23. F. Englert, *Phys. Rev. Letters* **5**, 102 (1960).
24. F. Keffer and R. Loudon, *Phys. Rev.*
25. T. Oguchi, *Phys. Rev.* **117**, 117 (1960).
26. T. Holstein and H. Primakoff, *Phys. Rev.* **58**, 1908 (1940).
27. F. Bloch, *Z. Physik* **57**, 545 (1929).
28. E. C. Stoner, *Proc. Roy. Soc.* **A165**, 372 (1938).
29. R. Kubo *et al.*, preprint, 1963.

30. M. Gutswiller, *Phys. Rev. Letters* **10**, 159 (1963).
31. H. Stern, private communication, 1963.
32. H. Stern, and R. Brout, *Physica* **30**, 1689 (1964).
33. P. A. Wolff, *Phys. Rev.* **120**, 814 (1960).
34. L. D. Landau, *Zh. Eksperim. i Teor. Fiz.* **30**, 1058 (1956); **32**, 59 (1957).
35. E. P. Wohlfarth, *Rev. Mod. Phys.* **24**, 211 (1953).
36. G. Horwitz, R. Brout, and F. Englert, *Phys. Rev.* **130**, 409 (1963).
36a. R. Brout and P. A. Carruthers, "Lectures on the Many Electron Problem." Wiley, New York, 1963.
37. F. Englert and R. Brout, *Phys. Rev.* **120**, 1085 (1960).
38. R. J. Elliott, private communication, 1962.
39. M. Coopersmith and R. Brout, *Phys. Chem. Solids* **17**, 254 (1961).
40. R. J. Elliott, B. R. Heap, D. J. Morgan, and G. S. Rushbrooke, *Phys. Rev. Letters* **5**, 366 (1960).
41. R. J. Elliott and B. R. Heap, *Proc. Roy. Soc.* **A265**, 264 (1962).
42. B. T. Matthias, H. Suhl, and E. Corenzwit, *Phys. Rev. Letters* **1**, 92 (1958).
43. M. A. Ruderman and C. Kittel, *Phys. Rev.* **96**, 99 (1954).
44. A. W. Overhauser, *Phys. Rev. Letters* **3**, 414 (1959).
45. W. Marshall, *Phys. Rev.* **118**, 1520 (1960).
46. M. Klein and R. Brout, *Phys. Rev.* **132**, 2412 (1963).
47. B. Kahn, Dissertation, Utrecht 1938 (N.V. Noord Hollandsche Uitgeversmaatschappij).
48. R. Brout and F. Englert, *Phys. Rev.* **120**, 1519 (1960).

3. Magnetic Symmetry

W. Opechowski[*] and Rosalia Guccione[**]

Department of Physics
University of British Columbia,
Vancouver, Canada

I. Introduction 105
II. Magnetic Groups; Definition and Method of Construction 106
    1. Definition of Magnetic Groups 106
    2. Fundamental Properties of Magnetic Groups 109
    3. Historical Digression on Magnetic Groups 111
III. Construction of the Magnetic Groups 114
    1. Construction of the Magnetic Point Groups and Magnetic Lattices . . . 114
    2. Construction of the Magnetic Space Groups 122
IV. Invariant Spin Arrangements 133
    1. Definition of Invariant Spin Arrangements 133
    2. Construction of All the Invariant Spin Arrangements 151
    3. Classification of All the Invariant Spin Arrangements 160
    4. Miscellaneous Remarks 162
    Appendix A. List of Some of the Symbols Used 163
    References 164

I. Introduction

Except for a very long but necessary digression on the theory of magnetic groups, the subject of this chapter is a group theoretical classification of all possible spin arrangements in perfect magnetic crystals. Here the term "spin" is an abbreviation for the "magnetic moment of an atom in a crystal," and the term "magnetic crystal" is used for any magnetically ordered crystal, that is, any antiferromagnetic, ferrimagnetic, or ferromagnetic crystal. (By saying that a

[*] Present address: Lorentz Institute, University of Leiden, Leiden, Holland.
[**] Present address: Department of Physics, University of Toronto, Toronto, Canada.

crystal is antiferromagnetic or ferrimagnetic we do not want to imply that its spin arrangement has necessarily a three-dimensional translational symmetry; the spin arrangement may very well be a "spiral" arrangement.) A "possible" spin arrangement means here a spin arrangement that does not violate any of the symmetry principles that are commonly accepted as valid in physics. Many of the spin arrangements which are in this sense possible may turn out to be impossible from some specific dynamical point of view.

The problem of classification of all possible spin arrangements is of course analogous to the problem of classification of all possible atom arrangements in crystals, and the two problems are not independent because the spins are located at the positions of paramagnetic atoms in a crystal.

The answer to the question what the possible atom arrangements in a crystal are has been known for many years. It is provided by the theory of space groups, and a catalog of all possible atom arrangements is given in "International Tables for X-Ray Crystallography" [1].

The answer to the analogous question about spin arrangements is provided by the theory of magnetic groups (not only magnetic space groups!), which is very recent. However, no catalog of all possible spin arrangements is yet available, so that the answer is known only "in principle." It is the principle of a classification of all possible spin arrangements based on the theory of magnetic groups which is described in this chapter. Since our classification of the spin arrangements is based on the theory of magnetic groups, and no systematic presentation of this theory has ever been published (to our best knowledge), a rather detailed outline of it had to be included in this chapter.

Many statements about magnetic groups and invariant spin arrangements are given in this chapter without proof. The proofs, and a detailed discussion of some points which here could only be mentioned, will be found in Opechowski and Guccione [2], from which we have borrowed freely.

II. Magnetic Groups; Definition and Method of Construction

1. Definition of Magnetic Groups

We consider a static arrangement of atoms regarded as mass points. If the arrangement is invariant under some space group F we call it a crystal. Since the arrangement is static it will also be invariant under the group χ consisting of time inversion and all time-displacements.

This invariance is trivial in the sense that it has nothing to do with the fact that the arrangement of atoms is a crystal: the position vector of each atom remains unchanged if any element of χ is applied to it. The symmetry group of the crystal is thus the direct product group **F** × χ, and this is also the most general group under which it is invariant (the crystal is assumed to be at rest).

If the atoms of the crystal are not simply mass points but have also nonvanishing magnetic moments, i.e., if the crystal is magnetic, its symmetry is necessarily lower. For, according to the standard assumptions of the electromagnetic theory, a magnetic moment changes its sign under time inversion. Hence the symmetry group of a magnetic crystal cannot have time inversion as an element.

We shall assume throughout this chapter that the magnetic moments of the atoms in a magnetic crystal are static. Hence a magnetic crystal will be trivially invariant under time displacements just as a nonmagnetic crystal. We will consistently disregard this trivial invariance (just as one usually disregards the trivial invariance of a nonmagnetic crystal under the group χ); this does not imply any loss of generality as far as the questions discussed in this chapter are concerned.

If **A** is the subgroup of χ which consists of the identity E and time inversion E' (i.e., the transformation $t \rightarrow -t$), we can thus simply say that the symmetry group of a magnetic crystal is necessarily one of those subgroups of the direct product group **F** × **A** which do not have time inversion combined with the identity of **F** as an element. Any such subgroup, for any choice of the space group **F** in **F** × **A**, will be called a magnetic group, and will be denoted by **m**. This is the definition of a magnetic group adopted in this chapter (see, however, Section III, 1b).

At this point it is important to realize that the elements of **F** × **A** are pairs FA where F is an element of **F** and A an element of **A**; we shall always write an element of the left-hand factor in a direct product to the left of an element of the right-hand factor. A magnetic group does not contain, according to the above definition, the element EE' of **F** × **A** where E is the identity element of **F**, but it may very well contain elements of the form FE' where F is not the identity element of **F**. A magnetic group which does not contain elements of the latter form will be called a "trivial magnetic group." All space groups, all lattices (i.e., groups of primitive translations), and all point groups are trivial magnetic groups.

One is led to the same definition of magnetic groups if one considers the invariance of a magnetic crystal from the quantum mechanical point of view. The magnetic moment of an atom in the crystal acquires, from this point of view, the meaning of the quantum statistical average of the magnetic moment operator **m** of the atom. The contribution of an energy level of the crystal to this average is proportional, at any given temperature, to $\Sigma \langle \mu | \mathbf{m} | \mu \rangle$, where the sum is taken over all the linearly independent eigenvectors $|\mu\rangle$ belonging to the energy level under consideration. The nonvanishing of this average value for the magnetic atoms of a crystal in the absence of an external magnetic field is an essential characteristic of the magnetically ordered state. But if the crystal is invariant under time inversion $\Sigma \langle \mu | \mathbf{m} | \mu \rangle$ necessarily vanishes. In fact, the time inversion invariance means that the eigenvectors $|\mu\rangle$ and $\theta |\mu\rangle$, where θ is the time inversion operator (see Chapter 26 of Wigner [3]), belong to the same energy eigenvalue. On the other hand, $\theta \mathbf{m} + \mathbf{m} \theta = 0$. Hence $\langle \mu | \mathbf{m} | \mu \rangle = - \langle \mu | \theta^{\dagger} \mathbf{m} \theta | \mu \rangle$, and $\Sigma \langle \mu | \mathbf{m} | \mu \rangle = 0$.

The nonvanishing of the average value of the magnetic moment operator of the atoms of a magnetic crystal is thus incompatible with the time inversion invariance of the crystal; and we conclude again that the symmetry group of a magnetic crystal cannot have time inversion as an element.

The existence of magnetic crystals should not, of course, be interpreted as a violation of the time inversion invariance of physical phenomena. From each magnetically ordered state one can obtain another one by time inversion, and both are compatible with the same Schroedinger equation which, in the absence of an external magnetic field, is time inversion invariant. Magnetically ordered states are metastable states, but from the experimental point of view they are in many cases stable for all practical purposes. Correspondingly, the theoretical point of view adopted in this chapter, and in most existing approximate theories of magnetic crystals, is equivalent to identifying the metastability of magnetically ordered states with absolute stability. The adoption of this point of view leads in most quantum mechanical calculations to keeping certain eigenvectors of the Schroedinger equation, and excluding the eigenvectors obtained from those by time inversion. In most cases the approximate procedure used in the calculations does it automatically.

No quantum mechanical applications of magnetic groups will be discussed in this chapter. That is why we need not consider the magnetic groups of quantum mechanical operators which correspond to the magnetic groups defined above. The elements of a magnetic group act on the space-time variables, while the elements of the corresponding magnetic group of quantum mechanical operators act on state vectors. Since the time inversion operator θ is antilinear and antiunitary, the transformations of state vectors under a nontrivial magnetic group of quantum mechanical operators do not generate representations of the group, but "corepresentations" (see Chapter 26 of Wigner [3]). Corepresentations of magnetic groups have been discussed by Dimmock and Wheeler [4, 5] and by Dimmock [6]; see also reference [2].

2. Fundamental Properties of Magnetic Groups

a. Some simple properties of magnetic groups. We assume here that the elements of the theory of space groups are known. However, to avoid misunderstandings as to terminology and notation we list here some definitions and theorems. For details see, for example, Koster [7], Lomont [8], or Opechowski and Guccione [2].

An element F of a space group **F** will be also denoted more explicitly by $(R \mid \tau(R) + \mathbf{t})$, where R is a proper or improper rotation, $\tau(R)$ is the nonprimitive translation associated with R, $\mathbf{t} = n_1\mathbf{a}_1 + n_2\mathbf{a}_2 + n_3\mathbf{a}_3$ is a primitive translation, $\mathbf{a}_1$, $\mathbf{a}_2$, $\mathbf{a}_3$ being three linearly independent basic primitive translations. and n_1, n_2, n_3 arbitrary integers. A space group is called symmorphic if for all its elements the nonprimitive translations are zero; it is called nonsymmorphic otherwise. Each space group **F** has a normal subgroup **T**, the group of all primitive translations $(E \mid \mathbf{t})$, where E is the identity rotation; this subgroup is also called the lattice of **F**, and its elements often will be denoted simply by $\mathbf{t}$ (rather than by $(E \mid \mathbf{t})$). The set of elements $(R \mid 0)$, where 0 stands for the zero translation, obtained from the elements of **F** by putting in the symbol $(R \mid \tau(R) + \mathbf{t})$ $\mathbf{t} = 0$, and replacing $\tau(R)$ by 0, is a group called the point group **R** of **F**. The elements of the point group **R** of **F** will be denoted simply by R [rather than by $(R \mid 0)$]. The point group **R** of **F** is a subgroup of **F** if **F** is symmorphic; it is not a subgroup of **F** if **F** is nonsymmorphic. In both cases **R** is isomorphic to the factor group **F/T**. The point group **R** of **F** leaves the lattice **T** of **F** invariant in the following sense: if $\mathbf{t}$ is an element of **T** then $R\mathbf{t}$ is also an element of **T** for any element $\mathbf{t}$, and any element R of **R**. The holohedry of a lattice is the largest point group which leaves it invariant.

An element FA of a magnetic group can accordingly be written as $(R \mid \tau(R) + \mathbf{t})A$. We shall call an element FA a primed element if $A = E'$, and an unprimed element if $A = E$. Correspondingly we shall often use a simplified notation: F' for FE', and F for FE. Obviously, the product of any two primed elements, or of any two unprimed elements, is an unprimed element; the product of a primed element and an unprimed one (in either order) is a primed element. From these simple rules some important properties common to all nontrivial magnetic groups follow readily.

Since a nontrivial magnetic group **m** is a subgroup of **F** × **A**, the set consisting of the unprimed elements of **m** and of the elements obtained from the primed elements of **m** by omitting the primes (that is, replacing FE' by FE) is a subgroup L_m of **F**; L_m is of course isomorphic to **m**. This property of magnetic groups leads to the following natural definitions: If L_m is a space group, **m** is called a "magnetic space group" and denoted by **M**; if L_m is a group of (some or all) primitive translations $(E \mid t)$, **m** is called a "magnetic lattice" and denoted by T_m; if L_m is a group of (some or all) rotations $(R \mid 0)$, **m** is called a "magnetic point group" and denoted by R_m. (N.B. The term "space group" will always mean in this chapter, unless qualified, a three-dimensional space-group; the same convention will be used for the term "magnetic space group").

The unprimed elements of a nontrivial magnetic group form a subgroup of index 2.

That they form a subgroup is obvious from the rule that the product of two unprimed elements is unprimed. That the subgroup is of index 2 is an immediate consequence of the rule that the product of two primed elements is unprimed, if the following theorem valid for an arbitrary group is used (we shall refer to this theorem as the fundamental lemma, apologizing for the pompous name):

A subgroup **H** of a group **G** is a subgroup of index 2 if and only if for any two elements G_i and G_k of **G** which are not elements of **H** the product $G_i G_k$ is an element of **H**.

b. *Prescription for constructing all magnetic groups.* We have proved that every magnetic group **m** has a subgroup of index 2 in common with the corresponding group L_m which is a subgroup of some space group. This implies that, conversely, all nontrivial magnetic groups can be constructed by means of a very simple prescription. We first formulate it for the magnetic space groups:

Consider all the space groups **F** which have subgroups **D** of index 2. For each **D** of each **F** combine the elements of **D** with the identity E of **A**, and the elements of the coset **F** − **D** with the element E' of **A**. Then, for each fixed **D**, the set of all elements of **D**E and (**F** − **D**)E' will necessarily constitute a nontrivial magnetic space group.

This prescription includes the case of magnetic lattices because any lattice is a space group of the class $P1$ (here, and in the rest of this chapter, we use the Hermann–Mauguin international symbols for the space groups and point groups, and for the Bravais classes of lattices).

Prescriptions for constructing all the nontrivial magnetic point groups, all the nontrivial two-dimensional magnetic space groups, and so forth are obtained from the above prescription by replacing in it the phrase "space group" by the appropriate phrase, and by changing the notation correspondingly.

A space group **F** and the magnetic space groups constructed from it by means of the above prescription from all the subgroups **D** of **F** will be said to form the "family" of **F**. We shall also use the term "family of a magnetic point group," and so forth in an analogous way.

The magnetic groups which belong to the same family are all isomorphic. This implies that they have all the same irreducible representations as the trivial magnetic group of the family.

3. Historical Digression on Magnetic Groups

Since there is only one abstract group of order 2, the problem of constructing all magnetic groups is clearly independent of the physical meaning assigned to the element E' of the group **A**. In fact, magnetic space groups were first considered under a different name, in a different way, and in quite a different connection more than 30 years ago by Heesch [9]; he refers to them as "four-dimensional groups of the three-dimensional space." The analogous two-dimensional problem (that is, the case in which **F** is a two-dimensional space group) was considered and solved even earlier by Hermann [10], Alexander and Herrmann [11], Weber [12], and Heesch [13]. The motivation for introducing such new groups was at that time either purely mathematical or crystallographical. The element E' of **A** was interpreted essentially as meaning the change of value of a coordinate capable of assuming two values, or as a change of color in a plane ornament (assuming that exactly two colors were available). Of course, no reference to the problem of magnetic ordering was possible at this early stage of the history of the problem.

Heesch [9] was not only the first to consider magnetic space groups in the sense just indicated, and to give a list of the magnetic space groups of the triclinic and monoclinic system, but also seems to have been the first to realize the importance of his groups from the physical point of view. He also gave the complete list of the magnetic point groups and he estimated the number of magnetic space groups to be close to 1800 (actually there are 1421 of them).

A complete list of magnetic space groups was published only in 1955 by Belov *et al.* [14]. These authors do not mention Heesch's work, nor do they make any reference to the problem of magnetic ordering and to the importance of time inversion in this connection. The same is true of Zamorzaev, who has also given a list of magnetic space groups in an unpublished thesis [15] unavailable to us, and who published [16] a short paper in which he outlines (without proofs) his method of deriving those magnetic space groups that are isomorphic to the symmorphic space groups. Magnetic space groups are called by Belov, Neronova, and Smirnova, and by Zamorzaev, "Shubnikov groups," a name now widely used but hardly justified from the historical point of view.

A clear realization that magnetic space groups are of importance for a systematic classification and discussion of magnetically ordered crystals, and that the element E' of **A** must be interpreted in this connection as time inversion, is due to Landau and Lifschitz [17]. In fact the term "magnetic space groups" was introduced by them.

At this point it is perhaps useful to give a brief dictionary of terms as used by various authors:

A = time inversion group = antisymmetry group = change-of-color-group.

M($\neq$ **F**) = nontrivial magnetic space group = black-and-white space group = minor Shubnikov group.

F × **A** = gray space group = major Shubnikov group.

F = space group = Fedorov group = colorless space group.

For magnetic point groups and magnetic lattices an equivalent terminology is used in a similar way. One speaks, for example, of black-and-white lattices, gray point groups, and so forth. Magnetic point groups are also called Heesch groups.

The concept of a "family of magnetic space groups" was first introduced by Zamorzaev [16], but in a somewhat different sense: his "family" consists of both minor and major Shubnikov groups.

The prescription given in Section II, 2b for constructing magnetic groups requires that one first find subgroups of index 2 of a space group, a point group, a lattice, whatever the case may be. This prescription is a straightforward adaptation of a well-known prescription for finding all those subgroups of the group of all proper and improper rotations which do not contain space inversion as an element (see, for example, Appendix B of Weyl [18]). Since the group of all proper and improper rotations is the direct product group $\mathscr{R} \times$ **I**, where $\mathscr{R}$ is the group of all proper rotations and **I** is the group consisting of the identity and space inversion, one has only to make **F** correspond to $\mathscr{R}$ and **A** to **I** in order to obtain the correspondence between the two prescriptions.

It is easy to see that the problem of finding all the subgroups **D**^{G} of index 2 of an arbitrary group **G** is equivalent to the problem of finding all alternating representations of **G**. In fact, the factor group

G/D^G has only two representations: the identical representation, and the alternating representation. Hence, each subgroup D^G of G will "engender" (the meaning of this term is explained in Lomont [8], p. 234) one alternating representation of G. Conversely, knowing all alternating representations of G, one can find all D^G's: for each alternating representation one finds the group D^G which engenders it by simply picking out from G all those elements to which $+1$ corresponds in the alternating representation in question. The equivalence of the two problems has been noticed for the case in which G is a point group or a space group by Indenbom [19] and Niggli [20].

Zamorzaev's method [16] for constructing the magnetic space groups is implicitly based on the prescription of Section II, 2b. His theorem 3 says essentially the same as our definition of magnetic space groups. The prescription has been used more explicitly by Tavger and Zaitsev [21] for constructing the magnetic point groups.

Although a list of all magnetic space groups is available [14, 22], and some indications how to prepare it are given in the above-quoted papers by Heesch [9], Belov *et al.* [14], and Zamorzaev [16], no rigorous and complete mathematical method of constructing all the magnetic space groups has been described in the existing literature until very recently.

A method which, we believe, satisfies the requirements of mathematical rigor and completeness is described in references [23, 24, 2]. The method is based on the prescription for constructing magnetic groups given above, and is outlined in Section III of this chapter.

It should be mentioned that there exists a still different method to construct all the magnetic space groups. It is a special case of a very general method, of investigating systematically the symmetries of the reciprocal space of a crystal, proposed by Bienenstock and Ewald [25].

Finally, we would like to emphasize that we have omitted from the above survey all reference to papers of purely crystallographic interest. Such papers are of two kinds: those which deal with problems in which the element E' of A is not interpreted as time inversion but as some other operation, for example, as change of color (hence the term "black-and-white" groups); and those which deal with obvious generalizations of the concept of black-and-white groups, like the so-called multicolor groups (here the group A of order 2 is replaced by a group of higher order). To obtain an idea of the former one may consult, for example the review articles by Mackay [26] and Neronova and Belov [27]; as a representative paper of the latter kind we may mention Zamorzaev's [28].

III. Construction of the Magnetic Groups

1. Construction of the Magnetic Point Groups and Magnetic Lattices

a. *Magnetic point groups.* According to our prescription (see Section II, 2b) for constructing nontrivial magnetic groups, we have first to find all subgroups of index 2 for all the 32 point groups. To find them it is sufficient to open "International Tables for X-Ray Crystallography" [1] to page 36, where the complete list of all subgroups of all point groups is given. The task of constructing all magnetic point groups is thus very simple. Except for point groups 1 (identity), 3 (trigonal), and 23 (tetrahedral), all point groups have subgroups of index 2.

It turns out that, in addition to the 32 trivial magnetic point groups, there are 58 nontrivial magnetic point groups [13, 21]. Table I gives a list of the 90 magnetic point groups arranged into families. (Strictly speaking, one should distinguish between a point group and a class of equivalent point groups. There are 90 classes of equivalent magnetic point groups. For brevity, we disregard this distinction whenever possible.) Each line in the list gives all magnetic point groups belonging to one family. The notation used in the list has been introduced by Belov *et al.* [14] and is a straightforward generalization of the international notation for ordinary point groups. The international symbol of a point group gives the generating elements of the group. In the symbol of a magnetic point group the symbols of those generating elements which are primed have a prime as a superscript on the right. By omitting the primes in the symbol of a magnetic point group one obtains the symbol of the ordinary point group to whose family the magnetic point group belongs. The symbols of the admissible magnetic point groups, to be defined in Section IV, 2, are marked with an asterisk.

The only peculiarity of magnetic point groups which is perhaps worth mentioning explicitly is that a rotation through 120° cannot be a primed rotation, i.e., $3'$ is never an element of a magnetic point group. This follows immediately from the fact that a primed element of any magnetic group must be of even order, for, if a primed element $F' = FE'$ of odd order n belonged to the group, then $(F')^n = F^n E'^n = EE'$ would also belong to the group, contrary to the definition of magnetic groups.

We will give now an example of how to construct the family of a given point group. Let this group belong to the class of point groups denoted by 422 and let it consist of the elements 1, 4_z, 2_z, 4_z^3, 2_x, 2_y, 2_{xy}, $2_{\bar{x}y}$. The symbol 1 stands for the identity of the group; the symbols

TABLE I
List of the Magnetic Point Groups

*1						
*$\bar{1}$	$\bar{1}'$					
*2	*2'					
*m	*m'					
*2/m	2'/m	2/m'	*2'/m'			
222	*2'2'2					
mm2	*m'm2'	*m'm'2				
mmm	m'mm	*m'm'm	m'm'm'			
*4	4'					
*$\bar{4}$	$\bar{4}'$					
*4/m	4'/m	4/m'	4'/m'			
422	4'22'	*42'2'				
4mm	4'm'm	*4m'm'				
$\bar{4}2m$	$\bar{4}'2'm$	$\bar{4}'2m'$	*$\bar{4}2'm'$			
4/mmm	4/m'mm	4'/mm'm	4'/m'm'm	*4/mm'm'	4/m'm'm'	
*3						
*$\bar{3}$	$\bar{3}'$					
32	*32'					
3m	*3m'					
$\bar{3}m$	$\bar{3}'m$	$\bar{3}'m'$	*$\bar{3}m'$			
*6	6'					
*$\bar{6}$	$\bar{6}'$					
*6/m	6'/m	6/m'	6'/m'			
622	6'2'2	*62'2'				
6mm	6'm'm	*6m'm'				
$\bar{6}m2$	$\bar{6}'m'2$	$\bar{6}'m2'$	*$\bar{6}m'2'$			
6/mmm	6/m'mm	6'/mm'm	6'/m'm'm	*6/mm'm'	6/m'm'm'	
23						
m3	m'3					
432	4'32'					
$\bar{4}3m$	$\bar{4}'3m'$					
m3m	m'3m	m3m'	m'3m'			

4_z, 2_z, and 4_z^3 stand for rotations through 90°, 180°, 270° about the z-axis and the remaining four symbols stand for rotations though 180° about the x- and y-axis and the two bisectors of the angles that the x- and y-axis form with each other.

As can easily be seen the point group under consideration has three subgroups of index 2. They are the group which consists of the elements 1, 4_z, 2_z, 4_z^3, the group which consists of the elements 1, 2_z, 2_x, 2_y, and that which consists of the elements 1, 2_z, 2_{xy}, $2_{\bar{x}y}$. The first group belongs to the class of point groups denoted by 4, the last two groups are equivalent and belong to the class of point groups denoted by 222.

Correspondingly the family of the given point group contains the three nontrivial magnetic point groups which consist of the unprimed elements 1, 4_z, 2_z, 4_z^3, and of the primed elements $2_x'$, $2_y'$, $2_{xy}'$, $2_{\bar{x}y}'$; of the unprimed elements 1, 2_z, 2_x, 2_y, and of the primed elements $4_z'$, $4_z^{3'}$, $2_{xy}'$, $2_{\bar{x}y}'$; and finally of the unprimed elements 1, 2_z, 2_{xy}, $2_{\bar{x}y}$, and of the primed elements $4_z'$, $4_z^{3'}$, $2_x'$, $2_y'$. The first of these magnetic point groups belongs to the class of magnetic point groups denoted by 42'2'; the last two magnetic point groups are equivalent, and belong to the class of magnetic point groups which is equally well denoted by 4'22' or by 4'2'2.

Disregarding the distinction between a class of equivalent magnetic point groups and individual magnetic point groups, we thus conclude that the family of the point group 422 consists of the magnetic point groups 422, 42'2', and 4'2'2.

b. Digression on noncrystallographic magnetic groups. Our definition of magnetic groups (as given in Section II, 1) excludes certain groups, which will be called "noncrystallographic magnetic groups."

A point group is a group of (proper, or proper and improper) rotations which leaves some lattice invariant. As is well known, if a group of rotations does not have this property it is sometimes called a noncrystallographic group of rotations, because point groups are often called crystallographic point groups. Similarly magnetic groups as defined in this chapter could also be called "crystallographic magnetic point groups."

If we now apply our prescription for constructing magnetic groups to a noncrystallographic group of rotations we shall obtain a nontrivial noncrystallographic magnetic group of rotations (if any).

Apart from the continuous group $\mathscr{R}$ of all proper rotations there exist five infinite continuous proper subgroups of the group $\mathscr{R} \times I$ of all proper and improper rotations. Four of these five have subgroups of index 2, which give rise to six nontrivial infinite noncrystallographic magnetic groups of rotations (see, for example Tavger [29], and reference [2]). Of these six one is of special importance here because it leaves invariant a magnetic field vector and a magnetic moment vector. It is denoted by $(\infty/m)(2'/m')$; its unprimed elements consist of all proper and improper rotations around a fixed axis; its primed elements are obtained from the unprimed elements by combining the latter with the rotation through 180° about an axis perpendicular to the fixed axis.

A magnetic moment is invariant under $(\infty/m)(2'/m')$ only if it is directed along the fixed axis.

There exist also other nontrivial finite and infinite discrete noncrystallographic magnetic groups of rotations.

Both crystallographic and noncrystallographic magnetic groups of rotations are, of course, those subgroups of the direct product group $(\mathscr{R} \times I) \times \mathbf{A}$ which do not have the identity of $\mathscr{R} \times I$ combined with E' as an element.

c. **Magnetic lattices.** Following our prescription for constructing nontrivial magnetic groups again we first look for subgroups of index 2 of ordinary lattices.

As is well known, two sets of three basic primitive translations generate the same lattice if and only if there exists an integral unimodular matrix (i.e., a matrix whose elements are integers, and whose determinant is $+1$ or -1) which transforms one set into the other. Using this criterion it is easy to prove that a given lattice has exactly seven subgroups of index 2. If the basic primitive translations of a lattice $\mathbf{T}$ are $\mathbf{a}_1$, $\mathbf{a}_2$, and $\mathbf{a}_3$, then the basic primitive translations of the seven subgroups can be chosen as follows:

$2\mathbf{a}_1, \mathbf{a}_2, \mathbf{a}_3$;

$\mathbf{a}_1, 2\mathbf{a}_2, \mathbf{a}_3$;

$\mathbf{a}_1, \mathbf{a}_2, 2\mathbf{a}_3$;

$2\mathbf{a}_1, \mathbf{a}_1 + \mathbf{a}_2, \mathbf{a}_3$;

$2\mathbf{a}_2, \mathbf{a}_2 + \mathbf{a}_3, \mathbf{a}_1$;

$2\mathbf{a}_3, \mathbf{a}_3 + \mathbf{a}_1, \mathbf{a}_2$;

$2\mathbf{a}_1, \mathbf{a}_1 + \mathbf{a}_2, \mathbf{a}_1 + \mathbf{a}_3$.

The seven subgroups have been listed by Zamorzaev [16].

As an example, let us apply the prescription for constructing magnetic groups to the case of the first and fourth of the seven subgroups. Evidently, the corresponding magnetic lattice consists of the unprimed translations

$$2n_1\mathbf{a}_1 + n_2\mathbf{a}_2 + n_3\mathbf{a}_3,$$

and of the primed translations

$$((2n_1 + 1)\mathbf{a}_1 + n_2\mathbf{a}_2 + n_3\mathbf{a}_3)',$$

where n_1, n_2, and n_3 are arbitrary integers.

Similarly, in the case of the fourth of the seven subgroups, the magnetic lattice consists of the unprimed translations

$$2n_1\mathbf{a}_1 + n_2(\mathbf{a}_1 + \mathbf{a}_2) + n_3\mathbf{a}_3,$$

and of the primed translations

$$((2n_1 + 1)\mathbf{a}_1 + n_2(\mathbf{a}_1 + \mathbf{a}_2) + n_3\mathbf{a}_3)'.$$

A magnetic lattice is thus uniquely characterized by specifying any three basic primitive translations of its subgroup of unprimed translations ($2\mathbf{a}_1$, $\mathbf{a}_2$, $\mathbf{a}_3$ in the first example; $2\mathbf{a}_1$, $\mathbf{a}_1 + \mathbf{a}_2$, $\mathbf{a}_3$ in the second example).

It is clear that the holohedry of a magnetic lattice, i.e., the largest magnetic point group which leaves it invariant, cannot be a nontrivial magnetic point group.

If the lattice *T* belongs to the triclinic system, i.e., if no relations exist between the three vectors $\mathbf{a}_1$, $\mathbf{a}_2$, $\mathbf{a}_3$, all its seven subgroups of index 2 belong to the same Bravais class, the only Bravais class of the triclinic system. If the lattice *T* belongs to any of the other six crystallographic systems, then not all the seven subgroups belong to the same Bravais class. Using the concept of "semidirect product" of a group with a group of its automorphisms (for a definition of the semidirect product, see Lomont [8, page 29]) one can define a Bravais class of (trivial or nontrivial) magnetic lattices as follows:

Two magnetic lattices *T*$_{m1}$ (holohedry *H*$_1$) and *T*$_{m2}$ (holohedry *H*$_2$) are said to belong to the same "magnetic Bravais class" if and only if (1) the semidirect product of *T*$_{m1}$ and *H*$_1$ is isomorphic to the semidirect product of *T*$_{m2}$ and *H*$_2$, and (2) the unprimed elements of the former correspond, under this isomorphism, to the unprimed elements of the latter (the phrase "a magnetic Bravais class" is an abbreviation of "a Bravais class of magnetic lattices").

In the case of two trivial magnetic lattices *T*$_1$ and *T*$_2$ this is equivalent to saying that *T*$_1$ and *T*$_2$ belong to the same Bravais class if and only if there exists an integral, unimodular matrix N such that $NH_1N^{-1} = H_2$, or, in other words, if and only if the holohedries *H*$_1$ and *H*$_2$ can be made identical by an appropriate choice of the basic primitive translations of *T*$_1$ and *T*$_2$; but not in general in the case of nontrivial magnetic lattices.

It turns out that there are 22 nontrivial magnetic Bravais classes so that there are altogether 36 magnetic Bravais classes. They were first derived by Belov et al. [14].

Instead of giving a list of the magnetic lattices by specifying for each representative lattice of a magnetic Bravais class three primitive translations of its subgroup of unprimed translations (such a list is given in references [16] and [2]), we characterize representative lattices by means of appropriate diagrams (Figs. 1 to 6).

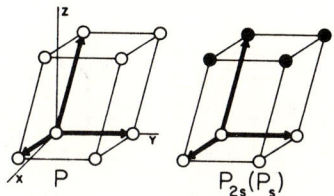

FIG. 1. Magnetic lattices of the triclinic system.

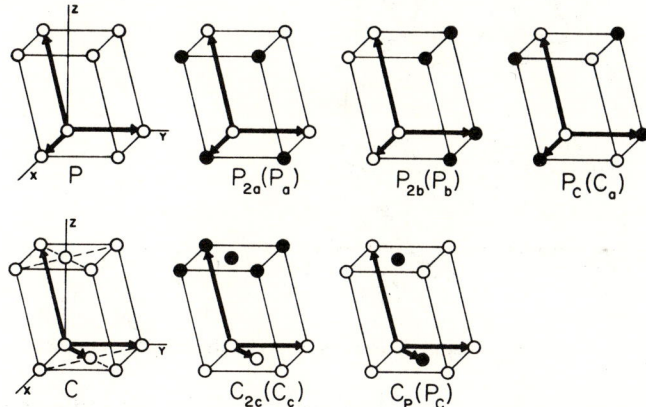

FIG. 2. Magnetic lattices of the monoclinic system (the twofold axis has been chosen as the y-axis).

A lattice is represented in the figures by indicating a few of its primitive translations. Open and full circles serve to indicate which translations are primed, and which are unprimed: a line joining two open or two full circles is an unprimed translation; a line joining an open and a full circle is a primed translation. We place always an open circle at the origin of the coordinate system which is chosen to be rectangular, except in Fig. 5. The arrows represent three basic primitive translations which generate the whole lattice: they define a primitive unit cell of the magnetic lattice. We could, of course, indicate only those three basic primitive

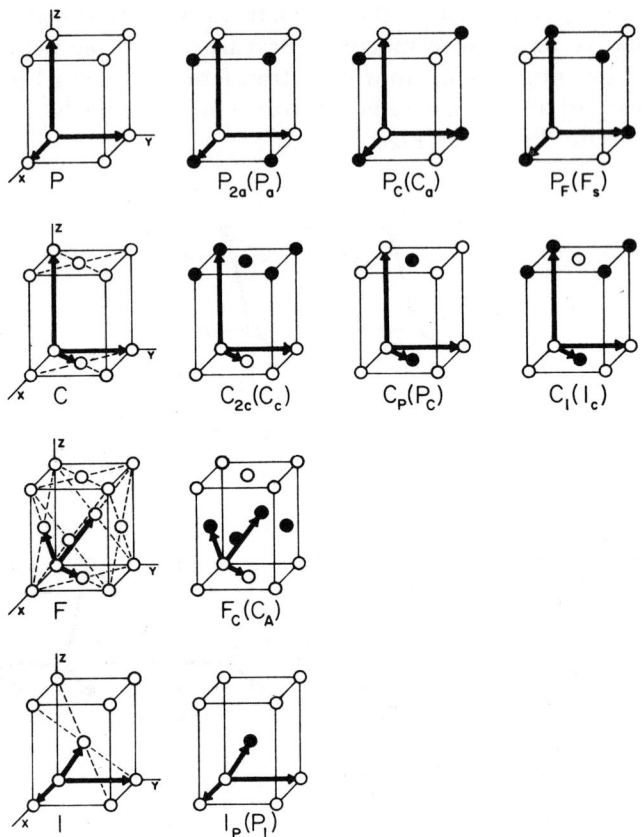

FIG. 3. Magnetic lattices of the orthorhombic system.

translations for each lattice; however, for clarity we indicate a few additional translations in each case.

In the case of a trivial magnetic lattice a primitive unit cell is defined as a parallelepiped with edges given by any three basic primitive translations of the lattice. The same definition is used here for a nontrivial magnetic lattice but some basic primitive translations will be primed in this case. In both cases (trivial, and nontrivial magnetic lattice) the whole magnetic lattice can be generated by translation of a primitive unit cell through primitive translations. However, in the case of a nontrivial magnetic lattice some of the primitive translations must be primed.

In physical applications a unit cell is usually chosen such that the whole magnetic lattice can be generated by translation of the cell through

3. MAGNETIC SYMMETRY 121

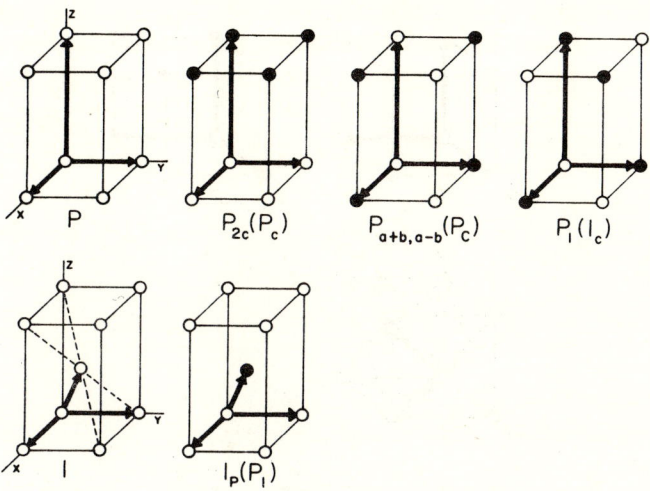

Fig. 4. Magnetic lattices of the tetragonal system.

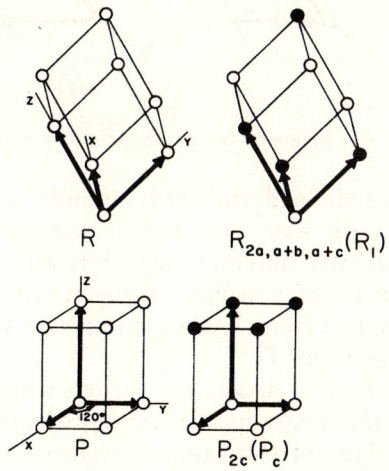

Fig. 5. Magnetic lattices of the trigonal and hexagonal systems.

unprimed translations only. A cell of this kind is called a "magnetic unit cell." The volume of a magnetic unit cell is in general a multiple of the volume of a primitive unit cell defined above. Some of the diagrams in Figs. 1 to 6 (but not all) represent magnetic unit cells.

In each figure lattices belonging to the same family of magnetic lattices are in the same row; in other words, each row of lattices represents

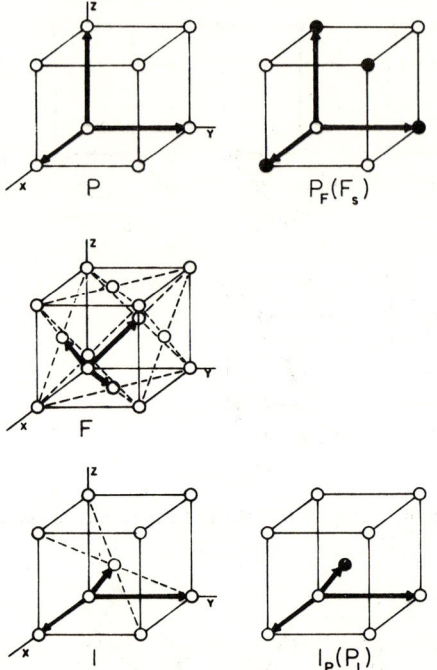

Fig. 6. Magnetic lattices of the cubic system.

one family. We have defined the term "family of magnetic groups" in Section II, 2b. In the case of magnetic lattices it is convenient to use that term in a slightly different way. A family of magnetic lattices will be taken to consist of a trivial magnetic lattice T, and all those magnetic lattices obtained from T whose unprimed subgroups of index 2 have the same holohedry as T.

The symbols used here to denote the various magnetic Bravais classes are given just below the diagrams of their representative lattices. Whenever the symbol is different from that introduced by Belov et al. [22], the latter is also given, in parentheses.

2. Construction of the Magnetic Space Groups

a. Introduction. We shall deal now in some detail with the problem of constructing all magnetic space groups, that is, all three-dimensional magnetic space groups. Nothing will be said about the construction of the two-dimensional magnetic space groups because the procedure is in principle exactly the same as in the three-dimensional case (and in practice simpler, and not so lengthy). The case of one-dimensional magnetic space groups is of course trivial.

Following our prescription (see Section II, 2b) we have to find all subgroups of index 2 of all space groups. How this can be done in a systematic and exhaustive way is described in this section (III, 2a and b). Once the subgroups of index 2 are known, the problem of obtaining the corresponding magnetic space groups is trivial according to our prescription, except for the following point: One may obtain in this way two or more magnetic space groups belonging to the same "class of magnetic space groups" to be defined in Subsection 2d, and one must then identify such redundant magnetic space groups. Some theorems which help deciding in practice whether or not two magnetic space groups belong to the same class can easily be proved, and are listed in reference [2]. An example of construction of a whole family of magnetic space groups is given below, in Subsection 2e.

We first state the following obvious fact:

If **D** is a subgroup of index 2 of a space group **F**, then **D** and **F** have either the same lattice **T** or the same point group **R**. In the former case a subgroup **D** will be denoted by $\boldsymbol{D_T}$, in the latter case by $\boldsymbol{D_R}$.

Later on we shall distinguish two kinds of subgroups $\boldsymbol{D_R}$: one kind will be denoted by $\boldsymbol{D_{R0}}$, the other by $\boldsymbol{D_{R\alpha}}$. It turns out that the rules for obtaining all subgroups of the kind $\boldsymbol{D_T}$ and $\boldsymbol{D_{R0}}$ of any given space group **F** can be easily derived. However the derivation, and the formulation of the rules, becomes rather cumbersome for the case of the subgroups $\boldsymbol{D_{R\alpha}}$ if **F** is nonsymmorphic. For that reason we shall give the derivation of the rules only for the case of subgroups $\boldsymbol{D_T}$ and $\boldsymbol{D_{R0}}$ of an arbitrary **F**, and for the case of subgroups $\boldsymbol{D_{R\alpha}}$ of an arbitrary symmorphic **F**. The derivation and formulation of the rules for constructing the subgroups $\boldsymbol{D_{R\alpha}}$ of a nonsymmorphic **F** will only be sketched in a few sentences.

The following convention will often be used in denoting the elements of the various groups considered in the sequel:

Let **G** be a group and $\boldsymbol{D^G}$ one of its subgroups of index 2. An element of **G** which belongs to $\boldsymbol{D^G}$ will be labeled with a Latin subscript, e.g., G_a, G_b, G_c, etc.; an element of **G** which does not belong to $\boldsymbol{D^G}$ (it belongs then to the coset $\boldsymbol{G-D^G}$) will be labeled with a Greek subscript, e.g., G_α, G_β, G_γ, etc.

b. Construction of all the subgroups **D** of index 2 for any given **F**. (1) *Subgroups of the kind* $\boldsymbol{D_T}$. The procedure for constructing all the subgroups $\boldsymbol{D_T}$ of a given **F** is particularly simple to prove.

Consider the set of all those elements $(R_a \mid \tau(R_a) + \mathbf{t})$ of **F** for which

R_a belongs to some specified subgroups $\mathbf{R}^D$ of index 2 of the point group $\mathbf{R}$ of $\mathbf{F}$, and $\mathbf{t}$ is any translation of the lattice $\mathbf{T}$ of $\mathbf{F}$. The set obviously constitutes a subgroup of $\mathbf{F}$ which has the lattice $\mathbf{T}$ in common with $\mathbf{F}$. It follows then immediately from the fundamental lemma (see Section II, 2a) that the subgroup is of index 2, i.e., of the kind $\mathbf{D}_T$. That one will obtain all the subgroups $\mathbf{D}_T$ of $\mathbf{F}$ in this way by taking in turn all the subgroups $\mathbf{R}^D$ follows from the theorem that the point group of any $\mathbf{D}_T$ is necessarily a subgroup $\mathbf{R}^D$ of the point group $\mathbf{R}$ of $\mathbf{F}$. This theorem can be proved easily using again the fundamental lemma.

A space group has thus no subgroups of the kind $\mathbf{D}_T$ if its point group has no subgroups of index 2, that is, if its point group is 1, 3, or 23.

(2) *Subgroups of the kind $\mathbf{D}_R$.* The point group of $\mathbf{D}_R$ is by definition identical with the point group of $\mathbf{F}$. Hence, the lattice $\mathbf{T}^D$ of $\mathbf{D}_R$ is a subgroup of index 2 of the lattice $\mathbf{T}$ of $\mathbf{F}$.

This does not imply, however, that translations which belong to $\mathbf{T} - \mathbf{T}^D$ cannot occur as translations in the elements of $\mathbf{D}_R$. It only follows that if they do occur they must be nonprimitive translations of $\mathbf{D}_R$. Hence the subgroups $\mathbf{D}_R$ may be of two kinds: those, denoted by $\mathbf{D}_{R0}$, in whose elements no translations of $\mathbf{T} - \mathbf{T}^D$ occur (the right-hand subscript in the symbol $\mathbf{D}_{R0}$ is "zero" for this reason); and those, denoted by $\mathbf{D}_{R\alpha}$, in whose elements the translations of $\mathbf{T} - \mathbf{T}^D$ do occur. Obviously, one can always choose as one of the three basic primitive translations of $\mathbf{T}$ a translation belonging to $\mathbf{T} - \mathbf{T}^D$. Let the translation so chosen be denoted by $\mathbf{t}_\alpha$. Since $\mathbf{T}^D$ is a subgroup of index 2 of $\mathbf{T}$, every translation of $\mathbf{T}$ belonging to $\mathbf{T} - \mathbf{T}^D$ can be expressed as a sum of $\mathbf{t}_\alpha$ and a translation belonging to $\mathbf{T}^D$.

An element of $\mathbf{D}_R$ is thus necessarily of the form $(R \mid \sigma(R) + \mathbf{t}_a)$ where R belongs to $\mathbf{R}$, $\mathbf{t}_a$ belongs to $\mathbf{T}^D$, and

$$\sigma(R) = \begin{cases} \tau(R) & \text{if } \mathbf{D}_R = \mathbf{D}_{R0} \\ \tau(R) + \mathbf{t}_\alpha(R) & \text{if } \mathbf{D}_R = \mathbf{D}_{R\alpha}; \end{cases}$$

$\mathbf{t}_\alpha(R)$ may be either zero or equal to $\mathbf{t}_\alpha$; the latter possibility must occur for at least one element of $\mathbf{R}$. (We assume here that by no choice of the coordinate system can $\mathbf{t}_\alpha$ be transformed away from all the elements of $\mathbf{D}_{R\alpha}$; for if this were possible the group in question would not be $\mathbf{D}_{R\alpha}$ but $\mathbf{D}_{R0}$).

Since $\mathbf{D}_R$ is a space group its lattice must be invariant under any rotation of its point group. Or, more precisely, for any R belonging to the point group $\mathbf{R}$ of $\mathbf{D}_R$ and any $\mathbf{t}_a$ belonging to the lattice $\mathbf{T}^D$ of $\mathbf{D}_R$, one must have

$$R\mathbf{t}_a = \mathbf{t}_b$$

where t_b belongs to T^D. But the point group R of D_R is also the point group of that F of which D_R is a subgroup of index 2. Hence, the lattice T^D of D_R and the lattice T of F have the same holohedry. Of course, the point group R need not be the holohedry of the two lattices.

Subgroups D_{R0}. To construct all the subgroups D_{R0} of a given space group F one has thus the following rule:

Take the set of all those elements $(R \mid \tau(R) + t_a)$ of F for which t_a belongs to some specified T^D which has the same holohedry as T and R is any rotation of the point group R of F. If F is symmorphic the set always forms a group; if F is nonsymmorphic it may or may not form a group; but, in both cases, the group (if it exists) is necessarily a subgroup D_{R0}. By taking all such T^D's in turn one obtains all the D_{R0}'s of F.

Subgroups $D_{R\alpha}$ (*F symmorphic*). Here we have the following theorem:

Those elements of $D_{R\alpha}$ of a symmorphic F for which $t_\alpha(R) = 0$ constitute a subgroup Q of index 2 of $D_{R\alpha}$, and, hence, Q is a subgroup of index 4 of F.

To prove this theorem we first observe that Q is a group; in fact, the product of two elements with $t_\alpha(R) = 0$ gives an element with $t_\alpha(R) = 0$ because (since $D_{R\alpha}$ is a space group) $Rt_a = t_b$. To show that Q is a subgroup of index 2 of $D_{R\alpha}$ let us multiply any two elements of $D_{R\alpha}$ which do not belong to Q, $(R_1 \mid t_\alpha + t_a)$ and $(R_2 \mid t_\alpha + t_b)$. The product is always of the form $(R_1R_2 \mid t_c)$ because $Rt_b = t_e$ and $Rt_\alpha = t_\beta$ (the last equation follows from the fact that if $Rt_\alpha = t_d$ one would have $t_\alpha = R^{-1}t_d$ which would violate the condition that the lattice of $D_{R\alpha}$ is invariant with respect to R), or in other words, the product belongs to Q. Hence, according to the fundamental lemma, the subgroup Q of $D_{R\alpha}$ is of index 2.

From the above proof it also follows that the lattice of Q is identical with the lattice T^D of $D_{R\alpha}$ and the point group of Q is a subgroup R^D of index 2 of R, that is, the general element of Q is of the form $(R_a \mid t_a)$ where R_a belongs to R^D and t_a to T^D. Hence, a symmorphic F whose point group does not have subgroups of index 2 cannot have subgroups of the kind $D_{R\alpha}$.

We now arrange the elements of F into left cosets relative to Q:

First coset: $(E \mid 0)\,(R_a \mid t_a)$,
Second coset: $(E \mid t_\alpha)\,(R_a \mid t_a)$,
Third coset: $(R_\alpha \mid 0)\,(R_a \mid t_a)$,
Fourth coset: $(R_\alpha \mid t_\alpha)\,(R_a \mid t_a)$.

It is immediately apparent that the first and the fourth coset constitute the group $D_{R\alpha}$ from which we have started out; the first and the third coset also constitute a group, which is of the kind D_{R0}; the first and the second coset also constitute a group, which is of the kind D_T.

The rule for constructing all the D_{Ra}'s of a given symmorphic F will then be as follows: Take the set Λ of all those elements $(R_a \mid t_a)$ of the symmorphic space group F for which R_a belongs to some specified R^D and t_a belongs to some specified T^D of T which has the same holohedry as T. The set Λ then will form a group which will be necessarily a subgroup of index 4 of F, and every subgroup Q of any subgroup D_{Ra} of F will be identical with one of the groups obtained in this way by taking all such pairs R^D, T^D. Also the set consisting of the elements of the first and fourth cosets will necessarily form a group. This is either a subgroup D_{Ra} or D_{R0} of F. The latter case occurs when all the translations t_α can be transformed away, or, in other words, when the group formed by the first and fourth cosets of F relative to Q belongs to the same class of space groups as the group formed by the first and third cosets of F relative to Q.

That Λ forms a group follows immediately by considering the product of two arbitrary elements of Λ. That the set constituted by the first and fourth cosets is a group follows similarly by considering the product of two arbitrary elements of the set. This product necessarily belongs to the set because $R_\alpha t_\alpha = t_\beta$, $R_\alpha t_\alpha = t_\gamma$ (as proved above).

(3) *Subgroups D of any symmorphic F.* The results just obtained in (1) and (2) can be summed up in one single rule for constructing all subgroups D of a symmorphic F whose point group has subgroups R^D (out of the 73 classes of symmorphic space groups only 6 have point groups which do not have subgroups of index 2):

Take the set of all those elements $(R_a \mid t_a)$ of the symmorphic space group F for which R_a belongs to some specified R^D and t_a belongs to some specified T^D of F, which has the same holohedry as T. This set is a group, which will be necessarily a subgroup Q of index 4 of F, and the elements of F can be arranged into left cosets relative to Q. The elements of the first and second coset will then necessarily form a subgroup D_T of F. Similarly, the elements of the first and third coset will form a subgroup D_{R0} and those of the first and fourth coset will form a subgroup D_{R0} or D_{Ra}. By taking all possible pairs R^D, T^D of F, one will obtain in this way all possible subgroups D_T, D_{R0}, and D_{Ra} of a symmorphic space group F.

(4) *Subgroups D_{Ra} of any nonsymmorphic F.* The procedure for constructing the subgroups D_{Ra} in the case of symmorphic space groups F is based on the validity of the theorem formulated at the beginning

of our discussion of that case. This theorem is not always valid in the case of nonsymmorphic space groups. Consequently the procedure cannot always be used for nonsymmorphic space groups without some modification.

If a nonsymmorphic space group has a point group which does not have elements of order higher than 2 the theorem in question is still valid although its proof is somewhat more complicated. For a nonsymmorphic space group whose point group has elements of higher order it is not valid in general, although it is valid in some cases. That is why it is necessary to treat the nonsymmorphic space groups belonging to the various crystallographic systems (i.e., tetragonal, rhombohedral, hexagonal, and cubic) separately. The discussion becomes rather lengthy. The result is that if the point group of a nonsymmorphic space group has subgroups of index 2 that is in 153 out of 157 cases) the problem can be reduced, as in the symmorphic case, to the problem of finding all subgroups, $Q_{\tau 0}$ and $Q_{\tau a}$, of index 4 of the space group F in question [23]. The latter problem is simple as one can construct all such subgroups by combining, loosely speaking, a "half" of the point group of F with a "half" of the lattice of F. For details, see reference [2].

c. **Historical digression.** A systematic discussion of the problem of finding and classifying the subgroups (not necessarily of index 2) of a given space group was given many years ago by Hermann [30]. In particular, he has introduced the division of all subgroups of a given space group into two classes: *klassengleiche Untergruppen* (which, in the case of the subgroups of index 2, are our subgroups D_R) and *zellengleiche Untergruppen* (which, in the case of the subgroups of index 2, are our subgroups D_T). However, Hermann was not specifically interested in the subgroups of index 2, and he has not considered the distinction between D_{R0} and D_{Ra} nor has he mentioned the important role played by subgroups of index 4 (Q, $Q_{\tau 0}$, and $Q_{\tau a}$).

The idea of combining a "half" of the elements of the point group of a space group with a "half" of the elements of its lattice for the case of symmorphic space groups (i.e., constructing a subgroup Q of index 4) is implicit in the work of Zamorzaev [16], who apparently has used this idea to obtain the magnetic space groups belonging to the families of the symmorphic space groups, without explicitly mentioning the role played by the subgroups of index 4.

d. **Magnetic space groups, their lattices and point groups.** Corresponding to the three kinds of subgroups D of index 2 of a space group F, i.e., D_T, D_{R0}, D_{Ra}, it is convenient to distinguish three kinds of nontrivial magnetic space groups: M_T, M_{R0}, $M_{R\alpha}$. The unprimed elements of M_T form a subgroup of the kind D_T, the unprimed elements of M_{R0} form a subgroup of the kind D_{R0}, and the unprimed elements of $M_{R\alpha}$ form a subgroup of the kind $D_{R\alpha}$.

We define magnetic lattice and magnetic point group of (or "belonging to") a nontrivial magnetic space group in exactly the same way as one does it for ordinary space groups (see Section II, 2a). In Table II, we give the elements of the magnetic lattice T_M and magnetic point group R_M belonging to a given magnetic space group M of each of the three kinds. [Table II does not include the case of those magnetic space groups $M_{R\alpha}$ which belong to the families of space groups whose point groups do not have subgroups of index 2. There are only two such magnetic space groups, $P_{2c}3_1$ and $P_{2c}3_2$; their magnetic point group is 3, and their magnetic lattice is P_{2c}].

TABLE II

Elements of Magnetic Space Groups, of their Magnetic Lattices, Their Magnetic Point Groups, and Their Standard Representations, for the Three Kinds of the Magnetic Space Groups[a]

Kind of magnetic space group M	Elements of magnetic space group M	Elements of its magnetic lattice T_M	Elements of its magnetic point group R_M	Matrices of its standard representation
M_T	$(R_a \mid \tau(R_a) + \mathbf{t})$ $(R_\alpha \mid \tau(R_\alpha) + \mathbf{t})'$	$\mathbf{t}$	R_a, R_α'	$\delta(R_a)R_a$ $-\delta(R_\alpha)R_\alpha$
M_{R0}	$(R \mid \tau(R) + \mathbf{t}_a)$ $(R \mid \tau(R) + \mathbf{t}_\alpha)'$	$\mathbf{t}_a$, $\mathbf{t}_\alpha'$	R	$\delta(R)R$ $-\delta(R)R$
$M_{R\alpha}$	$(R_a \mid \sigma(R_a) + \mathbf{t}_a)$ $(R_\alpha \mid \sigma(R_\alpha) + \mathbf{t}_\alpha)$ $(R_a \mid \sigma(R_a) + \mathbf{t}_\alpha)'$ $(R_\alpha \mid \sigma(R_\alpha) + \mathbf{t}_a)'$	$\mathbf{t}_a$, $\mathbf{t}_\alpha'$	R_a, R_α'	$\delta(R_a)R_a$ $\delta(R_\alpha)R_\alpha$ $-\delta(R_a)R_a$ $-\delta(R_\alpha)R_\alpha$

[a] The definition of the "standard representation" is given in Section IV, 1b; $\delta(R)$ is the determinant of the rotation matrix R. The meaning of the Greek and Latin subscripts is explained in Section III, 2a.

Except for two space groups, the family of each of the remaining 228 space groups contains nontrivial magnetic space groups. The two exceptions are the space groups $F23$ and $P2_13$; these are the only two space groups which do not have subgroups of index 2 at all.

As is well-known, two space groups F_1 and F_2 are regarded as not essentially different or as equivalent, and as belonging to the same class of space groups (denoted by the same international symbol) if the elements of F_1 can be made identical with the elements of F_2 by transforming appropriately the coordinate system relative to which the elements of F_1 are defined; moreover only those transformations of the coordinate system are allowed which do not change its "handedness." Exactly the same convention is used in the case of nontrivial magnetic space groups, and we speak of "equivalent magnetic space groups," and of "classes of magnetic space groups."

In Table III a list of all the 1421 classes of magnetic space groups (of which $1421 - 230 = 1191$ are nontrivial) is given. Each class is represented in the list by one arbitrarily chosen magnetic space group belonging to it (strictly speaking, the symbol standing for a class denotes a subclass of magnetic space groups, and not just one of them). For simplicity, we shall use the term "magnetic space group" instead of "class of magnetic space groups" whenever the context makes it possible.

The list in Table III differs in two respects from the list published by Belov *et al.* [22]. First, in our list, the magnetic space groups are arranged into families in the sense of definition given in Section II. Second, the symbols that we use for denoting the magnetic space groups have been chosen so as to indicate the family to which the magnetic space groups belong and the way in which they have been constructed following the prescription used in this chapter; also the symbol of a magnetic space group gives its magnetic lattice and its magnetic point group in the sense of the definitions given in Table II. In the case of the magnetic space groups of the kind M_T our symbol is identical with that of Belov, Neronova, and Smirnova. In the case of M_{R0} and M_{Ra} our symbols are always different from theirs, and both symbols are given, ours to the left of theirs.

The symbol of a magnetic space group consists of two parts (as in the case of the international symbols for ordinary space groups). The first part indicates the magnetic lattice of the magnetic space group in question. The symbols of all magnetic lattices have been given in Figs. 1 to 6. However, for brevity the symbol R_R is used instead of $R_{2a,a+b,a+c}$, and the symbol P_P instead of $P_{a+b,a-b}$. The second part gives the (primed and unprimed) rotations which generate its magnetic point group and the corresponding nonprimitive translations; the rotations and their nonprimitive translations are denoted in the same way as in the "International Tables for X-Ray Crystallography" [1]. (N.B. "Rotation" means a proper or an improper rotation.) Here it should be pointed out that in the case of a class of magnetic space groups of the kind M_{Ra} a magnetic point group cannot in general be assigned unambiguously to the class. The magnetic point groups of different magnetic space groups $M_{R\alpha}$ belonging to the same class may be different (but they will always belong to the same family of magnetic point groups). For example, the magnetic space group listed in Table III as $P_{2c}\,2/m'$ could have been listed as $P_{2c}2'/m'$.

In the second part of the symbol of a magnetic space group **M** the nonprimitive translations are in general referred to the lattice of the space group to whose family **M** belongs. If we referred them to the lattice of the unprimed subgroup of index 2 of **M** the second part of the symbol of **M** would become identical with the second part of the symbol used by Belov, Neronova, and Smirnova, and printed to the right of our symbol. For example, $P_{2c}4_2$ in our notation becomes P_c4_1 in the notation of Belov, Neronova, and Smirnova; $P_F nnn$ becomes $F_s ddd$; $P_I 4_2 nm$ becomes $I_c 4_1 md$. We depart from this convention in the case of magnetic space groups $P_{2c}3_2$, $P_{2c}3_1$, $P_{2c}3_212$, $P_{2c}3_221$, $P_{2c}3_112$, $P_{2c}3_121$.

If in the symbol of a magnetic space group **M** one replaces the symbols of screw axes and glide planes by the symbols of the corresponding ordinary axes and mirror planes then the second part of the symbol of **M** will become that of the magnetic point group of **M** as defined in Table II. For example, the magnetic point groups of $P2_1'$, $Pn'n'2$, $P_{2b}cca$, $I4_1'md'$ are $2'$, $m'm'2$, mmm, $4'mm'$.

The order in which the families of magnetic space groups are listed is the same as the order in which space groups (i.e., trivial magnetic space groups) belonging to these families are listed in the "International Tables for X-Ray Crystallography" [1].

Within one family the magnetic space groups are listed as follows: first the trivial magnetic space group (symbol underlined), then the magnetic space groups of the kind M_T, M_{R0}, and $M_{R\alpha}$ (in that order).

An asterisk to the left of the symbol of a magnetic space group indicates that the group is ferromagnetic in the sense to be defined in Section IV.

In connection with the difference in notation used by various authors we may refer to a recent paper by Zamorzaev [31].

A graphical representation of the monoclinic and orthorhombic magnetic space groups, similar to that of the space groups in "International Tables" [1], has been given by Spence [32].

A list of the two-dimensional magnetic space groups arranged into families can be found in reference [2].

e. *Example of constructing a family of magnetic space groups.* Since we have already constructed the family of the point group 422 whose generating elements are 4_z and 2_x [see Section III, 1a], we will construct the family of a space group **F** whose point group is generated by these two elements. Let this group **F** belong to the class of space groups *P*422. Its lattice **T** is primitive and belongs to the tetragonal system. It is characterized by three basic primitive translations $\mathbf{a}_1$, $\mathbf{a}_2$, $\mathbf{a}_3$ at right angles to each other and satisfying the relationship $|\mathbf{a}_1| = |\mathbf{a}_2| \neq |\mathbf{a}_3|$. The generating elements of **F** are thus:

$$(1 \mid n_1\mathbf{a}_1 + n_2\mathbf{a}_2 + n_3\mathbf{a}_3), (4_z \mid 0), (2_x \mid 0),$$

where n_1, n_2, and n_3 are arbitrary integers.

We have now to find all the subgroups of index 2 of **F**. In order to do that according to the rule given in Section III, 2*b*(3), we have first to consider all those sets of elements $(R_a \mid \mathbf{t}_a)$ of **F** for which R_a belongs to some subgroup of index 2 of the point group **R** of **F**, and $\mathbf{t}_a$ belongs to some subgroup of index 2 of **T** which has the same holohedry as **T**. As we have said earlier the only subgroups of index 2 of **R** are the groups consisting of the elements 1, 4_z, 2_z, 4_z^3, of the elements 1, 2_z, 2_x, 2_y and of the elements 1, 2_z, 2_{xy}, $2_{\bar{x}y}$. Of the seven subgroups of index 2 of **T** only three have the same holohedry as **T**. They are those whose basic primitive translations are

$\mathbf{a}_1, \mathbf{a}_2, 2\mathbf{a}_3$,

$2\mathbf{a}_1, \mathbf{a}_1 + \mathbf{a}_2, \mathbf{a}_3$,

$2\mathbf{a}_1, \mathbf{a}_1 + \mathbf{a}_2, \mathbf{a}_1 + \mathbf{a}_3$.

The first two lattices are again primitive, whereas the third one is a body-centered lattice. It follows that one has to consider nine sets of elements $(R_a \mid \mathbf{t}_a)$ of the given space group **F**. They are all groups, and are subgroups **Q** of index 4 of **F**. Their generating elements are:

Q₁: $(1 \mid n_1\mathbf{a}_1 + n_2\mathbf{a}_2 + 2n_3\mathbf{a}_3), (4_z \mid 0);$

Q₂: $(1 \mid 2n_1\mathbf{a}_1 + n_2(\mathbf{a}_1 + \mathbf{a}_2) + n_3\mathbf{a}_3), (4_z \mid 0);$

Q₃: $(1 \mid 2n_1\mathbf{a}_1 + n_2(\mathbf{a}_1 + \mathbf{a}_2) + n_3(\mathbf{a}_1 + \mathbf{a}_3)), (4_z \mid 0);$

Q₄: $(1 \mid n_1\mathbf{a}_1 + n_2\mathbf{a}_2 + 2n_3\mathbf{a}_3), (2_x \mid 0), (2_y \mid 0);$

Q₅: $(1 \mid 2n_1\mathbf{a}_1 + n_2(\mathbf{a}_1 + \mathbf{a}_2) + n_3\mathbf{a}_3), (2_x \mid 0), (2_y \mid 0);$

Q₆: $(1 \mid 2n_1\mathbf{a}_1 + n_2(\mathbf{a}_1 + \mathbf{a}_2) + n_3(\mathbf{a}_1 + \mathbf{a}_3)), (2_x \mid 0), (2_y \mid 0);$

Q₇: $(1 \mid n_1\mathbf{a}_1 + n_2\mathbf{a}_2 + 2n_3\mathbf{a}_3), (2_{xy} \mid 0), (2_{\bar{x}y} \mid 0);$

Q₈: $(1 \mid 2n_1\mathbf{a}_1 + n_2(\mathbf{a}_1 + \mathbf{a}_2) + n_3\mathbf{a}_3), (2_{xy} \mid 0), (2_{\bar{x}y} \mid 0);$

Q₉: $(1 \mid 2n_1\mathbf{a}_1 + n_2(\mathbf{a}_1 + \mathbf{a}_2) + n_3(\mathbf{a}_1 + \mathbf{a}_3)), (2_{xy} \mid 0), (2_{\bar{x}y} \mid 0).$

We then consider the second, third, and fourth coset of **F** relative to each **Q**, as has been explained in this section, [III, 2b(2)]. Each of these cosets together with the corresponding **Q** forms a group which is a subgroup **D** of index 2 of **F**. Combining the elements of each **D** with the identity of **A**, and the remaining elements of **F** with the element E' of **A** one obtains 27 magnetic space groups. According to our general theory, a representative of each class of magnetic space groups belonging to the family of $P422$ must be contained in the set of these 27 magnetic space groups. In fact, some of the classes will have more than one representative among the 27 magnetic space groups, as we shall see.

Let us consider first the cosets of **F** relative to $\mathbf{Q}_1$. They are:

First coset, $(1\mid 0)(1\mid n_1\mathbf{a}_1 + n_2\mathbf{a}_2 + 2n_3\mathbf{a}_3), (1\mid 0)(4_z\mid 0);$

Second coset, $(1\mid \mathbf{a}_3)(1\mid n_1\mathbf{a}_1 + n_2\mathbf{a}_2 + 2n_3\mathbf{a}_3), (1\mid \mathbf{a}_3)(4_z\mid 0);$

Third coset, $(2_x\mid 0)(1\mid n_1\mathbf{a}_1 + n_2\mathbf{a}_2 + 2n_3\mathbf{a}_3), (2_x\mid 0)(4_z\mid 0);$

Fourth coset, $(2_x\mid \mathbf{a}_3)(1\mid n_1\mathbf{a}_1 + n_2\mathbf{a}_2 + 2n_3\mathbf{a}_3), (2_x\mid \mathbf{a}_3)(4_z\mid 0).$

It is easy to see that $\mathbf{Q}_1$, which is the first coset, together with the second coset constitutes a group which belongs to the class of space groups $P4$. This is a subgroup $\mathbf{D}_T$ of **F** because it has the same lattice as **F** and because its point group is a subgroup of index 2 of **R**. The subgroup $\mathbf{Q}_1$, together with the third coset, constitutes a group which belongs to the class $P422$. This is a subgroup $\mathbf{D}_{R0}$ of **F** because it has the same point groups as **F**, a lattice which is a subgroup of index 2 of **T**, and contains no elements with $\mathbf{t}_\alpha$ which in this case is $\mathbf{a}_3$. Finally $\mathbf{Q}_1$, together with the fourth coset, forms a group which is equivalent to the previous one because the translation $\mathbf{a}_3$ is not a proper nonprimitive translation for the elements 2_x, 2_y, 2_{xy}, $2_{\bar{x}y}$ or, in other words, can be transformed away from these elements by choosing a different coordinate system.

The magnetic space groups whose unprimed subgroups are the subgroups of index 2 of **F** just found belong to the classes of magnetic space groups $P42'2'$ and $P_{2c}422$. The first is a class of magnetic space groups of the kind $\mathbf{M}_T$, the second is a class of magnetic space groups of the kind $\mathbf{M}_{R0}$.

Similarly, from $\mathbf{Q}_2$ one obtains a subgroup $\mathbf{D}_T$ of **F** which belongs to the class $P4$, a subgroup $\mathbf{D}_{R0}$ which belongs again to the class $P422$, and a subgroup $\mathbf{D}_{R\alpha}$ which is a space group of the class $P42_12$. The corresponding magnetic space groups belong to the classes $P42'2'$, $P_{a+b,a-b}422$, and $P_{a+b,a-b}42'2'$. It should be noticed that, although the unprimed subgroups of any two magnetic space groups of the classes $P_{2c}422$ and $P_{a+b,a-b}422$ are equivalent, the two magnetic space groups themselves are not equivalent. In other words the equivalence of the unprimed subgroups of two magnetic space groups is only a necessary condition for the two magnetic space groups being equivalent. In the case of magnetic point groups this condition is also sufficient.

From Q_3 one obtains again a subgroup D_T which belongs to the class $P4$, a subgroup D_{R0} which belongs to the class $I422$, and a $D_{R\alpha}$ which is equivalent to the subgroup D_{R0} just mentioned. Correspondingly the magnetic space groups belong to the classes $P42'2'$ and P_I422.

From Q_4 one obtains a magnetic space group of the kind M_T which belongs to the class $P4'22'$ and a magnetic space group of the kind $M_{R\alpha}$ which belongs to the class $P_{2c}4'22'$.

From Q_7 one obtains a magnetic space group of the kind M_T which belongs to the class $P4'2'2$.

Every one of the other magnetic space groups which can be obtained from the subgroups Q considered here and from the remaining subgroups Q not considered at all (Q_5, Q_6) is equivalent to some magnetic space group already obtained.

It should be noticed that the magnetic space group $M_{R\alpha}$ obtained from Q_5 belongs to the same class $P_{a+b,a-b}42'2'$ to which the group $M_{R\alpha}$ obtained from Q_2 belongs although according to the definition of magnetic point group belonging to a given magnetic space group given earlier they have different magnetic point groups.

In conclusion we can say that the nontrivial magnetic space groups of the family of any space group belonging to the class $P422$ belong to the following eight classes of magnetic space groups: $P4'22'$, $P42'2'$, $P4'2'2$, $P_{2c}422$, $P_{a+b,a-b}422$, P_I422, $P_{2c}4'22'$, $P_{a+b,a-b}4'22'$.

IV. Invariant Spin Arrangements

1. Definition of Invariant Spin Arrangements

a. Classification of all the simple crystals whose space group is **F**. We first state some definitions and well-known facts about the relation between a space group **F** and the crystals which have **F** as the symmetry group.

We shall call a crystal a "simple crystal" if it consists of identical atoms whose position vectors can all be obtained by applying all elements of a space group **F** to one position vector **r**. We shall say that the simple crystal **Fr** is generated by **F** from **r**, and we shall also use phrases like "the element F of **F** generates the position vector $F\mathbf{r}$."

Obviously every crystal can be regarded as consisting of a certain number of interlocking simple crystals, each of them being generated by the same space group from a different position vector, and no two of the

TABLE III
List of Magnetic Space Groups[a]

Triclinic system		$P_{2b}m$	P_bm	$P2_1/m'$	
1		P_Cm	C_am	*$P2'_1/m'$	
*$P1$		$P_{2c}m'$	P_cc	$P_{2a}2_1/m$	P_a2_1/m
		*Pc		$P_{2c}2_1/m'$	P_c2_1/c
$P_{2s}1$	P_s1			*$C2/m$	
$\bar{1}$		*Pc'			
*$P\bar{1}$		$P_{2a}c$	P_ac	$C2'/m$	
		$P_{2b}c$	P_bc	$C2/m'$	
$P\bar{1}'$		P_Cc	C_ac	*$C2'/m'$	
$P_{2s}\bar{1}$	$P_s\bar{1}$			$C_{2c}2/m$	C_c2/m
		*Cm		C_P2/m	P_C2/m
Monoclinic system				$C_{2c}2/m'$	C_c2/c
2		*Cm'		C_P2'/m	P_C2_1/m
*$P2$		$C_{2c}m$	C_cm	C_P2/m'	P_A2/c
		C_Pm	P_Cm	C_P2'/m'	P_A2_1/c
*$P2'$		$C_{2c}m'$	C_cc	*$P2/c$	
$P_{2a}2$	P_a2	C_Pm'	P_Ac		
$P_{2b}2$	P_b2			$P2'/c$	
P_C2	C_a2	*Cc		$P2/c'$	
$P_{2b}2'$	P_b2_1			*$P2'/c'$	
*$P2_1$		C_Pc	P_Cc	$P_{2a}2/c$	P_a2/c
		$2/m$		$P_{2b}2/c$	P_b2/c
*$P2'_1$		*$P2/m$		P_C2/c	C_a2/c
$P_{2a}2_1$	P_a2_1			$P_{2b}2'/c$	P_b2_1/c
*$C2$		$P2'/m$			
		$P2/m'$		*$P2_1/c$	
*$C2'$		*$P2'/m'$			
$C_{2c}2$	C_c2	$P_{2a}2/m$	P_a2/m	$P2'_1/c$	
C_P2	P_C2	$P_{2b}2/m$	P_b2/m	$P2_1/c'$	
C_P2'	P_C2_1	P_C2/m	C_a2/m	*$P2'_1/c'$	
m		$P_{2b}2'/m$	P_b2_1/m	$P_{2a}2_1/c$	P_a2_1/c
*Pm		$P_{2c}2/m'$	P_c2/c	*$C2/c$	
*Pm'		*$P2_1/m$		$C2'/c$	
$P_{2a}m$	P_am	$P2'_1/m$		$C2/c'$	

[a] Every page of Table III should be read first from the top to the bottom of the left-hand double column, then from the top to the bottom of the middle double column, and finally from the top to the bottom of the right-hand double column.

3. MAGNETIC SYMMETRY

TABLE III (continued)

Monoclinic system		$C222$		$P_{2c}m'm'2$	P_ccc2
$*C2/c$				$P_{2a}m'm'2$	P_ama2
$*C2'/c'$		$*C2'2'2$		$P_Am'm'2$	A_cbm2
C_P2/c	P_C2/c	$*C22'2'$			
C_P2'/c	P_C2_1/c	$C_{2c}222$	C_c222	$Pmc2_1$	
		C_P222	P_C222		
Orthorhombic system		C_I222	I_c222	$*Pm'c2_1'$	
222		$C_{2c}22'2'$	C_c222_1	$*Pmc'2_1'$	
$P222$		$C_P2'2'2$	$P_C2_12_12$	$*Pm'c'2_1$	
		$C_P22'2'$	P_A222_1	$P_{2a}mc2_1$	P_amc2_1
$*P2'2'2$		$C_I2'22'$	$I_c2_12_12_1$	$P_{2b}mc2_1$	P_bmc2_1
$P_{2s}222$	P_a222			P_Cmc2_1	C_amc2_1
P_C222	C_a222			$P_{2a}mc'2_1'$	P_amn2_1
P_F222	F_s222	$F222$		$P_{2b}m'c'2_1$	P_aca2_1
$P_{2c}22'2'$	P_c222_1				
		$*F2'2'2$		$Pcc2$	
$P222_1$		F_C222	C_A222		
		$F_C22'2'$	C_A222_1	$*Pc'c2'$	
$*P2'2'2_1$				$*Pc'c'2$	
$*P22'2_1'$		$I222$		$P_{2a}cc2$	P_acc2
$P_{2a}222_1$	P_a222_1			P_Ccc2	C_acc2
P_C222_1	C_a222_1	$*I2'2'2$		$P_{2b}c'c2'$	P_bnc2
$P_{2a}2'2'2_1$	$P_a2_12_12$	I_P222	P_I222		
		$I_P2'2'2$	$P_I2_12_12$	$Pma2$	
$P2_12_12$					
		$I2_12_12_1$		$*Pm'a2'$	
$*P2_1'2_1'2$				$*Pma'2'$	
$*P2_12_1'2'$		$*I2_1'2_1'2_1$		$*Pm'a'2$	
$P_{2c}2_12_12$	$P_c2_12_12$	$I_P2_12_12_1$	$P_I2_12_12_1$	$P_{2b}ma2$	P_bma2
$P_{2c}2_12_1'2'$	$P_a2_12_12_1$	$I_P2_1'2_1'2_1$	P_I222_1	$P_{2c}ma2$	P_cma2
				P_Ama2	A_cma2
$P2_12_12_1$		$mm2$		$P_{2b}m'a2'$	P_aba2
		$Pmm2$		$P_{2c}m'a2'$	P_cca2_1
$*P2_1'2_1'2_1$				$P_{2c}ma'2'$	P_cmn2_1
$C222_1$		$*Pm'm2'$		$P_{2c}m'a'2$	P_cnc2
		$*Pm'm'2$		$P_Am'a'2$	A_cba2
$*C2'2'2_1$		$P_{2c}mm2$	P_cmm2		
$*C22'2_1'$		$P_{2a}mm2$	P_amm2	$Pca2_1$	
C_P222_1	P_C222_1	P_Cmm2	C_amm2		
$C_P2'2'2_1$	$P_C2_12_12_1$	P_Amm2	A_cmm2	$*Pc'a2_1'$	
$C_P22'2_1'$	$P_A2_12_12$	P_Fmm2	F_smm2	$*Pca'2_1'$	
		$P_{2c}mm'2'$	P_cmc2_1		

TABLE III (continued)

Orthorhombic system		$Cmm2$		$A_Pm'm2'$	P_Bmn2_1
$Pca2_1$				$A_Pmm'2'$	P_Amc2_1
		$*Cm'm2'$		$A_Pm'm'2$	P_Anc2
$*Pc'a'2_1$		$*Cm'm'2$		$A_Im'm'2$	I_ama2
$P_{2b}ca2_1$	P_bca2_1	$C_{2c}mm2$	C_cmm2		
$P_{2b}c'a'2_1$	P_bna2_1	C_Pmm2	P_Cmm2	$Abm2$	
		C_Imm2	I_cmm2		
$Pnc2$		$C_{2c}m'm2'$	C_cmc2_1	$*Ab'm2'$	
		$C_{2c}m'm'2$	C_ccc2	$*Abm'2'$	
$*Pn'c2'$		$C_Pm'm2'$	P_Cma2	$*Ab'm'2$	
$*Pnc'2'$		$C_Pm'm'2$	P_Cba2	$A_{2a}bm2$	A_abm2
$*Pn'c'2$		$C_Im'm2'$	I_cma2	A_Pbm2	P_Bma2
$P_{2a}nc2$	P_anc2	$C_Im'm'2$	I_cba2	A_Ibm2	B_Ima2
$P_{2a}nc'2'$	P_ann2			$A_{2a}b'm'2$	A_aba2
		$Cmc2_1$		$A_Pb'm2'$	P_Bmc2_1
$Pmn2_1$				$A_Pbm'2'$	P_Bca2_1
		$*Cm'c2_1'$		$A_Pb'm'2$	P_Acc2
$*Pm'n2_1'$		$*Cmc'2_1'$		$A_Ib'm'2$	I_aba2
$*Pmn'2_1'$		$*Cm'c'2_1$			
$*Pm'n'2_1$		C_Pmc2_1	P_Cmc2_1	$Ama2$	
$P_{2b}mn2_1$	P_bmn2_1	$C_Pm'c2_1'$	P_Cca2_1	$*Am'a2'$	
$P_{2b}m'n2_1'$	P_ana2_1	$C_Pmc'2_1'$	P_Cmn2_1	$*Ama'2'$	
		$C_Pm'c'2_1$	P_Cna2_1	$*Am'a'2$	
$Pba2$				A_Pma2	P_Ama2
		$Ccc2$		$A_Pm'a2'$	P_Ana2_1
$*Pb'a2'$				$A_Pma'2'$	P_Amn2_1
$*Pb'a'2$		$*Cc'c2'$		$A_Pm'a'2$	P_Ann2
$P_{2c}ba2$	P_oba2	$*Cc'c'2$			
$P_{2c}b'a2'$	P_cna2_1	C_Pcc2	P_Ccc2	$Aba2$	
$P_{2c}b'a'2$	P_cnn2	$C_Pc'c2'$	P_Cnc2	$*Ab'a2'$	
		$C_Pc'c'2$	P_Cnn2	$*Aba'2'$	
$Pna2_1$				$*Ab'a'2$	
		$Amm2$		A_Pba2	P_Aba2
$*Pn'a2_1'$				$A_Pb'a2'$	P_Aca2_1
$*Pna'2_1'$		$*Am'm2'$		$A_Pba'2'$	P_Bna2_1
$*Pn'a'2_1$		$*Amm'2$		$A_Pb'a'2$	P_Bnc2
		$*Am'm'2$			
$Pnn2$		$A_{2a}mm2$	A_amm2	$Fmm2$	
		A_Pmm2	P_Amm2		
$*Pn'n2'$		A_Imm2	I_amm2	$*Fm'm2'$	
$*Pn'n'2$		$A_{2a}mm'2'$	A_ama2	$*Fm'm'2$	
P_Fnn2	F_sdd2				

3. MAGNETIC SYMMETRY

TABLE III (continued)

Orthorhombic system					
$Fmm2$		mmm		$P_{2c}ban$	P_cban
		$Pmmm$		$P_{2c}b'an$	P_bnna
				$P_{2c}b'a'n$	P_annn
F_Cmm2	C_Amm2	$Pm'mm$			
F_Amm2	A_Cmm2	*$Pm'm'm$		$Pmma$	
$F_Cmm'2'$	C_Amc2_1	$Pm'm'm'$			
$F_Cm'm'2$	C_Acc2			$Pm'ma$	
$F_Am'm2'$	A_Cbm2	$P_{2s}mmm$	P_ammm	$Pmm'a$	
$F_Amm'2'$	A_Cma2	P_Cmmm	C_ammm	$Pmma'$	
$F_Am'm'2$	A_Cba2	P_Fmmm	F_smmm	*$Pm'm'a$	
		$P_{2a}mmm'$	P_amma	*$Pmm'a'$	
$Fdd2$		$P_{2c}m'm'm$	P_cccm	*$Pm'ma'$	
		P_Cmmm'	C_amma	$Pm'm'a'$	
*$Fd'd2'$		$Pnnn$		$P_{2b}mma$	P_bmma
*$Fd\,d'2$				$P_{2c}mma$	P_cmma
		$Pn'nn$		P_Amma	C_amcm
$Imm2$		*$Pn'n'n$		$P_{2b}m'ma$	P_bbcm
		$Pn'n'n'$		$P_{2b}mma'$	P_ammn
*$Im'm2'$		P_Fnnn	F_sddd	$P_{2b}m'ma'$	P_amna
*$Im'm'2$				$P_{2c}m'ma$	P_abam
I_pmm2	P_Imm2	$Pccm$		$P_{2c}mm'a$	P_ebcm
$I_pmm'2'$	P_Imn2_1			$P_{2c}m'm'a$	P_ccca
$I_pm'm'2$	P_Inn2	$Pc'cm$		$P_Am'ma$	C_amca
		$Pccm'$			
$Iba2$		*$Pc'c'm$		$Pnna$	
		*$Pc'cm'$			
*$Ib'a2'$		$Pc'c'm'$		$Pn'na$	
*$Ib'a'2$		$P_{2a}ccm$	P_accm	$Pnn'a$	
I_pba2	P_Icc2	P_Cccm	C_accm	$Pnna'$	
$I_pba'2'$	P_Ica2_1	$P_{2a}ccm'$	P_acca	*$Pn'n'a$	
$I_pb'a'2$	P_Iba2	$P_{2a}c'c'm$	P_cmna	*$Pnn'a'$	
		$P_{2a}c'c'm'$	P_aban	*$Pn'na'$	
$Ima2$		P_Cccm'	C_acca	$Pn'n'a'$	
*$Im'a2'$		$Pban$		$Pmna$	
*$Ima'2'$					
*$Im'a'2$		$Pb'an$		$Pm'na$	
I_pma2	P_Ima2	$Pban'$		$Pmn'a$	
$I_pm'a2'$	P_Ina2_1	*$Pb'a'n$		$Pmna'$	
$I_pma'2'$	P_Imc2_1	*$Pb'an'$		*$Pm'n'a$	
$I_pm'a'2$	P_Inc2	$Pb'a'n'$		*$Pmn'a'$	

TABLE III (continued)

Orthorhombic system		$Pbcm$		$Pbca$	
$Pmna$					
		$Pb'cm$		$Pb'ca$	
$*Pm'na'$		$Pbc'm$		$*Pb'c'a$	
$Pm'n'a'$		$Pbcm'$		$Pb'c'a'$	
$P_{2b}mna$	P_bmna	$*Pb'c'm$			
$P_{2b}m'na$	P_cbcn	$*Pbc'm'$		$Pnma$	
$P_{2b}mna'$	P_annm	$*Pb'cm'$			
$P_{2b}m'na'$	P_anna	$Pb'c'm'$		$Pn'ma$	
		$P_{2a}bcm$	P_abcm	$Pnm'a$	
$Pcca$		$P_{2a}bc'm$	P_cnma	$Pnma'$	
		$P_{2a}bcm'$	P_abca	$*Pn'm'a$	
$Pc'ca$		$P_{2a}bc'm'$	P_bbcn	$*Pnm'a'$	
$Pcc'a$				$*Pn'ma'$	
$Pcca'$		$Pnnm$		$Pn'm'a'$	
$*Pc'c'a$					
$*Pcc'a'$		$Pn'nm$		$Cmcm$	
$*Pc'ca'$		$Pnnm'$			
$Pc'c'a'$		$*Pn'n'm$		$Cm'cm$	
$P_{2b}cca$	P_bcca	$*Pnn'm'$		$Cmc'm$	
$P_{2b}c'ca$	P_abcn	$Pn'n'm'$		$Cmcm'$	
$P_{2b}cca'$	P_accn			$*Cm'c'm$	
$P_{2b}c'ca$	P_cnna	$Pmmn$		$*Cmc'm'$	
				$*Cm'cm'$	
$Pbam$		$Pm'mn$		$Cm'c'm'$	
		$Pmmn'$			
$Pb'am$		$*Pm'm'n$		C_Pmcm	P_Amma
$Pbam'$		$*Pmm'n'$		$C_Pm'cm$	P_Cbcm
$*Pb'a'm$		$Pm'm'n'$		$C_Pmc'm$	P_Cmmn
$*Pb'am'$		$P_{2c}mmn$	P_cmmn	C_Pmcm'	P_Bnma
$Pb'a'm'$		$P_{2c}m'mn$	P_anma	$C_Pm'c'm$	P_Cnma
$P_{2c}bam$	P_cbam	$P_{2c}m'm'n$	P_cccn	$C_Pmc'm'$	P_Cnnm
$P_{2c}b'am$	P_bnma			$C_Pm'cm'$	P_Cbcn
$P_{2c}b'a'm$	P_cnnm	$Pbcn$		$C_Pm'c'm'$	P_Cnna
		$Pb'cn$			
$Pccn$		$Pbc'n$		$Cmca$	
		$Pbcn'$			
$Pc'cn$		$*Pb'c'n$		$Cm'ca$	
$Pccn'$		$*Pbc'n'$		$Cmc'a$	
$*Pc'c'n$		$*Pb'cn'$		$Cmca'$	
$*Pc'cn'$		$Pb'c'n'$		$*Cm'c'a$	
$Pc'c'n'$				$*Cmc'a'$	

TABLE III (continued)

Orthorhombic system		$C_{P}ccm$	$P_{C}ccm$	$Fm'm'm'$		
$Cmca$		$C_{P}c'cm$	$P_{C}mna$	$F_{C}mmm$	$C_{A}mmm$	
		$C_{P}ccm'$	$P_{C}ccn$	$F_{C}m'mm$	$C_{A}mcm$	
*$Cm'ca'$		$C_{P}c'c'm$	$P_{C}nnm$	$F_{C}mmm'$	$C_{A}mma$	
$Cm'c'a'$		$C_{P}cc'm'$	$P_{C}nna$	$F_{C}m'm'm$	$C_{A}ccm$	
$C_{P}mca$	$P_{C}bam$	$C_{P}c'c'm'$	$P_{C}nnn$	$F_{C}mm'm'$	$C_{A}mca$	
$C_{P}m'ca$	$P_{C}cca$			$F_{C}m'm'm'$	$C_{A}cca$	
$C_{P}mc'a$	$P_{C}nma$	$Cmma$				
$C_{P}mca'$	$P_{C}bcm$			$Fddd$		
$C_{P}m'c'a$	$P_{C}ccn$	$Cm'ma$				
$C_{P}mc'a'$	$P_{C}mna$	$Cmma'$		$Fd'dd$		
$C_{P}m'ca'$	$P_{C}bca$	*$Cm'm'a$		*$Fd'd'd$		
$C_{P}m'c'a'$	$P_{C}bcn$	*$Cmm'a'$		$Fd'd'd'$		
		$Cm'm'a'$				
$Cmmm$		$C_{2c}mma$	$C_{c}mma$	$Immm$		
		$C_{P}mma$	$P_{C}ccm$			
$Cm'mm$		$C_{I}mma$	$I_{c}bam$	$Im'mm$		
$Cmmm'$		$C_{2c}m'ma$	$C_{c}mca$	*$Im'm'm$		
*$Cm'm'm$		$C_{2c}m'm'a$	$C_{c}cca$	$Im'm'm'$		
*$Cmm'm'$		$C_{P}m'ma$	$P_{C}cca$	$I_{P}mmm$	$P_{I}mmm$	
$Cm'm'm'$		$C_{P}mm'a$	$P_{C}mma$	$I_{P}m'mm$	$P_{I}mmn$	
$C_{2c}mmm$	$C_{c}mmm$	$C_{P}mma'$	$P_{C}bcm$	$I_{P}m'm'm$	$P_{I}nnm$	
$C_{P}mmm$	$P_{C}mmm$	$C_{I}mm'a$	$I_{c}mma$	$I_{P}m'm'm'$	$P_{I}nnn$	
$C_{I}mmm$	$I_{c}mmm$	$C_{I}m'm'a'$	$I_{c}bca$			
$C_{2c}m'm'm$	$C_{c}ccm$			$Ibam$		
$C_{2c}mm'm'$	$C_{c}mcm$	$Ccca$				
$C_{P}m'mm$	$P_{C}mma$			$Ib'am$		
$C_{P}mmm'$	$P_{C}mmn$	$Cc'ca$		$Ibam'$		
$C_{P}m'm'm$	$P_{C}bam$	$Ccca'$		*$Ib'a'm$		
$C_{P}mm'm'$	$P_{C}mna$	*$Cc'c'a$		*$Iba'm'$		
$C_{P}m'm'm'$	$P_{C}ban$	*$Ccc'a'$		$Ib'a'm'$		
$C_{I}m'mm$	$I_{c}mma$	$Cc'c'a'$		$I_{P}bam$	$P_{I}ccm$	
$C_{I}m'm'm$	$I_{c}bam$	$C_{P}cca$	$P_{C}ban$	$I_{P}b'am$	$P_{I}bcm$	
		$C_{P}c'ca$	$P_{C}cca$	$I_{P}bam'$	$P_{I}ccn$	
$Cccm$		$C_{P}cca'$	$P_{C}bcn$	$I_{P}b'a'm$	$P_{I}bam$	
		$C_{P}cc'a'$	$P_{C}nna$	$I_{P}b'am'$	$P_{I}bcn$	
$Cc'cm$				$I_{P}b'a'm'$	$P_{I}ban$	
$Cccm'$		$Fmmm$				
*$Cc'c'm$				$Ibca$		
*$Ccc'm'$		$Fm'mm$				
$Cc'c'm'$		*$Fm'm'm$		$Ib'ca$		
				*$Ib'c'a$		

TABLE III (continued)

Orthorhombic system		*$P4_3$		*$P4_2/m$
Ibca				
		$P4'_3$		$P4'_2/m$
Ib'c'a'		P_P4_3	P_C4_3	$P4_2/m'$
I_Pbca	P_Ibca			$P4'_2/m'$
$I_Pb'ca$	P_Icca	*$I\bar{4}$		P_P4_2/m P_C4_2/m
				P_P4_2/m' P_C4_2/n
Imma		$I4'$		
		I_P4	P_I4	*$P4/n$
Im'ma		I_P4'	P_I4_2	
Imma'				$P4'/n$
Im'm'a		*$I4_1$		$P4/n'$
Imm'a'				$P4'/n'$
Im'm'a'		$I4'_1$		$P_{2c}4/n$ P_c4/n
I_Pmma	P_Imma	I_P4_1	P_I4_1	$P_{2c}4'/n$ P_c4_2/n
$I_Pm'm'a$	P_Inna	$I_P4'_1$	P_I4_3	*$P4_2/n$
$I_Pmm'a'$	P_Imna			
$I_Pm'ma'$	P_Inma	$\bar{4}$		$P4'_2/n$
		*$P\bar{4}$		$P4_2/n'$
Tetragonal system				$P4'_2/n'$
		$P\bar{4}'$		P_I4_2/n I_c4_1/a
4		$P_{2c}\bar{4}$	$P_c\bar{4}$	
*$P4$		$P_P\bar{4}$	$P_C\bar{4}$	*$I4/m$
		$P_I\bar{4}$	$I_c\bar{4}$	$I4'/m$
$P4'$				$I4/m'$
$P_{2c}4$	P_c4	*$I\bar{4}$		$I4'/m'$
P_P4	P_C4			I_P4/m P_I4/m
P_I4	I_c4	$I\bar{4}'$		I_P4'/m P_I4_2/m
$P_{2c}4'$	P_c4_2	$I_P\bar{4}$	$P_I\bar{4}$	I_P4/m' P_I4/n
				I_P4'/m' P_I4_2/n
*$P4_1$		$4/m$		
		*$P4/m$		*$I4_1/a$
$P4'_1$				
P_P4_1	P_C4_1	$P4'/m$		$I4'_1/a$
		$P4/m'$		$I4_1/a'$
*$P4_2$		$P4'/m'$		$I4'_1/a'$
		$P_{2c}4/m$	P_c4/m	422
$P4'_2$		P_P4/m	P_C4/m	
$P_{2c}4_2$	P_c4_1	P_I4/m	I_c4/m	$P422$
P_P4_2	P_C4_2	$P_{2c}4'/m$	P_c4_2/m	
P_I4_2	I_c4_1	P_P4/m'	P_C4/n	$P4'22'$
$P_{2c}4'_2$	P_c4_3			*$P42'2'$

3. MAGNETIC SYMMETRY

TABLE III (continued)

Tetragonal system		$P4_22_12$		$4mm$	
$P422$				$P4mm$	
		$P4'_22_12'$			
$P4'2'2$		$*P4_22'_12'$		$P4'm'm$	
$P_{2c}422$	P_c422	$P4'_22'_12$		$P4'mm'$	
P_P422	P_C422	$P_{2c}4_22_12$	$P_c4_12_12$	$*P4m'm'$	
P_I422	I_c422	$P_{2c}4'_22'_12$	$P_c4_32_12$	$P_{2c}4mm$	P_c4mm
$P_{2c}4'22'$	P_c4_22			P_P4mm	P_C4mm
$P_P4'22'$	P_C42_1	$P4_322$		P_I4mm	I_c4mm
				$P_{2c}4'm'm$	P_c4_2cm
$P42_12$		$P4'_322'$		$P_{2c}4'mm'$	P_c4_2mc
		$*P4_32'2'$		$P_{2c}4m'm'$	P_c4cc
$P4'2_12'$		$P4'_32'2$		$P_P4'mm'$	P_C4bm
$*P42'_12'$		P_P4_322	P_C4_322	$P_I4m'm'$	I_c4cm
$P4'2'_12$		$P_P4'_322'$	$P_C4_32_12$		
$P_{2c}42_12$	P_c42_12			$P4bm$	
$P_{2c}4'2'_12$	$P_c4_22_12$	$P4_32_12$			
				$P4'b'm$	
$P4_122$		$P4'_32_12'$		$P4'bm'$	
		$*P4_32'_12'$		$*P4b'm'$	
$P4'_122'$		$P4'_32'_12$		$P_{2c}4bm$	P_c4bm
$*P4_12'2'$				$P_{2c}4'b'm$	P_c4_2nm
$P4'_12'2$		$I422$		$P_{2c}4'bm'$	P_c4_2bc
P_P4_122	P_C4_122			$P_{2c}4b'm'$	P_c4nc
$P_P4'_122'$	$P_C4_12_12$	$I4'22'$			
		$*I42'2'$		$P4_2cm$	
$P4_12_12$		$I4'2'2$			
		I_P422	P_I422	$P4'_2c'm$	
$P4'_12_12'$		$I_P4'22'$	P_I4_222	$P4'_2cm'$	
$*P4_12'_12'$		$I_P42'2'$	P_I42_12	$*P4_2c'm'$	
$P4'_12'_12$		$I_P4'2'2$	$P_I4_22_12$	P_P4_2cm	P_C4_2mc
				$P_P4'_2cm'$	P_C4_2bc
$P4_222$		$I4_122$			
				$P4_2nm$	
$P4'_222'$		$I4'_122'$			
$*P4_22'2'$		$*I4_12'2'$		$P4'_2n'm$	
$P4'_22'2$		$I4'_12'2$		$P4'_2nm'$	
$P_{2c}4_222$	P_c4_122	I_P4_122	P_I4_122	$*P4_2n'm'$	
P_P4_222	P_C4_122	$I_P4'_122'$	P_I4_322		
P_I4_222	I_c4_122			P_I4_2nm	I_c4_1md
$P_{2c}4'_222'$	P_c4_322	$I_P4_12'2'$	$P_I4_12_12$	$P_I4_2n'm'$	I_c4_1cd
$P_P4'_222'$	$P_C4_22_12$	$I_P4'_12'2$	$P_I4_32_12$		

TABLE III (continued)

Tetragonal system	$I4'cm'$		$P\bar{4}'2_1m'$	
	$*I4c'm'$		$*P\bar{4}2'_1m'$	
$P4cc$	I_P4cm	P_I4bm	$P_{2c}\bar{4}2_1m$	$P_c\bar{4}2_1m$
	$I_P4'c'm$	P_I4_2cm	$P_{2c}\bar{4}'2_1m'$	$P_c\bar{4}2_1c$
$P4'c'c$	$I4'cm'$	P_I4_2bc		
$P4'cc'$	$I4c'm'$	P_I4cc	$P\bar{4}2_1c$	
$*P4c'c'$				
P_P4cc	P_C4cc		$P4'2'_1c$	
$P_P4'cc'$	P_C4nc		$P\bar{4}'2_1c'$	
	$I4_1md$		$*P\bar{4}2'_1c'$	
$P4nc$				
	$I4'_1m'd$		$P\bar{4}m2$	
$P4'n'c$	$I4'_1md'$			
$P4'nc'$	$*I4_1m'd'$		$P\bar{4}'m'2$	
$*P4n'c'$			$P\bar{4}'m2'$	
	$I4_1cd$		$*P\bar{4}m'2'$	
$P4_2mc$				
	$I4'_1c'd$		$P_{2c}\bar{4}m2$	$P_c\bar{4}m2$
	$I4'_1cd'$		$P_P\bar{4}m2$	$P_C\bar{4}2m$
$P4'_2m'c$	$*I4_1c'd'$		$P_I\bar{4}m2$	$I_c\bar{4}2m$
$P4'_2mc'$			$P_{2c}\bar{4}'m'2$	$P_c\bar{4}c2$
$*P4_2m'c'$	$\bar{4}2m$		$P_P\bar{4}'m2'$	$P_C\bar{4}2_1m$
P_P4_2mc	P_C4_2cm			
$P_P4'_2mc'$	P_C4_2nm	$P\bar{4}2m$	$P\bar{4}c2$	
$P4_2bc$	$P\bar{4}'2'm$		$P\bar{4}'c'2$	
	$P\bar{4}'2m'$		$P\bar{4}'c2'$	
	$*P\bar{4}2'm'$		$*P\bar{4}c'2'$	
$P4'_2b'c$	$P_{2c}\bar{4}2m$	$P_c\bar{4}2m$		
$P4'_2bc'$	$P_P\bar{4}2m$	$P_C\bar{4}m2$	$P_P\bar{4}c2$	$P_C\bar{4}2c$
$*P4_2b'c'$	$P_I\bar{4}2m$	$I_c\bar{4}m2$	$P_P\bar{4}'c2'$	$P_C\bar{4}2_1c$
	$P_{2c}\bar{4}2'm'$	$P_c\bar{4}2c$		
$I4mm$	$P_P\bar{4}'2m'$	$P_C\bar{4}b2$	$P\bar{4}b2$	
	$P_I\bar{4}'2m'$	$I_c\bar{4}c2$		
$I4'm'm$			$P\bar{4}'b'2$	
$I4'mm'$	$P\bar{4}2c$		$P\bar{4}'b2'$	
$*I4m'm'$			$*P\bar{4}b'2'$	
I_P4mm	P_I4mm		$P_{2c}\bar{4}b2$	$P_c\bar{4}b2$
$I_P4'm'm$	P_I4_2nm	$P\bar{4}'2'c$	$P_{2c}\bar{4}'b'2$	$P_c\bar{4}n2$
$I_P4'mm'$	P_I4_2mc	$P\bar{4}'2c'$		
$I_P4m'm'$	P_I4nc	$*P\bar{4}2'c'$	$P\bar{4}n2$	
		$P_P\bar{4}2c$	$P_C\bar{4}c2$	
$I4cm$		$P_P\bar{4}'2c'$	$P_C\bar{4}n2$	$P\bar{4}'n'2$
			$P\bar{4}'n2'$	
	$P\bar{4}2_1m$		$*P\bar{4}n'2'$	
$I4'c'm$	$P\bar{4}'2'_1m$		$P_I\bar{4}n2$	$I_c\bar{4}2d$

3. MAGNETIC SYMMETRY 143

TABLE III (*continued*)

Tetragonal system					
		$P4/m'm'm'$		$P4'/nn'c$	
$I\bar{4}m2$		$P_{2c}4/mmm$	P_c4/mmm	$P4'/nnc'$	
		P_P4/mmm	P_C4/mmm	$P4'/n'n'c$	
$I\bar{4}'m'2$		P_I4/mmm	I_c4/mmm	$*P4/nn'c'$	
$I\bar{4}'m2'$		$P_{2c}4'/mm'm$	P_c4_2/mcm	$P4'/n'nc'$	
$*I\bar{4}m'2'$		$P_{2c}4'/mmm'$	P_c4_2/mmc	$P4'/n'n'c'$	
$I_P\bar{4}m2$	$P_I\bar{4}m2$	$P_{2c}4/mm'm'$	P_c4/mcc		
$I_P\bar{4}'m'2$	$P_I\bar{4}n2$	$P_P4/m'mm$	P_C4/nmm	$P4/mbm$	
		P_P4'/mmm'	P_C4/mbm		
$I\bar{4}c2$		$P_P4'/m'mm'$	P_C4/nbm	$P4'/m'bm$	
		$P_I4/mm'm'$	I_c4/mcm	$P4'/mb'm$	
$I\bar{4}'c'2$				$P4'/mbm'$	
$I\bar{4}'c2'$		$P4/mcr$		$P4'/m'b'm$	
$*I\bar{4}c'2'$				$*P4/mb'm'$	
$I_P\bar{4}c2$	$P_I\bar{4}c2$	$P4/m'cc$		$P4'/m'bm'$	
$I_P\bar{4}c'2'$	$P_I\bar{4}b2$	$P4'/mc'c$		$P4'/m'b'm'$	
		$P4'/mcc'$			
$I\bar{4}2m$		$P4'/m'c'c$		$P_{2c}4/mbm$	P_c4/mbm
		$*P4/mc'c'$		$P_{2c}4'/mb'm$	P_c4_2/mnm
$I\bar{4}'2'm$		$P4'/m'cc'$		$P_{2c}4'/mbm'$	P_c4_2/mbc
$I\bar{4}'2m'$		$P4/m'c'c'$		$P_{2c}4/mb'm'$	P_c4/mnc
$*I\bar{4}2'm'$					
$I_P\bar{4}2m$	$P_I\bar{4}2m$	P_P4/mcc	P_C4/mcc	$P4/mnc$	
$I_P\bar{4}'2'm$	$P_I\bar{4}2_1m$	$P_P4/m'cc$	P_C4/ncc		
$I_P\bar{4}'2m'$	$P_I\bar{4}2c$	P_P4'/mcc'	P_C4/mnc	$P4/m'nc$	
$I_P\bar{4}2'm'$	$P_I\bar{4}2_1c$	$P_P4'/m'cc'$	P_C4/nnc	$P4'/mn'c$	
				$P4'/mnc'$	
$I\bar{4}2d$		$P4/nbm$		$P4'/m'n'c$	
		$P4/n'bm$		$*P4/mn'c'$	
$I\bar{4}'2'd$		$P4'/nb'm$		$P4'/m'nc'$	
$I\bar{4}'2d'$		$P4'/nbm'$		$P4'/m'n'c'$	
$*I\bar{4}2'd'$		$P4'/n'b'm$			
		$*P4/nb'm'$		$P4/nmm$	
$4/mmm$		$P4'/n'bm'$			
$P4/mmm$		$P4'/n'b'm'$		$P4/n'mm$	
		$P_{2c}4/nbm$	P_c4/nbm	$P4'/nm'm$	
$P4/m'mm$		$P_{2c}4'/nb'm$	P_c4_2/nnm	$P4'/nmm'$	
$P4'/mm'm$		$P_{2c}4'/nbm'$	P_c4_2/nbc	$P4'/n'm'm$	
$P4'/mmm'$		$P_{2c}4/nb'm'$	P_c4/nnc	$*P4/nm'm'$	
$P4'/m'm'm$				$P4'/n'mm'$	
$*P4/mm'm'$		$P4/nnc$		$P4'/n'm'm'$	
$P4'/m'mm'$		$P4/n'nc$		$P_{2c}4/nmm$	P_c4/nmm

TABLE III (continued)

Tetragonal system	$P_P4_2/m'cm$	P_C4_2/nmc	$*P4_2/mn'm'$
	P_P4_2'/mcm'	P_C4_2/mbc	$P4_2'/m'nm'$
$P4/nmm$	$P_P4_2'/m'cm'$	P_C4_2/nbc	$P4_2/m'n'm'$
$P_{2c}4'/nm'm$ P_c4_2/ncm	$P4_2/nbc$		$P4_2/nmc$
$P_{2c}4'/nmm'$ P_o4_2/nmc			
$P_{2c}4/nm'm'$ P_c4/ncc	$P4_2/n'bc$		$P4_2/n'mc$
	$P4_2'/nb'c$		$P4_2'/nm'c$
$P4/ncc$	$P4_2'/nbc'$		$P4_2'/nmc'$
	$P4_2/n'b'c$		$P4_2/n'm'c$
$P4/n'cc$	$*P4_2/nb'c'$		$*P4_2/nm'c'$
$P4'/nc'c$	$P4_2'/n'bc'$		$P4_2'/n'mc'$
$P4'/ncc'$	$P4_2/n'b'c'$		$P4_2/n'm'c'$
$P4'/n'c'c$			
$*P4/nc'c'$	$P4_2/nnm$		$P4_2/ncm$
$P4'/n'cc'$			
$P4/n'c'c'$	$P4_2/n'nm$		$P4_2/n'cm$
	$P4_2'/nn'm$		$P4_2'/nc'm$
$P4_2/mmc$	$P4_2'/nnm'$		$P4_2'/ncm'$
	$P4_2'/n'n'm$		$P4_2'/n'c'm$
$P4_2/m'mc$	$*P4_2/nn'm'$		$*P4_2/nc'm'$
$P4_2'/mm'c$	$P4_2'/n'nm'$		$P4_2'/n'cm'$
$P4_2'/mmc'$	$P4_2/n'n'm'$		$P4_2/n'c'm'$
$P4_2/m'm'c$	P_I4_2/nnm I_c4_1/amd		
$*P4_2/mm'c'$	$P_I4_2/nn'm$ I_c4_1/acd		$I4/mmm$
$P4_2'/m'm'c$			
$P4_2'/m'mc'$	$P4_2/mbc$		$I4/m'mm$
$P4_2/m'm'c'$			$I4'/mm'm$
P_P4_2/mmc P_C4_2/mcm	$P4_2/m'bc$		$I4'/mmm'$
$P_P4_2/m'mc$ P_C4_2/ncm	$P4_2'/mb'c$		$I4'/m'm'm$
$P_P4_2/mm'c'$ P_C4_2/mnm	$P4_2'/mbc'$		$*I4/mm'm'$
$P_P4_2'/m'mc$ P_C4_2/nnm	$P4_2/m'b'c$		$I4'/m'mm'$
	$*P4_2/mb'c'$		$I4/m'm'm'$
$P4_2/mcm$	$P4_2'/m'bc'$		
	$P4_2/m'b'c'$		I_P4/mmm P_I4/mmm
$P4_2/m'cm$			$I_P4/m'mm$ P_I4/nmm
$P4_2'/mc'm$	$P4_2/mnm$		$I_P4'/mm'm$ P_I4_2/mnm
$P4_2/mcm'$			I_P4'/mmm' P_I4_2/mmc
$P4_2'/m'c'm$	$P4_2/m'nm$		$I_P4'/m'm'm$ P_I4_2/nnm
$*P4_2/mc'm'$	$P4_2'/mn'm$		$I_P4/mm'm'$ P_I4/mnc
$P4_2'/m'cm'$	$P4_2'/mnm'$		$I_P4'/m'mm'$ P_I4_2/nmc
$P4_2/m'c'm'$	$P4_2'/m'n'm$		$I_P4/m'm'm'$ P_I4/nnc
P_P4_2/mcm P_C4_2/mmc			

3. MAGNETIC SYMMETRY

TABLE III (continued)

Tetragonal system		Trigonal system		$P3_121$	
		3			
$I4/mcm$		$*P3$		$*P3_12'1$	
				$P_{2c}3_221$	P_c3_221
$I4/m'cm$		$P_{2c}3$	P_c3	$P3_212$	
$I4'/mc'm$					
$I4'/mcm'$		$*P3_1$		$*P3_212'$	
$I4'/m'c'm$				$P_{2c}3_112$	P_c3_112
$*I4/mc'm'$		$P_{2c}3_2$	P_c3_2		
$I4'/m'cm'$		$*P3_2$		$P3_221$	
$I4/m'c'm'$					
I_P4/mcm	P_I4/mcc	$P_{2c}3_1$	P_c3_1	$*P3_22'1$	
$I_P4/m'cm$	P_I4/ncc			$P_{2c}3_121$	P_c3_121
$I_P4'/mc'm$	P_I4_2/mbc	$*R3$		$R32$	
I_P4'/mcm'	P_I4_2/mcm				
$I_P4'/m'c'm$	P_I4_2/nbc	R_R3		$*R32'$	
$I_P4'/mc'm'$	P_I4/mbm	$\bar{3}$		R_R32	
$I_P4'/m'cm'$	P_I4_2/ncm	$*P\bar{3}$		$3m$	
$I_P4/m'c'm'$	P_I4/nbm			$P3m1$	
		$P\bar{3}'$			
$I4_1/amd$		$P_{2c}\bar{3}$	$P_c\bar{3}$	$*P3m'1$	
		$*R\bar{3}$		$P_{2c}3m1$	P_c3m1
$I4_1/a'md$				$P_{2c}3m'1$	P_c3c1
$I4'_1/am'd$		$R\bar{3}'$		$P31m$	
$I4'_1/amd'$		$R_R\bar{3}$			
$I4'_1/a'm'd$		32		$*P31m'$	
$*I4_1/am'd'$		$P312$		$P_{2c}31m$	P_c31m
$I4'_1/a'md'$				$P_{2c}31m'$	P_c31c
$I4_1/a'm'd'$		$*P312'$			
		$P_{2c}312$	P_c312	$P3c1$	
$I4_1/acd$		$P321$		$*P3c'1$	
$I4_1/a'cd$		$*P32'1$		$P31c$	
$I4'_1/ac'd$		$P_{2c}321$	P_c321		
$I4_1/acd'$				$*P31c'$	
$I4'_1/a'c'd$		$P3_112$		$R3m$	
$*I4_1/ac'd'$		$*P3_112'$			
$I4'_1/a'cd'$				$*R3m'$	
$I4_1/a'c'd'$		$P_{2c}3_212$	P_c3_212		

TABLE III (continued)

Trigonal system				6/m	
$R3m$		$R_R\bar{3}m$		*$P6/m$	
		$R_R\bar{3}m'$	$R_I\bar{3}c$		
R_R3m		$R\bar{3}c$		$P6'/m$	
R_R3m'	R_I3c			$P6/m'$	
$R3c$		$R\bar{3}'c$		$P6'/m'$	
		$R\bar{3}'c'$		$P_{2c}6/m$	P_c6/m
*$R3c'$		*$R\bar{3}c'$		$P_{2c}6'/m$	P_c6_3/m
$\bar{3}m$		Hexagonal system		*$P6_3/m$	
$P\bar{3}1m$		6		$P6_3'/m$	
$P\bar{3}'1m$		*$P6$		$P6_3/m'$	
$P\bar{3}'1m'$		$P6'$		$P6_3'/m'$	
*$P\bar{3}1m'$		$P_{2c}6$	P_c6		
$P_{2c}\bar{3}1m$	$P_c\bar{3}1m$	$P_{2c}6'$	P_c6_3	622	
$P_{2c}\bar{3}1m'$	$P_c\bar{3}1c$	*$P6_1$		$P622$	
$P\bar{3}1c$		$P6_1'$		$P6'2'2$	
$P\bar{3}'1c$		*$P6_5$		$P6'22'$	
$P\bar{3}'1c'$		$P6_5'$		*$P62'2'$	
*$P\bar{3}1c'$		*$P6_2$		$P_{2c}622$	P_c622
$P\bar{3}m1$		$P6_2'$		$P_{2c}6'22$	P_c6_322
$P\bar{3}'m1$		$P_{2c}6_2$	P_c6_1	$P6_122$	
$P\bar{3}'m'1$		$P_{2c}6_2'$	P_c6_4	$P6_1'2'2$	
*$P\bar{3}m'1$		*$P6_4$		$P6_1'22'$	
$P_{2c}\bar{3}m1$	$P_c\bar{3}m1$	$P6_4'$		*$P6_12'2'$	
$P_{2c}\bar{3}m'1$	$P_c\bar{3}c1$	$P_{2c}6_4$	P_c6_2	$P6_522$	
$P\bar{3}c1$		$P_{2c}6_4'$	P_c6_5	$P6_5'2'2$	
$P\bar{3}'c1$		*$P6_3$		$P6_5'22'$	
$P\bar{3}'c'1$		$P6_3'$		*$P6_52'2'$	
*$P\bar{3}c'1$		$\bar{6}$		$P6_222$	
$R\bar{3}m$		*$P\bar{6}$		$P6_2'2'2$	
$R\bar{3}'m$		$P\bar{6}'$		$P6_2'22'$	
$R\bar{3}'m'$		$P_{2c}\bar{6}$	$P_c\bar{6}$	*$P6_22'2'$	
*$R\bar{3}m'$				$P_{2c}6_222$	P_c6_122
				$P_{2c}6_2'22$	P_c6_422

TABLE III (continued)

Hexagonal system		$\bar{6}m2$		$P6/mcc$	
$P6_422$		$P\bar{6}m2$			
				$P6/m'cc$	
$P6'_42'2$		$P\bar{6}'m'2$		$P6'/mc'c$	
$P6'_422'$		$P\bar{6}'m2'$		$P6'/mcc'$	
$*P6_42'2'$		$*P\bar{6}m'2'$		$P6'/m'c'c$	
$P_{2c}6_422$	P_c6_222	$P_{2c}\bar{6}m2$	$P_c\bar{6}m2$	$P6'/m'cc'$	
$P_{2c}6'_42'2$	P_c6_522	$P_{2c}\bar{6}'m'2$	$P_c\bar{6}c2$	$*P6/mc'c'$	
				$P6/m'c'c'$	
$P6_322$		$P\bar{6}c2$			
				$P6_3/mcm$	
$P6'_32'2$		$P\bar{6}'c'2$			
$P6'_322'$		$P\bar{6}'c2'$		$P6_3/m'cm$	
$*P6_32'2'$		$*P\bar{6}c'2'$		$P6'_3/mc'm$	
				$P6'_3/mcm'$	
$6mm$		$P\bar{6}2m$		$P6'_3/m'c'm$	
$P6mm$				$P6'_3/m'cm'$	
		$P\bar{6}'2'm$		$*P6_3/mc'm'$	
$P6'm'm$		$P\bar{6}'2m'$		$P6_3/m'c'm'$	
$P6'mm'$		$*P\bar{6}2'm'$			
$*P6m'm'$		$P_{2c}\bar{6}2m$	$P_c\bar{6}2m$	$P6_3/mmc$	
$P_{2c}6mm$	P_c6mm	$P_{2c}\bar{6}'2m'$	$P_c\bar{6}2c$		
$P_{2c}6'm'm$	P_c6_3cm			$P6_3/m'mc$	
$P_{2c}6'mm'$	P_c6_3mc	$P\bar{6}2c$		$P6'_3/mm'c$	
$P_{2c}6m'm'$	P_c6cc			$P6'_3/mmc'$	
		$P\bar{6}'2'c$		$P6'_3/m'm'c$	
$P6cc$		$P\bar{6}'2c'$		$P6'_3/m'mc'$	
		$*P\bar{6}2'c'$		$*P6_3/mm'c'$	
$P6'c'c$		$6/mmm$		$P6_3/m'm'c'$	
$P6'cc'$					
$*P6c'c'$		$P6/mmm$		Cubic system	
$P6_3cm$		$P6/m'mm$		23	
		$P6'/mm'm$			
$P6'_3c'm$		$P6'/mmm'$		$P23$	
$P6'_3cm'$		$P6'/m'm'm$			
$*P6_3c'm'$		$P6'/m'mm'$		P_F23	F_s23
$P6_3mc$		$*P6/mm'm'$		$F23$	
		$P6/m'm'm'$			
$P6'_3m'c$		$P_{2c}6/mmm$	P_c6/mmm	$I23$	
$P6'_3mc'$		$P_{2c}6'/mm'm$	P_c6_3/mcm		
$*P6_3m'c'$		$P_{2c}6'/mmm'$	P_c6_3/mmc	I_P23	P_I23
		$P_{2c}6/mm'm'$	P_c6/mcc		

TABLE III (continued)

Cubic system						
$P2_13$		$P4_23$2			$I\bar{4}3m$	
		$P4_2'32'$			$I\bar{4}'3m'$	
$I2_13$		$P_F4_23$2		$F_s4_13$2	$I_P\bar{4}3m$	$P_I\bar{4}3m$
					$I_P\bar{4}'3m'$	$P_I\bar{4}3n$
I_P2_13	P_I2_13	$F432$			$P\bar{4}3n$	
$m3$		$F4'32'$			$P\bar{4}'3n'$	
$Pm3$		$F4_13$2			$F\bar{4}3c$	
$Pm'3$		$F4_1'32'$			$F\bar{4}'3c'$	
P_Fm3	F_sm3	$I432$			$I\bar{4}3d$	
$Pn3$		$I4'32'$			$I\bar{4}'3d'$	
$Pn'3$		I_P432		P_I432	$m3m$	
P_Fn3	F_sd3	$I_P4'32'$		$P_I4_23$2	$Pm3m$	
$Fm3$		$P4_33$2			$Pm'3m$	
$Fm'3$		$P4_3'32'$			$Pm3m'$	
$Fd3$		$P4_13$2			$Pm'3m'$	
$Fd'3$		$P4_1'32'$			P_Fm3m	F_sm3m
$Im3$		$I4_13$2			P_Fm3m'	F_sm3c
$Im'3$					$Pn3n$	
I_Pm3	P_Im3	$I4_1'32'$			$Pn'3n$	
$I_Pm'3$	P_In3	$I_P4_13$2		$P_I4_33$2	$Pn3n'$	
$Pa3$		$I_P4_1'32'$		$P_I4_13$2	$Pn'3n'$	
$Pa'3$		$\bar{4}3m$			$Pm3n$	
$Ia3$		$P\bar{4}3m$			$Pm'3n$	
$Ia'3$		$P\bar{4}'3m'$			$Pm3n'$	
I_Pa3	P_Ia3	$P_F\bar{4}3m$		$F_s\bar{4}3m$	$Pm'3n'$	
432		$P_F\bar{4}'3m'$		$F_s\bar{4}3c$	$Pn3m$	
$P432$		$F\bar{4}3m$			$Pn'3m$	
$P4'32'$					$Pn3m'$	
P_F432	F_s432	$F\bar{4}'3m'$			$Pn'3m'$	

TABLE III (continued)

Cubic system	Fm3c'	Im3m	
	Fm'3c'		
Pn3m		Im'3m	
		Im3m'	
$P_F n3m$	$F_s d3m$	Fd3m	
$P_F n3m'$	$F_s d3c$		Im'3m'
		$I_P m3m$	$P_I m3m$
Fm3m		Fd'3m	$I_P m'3m$	$P_I n3m$
	Fd3m'	$I_P m3m'$	$P_I m3n$
Fm'3m	Fd'3m'	$I_P m'3m'$	$P_I n3n$
Fm3m'			
Fm'3m'	Fd3c	Ia3d	
Fm3c	Fd'3c	Ia'3d	
	Fd3c'	Ia3d'	
Fm'3c	Fd'3c'	Ia'3d'	

simple crystals having atoms in common. That is why we shall consider only simple crystals; the generalization of the discussion given in the rest of this section to the case of an arbitrary crystal is trivial.

The invariance of a simple crystal generated by **F** under **F** is implicit in the above definition. By applying an arbitrary element F of **F** to all position vectors which form the set **Fr** one changes only the order in which the position vectors are arranged in the set: $F\mathbf{Fr} = \mathbf{Fr}$. It is in this sense that a simple crystal is invariant under **F**.

If no two position vectors of a simple crystal C generated by **F** from **r** are equal, the position vectors of C are called "general position vectors"; they are called "special position vectors" otherwise.

A position vector **r** from which a simple crystal is generated can be characterized by specifying its "site space group" **F(r)**, which is defined as the group formed by all those elements of **F** which generate the set of position vectors $\mathbf{Tr} = \mathbf{r} + \mathbf{t}$. The point group **R(r)** of **F(r)** is called the "site point group of **r**."

Let us decompose **F** into left cosets relative to **F(r)**,

$$\mathbf{F} = \mathbf{F(r)} + F_2\mathbf{F(r)} + \cdots + F_n\mathbf{F(r)},$$

where F_k stands for $(R_k \mid \tau_k)$, and τ_k for $\tau(R_k)$. Each of these n left cosets generates a different set of position vectors:

$$\mathbf{r} + \mathbf{t}, R_2\mathbf{r} + \tau_2 + \mathbf{t}, \cdots, R_n\mathbf{r} + \tau_n + \mathbf{t},$$

where **t** is any primitive translation belonging to **T**. If **t** is fixed, we get in this way a set of equivalent position vectors whose components are given for each **F** and each **r** in the "International Tables for X-Ray Crystallography" and are called there the "coordinates of equivalent positions," while the site point group **R(r)** is called the "point symmetry" of each of the equivalent positions. Each set of equivalent positions defines one simple crystal. In particular, if **r** is a general position vector, **F(r) = T**, and **R(r)** consists of the identity element only.

The atoms whose position vectors are **r + t** (fixed **r**, variable **t**) form an "atom lattice," those whose position vectors are $R_2\mathbf{r} + \boldsymbol{\tau}_2 + \mathbf{t}$ form another atom lattice, etc. (N.B. A lattice is a group, an atom lattice is an arrangement of atoms.) A simple crystal can always be regarded as consisting of such n interlocking atom lattices.

b. *Definition of invariant spin arrangements.* We now consider a magnetic crystal, that is, a crystal each atom of which has a nonvanishing magnetic moment. For the purpose of this discussion those atoms of a crystal which are nonmagnetic may be regarded as nonexistent provided it is remembered that the arrangement of both magnetic and nonmagnetic atoms determines the space group **F** of the crystal. We shall call, for brevity, the magnetic moment of an atom its "spin" (by this we do not imply, of course, that there may not be any orbital contribution to the magnetic moment). A "spin arrangement" in a magnetic crystal is given if the spin of each atom is specified. In other words, a spin arrangement is a vector function $\mathbf{S}(\mathbf{r}_i)$ defined on the set $\mathbf{r}_i$ ($i = 1, 2, 3, ...$) of the position vectors of all atoms of the crystal.

We shall now define invariant spin arrangements. To do that we must first introduce an assumption as to the way an element of a magnetic group acts on such a spin vector function. An element of a magnetic group **m** is always of the form $m = FA$ where $F = (R \mid \tau(R) + \mathbf{t})$ and $A = E$ or $A = E'$. We now assume that m applied to a given spin arrangement $\mathbf{S}(\mathbf{r}_i)$, $i = 1, 2, 3, ...$, has the triple effect of

(1) translating $\mathbf{S}(\mathbf{r}_i)$ from $\mathbf{r}_i$ to the position

$$F\mathbf{r}_i = R\mathbf{r}_i + \tau(R) + \mathbf{t};$$

(2) rotating $\mathbf{S}(\mathbf{r}_i)$ in the same way as $\mathbf{r}_i$ is rotated, except for the difference which may arise from the fact that the spin is an axial vector while a position vector is polar;

(3) changing the sign of $\mathbf{S}(\mathbf{r}_i)$ if the element m is primed (because the spin changes sign under time inversion);

that is,

$$m\mathbf{S}(F\mathbf{r}_i) = \epsilon\delta\, R\mathbf{S}(\mathbf{r}_i), \qquad i = 1, 2, 3, ...,$$

where δ is the determinant of R, and $\epsilon = +1$ if $A = E$ and $\epsilon = -1$ if $A = E'$ (we shall call ϵ the "signature" of the element m).

Consequently we shall say that the spin arrangement $\mathbf{S}(\mathbf{r}_i)$, $i = 1, 2, 3, \ldots$, is invariant under m if for all i's

$$m\mathbf{S}(\mathbf{r}_i) = \mathbf{S}(\mathbf{r}_i);$$

if this is also true for all elements m of **m** we shall say that the spin arrangement is "invariant under **m**." The largest magnetic group under which a spin arrangement is invariant will be called its magnetic group.

The set of matrices $\epsilon \delta R$, or more explicitly $\epsilon(R)\delta(R)R$, evidently constitutes a representation of the magnetic group **m**. It will be called the "standard representation" of **m**. A standard representation is thus a representation of **m** generated by the transformations of a spin vector when elements of **m** are applied to it.

2. Construction of All the Invariant Spin Arrangements

a. Introduction. We want to classify all invariant spin arrangements which could possibly exist in an arbitrary magnetic crystal. We shall do that by showing how to construct for a given magnetic crystal C whose space group is **F** all spin arrangements which are invariant under an arbitrary magnetic subgroup **m** of **F** × **A**. It will turn out that each spin arrangement invariant under **m** in C can be unambiguously characterized by specifying the location, magnitude, and orientation of a certain number (possibly infinite) of spins. The number, and the possible location and orientation of those spins, will depend on **F**, **r**, and **m**.

Let **L** be that subgroup of **F** to whose family a given but arbitrary magnetic group $\mathbf{m_L}$ belongs. There are the six following cases to be distinguished:

(1) **L** is identical with **F**,
(2) **L** belongs to the same class of space groups as **F** (i.e., **L** and **F** would be denoted by the same international symbol),
(3) **L** is a space group that does not belong to the same class of space groups as **F**,
(4) **L** is a two-dimensional space group,
(5) **L** is a one-dimensional space group,
(6) **L** is a finite group (i.e., a group of rotations).

We shall first discuss case (1) in some detail. Next we shall show that the problem of constructing all invariant spin arrangements for cases

(2) to (5) can be reduced to that for case (1), and that the problem for case (6) is trivial.

b. Case (1). Since any crystal whose space group is **F** can always be regarded as consisting of interlocking simple crystals each of which has **F** as its space group, and since an element of **F** permutes the atoms of each simple crystal among themselves, it is sufficient to consider the problem of constructing all invariant spin arrangements for a simple crystal.

Let us then consider a simple crystal generated by **F** from an atom located at **r**, and a magnetic space group **M** belonging to the family of **F** [case (1)].

If **r** is a general position vector, no two position vectors in the set obtained by applying to **r** all elements of **F** are equal. Hence, by assigning to the atom at **r** an arbitrary spin **S(r)**, and to the remaining atoms (position vectors: $F\mathbf{r}$) the spins

$$\mathbf{S}(F\mathbf{r}) = \epsilon\delta\, R\mathbf{S}(\mathbf{r}),$$

where the notation is the same as previously (see this section, IV, 1b), we unambiguously define a spin arrangement which is evidently invariant under **M**. Moreover, every spin arrangement invariant under **M** must be of this form. We shall call this prescription for assigning spin to atoms the "standard prescription."

If **r** is a special position vector the standard prescription may not be unambiguous; it may lead to contradictions. For, since **r** is a special position vector, there exist pairs M_k, M_l of elements of **M** for which $F_k\mathbf{r} = F_l\mathbf{r}$, and it may very well happen that the standard prescription will lead to $\mathbf{S}(F_k\mathbf{r}) \neq \mathbf{S}(F_l\mathbf{r})$ if $\mathbf{S}(\mathbf{r})$ is chosen arbitrarily. In fact, it may not be possible to avoid this kind of contradiction with any choice of $\mathbf{S}(\mathbf{r})$: if **r** is a special position vector, a spin arrangement which is invariant under **M** may not exist at all.

We shall now give without proof the conditions under which a spin arrangement invariant under **M** does exist, and is then obtained by the standard prescription (for a formal proof, which is straightforward, see reference [2]).

Let $R_k(\mathbf{r})$, $k = 1, 2, ..., h$, be the matrices of the site point group of **r**, and τ_k the nonprimitive translation associated with the rotation $R_k(\mathbf{r})$ in the space group **F**. Further, let $\mathbf{t}_k$ be the primitive translation defined by

$$\mathbf{t}_k = R_k(\mathbf{r})\mathbf{r} + \tau_k - \mathbf{r} \qquad (k = 1, 2, ..., h).$$

Then the set of h matrices

$$\epsilon(\mathbf{t}_k)\,\epsilon(R_k)\,\delta(R_k)\,R_k(\mathbf{r}),$$

where $\epsilon(\mathbf{t}_k)$ and $\epsilon(R_k)$ are the signatures of the elements $(E \mid \mathbf{t}_k)A$ and $(R_k(\mathbf{r}) \mid \boldsymbol{\tau}_k)A$, respectively, and $\delta(R_k)$ is the determinant of $R_k(\mathbf{r})$, constitutes a group. The magnetic point group belonging to the family of **R(r)** of which this group of matrices is the standard representation will be called the "magnetic site point group of **r**," and will be denoted by **R_M(r)**.

A necessary and sufficient condition for the existence of a spin arrangement invariant under **M** belonging to the family of **F** in a simple crystal generated by **F** from an atom at **r** is the existence of at least one spin vector **S(r)** invariant under **R_M(r)**. For each such invariant spin vector the standard prescription defines unambiguously a spin arrangement invariant under **M**.

It is thus essential to know which magnetic point groups leave at least one spin vector invariant. Those which have this property will be called "admissible magnetic point groups," and their invariant spin vectors will be called admissible.

The number n_1 of linearly independent spin vectors which are invariant under a magnetic point group **R_m** is equal to the number of times the identical representation occurs in the standard representation of **R_m**, that is,

$$n_1 = (1/h) \sum_{R_m} \chi(R_m),$$

where $\chi(R)$ is the trace of the matrix $\epsilon(R)\delta(R)R$, h is the order of the group **R_m**, and the sum is extended over all its h elements. The magnetic point group is admissible if $n_1 > 0$, i.e., if $n_1 = 1, 2$, or 3.

It turns out that 31 of the 90 magnetic point groups are admissible. Their list (first published by Tavger [33]) and their invariant spin vectors are given in Table IV. No magnetic point group of the cubic system is admissible. Each of the 31 admissible magnetic point groups is a subgroup of the infinite noncrystallographic magnetic group $(\infty/m)(2'/m')$, defined in Section III, 1b.

We thus conclude that if the magnetic site point group **R_M(r)** of **r** is not admissible no spin arrangement invariant under **M** in a simple crystal generated by **F** from an atom located at **r** can exist. If it is admissible, one obtains all invariant spin arrangements by using the standard prescription, and by choosing for **S(r)** all admissible spin vectors in turn. Each invariant spin arrangement is thus uniquely characterized by specifying **M** and **S(r)**.

TABLE IV
List of the Admissible Magnetic Point Groups

Magnetic point groups	Admissible spin directions
	$n_1 = 3$
1 $\bar{1}$	Any direction
	$n_1 = 2$
2' 2'/m' m'm2'	Perpendicular to the axis
m'	Any direction in the plane
	$n_1 = 1$
m	Perpendicular to the plane
m'm'm	Perpendicular to the unprimed plane
2'2'2	Along the unprimed axis
2 2/m m'm'2	Along the axis
4 $\bar{4}$ 4/m 42'2'	Along the axis of higher order
4m'm' $\bar{4}$2'm' 4/mm'm'	Along the axis of higher order
3 $\bar{3}$ 32' 3m' $\bar{3}$m'	Along the axis of higher order
6 $\bar{6}$ 6/m 62'2'	Along the axis of higher order
6m'm' $\bar{6}$m'2' 6/mm'm'	Along the axis of higher order

c. *Ferro-, ferri-, and antiferromagnetic crystals.* As we have seen, a spin arrangement invariant under a magnetic space group **M** which belongs to the family of the space group **F** of a simple crystal [case (1)] is given, according to the standard prescription, by

$$\mathbf{S}(F\mathbf{r}) = \epsilon \delta R \mathbf{S}(\mathbf{r}),$$

where $\mathbf{S}(\mathbf{r})$ is an admissible spin vector located at $\mathbf{r}$. Summing both sides of this equation over all elements of **M** we obtain

$$\sum_M \mathbf{S}(F\mathbf{r}) = \left(\sum_M \epsilon \delta R \right) \mathbf{S}(\mathbf{r}).$$

The last column of Table II shows that the sum between brackets vanishes whenever the magnetic space group is of the kind $\mathbf{M_{R0}}$ or $\mathbf{M_{R\alpha}}$. This means that the resultant spin vector (i.e., the macroscopic magnetization) of a spin arrangement which is invariant under $\mathbf{M_{R0}}$ or $\mathbf{M_{R\alpha}}$ is necessarily zero; in other words, the magnetic simple crystal is necessarily antiferromagnetic.

We have thus proved that in a ferromagnetic or a ferrimagnetic crystal the spin arrangement can only be invariant under a trivial magnetic space group or a nontrivial magnetic space group of the kind $\mathbf{M_T}$. The invariance under a group of the kind $\mathbf{M_T}$ is, however, not a sufficient condition for a magnetic crystal to be ferromagnetic or ferrimagnetic. The sum $\Sigma_M \epsilon \delta R$ may very well vanish for a group of the kind $\mathbf{M_T}$.

The sufficient and necessary condition for a magnetic space group to be the symmetry group of a ferromagnetic crystal is that the group be either trivial or of the kind $\mathbf{M_T}$, and that its magnetic point group be one of the 31 admissible magnetic point groups. The magnetic space groups which satisfy this condition have been called ferromagnetic space groups by Neronova and Belov [34]. The ferromagnetic space groups are marked in Table III by an asterisk. There are 275 ferromagnetic space groups.

The fact that a group is ferromagnetic does not mean that it is incompatible with an antiferromagnetic spin arrangement. It only means that the matrix $\Sigma_M \epsilon \delta R$ does not vanish (i.e., that the standard representation of a ferromagnetic group contains the identical representation). The resultant spin vector ($\Sigma_M \epsilon \delta R)\mathbf{S}(\mathbf{r})$ may still be zero for some choice of $\mathbf{S}(\mathbf{r})$.

The macroscopic symmetry of a ferromagnetic or ferrimagnetic crystal is, of course, characterized by the magnetic point group belonging to the magnetic space group $\mathbf{M_T}$ under which the spin arrangement in the crystal is invariant.

d. *Example.* As an example of spin arrangements in case (1) let us consider a simple crystal generated by the same space group $\mathbf{F}$ of the class $P422$ that we have considered in Section III, 2e, and let it be generated from the special position $\mathbf{r} = (\frac{1}{2}, \frac{7}{8}, 0)$ (see "International Tables" [1], p. 179). The site space group $\mathbf{F(r)}$ in this case is the space group $P2_y$; the site point group $\mathbf{R(r)}$ is 2_y. We will construct a few spin arrangements invariant under the magnetic space groups $\mathbf{M_1}$ and $\mathbf{M_2}$ of the family of $\mathbf{F}$ which belong to the classes $P42'2'$ and $P_{a+b,a-b}422$, respectively. The group $\mathbf{M_1}$ is a ferromagnetic group, the group $\mathbf{M_2}$ is of the kind $\mathbf{M_{R0}}$.

According to the definitions given in this section (IV,2b), the magnetic site point group of $\mathbf{r}$, $\mathbf{R_M(r)}$, for the magnetic space group $\mathbf{M_1}$ is $2_y'$. This is an admissible magnetic point group and its admissible spin vectors lie in the xz-plane. One thus obtains all the spin arrangements invariant under $\mathbf{M_1}$ by choosing for $\mathbf{S(r)}$ all the admissible spin vectors of $2_y'$ in turn and by using the standard prescription given in this section (IV, 2b). In Fig. 7 we give two of the infinity of the spin arrangements invariant under $\mathbf{M_1}$. In Fig. 7a we have chosen the spin $\mathbf{S(r)}$ to be parallel to the x-axis; in Fig. 7b, parallel to the z-axis. In the first case the crystal is antiferromagnetic; in the second case it is ferromagnetic.

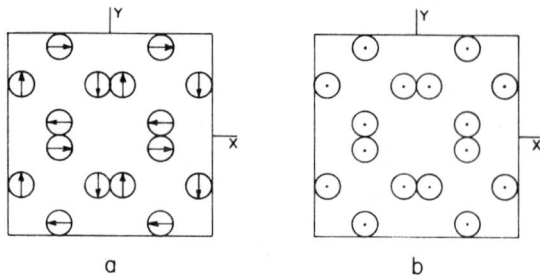

Fig. 7. (a) An "antiferromagnetic" and (b) a "ferromagnetic" spin arrangement, both invariant under the magnetic space group $P42'2'$.

For the magnetic space group $\mathbf{M_2}$ the magnetic site point group $\mathbf{R_M(r)}$ is again $2_y'$. In fact $2_y\mathbf{r} = \mathbf{r} - \mathbf{a}$ and hence the matrices $\epsilon(\mathbf{t}_k)\,\epsilon(R_k)\,\delta(R_k)R_k(\mathbf{r})$ as defined in this section (IV, 2b) constitute the standard representation of the magnetic point group $2_y'$. One can therefore construct spin arrangements invariant under $\mathbf{M_2}$ only if the spin in $\mathbf{r}$ is again in the xz-plane. In Fig. 8a we have chosen again $\mathbf{S(r)}$ to be along the x-axis; in Fig. 8b, along the z-axis; and we obtain in this way two different antiferromagnetic crystals.

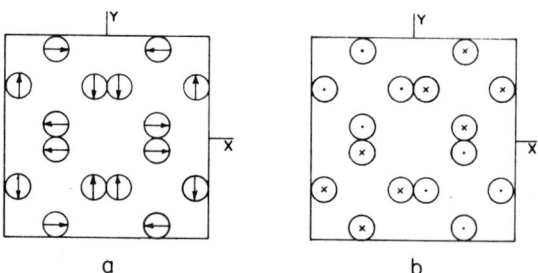

Fig. 8. Two "antiferromagnetic" spin arrangements, both invariant under the magnetic space group $P_{a+b,a-b}422$.

e. *Cases (2) and (3)*. We consider again a simple crystal generated by $\mathbf{F}$ from an atom located at $\mathbf{r}$. As a magnetic crystal it is supposed to be invariant under a magnetic space group $\mathbf{M_L}$ which, this time, belongs to the family of a space group $\mathbf{L}$, $\mathbf{L}$ being a proper subgroup of $\mathbf{F}$ [case (2) or (3)]. This invariance means that the equation

$$M_L\mathbf{S}(L\mathbf{r}) = \mathbf{S}(L\mathbf{r})$$

must be valid for all elements M_L of $\mathbf{M_L}$, but it also means that the equation

$$S(Fr) = \delta R S(r)$$

need not be valid for those elements F of $\mathbf{F}$ which do not belong to the subgroup $\mathbf{L}$.

To construct all the spin arrangements invariant under $\mathbf{M_L}$ we first decompose $\mathbf{F}$ into the right cosets relative to $\mathbf{L}$,

$$\mathbf{F} = \mathbf{L} + \mathbf{L}F_2 + \cdots + \mathbf{L}F_l,$$

where l is the index of the subgroup $\mathbf{L}$, and write correspondingly

$$\mathbf{F}\mathbf{r} = \mathbf{L}\mathbf{r} + \mathbf{L}F_2\mathbf{r} + \cdots + \mathbf{L}F_l\mathbf{r}.$$

It can be shown (see reference [2]) that the sets of position vectors generated by two different right cosets from the position vector $\mathbf{r}$ are either identical or have no position vectors in common. Hence we can write, by omitting appropriate cosets in the above decomposition,

$$\mathbf{F}\mathbf{r} = \mathbf{L}\mathbf{r} + \cdots + \mathbf{L}F_{l'}\mathbf{r},$$

where $l' \leqslant l$, and where no two sets of position vectors any longer have position vectors in common. This means that a simple crystal generated by $\mathbf{F}$ from an atom located at $\mathbf{r}$ can be regarded as composed of l' interlocking simple crystals generated by a subgroup $\mathbf{L}$ of $\mathbf{F}$ from the atoms located at $\mathbf{r}, F_2\mathbf{r}, ..., F_{l'}\mathbf{r}$. Making use of this fact we can immediately reduce the problem of constructing all spin arrangements invariant under $\mathbf{M_L}$ in a simple crystal generated by $\mathbf{F}$ to the problem discussed and solved for case (1). We simply assign spins to the atoms of any one of the l' simple crystals quite independently of how we assign spins to the atoms of the others; but for each simple crystal we follow the procedure described for case (1). The site point groups $\mathbf{R}(F_k\mathbf{r})$ of the l' simple crystals will not, in general, be the same. Hence, the corresponding magnetic site point groups and the sets of admissible spin vectors for the l' simple crystals will not be the same either.

Each spin arrangement invariant under $\mathbf{M_L}$ in a simple crystal generated by $\mathbf{F}$ from an atom located at $\mathbf{r}$ is thus uniquely characterized by specifying the l' admissible spin vectors $\mathbf{S}(\mathbf{r}), \mathbf{S}(F_2\mathbf{r}), ... \mathbf{S}(F_{l'}\mathbf{r})$. One obtains then all invariant spin arrangements by choosing these l' admissible spin vectors in all possible ways.

f. Example. Let us consider the simple crystal generated by the space group $\mathbf{F}$ of the class $P422$ from the special position $\mathbf{r} = (\frac{1}{2}, \frac{7}{8}, 0)$

already considered in this section (IV, 2d). We want to construct all the spin arrangements invariant under the magnetic space group $P2_y'$, which has the same lattice as $P422$.

According to the procedure explained in this section (IV, 2e), we have first to decompose F into the right cosets relative to the space group $P2_y$ to whose family $P2_y'$ belongs:

$$F = P2_y + P2_y(2_z \mid 0) + P2_y(4_z \mid 0) + P2_y(4_z{}^3 \mid 0).$$

Correspondingly, we obtain

$$\mathbf{Fr} = P2_y\mathbf{r} + P2_y(2_z \mid 0)\mathbf{r} + P2_y(4_z \mid 0)\mathbf{r} + P2_y(4_z{}^3 \mid 0)\mathbf{r}.$$

It is very easy to show that the sets of position vectors generated by the first and second coset have no position vectors in common, whereas the sets of position vectors generated by the third and fourth coset are identical. We can therefore consider the crystal as composed of three interlocking simple crystals generated by $P2_y$ from $\mathbf{r}$, $(2_z \mid 0)\mathbf{r}$, and $(4_z \mid 0)\mathbf{r}$. These three simple crystals are represented in Fig. 9 by dotted, heavy, and thin circles, respectively. In order to obtain all the spin arrangements invariant under $P2_y'$ we now assign spins to the atoms of each simple crystal using the standard prescription.

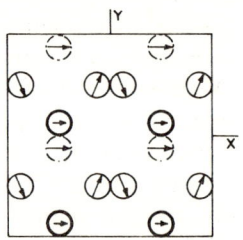

Fig. 9. A spin arrangement invariant under the magnetic space group $P2'$ in a crystal whose space group is $P422$.

The site point group $\mathbf{R}(\mathbf{r})$ is the group 2_y, and the site point group $\mathbf{R}((2_z \mid 0)\mathbf{r})$ is also 2_y, but the site point group $\mathbf{R}((4_z \mid 0)\mathbf{r})$ is 2_x. It follows that the site magnetic point group is $2_y'$ in the first two cases, and 1 in the third case. The admissible spin vectors of $2_y'$ lie in the xz-plane; those of 1 are arbitrary. One thus obtains all the spin arrangements invariant under $P2_y'$ by choosing for $\mathbf{S}(\mathbf{r})$ and $\mathbf{S}((2_z \mid 0)\mathbf{r})$ all the admissible spin vectors of $2_y'$ in turn, and for $\mathbf{S}((4_z \mid 0)\mathbf{r})$ a spin along all directions in turn. In Fig. 9 we have chosen $\mathbf{S}(\mathbf{r})$ to be parallel to the x-axis, $\mathbf{S}((2_z \mid 0)\mathbf{r})$ to be in the xz-plane, and $\mathbf{S}((4_z \mid 0)\mathbf{r})$ to be in the

xy-plane. The arrows representing the spins are shorter in the case of heavy circles, to indicate that the spin $\mathbf{S}((2_z \mid 0)\mathbf{r})$ is not parallel to the x-axis.

g. **Cases (4), (5), and (6).** In these three cases the number of right cosets in the decomposition

$$\mathbf{F} = \mathbf{L} + \mathbf{L}F_2 + \mathbf{L}F_3 + \cdots$$

is infinite, and, correspondingly, the number of sets of different position vectors, generated by these cosets,

$$\mathbf{Fr} = \mathbf{Lr} + \mathbf{L}F_2\mathbf{r} + \mathbf{L}F_3\mathbf{r} + \cdots$$

is also infinite (as previously, we keep in this sum only one coset from each set of cosets which generate the same set of position vectors).

Let us first consider case (4), where $\mathbf{L}$ is a two-dimensional space group, and $^2\mathbf{M_L}$ is a two-dimensional magnetic space group belonging to the family of $\mathbf{L}$. We decompose a simple crystal generated by $\mathbf{F}$ from an atom located at $\mathbf{r}$ into an infinity of two-dimensional simple crystals generated by $\mathbf{L}$ from the atoms located at $\mathbf{r}, F_2\mathbf{r}, F_3\mathbf{r}, \ldots$. In each of these simple crystals we assign spins to atoms quite independently, following a procedure immediately obtained from the procedure described for case (1) by appropriately adapting the terminology to the present two-dimensional case. The conclusion is thus that each spin arrangement invariant under $^2\mathbf{M_L}$ in a simple crystal generated by $\mathbf{F}$ from an atom located at $\mathbf{r}$ is uniquely characterized by specifying an infinity of admissible spin vectors $\mathbf{S}(\mathbf{r}), \mathbf{S}(F_2\mathbf{r}), \mathbf{S}(F_3\mathbf{r}), \ldots$. One obtains then all invariant spin arrangements by choosing these admissible spin vectors in all possible ways. If the admissible spin vectors are chosen in an arbitrary manner, the magnetic crystal will have in general only a two-dimensional translatory symmetry.

It is easy to adapt the procedure just described for case (4) to cases (5) and (6), although we will not do this here. It is clear that in case (5) a magnetic crystal will have in general only one-dimensional translatory symmetry, and in case (6) no translatory symmetry at all. In fact, case (6) is compatible with a completely random spin arrangement.

h. **Example.** Let us take once more the simple crystal already considered in this section (IV, 2d and 2f). We want now to construct all the spin arrangements invariant under the two-dimensional magnetic space group $^2\mathbf{M_L}$ of the class $p4$. This is a trivial magnetic space group whose primitive translations are identical with those primitive translations of the space group **F** (which generates the crystal) which are parallel to the xy-plane.

According to the procedure explained in this section (IV, 2g), we have first to decompose **F** into the right cosets relative to $p4_z$. In this case **L** and ^2M_L are identical.

$$\mathbf{F} = p4_z + p4_z(2_x \mid 0) + p4_z(1 \mid \mathbf{a}_3) + \cdots.$$

Correspondingly, we obtain

$$\mathbf{Fr} = p4_z\mathbf{r} + p4_z(2_x \mid 0)\mathbf{r} + p4_z(1 \mid \mathbf{a}_3)\mathbf{r} + \cdots.$$

It can easily be shown that the sets of position vectors generated by the first and second coset of **F** are identical, whereas all the other cosets have no position vector in common. This means that the given crystal consists of an infinity of two-dimensional simple crystals generated by $p4$ from atoms located at $\mathbf{r}$, $(1 \mid \mathbf{a}_3)\mathbf{r}$, $(1 \mid 2\mathbf{a}_3)\mathbf{r}$ .... To obtain all the spin arrangements invariant under $p4_z$ we now have to assign spins to the atoms of each two-dimensional simple crystal using the standard prescription. This time the site point group for all atoms is the point group 1; thus the site magnetic point group is also 1. Its admissible spin vectors are arbitrary. We thus obtain all the spin arrangements invariant under $p4$ by choosing for $\mathbf{S}(\mathbf{r})$, $\mathbf{S}((1 \mid \mathbf{a}_3)\mathbf{r})$ ... all possible vectors in turn. In particular, we can choose $\mathbf{S}(\mathbf{r})$, $\mathbf{S}((1 \mid \mathbf{a}_3)\mathbf{r})$, $\mathbf{S}((1 \mid 2\mathbf{a}_3)\mathbf{r})$ ... in such a way as to obtain a so-called "spiral" invariant spin arrangement. If for example $\mathbf{S}(\mathbf{r})$ is parallel to the x-axis, $\mathbf{S}((1 \mid \mathbf{a}_3)\mathbf{r})$ can be chosen to make an angle ψ with the x-axis, $\mathbf{S}((1 \mid 2\mathbf{a}_3)\mathbf{r})$ an angle 2ψ with the x-axis, and so on. A spiral invariant spin arrangement of this kind can thus be characterized more generally by its being invariant under $p4_z$, and by the relation

$$\mathbf{S}((1 \mid n\mathbf{a}_3)\mathbf{r}) = (R_\psi)^n \mathbf{S}(\mathbf{r})$$

where R_ψ is the rotation through an angle ψ around the z-axis, $\mathbf{S}(\mathbf{r})$ an arbitrarily chosen spin, and n an arbitrary integer. The rotation matrices E, R_ψ, R_ψ^{-1}, R_ψ^2, R_ψ^{-2}, ... form a group.

3. Classification of All the Invariant Spin Arrangements

We have described in this section a systematic procedure for constructing all the possible invariant spin arrangements in magnetic crystals. This procedure implies that each invariant spin arrangement in a magnetic crystal can be uniquely characterized by the following data:

3. MAGNETIC SYMMETRY

(1) The space group $\mathbf{F}$ of the crystal C.

(2) The position vector $\mathbf{r}^{(k)}$ of one magnetic atom (i.e., an atom with nonvanishing magnetic moment) for each simple crystal C_k generated by $\mathbf{F}$ from $\mathbf{r}^{(k)}$.

(3) The magnetic group $\mathbf{m}_L^{(k)}$ of each C_k (i.e., the magnetic group under which the spin arrangement in C_k is invariant), where the subscript $\mathbf{L}$ denotes the subgroup $\mathbf{L}^{(k)}$ of $\mathbf{F}$ to whose family $\mathbf{m}_L^{(k)}$ belongs.

(4) A minimum number of position vectors $\mathbf{r}_1^{(k)}$, $\mathbf{r}_2^{(k)}$, $\mathbf{r}_3^{(k)}$ ... of the magnetic atoms from which each C_k can be generated by $\mathbf{L}^{(k)}$.

(5) The magnitude and direction of the spins $\mathbf{S}(\mathbf{r}_1^{(k)})$, $\mathbf{S}(\mathbf{r}_2^{(k)})$, $\mathbf{S}(\mathbf{r}_3^{(k)})$, ... located at $\mathbf{r}_1^{(k)}$, $\mathbf{r}_2^{(k)}$, $\mathbf{r}_3^{(k)}$, ... for each C_k. (The magnitude of the spins for a fixed k must be the same, because the atoms of a simple crystal are identical).

All the invariant spin arrangements can thus be divided first into classes using (1), then each class can be divided into subclasses using (2), then each subclass can be divided into sub-subclasses using (3), and so on.

The classification into classes and subclasses [i.e., using (1) and (2)] is of course identical with the usual classification of all the possible arrangements of atoms in crystals. A complete catalog of all classes and subclasses is given in "International Tables for X-Ray Crystallography" [1]; each subclass corresponds to a set of "special positions" listed there. Further classification of invariant spin arrangements [using (3), (4), and (5)] has not yet been carried out in detail. However, a first step in this direction has been made by Belov *et al.* [35]. They consider the most important case in which $\mathbf{L}^{(k)} = \mathbf{L} = \mathbf{F}$, and give, as an example, a catalog of all "magnetic special positions" (i.e., all the invariant spin arrangements) for all the eight magnetic space groups of the kind $\mathbf{M}_T$ belonging to the family of *P4/mmm*.

In the case of some invariant spin arrangements it may not be necessary to specify all the spins $\mathbf{S}(\mathbf{r}_1^{(k)})$, $\mathbf{S}(\mathbf{r}_2^{(k)})$, ... explicitly as is required in (5). There may be relationships among these spins so that, for example, only a few spins and these relationships would have to be specified. This is so, in particular, in the case of some of the "spiral" invariant spin arrangements, of which a simple example has been given in this section (IV, 2h). In the case of that example it was sufficient to specify one spin, $\mathbf{S}(\mathbf{r})$, and a group which generates the spins located at $\mathbf{r} + n\mathbf{a}_3$ from $\mathbf{S}(\mathbf{r})$. There is of course a whole class of invariant spin arrangements for which the specification of the spins $\mathbf{S}(\mathbf{r}_1^{(k)})$, $\mathbf{S}(\mathbf{r}_2^{(k)})$, ... can be replaced by the specification of a few of them, and by defining a group which generates them from these few.

It is not our intention here to formulate practical rules for determining the characteristics (3), (4), and (5) of an invariant spin arrangement from some given experimental data concerning it. A survey of the experimental data on invariant spin arrangements is outside the scope of this chapter. We want to mention, however, that the first to use magnetic space groups to characterize invariant spin arrangements as obtained from neutron diffraction in magnetic crystals were Donnay *et al.* [36] and Le Corre [37]. Magnetic space groups have also been used, by Riedel and Spence [38] and van der Lugt [39], to interpret experimental data on invariant spin arrangements as obtained from the orientation dependence of nuclear magnetic resonance spectra in antiferromagnetic single crystals. Dimmock [39a] has used the theory of magnetic space groups to discuss the part played by the symmetry in determining the spin arrangement in a crystal which undergoes a second-order phase transition from the disordered paramagnetic state to an ordered magnetic state.

4. Miscellaneous Remarks

There are three additional remarks which we want to make in connection with the subject of this section. One remark has to do with the spin arrangements which are not invariant under any magnetic group. It may be convenient in some cases to consider such spin arrangements. They will generate representations of magnetic groups. The various (in general, reducible) representations obtained in this way have been discussed in detail for the case of space groups (i.e., trivial magnetic space groups) by Alexander [40], and by Bertaut [41], who also gives a survey of the work of others. The conclusions from such discussions have, however, only limited validity since no nontrivial magnetic space groups were considered.

The second remark is about the macroscopic symmetry of a magnetic crystal. We have seen that only a spin arrangement which is invariant under a magnetic space group which is of the kind $\mathbf{M}_T$ and whose point group is admissible can have a nonvanishing resultant spin vector, i.e., is compatible with a nonvanishing magnetization of the crystal; and Table IV specifies, for each admissible magnetic point group, which components of the magnetization can be different from zero. The question of which components of the magnetization (if any) are different from zero if the magnetization is to be invariant under a given magnetic point group could of course be answered without any reference to invariant spin arrangements. Similar questions concerning the nonvanishing of components of the invariant tensors of higher rank which are also of importance in the theory of magnetic properties of crystals, like the magnetoelectric tensor, or the piezomagnetic tensor, can be answered using the standard theory of group representations as applied to magnetic point groups. A systematic and detailed discussion of these questions is given in a review article by Birss [42]. His article also contains extensive tables which describe the properties of various invariant tensors for all the magnetic point groups.

While a discussion of the macroscopic symmetry of magnetic crystals based on the theory of magnetic point groups is perfectly general, a discussion of their microscopic symmetry (i.e., a discussion of invariant spin arrangements) which is not based on the theory of magnetic space groups but only on the theory of magnetic point groups is necessarily of limited validity. This third and final remark may seem superfluous. Careful reading of the existing literature on this subject (see, for example, Dzialoshinskii [43] and Birss [42], §4.1) makes us think, however, that perhaps it is not.

Appendix A. List of Some of the Symbols Used

$\mathscr{R}$	Group of all proper rotations
F	Space group
L	Any finite or infinite, proper or improper subgroup of a space group
χ	Group of all time displacements and time inversion
A	Abstract group of order 2; also the time inversion group
E	Identity element of a group
E'	Element of order 2 of A; also time inversion
D^G	Subgroup of index 2 of any group G
$D = D^F$,	Subgroup of index 2 of a space group F
D_T	D^F which has the same lattice T as F
D_R	D^F which has the same point group R as F
D_{R0} and $D_{R\alpha}$	Two kinds of D_R
T	Lattice (that is, group of primitive translations)
$T^D = D^T$	
R	Point group
$R^D = D^R$	
Q	Subgroup of index 4 of a symmorphic F
Q_τ	Subgroup of index 4 of a nonsymmorphic F
$Q_{\tau 0}$ and $Q_{\tau\alpha}$	Two kinds of Q_α
m	Any magnetic group
L	Subgroup of F to whose family m belongs
m_L	Magnetic group belonging to the family of L

M	Magnetic space group
M_T	Magnetic space group whose unprimed subgroup is D_T
M_R	Magnetic space group whose unprimed subgroup is D_R
M_{R0} and $M_{R\alpha}$	Two kinds of M_R
T_M	Magnetic lattice of M
T_m	Any magnetic lattice
R_M	Magnetic point group of M
R_m	Any magnetic point group
$F(\mathbf{r})$	Site space group of $\mathbf{r}$
$R(\mathbf{r})$	Site point group of $\mathbf{r}$
$R_M(\mathbf{r})$	Magnetic site point group of $\mathbf{r}$.

References*

1. "International Tables for X-Ray Crystallography" (N.F.M. Henry and Kathleen Lonsdale, eds.), Vol. 1. Kynoch Press, Birmingham, England, 1952.
2. W. Opechowski and R. Guccione, "Magnetic Groups." North-Holland Publ. Amsterdam (to be published).
3. E. Wigner, "Group Theory and its Applications to the Quantum Mechanics of Atomic Spectra." Academic Press, New York, 1959.
4. J. O. Dimmock and R. G. Wheeler, *Phys. Chem. Solids* **23**, 729 (1962).
5. J. O. Dimmock and R. G. Wheeler, *Phys. Rev.* **127**, 391 (1962).
6. J. O. Dimmock, *J. Math. Phys.* **4**, 1307 (1963).
7. G. F. Koster, *Solid State Phys.* **5**, 173 (1957).
8. J. S. Lomont, "Applications of Finite Groups." Academic Press, New York, 1959.
9. H. Heesch, *Z. Krist.* **73**, 325 (1930).
10. C. Hermann, *Z. Krist.* **69**, 250 (1928).
11. E. Alexander and K. Herrmann, *Z. Krist.* **69**, 295 (1928).

* *Note added in proof:* Several books have appeared containing some new material on magnetic symmetry since this chapter was submitted for publication. These are:

A. V. Shubnikov, N. V. Belov, and others, "Colored Symmetry" (W. J. Holser, ed.). (A series of publications from the Institute of Crystallography, Academy of Sciences of the U.S.S.R., Moscow, 1951–1958.) Macmillan, New York, 1964. In particular, this book contains an English translation of the first paper by N. V. Belov *et al.* [14] on Shubnikov groups.

R. R. Birss, "Symmetry and Magnetism." North-Holland Publ., Amsterdam, 1964. This book does not much differ in contents from reference [42].

H. Margenau and G. M. Murphy, "The Mathematics of Physics and Chemistry," Vol. 2. Van Nostrand, Princeton, New Jersey, 1964—see in particular Chapter 12, "Symmetry Properties of Magnetic Crystals" by J. O. Dimmock and R. G. Wheeler. This is an extended and improved version of the contents of references [4, 5, 6, 39a].

"American Institute of Physics Handbook," 2nd ed. (D. E. Gray, ed.). McGraw-Hill, New York, 1963. Table 5g-22, compiled by L. M. Corliss and J. M. Hastings, gives a list and some characteristics of invariant spin arrangements as obtained from neutron diffraction studies of various magnetically ordered crystals.

12. L. Weber, *Z. Krist.* **70**, 309 (1929).
13. H. Heesch, *Z. Krist.* **71**, 95 (1929).
14. N. V. Belov, N. N. Neronova, and T. S. Smirnova, *Tr. Inst. Kristallogr. Akad. Nauk. SSSR* **11**, 33 (1955).
15. A. M. Zamorzaev, dissertation (in Russian), Univ. of Leningrad, 1953.
16. A. M. Zamorzaev, *Kristallografiya* **2**, 15 (1957); *Soviet Phys. Cryst. (English Transl.* **3**, 401 (1958).
17. L. L. Landau and E. M. Lifschitz, "Statisticheskaya Fizika." GITTL (State Tech. Lit. Press), 1951; "Statistical Physics." Addison-Wesley, Reading, Massachusetts, 1958.
18. H. Weyl, "Symmetry." Princeton Univ. Press, Princeton, New Jersey, 1952.
19. V. L. Indenbom, *Kristallografiya* **4**, 619 (1959); *Soviet Phys. Cryst. (English Transl.)* **4**, 578 (1959).
20. A. Niggli, *Z. Krist.* **111**, 288 (1959).
21. B. A. Tavger and V. M. Zaitsev, *Zh. Eksperim. i Teor. Fiz.* **30**, 564 (1956); *J. Exptl. Theoret. Phys. (U.S.S.R.)* **3**, 430 (1956).
22. N. V. Belov, N. N. Neronova, and T. S. Smirnova, *Kristallografiya* **2**, 3 (1957); *Soviet Phys. Cryst. (English Transl.)* **2**, 311 (1957).
23. R. Guccione, *Phys. Letters* **5**, 105 (1963).
24. R. Guccione, Ph.D. Thesis, Univ. of British Columbia, Vancouver, Canada, 1963.
25. A. Bienenstock and P. P. Ewald, *Acta Cryst.* **15**, 1253 (1962).
26. A. L. Mackay, *Acta Cryst.* **10**, 543 (1957).
27. N. N. Neronova and N. V. Belov, *Kristallografiya* **6**, 3 (1961); *Soviet Phys. Cryst. (English Transl.)* **6**, 1 (1961).
28. A. M. Zamorzaev, *Kristallografiya* **3**, 399 (1958); *Soviet Phys. Cryst. (English Transl.)* **3**, 401 (1958).
29. B. A. Tavger, *Kristallografiya* **5**, 677 (1960); *Soviet Phys. Cryst. (English Transl.)* **5**, 646 (1961).
30. C. Hermann, *Z. Krist.* **69**, 546 (1929).
31. A. M. Zamorzaev, *Kristallografiya* **7**, 813 (1962); *Soviet Phys. Cryst. (English Transl.)* **7**, 661 (1963).
32. R. D. Spence, unpublished report, 1963.
33. B. A. Tavger, *Kristallografiya* **3**, 340 (1958); *Soviet Phys. Cryst. (English Transl.)* **3**, 341 (1958).
34. N. N. Neronova and N. V. Belov, *Kristallografiya* **4**, 807 (1959); *Soviet Phys. Cryst. (English Transl.)* **4**, 769 (1960).
35. N. V. Belov, N. N. Neronova, J. D. H. Donnay, and G. Donnay, Proc. Intern. Conf. Magnetism Cryst., Kyoto, 1961, *J. Phys. Soc. Japan* **17** (Suppl. B-II), 332 (1962).
36. G. Donnay, L. M. Corliss, J. P. H. Donnay, N. Elliott, and J. M. Hastings, *Phys. Rev.* **112**, 1917 (1958).
37. Y. Le Corre, *J. Phys. Radium* **19**, 750 (1958).
38. E. P. Riedel and R. D. Spence, *Physica* **26**, 1174 (1960).
39. W. van der Lugt, thesis, Univ. of Leyden, Leyden, 1961.
39a. J. O. Dimmock, *Phys. Rev.* **130**, 1337 (1963).
40. S. Alexander, *Phys. Rev.* **127**, 420 (1962).
41. E. F. Bertaut, "Magnetism" (G. T. Rado and H. Suhl, eds.), Vol. 3, p. 150. Academic Press, New York, 1963.
42. R. R. Birss, *Rept. Progr. Phys.* **26**, 307 (1963).
43. I. E. Dzialoshinskii, *Zh. Eksperim. i Teor. Fiz.* **32**, 1547 (1957); *Soviet Phys. JETP (English Transl.)* **5**, 1259 (1957).

4. Hyperfine Interactions in Magnetic Materials

Arthur J. Freeman*

National Magnet Laboratory,**
Massachusetts Institute of Technology,
Cambridge, Massachusetts

Richard E. Watson

Bell Telephone Laboratories,
Murray Hill, New Jersey, and
Quantum Chemistry Group,
University of Uppsala,
Uppsala, Sweden

I. Introduction	168
1. One-Electron Theory of Hyperfine Interactions	169
2. Multi-Electron Contributions to Hyperfine Interactions	175
II. Experimental Methods	182
1. Introduction	182
2. Nuclear Orientation	183
3. Nuclear Specific Heats	186
4. Recoilless Emission and Absorption of γ-Rays; the Mössbauer Effect	194
5. Angular Correlation of γ-Rays	215
6. Interaction of Polarized Neutrons with Polarized Nuclei	221
III. The Hartree–Fock Method in Its Conventional and "Unrestricted" Forms	223
1. The Hartree–Fock Formalism and the Fermi Exchange Hole	224
2. Correlation in Atoms	229
3. Restrictions Associated with the Conventional Hartree–Fock Scheme	233
4. Exchange Polarization	237
5. The Spin-, or Exchange-, Polarized Hartree–Fock Method	239
6. Aspherical Distortions and Hyperfine Interactions	254
7. Extended Hartree–Fock Functions for Closed-Shell Systems	257
8. Concluding Comments	258
IV. Interpretation of Measured Hyperfine Interactions	259
1. Hyperfine Fields in Iron Series Salts	259
2. Hyperfine Fields in Rare Earth Salts	266
3. Hyperfine Fields in Ferromagnetic Metals	270
4. Hyperfine Fields in Nonmagnetic Ions in Magnetic Materials	274
5. Hyperfine Fields (Knight Shifts) in Metals	280
6. Mössbauer Effect in an External Magnetic Field	282

* Part of the work of this author was done at the Ordnance Materials Research Office, Watertown, Massachusetts. Support is gratefully acknowledged.
** Supported by the Air Force Office of Scientific Research.

7. The Mössbauer Effect and Hyperfine Fields in Iron Metal with Dilute Impurities 285
Appendix A. Comments on the Solution of Conventional Hartree–Fock Radial Equations for Atoms 287
Appendix B. Hartree–Fock $\langle r^n \rangle$ Values for $3d$, $4d$, and $4f$ Ions 290
Acknowledgments 292
References 292

I. Introduction

Hyperfine interactions have been studied for many years. Indeed, as early as 1924, it was the hyperfine structure of spectral lines which led Pauli [1] to postulate the existence of a spin-angular momentum and an associated magnetic moment for the nucleus. Hyperfine interactions, involving as they do the interaction between electrons and nuclei, form the bridge between the otherwise separated disciplines called solid state physics and nuclear physics. Such measurements provide the basis for (1) the determination of nuclear moments (magnetic and electric quadrupole) and (2) a better understanding of the electronic structure of matter (atoms, molecules, and solids).

It has only been in recent years that the discovery of a number of new techniques for studying hyperfine interactions in magnetic materials brought with it the possibility of improving our understanding of the origin of electronic magnetism. This was quickly followed by a growing interest in the use of these techniques; the development and application of these methods, by a large number of workers, has led rapidly to the accumulation of a good many results on a wide variety of materials.

In this chapter we review and discuss some of these techniques with particular emphasis on the Mössbauer effect (since this method is not being discussed elsewhere in this treatise), some of the information obtained to date, and some theoretical attempts at understanding the origin of these phenomena in magnetic materials. This is by no means a closed field, as much more theoretical and experimental work is needed to help our understanding of the origin of the hyperfine interactions. Such an understanding will be of invaluable importance for the study of magnetism because the basic mechanisms responsible for observed hyperfine effects are intimately related to, and involved with, some of the least understood properties of the electronic structure of magnetic materials.

Some of the information obtained in this way is rather unique and is very helpful in understanding a wide range of phenomena in magnetic materials. Among these, one may cite determinations of electronic charge and spin distributions, magnetization and magnetic moments, ionicities (or valencies), electron configurations in metals and alloys, covalency effects in magnetic salts, localized magnetic moments in metals and alloys, crystal-field parameters and splittings of crystal-field states, ordering mechanisms in alloys, magnetic (and crystallographic) transitions and magnetic ordering processes, magnetic impurities, and spin structures, to mention but a few. In short, the measurement of hyperfine interactions covers such a broad range and variety of phenomena that it has become an indispensable tool for the study of magnetism.

1. One-Electron Theory of Hyperfine Interactions

We are concerned with the interaction between electrons and the nucleus, which, for free atoms, arises from two sources: the magnetic interaction with the magnetic moment of the nucleus (linear in the nuclear spin I); and the electrostatic interaction with the electric quadrupole moment of the nucleus (quadratic in I). The effects of higher electric and magnetic poles are negligibly small and so will not be considered.

a. *Magnetic interaction between an atomic nucleus and its electrons.* The Hamiltonian for the interaction between a single electron's spin and orbital moments and the nuclear magnetic moment was first derived by Fermi from Dirac relativistic theory for the electron [2]. If $\mathbf{L}$, $\mathbf{S}$, and $\mathbf{I}$ represent, respectively, electron orbital, electron spin, and nuclear-spin angular momentum operators, μ_0 and μ_N Bohr and nuclear magnetons, and g and g_I the electronic and nuclear spectroscopic splitting factors, then the magnetic hyperfine interaction can be written as

$$\mathscr{H}_D = -gg_I\mu_0\mu_N \left\{ \frac{8\pi}{3}\delta(\mathbf{r})\mathbf{S}\cdot\mathbf{I} + \frac{(\mathbf{L}-\mathbf{S})\cdot\mathbf{I}}{r^3} + \frac{3(\mathbf{S}\cdot\mathbf{r})(\mathbf{I}\cdot\mathbf{r})}{r^5} \right\} \quad (1.1)$$

The delta function term,* which is called the Fermi contact term, is non-zero only for s electrons, in which case the last two terms are zero. These two terms are dipolar interaction terms and are analogous to the classical expression for the mutual potential energy of two point dipoles $g_I\mu_N\mathbf{I}$ and $g\mu_0\mathbf{S}$ plus the interaction of a point dipole $g_I\mu_N\mathbf{I}$ with a circulating charge of angular momentum $\mathbf{L}$. Equation (1.1) has been derived more recently by the powerful methods of quantum field theory [4] and a number of higher-order corrections were included [5]. As already

---

* The delta function notation was apparently first introduced by Abragam and Pryce [3].

noted, the anomalous magnetic moment of the electron was, in fact, indicated by the exact determination of the hyperfine splitting for the hydrogen atom.

If the atom has several electrons outside closed shells, the operator in the Hamiltonian leading to the magnetic hyperfine interaction is taken to be a sum of terms like Eq. (1.1), which gives the interaction of each electron with the nuclear magnetic moment [6]. This sum of one-electron operators for the many-electron case has been treated extensively in the literature in terms of the vector model, using the methods of Racah, as has been done by Schwartz [7], Trees [8], and others [9]. Additional factors, such as relativistic effects, perturbations due to the mutual interaction between hyperfine states, and effects of finite nuclear size have also been investigated [6].

Equation (1.1) can be written in the form

$$\mathcal{H}_D = -\mu_I \cdot \mathbf{H}_J \tag{1.2}$$

where μ_I is the nuclear magnetic moment and $\mathbf{H}_J$ is the magnetic field at the nucleus arising from the rest of the atom having total angular momentum J. As in Eq. (1.1), the magnetic moment can be taken as proportional to its spin angular momentum I and written as

$$\mu_I = (\mu_I/I)\mathbf{I} = g_I \mu_N \mathbf{I} \tag{1.3}$$

and therefore

$$\mathcal{H}_D = -g_I \mu_N \mathbf{I} \cdot \mathbf{H}_J \tag{1.4}$$

Since H_J can be taken as proportional to J for matrix elements diagonal in J the interaction assumes the familiar form

$$\mathcal{H}_D = A \mathbf{I} \cdot \mathbf{J} \tag{1.5}$$

where A, the hyperfine, or magnetic dipole interaction, constant, is given by

$$A = \frac{-g_I \mu_N \mathbf{H}_J \cdot \mathbf{J}}{\mathbf{J} \cdot \mathbf{J}} \tag{1.6}$$

(1) *Electrons which have $L \neq 0$.* For a single electron (or a single electron or single hole in an otherwise closed shell system) with $L \neq 0$, Eq. (1.1) can be replaced by an equivalent operator, provided we are interested only in the diagonal matrix components of $\mathcal{H}_D$ with respect to wave functions in which J is a good quantum number. In this case the diagonal matrix elements of the operator L equal those of $(\mathbf{L} \cdot \mathbf{J})\mathbf{J}/J(J+1)$, i.e., it is only the component of $\mathbf{L}$ (and similarly of $\mathbf{S}$)

along their resultant **J** which comes into the average value. Then $\mathcal{H}_D$ of Eq. (1.1) can be replaced by the operator

$$\mathcal{H}_D' = -gg_I\mu_0\mu_N \frac{\mathbf{I}\cdot\mathbf{J}}{J(J+1)} \left[\frac{\mathbf{L}\cdot\mathbf{J}}{r^3} - \frac{\mathbf{S}\cdot\mathbf{J}}{r^3} - \frac{3(\mathbf{S}\cdot\mathbf{r})(\mathbf{r}\cdot\mathbf{J})}{r^5}\right] \quad (1.7)$$

which is in the form of Eq. (1.5). Since we know $\mathbf{L}\cdot\mathbf{r} = 0$ and we can let $\mathbf{J} = \mathbf{L} + \mathbf{S}$, we have

$$\mathcal{H}_D' = -gg_I\mu_0\mu_N \frac{\mathbf{I}\cdot\mathbf{J}}{J(J+1)} \left[\frac{L^2}{r^3} - \frac{S^2}{r^3} + \frac{3(\mathbf{S}\cdot\mathbf{r})^2}{r^5}\right] \quad (1.8)$$

For a single electron (or a single hole in a closed shell), the last two terms cancel, leaving only the term involving L^2.

As an example, consider the case when the expection value of $\mathcal{H}_D'$ with respect to the electronic wave function is taken in the representation $\mathbf{F} = \mathbf{I} + \mathbf{J}$ where F, I, L, J are good quantum numbers. Using the operator relation $2\mathbf{I}\cdot\mathbf{J} = F^2 - I^2 - J^2$, Eq. (1.8) leads to

$$\langle\mathcal{H}_D'\rangle = -gg_I\mu_0\mu_N \frac{F(F+1) - I(I+1) - J(J+1)}{J(J+1)} L(L+1)\langle r^{-3}\rangle \quad (1.9)$$

or

$$\langle\mathcal{H}_D'\rangle = -gg_I\mu_0\mu_N \frac{KL(L+1)}{J(J+1)} \langle r^{-3}\rangle \quad (1.10)$$

with $K = F(F+1) - I(I+1) - J(J+1)$. Note that here the electronic spin contributes to the $\mathcal{H}_D'$ of Eqs. (1.9) and (1.10) despite the fact that **S** does not appear explicitly.

Correction factors due to relativity, finite size of nucleus, etc. must be included in these expressions as they can become quite appreciable, particularly for the high-Z elements. Kopfermann [6] gives these corrections tabulated as functions of Z and J and recently some of these matters have been the subject of a study by Stroke et al. [10]. Equation (1.9) or (1.10) gives the magnetic hyperfine interaction energy from which $\mathbf{H}_J$, the effective magnetic field at the nucleus, can be found if the nuclear magnetic moment (i.e., g_I) is known; or, reciprocally, g_I can be calculated from the observed interaction from Eq. (1.10) if one assumes the interaction to be correct. We shall return to this point in some detail later (Section IV).

(2) *s Electrons.* For ions with unpaired *s* electrons, the field at the nucleus arises from just the contact part of the Hamiltonian. The field $\mathbf{H}_c$ is of the form

$$\mathbf{H}_c = \frac{8\pi}{3} g\mu_0 \mathbf{S} |\psi_s(0)|^2 \quad (1.11)$$

where $\rho(0) = |\psi_s(0)|^2$ is the density of the *s* electron at the nucleus.

Historically, the Fermi contact term has been generally accepted as the origin of the hyperfine structure observed in certain free-atom spectra and molecular beam measurements and more recently has been used to explain the observed Knight shifts in metals [11]. In all these systems the density at the nucleus of an outer unpaired s electron is considered to be responsible for the observed effective magnetic field.

For ions with a net spin but no unpaired s electrons, conventional one-electron theory predicts zero values for H_c, whereas large non-zero fields do, in fact, occur for such cases. A good example of this is the Mn^{2+} ion in the $3d^5, {}^6S$ state; since it is spherical, it does not have the orbital and spin dipolar hyperfine terms of Eq. (1.1); yet, it has a hyperfine field of ~ 700 kgausses.* These cases point out the inadequacy of the one-electron theory. As we shall see, recent theoretical efforts have centered almost entirely on many-electron polarization effects involving core electrons; for this reason most of the theoretical discussion in the remainder of this chapter will be concerned with these effects.

b. *Electrostatic interactions between electrons and nuclei.* The theory of the electrostatic interaction between electrons and nuclei was worked out by Casimir [13], Pound [14], and others; nuclear quadrupole effects in solids have recently been reviewed at length by Das and Hahn [15], Cohen and Reif [16], and Abragam [17]. We shall present here only a brief outline of some of the results pertinent to our later discussions; our concern is only with the quadrupole part of the interaction.

If the finite volume of the charged nucleus is considered then the electrostatic interaction between a nucleus and its electrons is given by

$$H_E = \int\int \frac{\rho_e(\mathbf{r}_e)\rho_n(\mathbf{r}_n)\,d\tau_e\,d\tau_n}{|\mathbf{r}_e - \mathbf{r}_n|} \tag{1.12}$$

where $\mathbf{r}_e$ and $\mathbf{r}_n$ denote electronic and nuclear coordinates, respectively. The general procedure is to use the expansion

$$\frac{1}{|\mathbf{r}_e - \mathbf{r}_n|} = 4\pi \sum_{k=0}^{\infty} \sum_{m=-k}^{+k} \frac{1}{2k+1} \frac{r_<^k}{r_>^{k+1}} Y_k^{m*}(\theta_e, \phi_e) Y_k^m(\theta_n, \phi_n) \tag{1.13}$$

where $r_<$ and $r_>$ are the smaller and larger, respectively, of the two distances r_e and r_n. Assuming that the charge distribution of the nucleus is completely inside the electronic distribution one can equate r_n with

---

* For a discussion and review of these data, see Watson and Freeman [12].

4. HYPERFINE INTERACTIONS IN MAGNETIC MATERIALS

$r_<$ and r_e with $r_>$, arriving at a simple expression for H_E which is separable into electronic and nuclear parts.

$$H_E = 4\pi \sum_k \sum_m \frac{1}{2k+1} \int \rho_n(\mathbf{r}_n) r_n^k Y_k^m(\theta_n \phi_n) \, d\tau_n \int \rho_e(\mathbf{r}_e) r_e^{-(k+1)} Y_k^{m*}(\theta_e, \phi_e) \, d\tau_e \quad (1.14)$$

The first integral represents a multipole component of the nuclear charge, and the second that part of the electronic density at the nucleus which has the appropriate symmetry to give a non-zero interaction with this multipole. Odd multipoles (k odd) are forbidden by symmetry (for nuclear states of well-defined parity), as is observed experimentally for $k = 1$. The electric quadrupole interaction comes from $k = 2$; we shall not be concerned with higher ($k = 4, 6, \ldots$) multipoles.

Equation (1.14) is conveniently treated by Racah's tensor operators [7–9]. We can only summarize some of the results of interest here. For a single electron with angular momentum J, retaining only diagonal terms, we have

$$H_{EQ} = e^2 q_J Q \left[\frac{3(\mathbf{I} \cdot \mathbf{J})^2 + (\tfrac{3}{2})(\mathbf{I} \cdot \mathbf{J}) - I(I+1)J(J+1)}{2I(2I-1)J(2J-1)} \right] \quad (1.15)$$

where Q is a scalar quantity called the quadrupole moment defined by

$$eQ = \int \rho_n(r_n)_{m_I=I} (3z_n^2 - r_n^2) \, dv_n, \quad (1.16)$$

the integral being carried out for the nuclear state with $m_I = I$. Similarly q_J is given by

$$eq_J = \int \rho_e(r_e)_{m_J=J} \left(\frac{3\cos^2\theta - 1}{r_e^3} \right) dv_e \quad (1.17)$$

Equation (1.17) is the expectation value of the operator $(3\cos^2\theta - 1)/r^3$ with respect to the appropriate eigenfunction for the state $m_J = J$. For wave functions separable into a radial and angular part

$$eq_J = -\frac{2J-1}{2J+2} \langle r^{-3} \rangle \quad (1.18)$$

As in the preceding section, in the F, m representation ($\mathbf{F} = \mathbf{I} + \mathbf{J}$), the $\mathbf{I} \cdot \mathbf{J}$ factors can be worked out to give

$$H_{EQ} = e^2 q_J Q \left[\frac{\tfrac{3}{4}K(K+1) - I(I+1)J(J+1)}{2I(2I-1)J(2J+1)} \right] \quad (1.19)$$

Generally the electric quadrupole interaction for any atomic or molecular system is written as

$$H_{EQ} = e^2 \xi Q \,[3(\mathbf{I} \cdot \mathbf{J})^2 + \tfrac{3}{2}(\mathbf{I} \cdot \mathbf{J}) - J(J+1)I(I+1)] \qquad (1.20)$$

where ξ contains the pertinent information regarding the electronic distribution.

For solids, it is customary to treat the electric field gradient at the nucleus classically, i.e., the limiting case of the molecular system with $J \to \infty$. The magnitude of the electric field gradient at the nucleus is described by the tensor V_{ij}, which has in general five independent components, since it is symmetric and has zero trace. By choosing a set of orthogonal axes (the principal axes), this symmetric tensor may be reduced to diagonal form. In this system there are only three non-vanishing components, V_{xx}, V_{yy}, V_{zz}, which must satisfy

$$\nabla^2 V = V_{xx} + V_{yy} + V_{zz} = 0 \qquad (1.21)$$

These components are therefore not independent and in general there are but two parameters which describe the electric field gradient. Usually one chooses the principal axes such that

$$|V_{zz}| \geqslant |V_{yy}| \geqslant |V_{xx}| \qquad (1.22)$$

and defines

$$eq = V_{zz} = \left(\frac{\partial^2 V}{\partial z^2}\right)_0; \qquad \eta = (V_{xx} - V_{yy})/V_{zz} \qquad (1.23)$$

From Eqs. (1.21) and (1.22) it follows that V_{xx} and V_{yy} have the same sign, which implies that $0 \leqslant \eta \leqslant 1$. The quadrupole Hamiltonian can therefore be written

$$H_{EQ} = \frac{e^2 q Q}{4I(2I-1)} [3I_z^2 - I(I+1) + \tfrac{1}{2}\eta(I_+^2 + I_-^2)] \qquad (1.24)$$

where $I_{\pm} = I_x \pm i I_y$. If the field gradient has axial symmetry, then $V_{xx} = V_{yy}$ and $\eta = 0$; η thus measures the degree of asymmetry or the departure from axial symmetry. For cubic or spherical symmetry each of the $V_{ii} = 0$ and the quadrupole interaction is zero.

The determination of Q from experiment (which gives the product qQ) requires some independent determination of the parameters q and η, either theoretically or from some other experiment. However, unlike the case of magnetic hyperfine interactions, where in certain cases the internal magnetic field is smaller than the externally applied field and precise nuclear magnetic moments can be determined, appreciable electric field gradients cannot be produced by external means. They are generally produced internally and are very sensitive to details of the electronic charge distributions and, as has been emphasized by Sternheimer [18, 18a, b, 19], to the large polarizations of ion cores induced by the field gradients. Lack of accurate values of q provides the major uncertainty in the measurement of Q's and is the reason that values of nuclear quadrupole moments are not precisely known. This in turn explains why quadrupole interactions have not given detailed quantitative information about internal fields in solids despite the fact that the nuclear quadrupole moment provides a sensitive probe for such studies.

2. Multi-Electron Contributions to Hyperfine Interactions

We have already had several occasions to point out the inadequacy of the conventional one-electron theory for the determination of both magnetic and electric hyperfine fields. Here we shall indicate but briefly several ways that many-electron effects are included in the theoretical framework; some of the most recent developments (i.e., unrestricted Hartree–Fock theory) will be discussed at length later on. Our purpose is to introduce some of these matters here so as to ensure the reader's awareness of them when he reads the sections that follow.

The inadequacy of the one-electron theory outlined above was recognized many years ago. As early as 1933, Fermi and Segrè [20] attempted to improve the conventional approach to magnetic hyperfine interactions by applying configuration interaction techniques to the outer electronic shells of Tl. They considered the contact term associated with the closed $6s$ shell arising from the influence of a singly occupied $6p$ electron; it seemed physically reasonable (and was also computationally necessary) to limit the investigation to the $6s$ shell alone. There have been other estimates of the H_c associated with an outermost s shell, such as that of Koster [21] for Ga and of Abragam et al. (AHP) [22] for Mn^{2+}. More recently it has become known that all closed s shells normally contribute significantly to an H_c. This was first shown by Sternheimer in his perturbation treatment [23] of Cl. We shall return to some of these matters in a later section (Section III) where we also discuss the spin (or exchange) polarized Hartree–Fock method for treating these problems.

a. *Quadrupole polarizabilities and Sternheimer antishielding factors.* As emphasized by Sternheimer and co-workers [18, 18a,b, 19], contributions to magnetic dipole and electric quadrupole interactions arising from the distortion of the otherwise spherical closed electronic shells of the system play an important role in observations of hyperfine interactions. A striking example is the polarization induced by electric field gradients in either free atoms or solids; the induced distortions can be very large, and give rise to additional electric field gradients which are sometimes several orders of magnitude larger than the source gradients. While these additional contributions are a hindrace to determinations of nuclear quadrupole moments, studies of these effects offer a means of investigating the nature of electron distributions in certain materials and may therefore prove valuable.

Two sources of the field gradient must be considered: (1) the field gradient coming from the aspherical outer (or valence) electrons, q_v, and (2) the field gradient due to the crystalline field, q_c. The induced field gradients are, to second order, proportional to the source gradients, and the quadrupole Hamiltonian, given by Eq. (1.24), must be modified by certain proportionality factors, now called Sternheimer antishielding factors. In Eq. (1.24), q is replaced by $q_v(1 - R_q)$ and $q_c(1 - \gamma_\infty)$ for the two contributions. Computed values of $|R_q|$ are <1.0 whereas γ_∞ ranges from -10 to -100, except for small ions consisting only of s shells, in which case γ_∞ is small (<1) and positive (i.e., shielding). The reason for the difference in magnitude between R_q and γ_∞ will become apparent later in our discussion.

One may calculate either (a) the electronic distortion due to the electric field gradient and thence the interaction of this distorted distribution with the nuclear quadrupole moment, Q, or (b) the electronic distortion due to Q and in turn its interaction with the external field [18, 18a,b]. Both approaches give identical results within second-order perturbation theory [24], but the situation would become more complicated if one were to go beyond it.

Let us now turn to a description of the way in which these calculations are generally carried out, using the perturbation theory method of Sternheimer and his collaborators. This approach to the problem of the polarization of electron shells of atoms and ions has been discussed previously at length [18, 18a,b, 19, 24]; some brief details will be given here in order to illustrate these ideas and to provide a framework for later discussions.

Writing the Hamiltonian in the usual form, $H = H_0 + H_1$, with H_0 the unperturbed Hamiltonian and H_1 a perturbing potential, the first-order perturbation $\psi_1(\mathbf{r})$ to the unperturbed solution $\psi_0(\mathbf{r})$, is determined from the usual second perturbation theory relation

$$(H_0 - E_0)\psi_1(\mathbf{r}) = -(H_1 - E_1)\psi_0(\mathbf{r}) \tag{1.25}$$

Here

$$E_0 = \langle \psi_0 | H_0 | \psi_0 \rangle, \quad E_1 = \langle \psi_0 | H_1 | \psi_0 \rangle, \quad \text{and} \quad \langle \psi_0 | \psi_1 \rangle = 0$$

Equation (1.25) has been solved in two distinct ways: (a) directly by exact numerical solution, as was done by Sternheimer and his collaborators [18, 18a,b, 19], and (b) by the analytic approach of Das and Bersohn [25, 26] in which the radial part of $\psi_1(\mathbf{r})$ is assumed to be related to the radial part of $\psi_0(\mathbf{r})$ by

$$u_1(\mathbf{r}) = u_0(\mathbf{r}) \sum_m a_m r^m \tag{1.26}$$

and the parameters a_m are determined by minimizing the second-order perturbation energy with respect to variation of these parameters. This technique has the advantage of being easier to carry out than Sternheimer's but is inferior because full variational freedom is not accorded to u_1 (e.g., if u_0 is a noted function then u_1 is constrained to have the same nodes). A detailed discussion of the Sternheimer and other perturbation approaches has been given recently by Allen [27], Kaneko [28], and Dalgarno [24].

Since for quadrupole interactions the perturbing potential H_1 has $Y_2^0(\theta, \varphi)$ symmetry, perturbed orbital character will be mixed into the unperturbed orbitals, ψ^0, in the following ways:

$$\psi_s^0 \to N_s(\psi_s^0 + \psi_d')$$
$$\psi_p^0 \to N_p(\psi_p^0 + \psi_p' + \psi_f')$$
$$\psi_l^0 \to N_l(\psi_l^0 + \psi_l' + \psi_{l-2}' + \psi_{l+2}'); \quad l > 1 \tag{1.27}$$

where the N_i are normalization constants. The mixing of ψ' of l in common with ψ^0 is called a "radial" excitation whereas the ψ' components having $l \pm 2$ are called "angular" excitations. The "radial" and "angular" effects on the charge distribution are shown schematically [16] in Fig. 1. As examples of numerical results we list in Table I the various angular and radial contributions to γ_∞ for He, Cl$^-$, and Cu$^+$ obtained by Sternheimer and co-workers [18, 18a,b, 19]. The most prominent feature of these results is that the radial $\gamma_{l \to l}$'s dominate and are antishielding.

The Sternheimer perturbation method is essentially a Hartree perturbation theory. This raises the question of what repercussions occur when we go to a Hartree–Fock theory. Sternheimer and Foley [18, 18a,b, 19], Allen [27], Kaneko [28], Dalgarno [29], and Burns [30] have discussed the roles of exchange and higher-order effects. Dalgarno

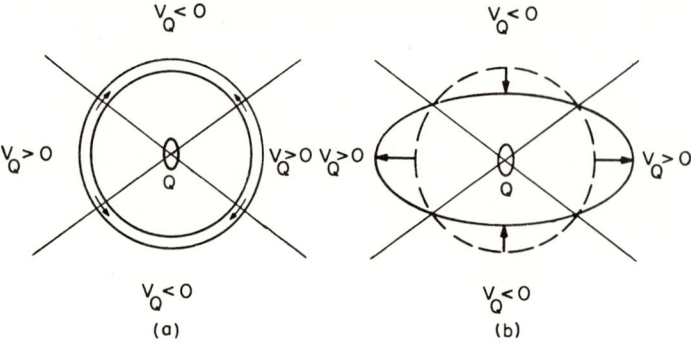

Fig. 1. Shielding and antishielding. (a) Angular excitation. Electrons within each radial shell move into regions of negative potential V_Q, from the regions of positive V_Q, thus piling negative charge along the axis of Q and shielding it. (b) Radial excitation. Electrons move inward where V_Q is negative and outward where V_Q is positive. The distortion of the electron density is opposite to that of the nuclear charge distribution producing Q. Hence, the radial excitation of the negative electrons produces antishielding of Q. (After Cohen and Reif [16].)

TABLE I

Individual Contributions to γ_∞ for He, Cl$^-$, and Cu$^+$ Obtained from Numerical Integration of

	He [a]	Cl$^-$ [b]	Cu$^+$ [b]
$\gamma_{1s\to d}$	0.41	0.04	—
$\gamma_{2s\to d}$	—	0.11	—
$\gamma_{3s\to d}$	—	0.44	0.18
$\gamma_{2p\to p}$	—	−1.5	−0.62
$\gamma_{3p\to f}$	—	0.16	—
$\gamma_{3p\to p}$	—	−56.5	−7.9
$\gamma_{3p\to f}$	—	0.73	0.25
$\gamma_{3d\to s}$	—	—	−0.86
$\gamma_{3d\to d}$	—	—	−8.5
$\gamma_{3d\to g}$	—	—	0.37
γ_∞ [c]	0.41	−56.5	−17.1
γ_∞ [d]	—	−88.9	−17.6

[a] See Sternheimer [31].
[b] For $\gamma_{l\to l}$'s see Sternheimer and Foley [18a]; and for $\gamma_{l\to l\pm 2}$ see Sternheimer [31a].
[c] Using these values.
[d] Using the unrestricted Hartree–Fock $\gamma_{l\to l}$'s of Freeman and Watson [33].

[29] has derived two methods which include exchange; his first (his Method I) includes means for self-consistency and thus is related to the unrestricted Hartree–Fock methods we will discuss later. Dalgarno and co-workers [24] have investigated the effect of including exchange in the calculation of γ_∞ for He and have found it to be important. In addition they have also applied his Method I to He and have found self-consistency to be unimportant. Inspection of Table I suggests that this is not surprising. More recently, Khubchandani et al. [32] have applied an approximate form of Dalgarno's Method II (which does not include self-consistency) to the $\gamma_{p \to p}$'s of K^+ and Cl^-. They again found the inclusion of exchange to be significant but they did not investigate self-consistency. Utilizing the unrestricted Hartree–Fock method to be discussed later, the authors [33] have obtained values of γ_∞ for Cu^+ and Cl^- and have discovered self-consistency to be important for Cl^- (see Table I).

Another computational problem which arises concerns the maintenance of orbital orthogonality in the perturbation calculations and is associated with the application of the variation principle. There is considerable formal justification for ignoring this matter on the grounds that the resulting errors are largely self-canceling. Sternheimer does this with great success. Recent calculations for Fe^{3+} by Ingalls [34] and for Cl^- by the authors [33], using the standard analytic variation–perturbation scheme (which partially accounts for orthogonalization), show the resulting γ_∞ to be very sensitive to the way orthogonalization is maintained between the $2p \to p$ and $3p \to p$ distortions.

In addition to the problems associated with accurately computing a γ_∞ factor, there is also the problem of relating it to a γ factor appropriate for the ion in a specific crystal or molecule. It has been suggested [35] that theoretical values of γ_∞ are in rough agreement with experimental values of γ for positive ions but are badly out of line for negative ions. For example, in contrast to the theoretical results given in Table I, Cl^- and Cu^+ are both considered to have [35] experimental values of γ of roughly -10 to -15. The discrepancy between theory and experiment arises from limitations of what is essentially a tight binding approach, repercussions from other effects of the crystalline environment (e.g., if the environment causes an over-all contraction of the ion's charge distribution, this will affect the value of γ_∞ [35]), and difficulties in defining the external field. A very serious problem is the matter of the perturbing field, for while a field such as that described by a lattice of point charges external to the ion may be adequate for describing the interaction with the nucleus, it is naïve to presume that it remains adequate for the ion's electrons. There are considerable complications associated with defining an appropriate crystal field for an ion. One obvious point ignored by such a potential is that the charge of the rest of the crystal does overlap that of the ion. Unfortunately, the crystal-field theory necessary for the problem at hand is not in a completely satisfactory state [36, 37]. In any case, when one is comparing theory and experiment, the problems related to the crystal field appear to pose greater difficulties than do details in the mechanics of γ_∞ calculations.

Recent Mössbauer effect measurements for rare earth ions have shown that Sternheimer γ_∞ and R_q factors can be determined for these ions, particularly when independent Coulomb excitation data are available.*
In these cases $\gamma_\infty \sim -100$ and $R_q \simeq 0.3$, in quite good agreement with theory (see discussion below and in Section IV, 3).

b. *Magnetic and electric hyperfine interactions due to closed-shell distortions in open-shell ions.* Most of our remarks have been concerned chiefly with the distortion of a spherical ion due to an external crystalline electric field gradient. Suppose we now consider a free ion with an unfilled shell (or shells) and $L \neq 0$ and ask how that shell distorts the closed shells and, in turn, causes an interaction with the nuclear electric quadrupole or magnetic dipole moment.

In a conventional Hartree–Fock description the closed shells contribute nothing to the hyperfine interaction and the unfilled shell has an electric quadrupole interaction which can be written as

$$H_Q = \frac{e^2 Q}{2I(2I-1)} \langle r^{-3} \rangle \sum_i \left[I(I+1) - \frac{3(\mathbf{r}_i \cdot \mathbf{I})^2}{r_i^2} \right] \quad (1.28)$$

where the sum is over open-shell electrons and the operator inside the sum is an angular operator. Given conventional one-electron wave functions, one can utilize perturbation techniques similar to the γ_∞ case and obtain distorted closed-shell contributions to H_Q as was done originally by Sternheimer [18, 18a,b, 19]. The symmetry of the problem is such that the closed-shell contributions can be included in Eq. (1.28) by replacing $\langle r^{-3} \rangle$ by an effective $\langle r^{-3} \rangle$ parameter, $\langle r^{-3} \rangle_{\text{eff},q}$, or by introducing the shielding factor R_q where

$$\langle r^{-3} \rangle_{\text{eff},q} = \langle r^{-3} \rangle (1 - R_q) \quad (1.29)$$

Unfortunately, as soon as the closed shells are distorted, the many-electron wave function ceases to be an eigenfunction of L^2. When

---

* See discussions in Section II and reference [38].

4. HYPERFINE INTERACTIONS IN MAGNETIC MATERIALS

dealing with the γ_∞ factors, we found of course, that the ions ceased to be of their original symmetry when distorted; but this was all right, since the source of the distortion was external to the ion. When unfilled shells are the source of the distortion, one still wants proper eigenfunctions of L^2 and the implications of either ignoring the symmetry failure or of trying to correct it are serious [39]. We will return to this problem in Section III.

For an ion with a single [non-s] open shell, the conventional Hartree–Fock description would predict a magnetic hyperfine interaction of the form

$$H_m = \frac{\beta\mu}{I}\langle r^{-3}\rangle \sum \mathbf{I}\cdot \mathbf{l}_i + \frac{\beta\mu}{I}\langle r^{-3}\rangle \sum [\mathbf{I}\cdot \mathbf{s}_i - 3(\mathbf{I}\cdot \mathbf{r}_i/r_i)(\mathbf{s}_i \cdot \mathbf{r}_i/r_i)] \quad (1.30)$$

where again the sums are over the unclosed shell. If one wishes to include the effect of closed-shell distortions one can again replace $\langle r^{-3}\rangle$ by $\langle r^{-3}\rangle_{\text{eff},m}$. Separate $\langle r^{-3}\rangle_{\text{eff},m}$ terms are required for the spin dipolar and orbital terms and these, in turn, will differ with $\langle r^{-3}\rangle_{\text{eff},q}$. Unlike the case of $\langle r^{-3}\rangle_{\text{eff},q}$ parameters, Coulomb interactions alone (between the open and closed shells) will not lead to closed-shell contributions to the magnetic $\langle r^{-3}\rangle_{\text{eff}}$ values. In the case of the spin dipolar term, not only must the closed shell be aspherically distorted (whether by Coulomb and/or by exchange interactions) but pairs of orbitals differing in m_s value must have differing spatial behavior in order that the spin dipolar sum of Eq. (1.30) be different from zero. This requires an imbalance in the distorting open-shell exchange terms, for such a pair. As for the $\mathbf{L} \cdot \mathbf{I}$ term, Coulomb interactions treat a pair of orbitals of $+m_l$ and $-m_l$ identically and the resulting contributions to this term will thus cancel identically. Exchange interactions, on the other hand, can be non-symmetric in the treatment of $+m_l$ and $-m_l$ and, in turn, yield non-zero $\mathbf{L}\cdot\mathbf{I}$ contributions [39].

In addition to Sternheimer's calculations for the R shielding factors, Ingalls [34] has obtained an R_q for divalent iron, and "unrestricted" Hartree–Fock results have been obtained by the authors [39, 40] for Cl, Ce^{3+}, and Fe^{2+}; using related methods, Bessis et al. have obtained results for F[41] and O[42]. We shall discuss some of these results in Section III. Two prominent features emerge from these calculations. First, the values of R are much smaller than the values of γ_∞ and generally fall within a range of ± 0.3. The reduced magnitude is due largely to the fact that the perturbing shell overlaps the closed shells and thus reduces the effectiveness of a distortion $r_<^2/r_>^3$ radial operator. Second, the magnetic and quadrupole $\langle r^{-3}\rangle_{\text{eff}}$ terms are similar (though by no means identical). This similarity is somewhat surprising, since one arises from Coulomb distorting effects, the other from differing aspects of exchange effects. Sternheimer's early results emphasized the similarity in $\langle r^{-3}\rangle_{\text{eff}}$ values; our calculations suggest that differences of as much as 10% can occur. Recent atomic beam studies [43]* of oxygen and fluorine suggest distinct values of $\langle r^{-3}\rangle_{\text{eff}}$ for the magnetic spin and orbital dipolar terms. For oxygen this involved, with certain assumptions, a three-parameter fit to four independent measured hyperfine parameters, while for fluorine the three parameters were obtained from three measured numbers. These are given in Table II. Their estimated error is 2%. It is seen that in both cases the two $\langle r^{-3}\rangle_{\text{eff}}$ values differ by slightly more than 10% with the spin dipolar terms being larger. These results have been the subject of recent theoretical investigations by Bessis et al. [41, 42].

TABLE II

Spin Dipolar and Orbital $\langle r^{-3}\rangle_{\text{eff}}$ Values for Atomic F and O as Obtained by Harvey [43]

	Spin dipolar $\langle r^{-3}\rangle_{\text{eff}}$	Orbital $\langle r^{-3}\rangle_{\text{eff}}$
O	5.19 au	4.58 au
F	8.14 au	7.35 au

One would like to compare experimental magnetic and quadrupolar $\langle r^{-3}\rangle_{\text{eff}}$ values, but this is, in general, difficult, because the standard method of obtaining values of Q is to combine a known nuclear magnetic moment with experimental quadrupole and magnetic hyperfine data and to assume that $\langle r^{-3}\rangle_{\text{eff},q} = \langle r^{-3}\rangle_{\text{eff,mag}}$. The above observations suggest that it is optimistic to assume an accuracy of better than 10% for the resulting values. While many workers at present would be pleased to have the moments known to this great an accuracy, one would like to be more exact in the future. This, in turn, requires a better understanding of $\langle r^{-3}\rangle_{\text{eff}}$ behavior.

II. Experimental Methods

1. Introduction

The conventional methods for observing hyperfine interactions (hyperfine structure of spectral lines, deflection of atomic and molecular

---

* We are grateful to P. G. H. Sandars for informing us of this work.

4. HYPERFINE INTERACTIONS IN MAGNETIC MATERIALS

beams) have been described in the literature. The experimental methods which are important for measuring hyperfine fields in magnetic materials are (1) nuclear magnetic resonance (NMR), (2) electron spin resonance (EPR), (3) nuclear specific heats, (4) recoilless emission and absorption of γ-rays (the Mössbauer effect), (5) angular correlation of γ-rays, and (6) the interaction of polarized neutrons with polarized nuclei. In this section we shall discuss methods (3) to (6) inclusively, with particular emphasis on the Mössbauer effect; two other chapters in this volume (Chapters 5 and 6) are devoted to NMR and one in Volume I of this treatise (Chapter 1) takes up the subject of EPR—a subject which has been treated extensively by Low [44].

2. Nuclear Orientation

Since all of the experimental methods to be referred to are based on the existence of oriented nuclei, we shall give here some of the basic definitions and then cite some of the methods which have been proposed, and used, to produce nuclear orientation [45].

a. *Basic definitions.* A collection of oriented nuclei, each with spin I, has an axis of rotational symmetry η. The different states with magnetic quantum numbers $m = -I, ..., +I$ with respect to the axis η have populations a_m which characterize the orientation of the nuclei. The a_m are $2I + 1$ independent parameters and are normalized, i.e., $\sum_{m=-I}^{+I} a_m = 1$.

One speaks of orientation of nuclei if not all of the a_m are equal. The degree of orientation, of order k, can be characterized by the percentage excess of the expectation value of $\mathbf{M}_k$ (where $\mathbf{M}$ is the component of $\mathbf{I}$ along η) over its isotropic value. For our purposes we need mention only the first two parameters, given by

$$f_1 = \frac{1}{I} \sum_m m a_m \tag{2.1}$$

$$f_2 = \frac{1}{I^2} \left[\sum_m m^2 a_m - \tfrac{1}{3} I(I+1) \right] \tag{2.2}$$

If the positive and negative η directions are not equivalent, then $f_1 \neq 0$ and one speaks of "polarization" of the nuclei. If these directions are equivalent, then all odd $f_k = 0$ and one speaks of "alignment" (here $f_1 = 0$ but $f_2 \neq 0$). In the former case, the nuclei are so oriented that there is a resultant spin along the axis of symmetry, whereas in the latter case there is no resultant spin.

The experiments to be described thus fall into two groups. In one category we have experiments, such as the angular distribution of radiation emitted from nuclei, in which only the alignment of the nuclear moments along an axis is of importance, and *not* the direction in which they point, since this distribution is symmetrical about the plane perpendicular to the dipole. In the other, it is vital that the nuclei should be polarized, i.e., should give rise to a net magnetic moment; an example of this is the dependence of the cross section of nuclei for polarized neutrons as a function of their relative orientation.

b. Methods for achieving nuclear orientation. Nuclear orientation is usually produced by low-temperature methods in which the nuclear spin system is coupled to a fixed spatial direction by electric or magnetic fields. Each substate m thus has a different energy E_m with equilibrium populations given by a Boltzmann function $a_m \alpha \exp(-E_m/kt)$. At temperatures where $(E_m - E_{m+1})/kT$ is small the populations are approximately equal and the nuclear spins are oriented at random; when $E_m - E_{m+1} \approx kT$ then the populations are unequal and there is a resultant degree of nuclear orientation.

External field polarization was suggested by Gorter [46] and Kurti and Simon [47]. In this "brute force" method the direct coupling between a high external magnetic field and the nuclear moment is used to polarize nuclei at very low temperatures. However, for any appreciable polarization the necessary combination of field and temperature is very hard to achieve (e.g., about 20,000 gausses at 0.01°K are needed to get a 20% ($f_1 = 0.2$) polarization of protons).

The method which most concerns us here is the use of magnetic hyperfine splittings in paramagnetic ions to orient nuclei, as proposed by Gorter [48] and Rose [49]. For these ions, $H_{\text{eff}} \approx 10^5$–$10^6$ gausses and a significant degree of polarization can be produced at 0.01°K, if the ions have their own electronic moments aligned. An external field of several hundred gausses is sufficient at these low temperatures. Since then, a number of other suggestions have been made, corresponding to the particular crystalline environment at hand, for producing nuclear orientation by means of hyperfine interactions. Among these [45] are magnetic polarization in antiferromagnets or ferromagnets; magnetic hyperfine alignment [50] (in which no external magnetic field is used), and electric quadrupole alignment [51].

The explanation of these methods is generally given in terms of the familiar phenomenological spin Hamiltonian [50, 3], which describes the various interactions of the nuclear spin, electron spin, and nuclear quadrupole moment of the ion or atom, the crystalline electric field, and an external magnetic field. Explicitly,

$$\mathcal{H} = [-\mu \cdot \mathbf{H}] + [\mu_0\{g_\| H_z S_z + g_\perp(H_x S_x + H_y S_y)\}]$$
$$+ [AS_z I_z + B(S_x I_x + S_y I_y)] + [D\{S_z{}^2 - \tfrac{1}{3}S(S+1)\}]$$
$$+ [P\{I_z{}^2 - \tfrac{1}{3}I(I+1)\}] \tag{2.3}$$

Equation (2.3) has been written assuming axial symmetry about the z-axis. The effective electron spin S is defined by setting the multiplicity of the electronic energy levels equal to $2S + 1$; g is a tensor with pincipal values $g_\|$ and $g_\perp$, the D term represents the effect of the crystalline field; and the P term is the nuclear quadrupole interaction term [see Section I].

The first term gives the "brute force" method; terms 2 and 3 coupled together for sufficiently large H ($\sim$1000 gausses) produce a hyperfine field and a resulting nuclear polarization (Gorter–Rose method).

For Bleaney's method [50] the magnetic hyperfine coupling alone is used ($H = 0$) so that terms 3 and 4 are used. If $S = \tfrac{1}{2}$, there is no D term and the method requires the existence of an anistropy in the hyperfine interaction [$A \neq B$]. For $A \gg B$, the nuclear and electronic moments are aligned along the crystalline symmetry axis; if $B \gg A$, the moments will be normal to this axis. When $S \neq \tfrac{1}{2}$, the D term lifts the electronic orbital degeneracy and at low temperatures the electron spins are aligned with respect to the crystal field symmetry axis. The hyperfine coupling then produces the nuclear alignment (even if $A = B$).

The quadrupole coupling (term 5) of Eq. (2.3) can by itself produce alignment, rather than polarization, in the rare case that P is large enough (as in covalently bonded complexes, e.g., UO_2). In this method, however, the sign of the magnetic moment is not determined, since orientations parallel and antiparallel to the axis (of the homopolar bond) are equally favored.

In ferro- or antiferromagnets the alignment of electronic spins produces large internal magnetic fields and so is a natural and obvious extension of the Gorter–Rose method for producing nuclear polarization.

There are also a number of "dynamic" methods for producing nuclear orientation, all of which have in common the saturation of certain resonances of a system of coupled nuclear and electron spins in a steady external magnetic field.* These include optical pumping, rf and microwave pumping in metals (Overhauser's method), double resonance, etc.* Details of these methods are also given in the references cited.

---

* See the detailed review articles by Ambler [52] and Jeffries [52a].

3. Nuclear Specific Heats

a. *Contribution of nuclear spin orientation to low-temperature specific heats.* It is now well known that the measurement of specific heats of magnetic materials at low temperatures ($T < 1°K$) can be interpreted to yield information about magnetic and electric hyperfine interactions [48]. By this means the large electron-nuclear interactions arising from the unfilled $4f$ shell electrons in the rare earths have been measured. Since other methods, until very recently, have been difficult to carry out, specific heat measurements have been the major source for this information.

Consider a collection of nuclei, each seeing some effective magnetic field H_{eff} [arising from a term in the Hamiltonian given by Eq. (1.1)]. If the nucleus has a magnetic moment μ_I, then the $(2I + 1)$-fold degeneracy in the absence of H_{eff} is removed and the distribution of spin orientations among these Zeeman energy levels is given by the Boltzmann factor. At high temperatures the distribution will be uniform, i.e., random orientation for the nuclear spins, but at low temperatures those spin orientations having lower energy will be preferred. Therefore, this redistribution among the energy levels (in the temperature region in which this can occur) causes a contribution C_N to the specific heat. Since the other contributions to the specific heat (electronic, lattice, spin wave) have different temperature dependences and become very small for $T < 1°K$, one can easily determine C_N (and therefore H_{eff}). One finds C_N is a maximum when $kT \approx \mu_I H_{\text{eff}}/I$; and at sufficiently high temperatures, i.e., the tail of the anomaly, where the Boltzmann exponential can be expanded in a power series, one obtains a general power series expansion

$$\frac{C_N}{R} = \sum_{n=2}^{\infty} c_n (kT)^{-n}. \tag{2.4}$$

Bleaney [53] and Bleaney and Hill [54] have recently discussed the various terms in this expansion, including effects due to quadrupole coupling, which are quite often important for the rare earths. The quadrupole interaction is represented by an additional contribution to the Hamiltonian (see also Section II, 2):

$$\mathcal{H}_{EQ} = P[I_z^2 - \tfrac{1}{3}I(I+1)] + P'(I_x^2 - I_y^2) \tag{2.5}$$

where P varies as $J_z^2 - \tfrac{1}{3}J(J+1)$ (and gives a measure of the electronic quadrupole distribution) and P' varies as $\tfrac{1}{3}(J_x^2 - J_y^2)$ (and is zero when the electric field gradient lies along an axis of symmetry). With these definitions [53],

$$c_2 = \frac{(I+1)}{3I}(\mu_n H_{\text{eff}})^2 + \frac{1}{45}[P^2 + 3(P')^2]I(I+1)(2I-1)(2I+3) \quad (2.6)$$

$$c_3 = -\frac{1}{15}P\frac{(I+1)}{I}(2I-1)(2I+3)(\mu_n H_{\text{eff}})^2 \quad (2.7)$$

and

$$c_4 = -\frac{1}{30}\frac{(I+1)}{I^3}(2I^2+2I+1)(\mu_n H_{\text{eff}})^4. \quad (2.8)$$

A term in P^3 has been omitted in c_3 and terms in $P^2(\mu_n H_{\text{eff}})^2$ and P^4 have been omitted in c_4.

Marshall [55] has discussed the specific heat anomaly for ferromagnets, using the effective-field concept and the simple Heisenberg model with spin S on each atom. Whereas in the usual treatment in terms of the spin Hamiltonian for paramagnetic salts

$$H = AS_zI_z + B(S_xI_x + S_yI_y) \quad (2.9)$$

and [56]

$$C_N/R = S(S+1)I(I+1)(A^2 + 2B^2)/9k^2T^2 \quad (2.10)$$

Marshall finds that for zero applied magnetic field the nuclear specific heat for ferromagnets is

$$C_N/R = I(I+1)A^2S^2/3kT^2 \quad (2.11)$$

[where S_z in Eq. (2.6) is replaced by S, and S_x and S_y by zero because of the alignment of the electronic spins by the strong exchange interaction].

If, as in the case of holmium metal, the hyperfine interaction is very strong, then the exact formula

$$\frac{C_N}{R} = \frac{\sum_{i=-I}^{I}\sum_{j=-I}^{I}(E_i^2 - E_iE_j)\exp[(-E_i - E_j)/kT]}{(kT)^2\sum_{i=-I}^{I}\sum_{j=-I}^{I}\exp[(-E_i - E_j)/kT]} \quad (2.12)$$

must be used instead of the series expansion. Equation (2.12) is obtained from the partition function in the usual way; E_i are the energies of the various nuclear spin states, i.e., eigenvalues of a Hamiltonian which includes the magnetic and electric quadrupole hyperfine interactions.

b. **Experimental results.** Bleaney [56] analyzed measurements of nuclear specific heats for a number of copper, manganese, and cobalt salts [57] using Eq. (2.7) and the second term of Eq. (2.3). He determined A, B, and P from paramagnetic experimental data on C_N then available but did not include any Sternheimer corrections.

Stimulated by measurements of the anisotropy of γ-ray emission (see Section II, 5), Heer and Erickson suggested [58] that the hyperfine coupling in transition metals be found by measurements of low-temperature specific heats. These were carried out by Heer and Erickson [59] for Co metal (and some cobalt salts) and by Arp et al. [60] for Co metal (and a Co-Ni alloy) and gave a field in Co metal of 183 kgausses and 200 kgausses, respectively (the nuclear orientation experiment gave 193 kgausses for the metal). Figure 2 presents Heer and Erickson's results and shows also the separate contributions, i.e., lattice, electronic, and nuclear, to the observed specific heat. Later, NMR measurements showed that these experiments underestimated the actual field. More recent measurements were reported by Arp et al. [61] for pure Co and

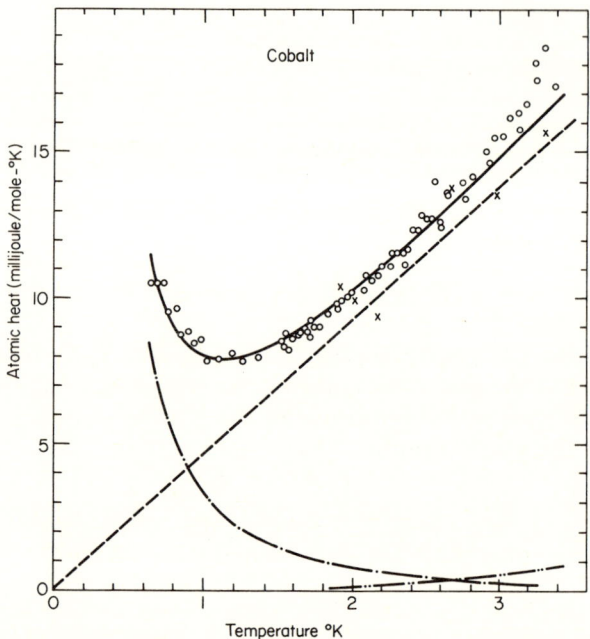

FIG. 2. Specific heat cobalt metal. The solid line (—) is the sum of the lattice specific heat (—··—), the electronic specific heat (— — —), and the nuclear specific heat (—·—·—). (After Heer and Erickson [59].)

some Co-Ni and Co-Fe alloys, and by Beck *et al.** [62–65] for alloys of the transition series. Some of these results are given in Table III.

In the rare earths, the unfilled shell of $4f$ electrons produces at the nucleus a magnetic field of several million gausses. At temperatures below 1°K, the Schottky anomaly arising from the interaction of this field with the nuclear magnetic moment can easily be observed because it is the dominant contribution to the heat capacity. For the rare earth metals, these data are the main source of information about hyperfine interactions. Since the original measurements by Kurti and Safrata [66] on Gd and Tb, a large number of nuclear specific heats have been determined for the rare metals [67–69]. The separation of the nuclear term from the other contributions (electronic, magnetic, and lattice) varies in accuracy from case to case, requiring accurate data over a wide range of temperature; the least reliable term, in general, is the magnetic specific heat as this requires a separate assumption (or belief

TABLE III

Hyperfine Fields, H_{eff} in kGausses for Various Alloys as Determined by Specific Heat Measurements

Nucleus	Host	H_{eff} (in kgausser)	Ref.
Co	Co	219 ± 4	[61, 63]
	$Fe_{0.90}Co_{0.10}$	395	[63]
	$Fe_{0.86}Co_{0.14}$	367	[63]
	$Fe_{0.26}Co_{0.24}$	308	[63]
	$Fe_{0.70}Co_{0.30}$	312	[63, 64]
	$Fe_{0.66}Co_{0.34}$	292	[63]
	$Fe_{0.62}Co_{0.38}$	306	[63]
	$Fe_{0.49}Co_{0.51}$	297	[63]
	$Fe_{0.40}Co_{0.60}$	270	[63]
	$Fe_{0.31}Co_{0.69}$	309	[63]
	$Fe_{0.28}Co_{0.72}$	238	[63]
	$Fe_{0.25}Co_{0.75}$	227	[63]
	$Fe_{0.07}Co_{0.93}$	236	[63]
Fe	$Co_{0.915}Fe_{0.085}$	223 ± 4	[61]
	$Co_{0.587}Fe_{0.413}$	25 ± 3	[61]
	$Co_{0.172}Fe_{0.828}$	293 ± 10	[61]
	$Co_{0.048}Fe_{0.952}$	314 ± 9	[61]
	$Fe_{0.90}Re_{0.10}$	610 ± 35	[65]
	$Fe_{0.946}Sb_{0.054}$	169 ± 8	[65]

* A large program for measuring specific heats of transition metal alloys is being carried out by Beck and his collaborators. It cannot be discussed in detail here. See Beck [62] for a review of the work on the electronic specific heats of some of these alloys.

in theory) regarding its temperature dependence. (For example, in samarium metal, antiferromagnetic below 14°K, it is assumed that spin wave theory may be used; since this gives a T^3 dependence for the magnetic contribution, its separation from the lattice contribution, which is also given by a T^3 term, is not possible.) However, if measurements can be done at low enough temperatures, the separation of C_N from the remainder can be done accurately. A far more important limit on

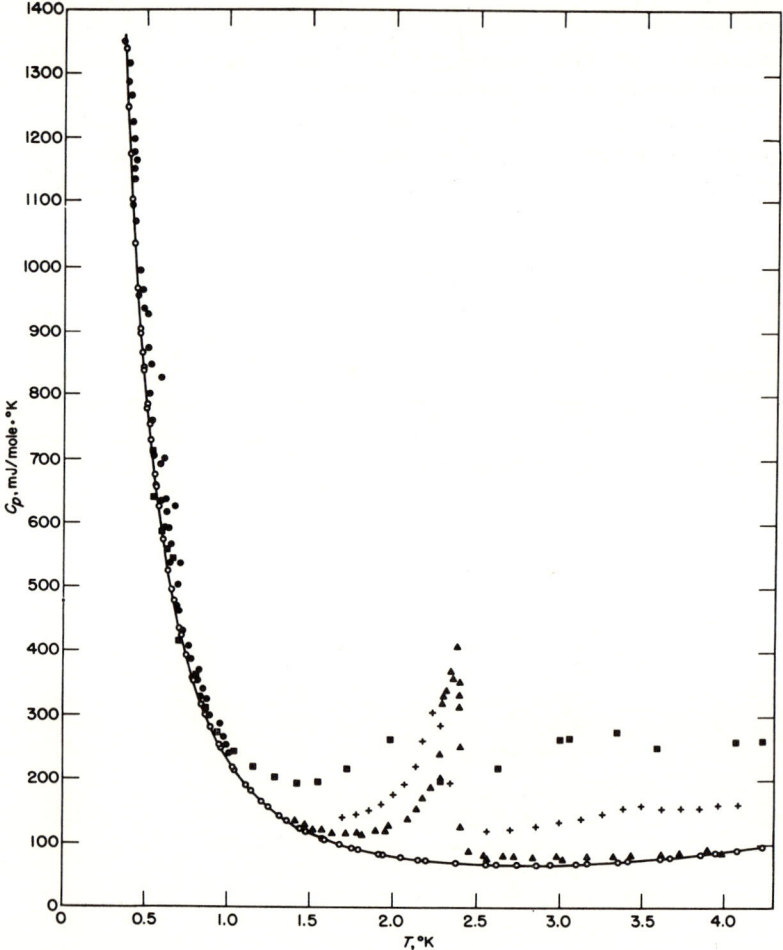

FIG. 3. The specific heat of terbium metal: ○, Lounasmaa and Roach [76]; □, Kurti and Safrata [66]; ▲, Stanton *et al.* [81a]; ●, Heltemes and Swenson [70]; +, Bailey [81b].

experimental accuracy is the effect of impurities (mostly oxides); these are believed to be the major source of differences between the results of various investigators.

From the available measurements of C_N for the rare earth metals we list in Table IV some recent data obtained by Lounasmaa and collaborators [74-80],* since these represent results obtained with samples of presumably higher purity than were previously available. The magnetic hyperfine constant $a(=\mu H_{\text{eff}}/kT)$, the quadrupole coupling constant P, the coefficient D in the T^2 term in C_N, and H_{eff} (based on an assumed nuclear moment, which is also listed) are given for Sm, Gd, Tb, Dy, Ho, and Yb metals. Also listed are values of D, a, and P calculated by Bleaney [81] for trivalent rare earth ions from electron paramagnetic resonance data in dilute salts. Except for Yb, agreement with specific heat (and NMR and Mössbauer) measurements on metals (and garnets) was found to be within experimental errors. Figures 3 and 4 illustrate some typical observations. Figure 3 summarizes various data for Tb metal [76]; the agreement below 1°K is seen to be very good. The anomaly between 2° and 3°K had been attributed to oxygen impurities† and this has recently been verified by Gerstein et al. [82], who measured the magnetic susceptibility of Tb_2O_3 and found a maximum at 2.42°K. A similar anomaly is seen in Fig. 4 for Dy metal‡ and here, too, this may arise from

TABLE IV

Comparison of Specific Heat Measurements of Lounasmaa (L) and Collaborators [74-80] with Data of [53] Bleaney (B) Compiled from NMR and EPR Experiments

		Sm	Gd	Tb	Dy	Ho	Yb
D (millijoule °K/mole)	L	8.6	—	238	26.4	4480	0.012
	B	8.9	0.02	250	26.6	4200	16[a]
a' (°K)	L	0.026	—	0.150	0.048	0.320	0.002
	B	0.027	0.002	0.153	0.048	0.312	0.09 [a]
P (°K)	L	—	—	0.021	0.008	0.007	—
	B	<0.001	—	0.023	0.008	0.001	—
H_{eff} [b]		3.3	—	4.1	7.1	9.3	0.14

[a] Based on measurements with trivalent Yb.
[b] For nuclear moments see references [53, 74-80].

* Reference [79] contains a convenient summary of the data for the rare earth metals.
† See reference [76] for a discussion of the effect of impurities.
‡ See the compilation of Lounasmaa [75] and references therein.

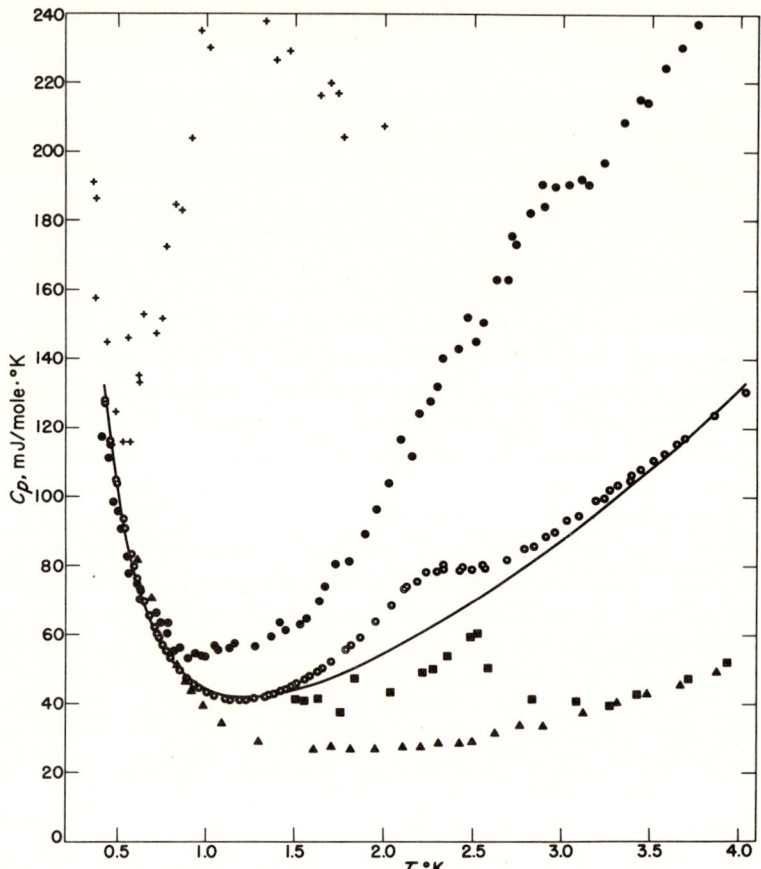

FIG. 4. The specific heat of dysprosium metal: ○, results of Lounasmaa and Guenther [75]; +, Dash et al. [67]; ●, Parks [71]; ▲, Dreyfus et al. [69], sample A; ■ Dreyfus et al. [69], sample B.

impurities. The large differences seen above 1°K are probably due to differences in the magnetic ($T^{3/2}$) specific heat [75].

For Dy there is evidence for a quadrupole interaction and for Tb such a term is necessary to explain the experimental data. An exceptionally strong magnetic hyperfine interaction was found for Ho; as stated earlier, for this case the exact formula for C_N is needed to represent the Schottky anomaly. For Yb metal, the $4f$ shell is apparently full; hence a very small C_N is expected and observed. Similarly, for Gd metal one expects a small C_N because the $4f$ shell forms a half-closed shell (^8S state). We shall return to these matters later on (cf. Section IV). The anomalous humps observed for Gd are tentatively attributed to the magnetic ordering of Gd^{3+} ions in the Gd_2O_3 impurity [77]. As an instructive comparison Fig. 5 shows the relative specific heat data for all these metals [79, 80].* Lounasmaa [83] has recently measured the specific heat of Ce at low temperatures. However, because of the large magnetic contributions no quadrupole interactions were detected for this case.

In none of the above comparisons between theory and experiment has the Sternheimer antishielding correction [cf. Section I, 2] been made. Lounasmaa [84] has just studied the specific heat of luteticium metal between 0.4° and 4°K. Comparing his data with electric field gradients calculated by de Wette [85] he finds an antishielding factor (γ_∞) of about -140 which appears to be in reasonable agreement with γ_∞ calculated theoretically [40, 18, 18a, b].

In addition to both supplementing and complementing the information gotten from other measurements, specific heat determinations of H_{eff} are valuable because they also provide a reliable estimate of the position

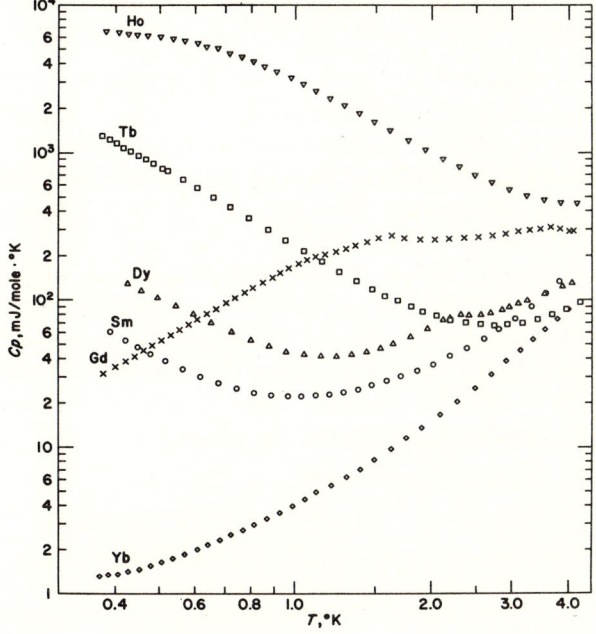

FIG. 5. Specific heats of Sm, Gd, Tb, Dy, Ho, and Yb metals. (After Lounasmaa [79, 80].)

* Reference [79] contains a convenient summary of the data for the rare earth metals.

of the resonant frequency for, say, NMR experiments, which can then give much more accurate values of H_{eff}. This has already occurred a number of times [i.e., Co and Tb], making otherwise hopelessly difficult searches for resonance fairly simple to achieve.

4. Recoilless Emission and Absorption of γ-rays; the Mössbauer Effect

The discovery by R. L. Mossbauer [86, 87] of the recoilless emission and absorption of γ-rays has brought with it a diversity of unusual applications in the fields of relativity, nuclear physics, and solid state physics [88a],[*] including the determination of a number of physical properties which were formerly thought unmeasurable[†] [90]. While the effect originally belonged to nuclear physics, solid state applications are now dominant and will very likely remain so. Of the very large number of such papers, the majority have been concerned with investigations of hyperfine fields. For the study of hyperfine fields in magnetic materials the Mössbauer effect has proved to be a powerful tool augmenting the other experimental methods and extending the information so obtained to many materials by means of relatively easy measurements.

a. *Resonance fluorescence.* Mössbauer showed that nuclei that are embedded in solids can emit and absorb low-energy γ-rays which have the natural line width, i.e., no recoil energy is transferred to the lattice. This process of emission of radiation by one system and the absorption and re-emission by another identical one is well known for optical wavelengths as resonance fluorescence.

Consider a system, such as a free atom or nucleus, of mass M in an excited energy level which can decay to the ground state by the emission of a photon. If the system (and we shall henceforth consider it to be a nucleus) is initially at rest, then after the emission of the photon of momentum $\mathbf{p} = \hbar\mathbf{k}$ the nucleus must have a recoil momentum $\mathbf{p} = -\hbar\mathbf{k}$ and hence a recoil energy E_R which is given by

$$E_R = \frac{\hbar^2 k^2}{2M} = \frac{E_\gamma^2}{2Mc^2} \qquad (2.13)$$

where E_γ is the energy of the γ-ray. Conservation of energy requires

$$E_\gamma = E_0 - E_R \qquad (2.14)$$

---

[*] For a review with extensive references, and a collection of early reprints, see reference [88].

[†] Among these are a terrestial measurement of the gravitational red shift [89], the observation of the Zeeman splitting of excited nuclear levels [89a], and a test of the equivalence principle for rotating systems [89b].

where E_0 is the nuclear excitation energy. Since $E_R \ll E_\gamma$ it is a good approximation to write $E_R \approx E_0^2/2Mc^2$ and to replace Eq. (2.14) by

$$E_\gamma = E_0 - E_R = E_0 - E_0^2/2Mc^2 \tag{2.15}$$

Hence the emitting nucleus takes up a recoil energy E_R at the expense of the emitted γ-ray. Quite similarly, in absorption, the incident photon must have an energy given by $E_0 + E_R$. There therefore exists an energy deficit $2E_R$ between the emitted and absorbed photons. Resonance absorption can occur only if $2E_R$ is not larger than the width Γ of the excited nuclear state, i.e., only if there is appreciable overlap of the emission and absorption lines ($\Gamma/2E_R \geqslant 1$). This is shown in Fig. 6.

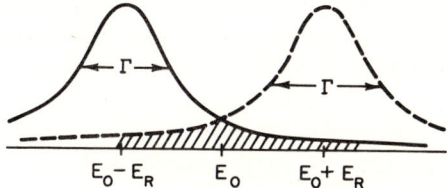

FIG. 6. Overlap of emission and absorption lines showing the condition for resonance absorption.

The width Γ is the natural line width of the emitted photon, and is given by the Heisenberg uncertainty relation as $\Gamma\tau = \hbar$, where τ is the lifetime of the excited state. For nuclear systems Γ/E_R has values of 10^{-5} or smaller, whereas for atomic systems Γ/E_R is about 10^3. Thus, while the recoil energy plays an insignificant role in atomic spectra, it is of great importance for γ-ray transitions.

One other feature must be considered. If the nucleus is not initially at rest, then the energy of the emitted photon is shifted, because of the Doppler effect, by an amount equal to $(v/c)E_0$, where v is the velocity of the nucleus. In absorption the energy of the nucleus is similarly shifted. This produces an additional broadening of the lines and a width Δ which is much greater than Γ and affects the resonance fluorescence in two ways: (1) It increases the resonance fluorescence by increasing the overlap between emission and absorption lines; (2) it decreases the maximum absorption cross section (see discussion and references in [88]). Again, for nuclear transitions the recoil energy is comparable to or greater than the Doppler broadening and the maximum absorption cross section is thus small.*

---

* See reference [88] and Heitler [91] for a discussion of cross sections and their role in resonance fluorescense.

Techniques have been developed for observing resonance fluorescence, e.g., by compensating for the recoil shift (by use of a rotor) or increasing the temperature of source and absorber (and thus the Doppler width and hence the overlap between emission and absorption lines). These methods are described in the literature [92].

R.L. Mössbauer discovered the effect which bears his name while working on the nuclear resonance scattering of the 129 kev γ-ray from Ir^{191}. The recoil energy for this transition is 0.05 ev with a Doppler broadening at room temperature of about 0.1 ev. The overlap of the emission and absorption lines is therefore appreciable and, to decrease the resonance scattering, Mössbauer cooled the absorber and the source. Instead of observing a decrease, he found that the resonance fluorescence increased. To explain the effect, Mössbauer adapted an earlier theory by Lamp [93] for the resonance absorption of neutrons in solids to the case of γ-rays. What he had found was that for a certain fraction of the γ-rays emitted from the cooled source the transfer of recoil energy is prevented. These γ-rays also do not show a Doppler broadening but have a natural line width. In other words, at low temperatures, nuclei of not too high an excitation energy can be tightly bound into a solid so that the recoil momentum cannot go into the nucleus but goes instead into the whole crystal. In the ideal case of infinite binding, the energy of the emitted and absorbed γ-rays must be the nuclear excitation energy, $E_\gamma = E_0$. Saying that nuclear recoil is absent does not imply a violation of the law of conservation of momentum; the whole crystal recoils as a unit. Hence, as a result of the large mass of the solid, the energy of recoil is very small indeed.

A real crystal, of course, does not provide infinite binding to nuclei; it possesses internal degrees of freedom which may be excited by the emission or absorption process. The nuclear excitation energy is shared between the γ-ray and the lattice vibrations (phonons) with a resulting loss of intensity of Mössbauer (or recoilless) transitions (provided the recoil energy is small enough so that the atom is not removed from its lattice site). The fraction f of γ-rays emitted or absorbed without recoil is given by the Debye–Waller factor, $f = e^{-2w}$, well known from x-ray elastic scattering measurements. For a Debye model of the lattice vibrations,

$$f = \exp\left[-\frac{3}{2}\frac{E_R}{k\theta}\left\{1 + 2(T/\theta)^2 \int_0^{\theta/T} \frac{x\,dx}{e^x - 1}\right\}\right] \qquad (2.16)$$

where θ is the Debye temperature. Hence, for $T \approx \theta$, f decreases rapidly with T, whereas f is large when $E_R \ll k\theta$ and $T \ll \theta$. The theory of the Mössbauer effect has been worked out by Lamb [93], Mössbauer

[86, 87], Visscher [94], Singvi and Sjölander [95], and others [88], but will not be discussed here.

b. **Method of measurement.** Experiments using the Mössbauer effect can be done quite simply and relatively inexpensively. As is usual in resonance fluorescence, a part of the incoming radiation is absorbed and then re-emitted in all directions. The Mössbauer effect can therefore be observed either by an increase in scattered radiation or by a decrease in the transmitted intensity. Most experiments are done by the latter method, the detection of the effect being based on the destruction of the resonance condition between source and absorber. This is done most easily by Doppler shifting the source relative to the absorber, e.g., by moving the source with a velocity with respect to the absorber, to produce an energy shift greater than the natural line width. Typical velocities are of the order of a fraction of a centimeter per second.

The basic experimental apparatus for transmission experiments consists of a source attached to some sort of device for producing the velocity required, an absorber, and a counter (placed beyond the absorber, to measure the reduced intensity [88]). The transmitted intensity is plotted as a function of velocity; at large velocity no resonance absorption occurs. A typical spectrum is shown in Fig. 7. Since the change ΔE

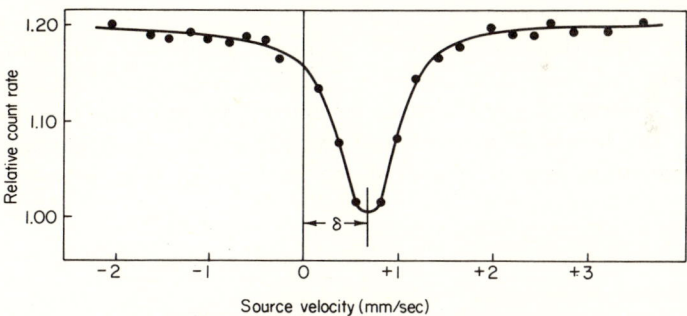

FIG. 7. Typical Mössbauer absorption spectrum. The shift δ is a measure of the isomer shift.

in the γ-ray's energy is $\Delta E = (v/c)E_\gamma$, the spectrum can be plotted as a function of ΔE or v. (Details of experimental apparatus, associated problems, relevant information about isotopes as sources and absorbers, etc., are given in reference [88].)

c. **Nuclear properties and hyperfine effects.** Although Mössbauer's discovery was made with Ir^{191}, the great interest in the effect stems from the existence of many other Mössbauer nuclides and the corresponding variety of experiments which can be performed. Fortunately for the study of magnetism, one of the best Mössbauer nuclides is Fe^{57}, which permits important studies to be made in its natural environment or otherwise. In addition, a good deal of work has also been done with rare earth nuclides and with Au^{197} and Sn^{119}.

We shall illustrate the study of hyperfine interactions by a brief description of the nuclear properties of Fe^{57} in the presence of electric and magnetic fields. The decay scheme of Co^{57} and Fe^{57} is shown in Fig. 8. Cobalt-57 is converted by electron capture of Fe^{57} in its 136 kev excited state, which then decays by emission of γ-radiation to the first excited state (91%) and the stable ground state (9%). The first excited state (at 14.4 kev) has half-life of 1.4×10^{-7} sec and is an ideal Mössbauer nuclide, as is evident from the following data:

Natural line width, $\Gamma = 4.7 \times 10^{-9}$ ev

Shift due to recoil, $2E_R = 3.9 \times 10^{-3}$ ev

Doppler velocity corresponding to natural width,
$$v_\Gamma = (\Gamma/E_0)c = 0.095 \text{ cm/sec}$$

Debye temperature, $\theta \approx 420°$ to $470°K$

Recoilless fraction (with $\theta = 420°K$, $f(0°K) = 0.92$; $f(300°K) = 0.79$

Even at room temperatures, f is still quite large because of the high θ and low E_R, and this obviates the general need for low-temperature equipment.

In the absence of any magnetic or electric fields at the position of the nucleus, the levels of ground and excited states are degenerate and a single transition line is observed (as in Fig. 7). The interaction of a

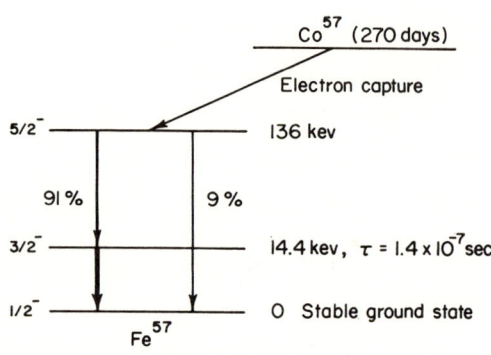

FIG. 8. Decay scheme of Co^{57} and Fe^{57}.

magnetic field H_{eff} with the magnetic moments of the ground state μ_0 and excited state μ_1 causes a Zeeman splitting of the nuclear levels (as in Fig. 9a). There will then be six possible transitions governed by the selection rules for dipole radiation and the Mössbauer transition spectrum observed will be as in Fig. 10 (which is the transmission

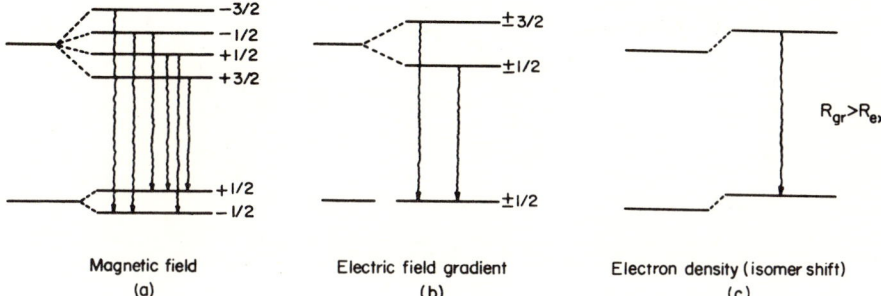

FIG. 9. Hyperfine splittings of nuclear levels in Fe^{57} showing the expected spectrum: (a) in a magnetic field; (b) under influence of an electric gradient; (c) varying electron density (isomer shift).

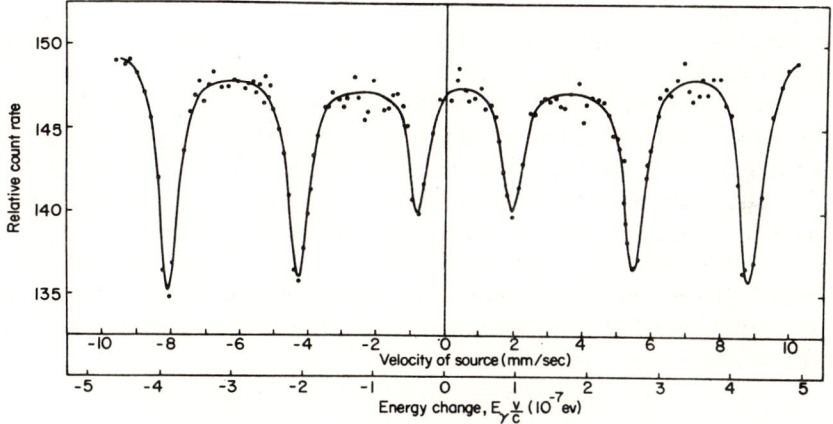

FIG. 10. Six-line transmission spectrum for Co^{57} in ferromagnetic iron (cf. Fig. 9a).

pattern for Co^{57} in ferromagnetic iron as source and for an absorber with $H_{\text{eff}} = 0$ at its nuclei). Similarly, for the case of a non-zero electric field gradient acting on the nucleus, there will be an electric quadrupole interaction with the quadrupole moment Q of the nucleus (in the case at hand, only the excited state has $Q \neq 0$), and again a splitting of the nuclear levels as shown in Fig. 9b. The observed transmission spectrum for this case (with $H_{\text{eff}} = 0$) is a simple two-line pattern. It should be said here that the Mössbauer effect presents some unique information about the nuclear quadrupole interaction, because conventional methods fail since there are no stable isotopes of iron with a ground state spin $>\frac{1}{2}$. A typical Mössbauer spectrum for this case is shown in Fig. 11.

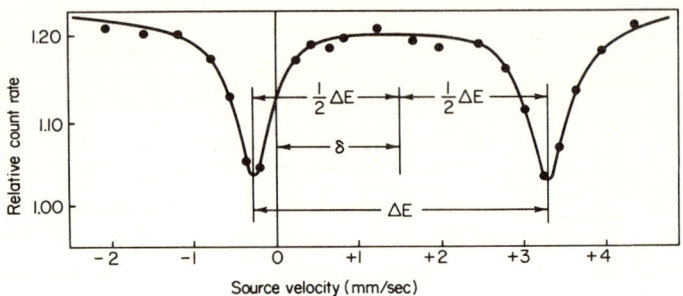

Fig. 11. Mössbauer spectrum showing the splitting due to an electric field gradient (ΔE) and the isomer shift ($\grave{e}$).

When one has combined magnetic and electric interactions the observed spectrum is a superposition of the combined magnetic and electric effects. Absorption of the "unsplit" emission line results in the usual six possible absorption energies, but these show an additional energy shift, positive or negative, of the individual excited states.

There is one additional interaction, called the isomer (or chemical) shift, which may be observed by means of the Mössbauer effect [96–99]. This is a nuclear volume effect, resulting from the fact that the ground and excited states of the nucleus have different effective charge radii. The electrostatic interaction due to the overlap of electronic and nuclear charge is therefore different for the two levels, and the γ-ray energy is changed. If the density of s electrons at the nucleus is different for the source and absorber, the difference in γ-ray energies is defined as the isomer shift,

$$\delta = \tfrac{2}{3}\pi Z e^2 [R_{\text{ex}}^2 - R_{\text{gr}}^2][|\psi(0)|^2 A - |\psi(0)|^2 E] \qquad (2.17)$$

Here Z is the nuclear charge, R_{ex}^2 and R_{gr}^2 are the average of the square of the charge radii of the excited and ground states, and $|\psi(0)|^2 A$ and $|\psi(0)|^2 E|$ are the total s electron densities for absorber and emitter. The effect of δ on a Mössbauer spectrum is shown in Figs. 7 and 11.* It should be

---

* The observed isomer shift actually consists of several additional parts: (a) a second-order Doppler shift which may exist between source and absorber; and (b) the difference in zero-point energy between source and absorber. These parts are, however, usually small.

emphasized that the measurement of δ gives some rather unique information about electron densities obtainable only in this way.

The above was a brief description of hyperfine interactions and their effect on Mössbauer measurements. A good deal of such experimental data are now available; some of these will be presented below and their interpretation will be discussed in Section IV.

d. *Experimental results.* By far the largest application of the Mössbauer effect has been to the study of hyperfine interactions. Of these, the magnetic hyperfine interaction has been most intensively investigated. Although NMR may be used to obtain some of the same information as obtained with the Mössbauer effect, NMR requires the use of rf fields, which limits its applicability (e.g., in superconductors and in single metal crystals), depends on domain-wall enhancement in ferromagnets (in order to have a high enough signal), and is too sensitive a probe for many applications (e.g., for studying the distribution of magnetization about localized magnetic moments, or when quadrupole interactions are such as to strongly broaden the lines). Because of these factors, and others, Mössbauer measurements will continue to be a useful tool for solid state studies.

(1) *Hyperfine fields at nuclei in magnetic materials.* The Mössbauer effect has already been used to determine magnetic hyperfine fields in a wide variety of magnetic materials: ferromagnets, antiferromagnets, and, most recently, paramagnetic solids (using external magnetic fields) have been studied. The dependence of hyperfine fields on such parameters as crystalline environment, pressure, temperature, and external fields have been investigated. Most of these experiments have been done with Fe^{57} or with Sn^{119} as probe. The number and diversity of published papers on these subjects are so vast that only a brief summary can be considered here [88, 90, 100].

Table V summarizes some results obtained to date [101-157].* The striking features of this table are that (1) the hyperfine fields are large, 200 to 1000 kgausses for iron series nuclides and several million gausses for rare earth nuclides, and (2) for transition metal ferromagnets (and some antiferromagnets) the fields are negative (i.e., antiparallel to the direction of magnetization) [101, 102]. The interpretation given to these results will be discussed in Section IV.

Clearly, the papers referred to in Table V contain a good deal of

---

* Roberts and Thompson [134] use $\mu = 0.1439$ for the ground state nuclear moment.

TABLE V

Magnetic Hyperfine Fields, H_{eff}, for Nuclei in Various Host Materials as Obtained from Mössbauer Measurements

Nucleus	Host	T (°K)	H_{eff} (kgausser)	Ref.
$^{57}\text{Fe}^{2+}$(II)	Fe_3O_4 [a]	30, 85, 300	470 ± 20	[108, 140, 141, 145]
$^{57}\text{Fe}^{3+}$(II)	Fe_3O_4 [a]	300	470 ± 20	[108, 140, 141, 145]
		85	510 ± 20	[108, 140, 141, 145]
		30	510 ± 20	[108, 140, 141, 145]
$^{57}\text{Fe}^{3+}$(II)	Fe_3O_4 [a]	300	495 ± 20	[108, 140, 141, 145]
		85	510 ± 20	[108, 140, 141, 145]
		30	510 ± 20	[108, 140, 141, 145]
^{57}Fe	Fe	0	−342	
		Room	330	[101, 102, 139]
^{57}Fe	Co	0	312	[137]
		Room	−310	[105, 126]
	$\text{Fe}_{0.3}\text{Co}_{0.7}$	0	365	[103]
	$\text{Co}_{0.03}\text{Pd}_{0.97}$	4	−310 ± 9	[105]
	$\text{Co}_{0.08}\text{Pd}_{0.92}$	4	−315 ± 4	[105]
		88	330	[104]
	$\text{Co}_{0.15}\text{Pd}_{0.85}$	4	−296 ± 8	[105]
	$\text{Co}_{0.30}\text{Pd}_{0.70}$	80	−305 ± 10	[105]
^{57}Fe	Ni	0	283	[125, 137]
		300	265	[126, 137]
	$\text{Cu}_{0.05}\text{Ni}_{0.95}$	0	276 ± 3	[125]
	$\text{Cu}_{0.1}\text{Ni}_{0.9}$	0	280 ± 3	[125]
	$\text{Cu}_{0.2}\text{Ni}_{0.8}$	0	273 ± 3	[125]
	$\text{Cu}_{0.3}\text{Ni}_{07}$	0	266 ± 3	[125]
	$\text{Cu}_{0.45}\text{Ni}_{0.55}$	0	246 ± 4	[125]
^{57}Fe (corner)	$(\text{Fe}_{3.6}\text{Ni}_{0.4})\text{N}$	300	363 ± 15	[127]
	Fe_4N	300	345 ± 10	[127]
^{57}Fe (face center)	$(\text{Fe}_{3.6}\text{Ni}_{0.4})\text{N}$	300	220 ± 15	[127]
	Fe_4N	300	215 ± 10	[127]
	$(\text{Fe}_3\text{Ni})\text{N}$	300	205 ± 15	[127]
$^{57}\text{Fe}^{3+}$	NiFe_2O_4	Room	510 ± 20	[108]
$^{57}\text{Fe}^{3+}$	$(\text{Li}_{0.5}\text{Fe}_{2.5})\text{O}_4$ (ordered)	Room	508 ± 20	[110]
	$(\text{Li}_{0.5}\text{Fe}_{2.5})\text{O}_4$ (disordered)	Room	510 ± 20	[110]
$^{57}\text{Fe}^{2+}$	$(\text{Fe}_{0.77}\text{Al}_{0.23})_2\text{O}_3$	Room	495	[109]
		0	520	[109]
	$(\text{Fe}_{0.5}\text{Cr}_{0.5})_2\text{O}_3$	Room	435	[109]
		0	540	[109]
	$(\text{Fe}_{0.5}\text{V}_{0.5})_2\text{O}_3$	Room	430	[109]
		0	540	[109]
$^{57}\text{Fe}^{3+}$	FeTiO_3	300	520	[144]
$^{57}\text{Fe}^{2+}$	FeTiO_3	20	70	[144]

TABLE V *(continued)*

Nucleus	Host	T (°K)	H_{eff} (kgausser)	Ref.
$^{57}\text{Fe}^{3+}$	$\alpha\text{-Fe}_2\text{O}_3$	Room	520	[96, 109, 119, 141]
		78	525 ± 5	[119]
		0	535	[109]
	$\gamma\text{-Fe}_2\text{O}_3$	Room	496 ± 20	[108, 110]
		85	515 ± 20	[108]
$^{57}\text{Fe}^{2+}$	CoO	298	0	[119]
		243	169 ± 3	[119]
		169	200 ± 3	[119]
		78	180 ± 10	[119]
$^{57}\text{Fe}^{3+}$	CoO	298	0	[119]
		243	426 ± 5	[119]
		169	544 ± 6	[119]
		78	557 ± 6	[119]
$^{57}\text{Fe}^{2+}$	FeF_2	0	340	[111, 137]
$^{57}\text{Fe}^{3+}$	FeF_3	4	622 ± 6	[123]
$^{57}\text{Fe(I)}$	Fe_3Al (ordered)	300	229 ± 10	[135, 149]
		30	246 ± 10	[135]
		4.2	242	[149]
$^{57}\text{Fe(II)}$	Fe_3Al (ordered)	300	299 ± 10	[135, 149]
		30	336 ± 10	[135]
		4.2	334	[149]
^{57}Fe	$\text{Fe}_{1-x}\text{Si}_x$ alloys ($0.14 \leqslant x \leqslant 0.27$)	Room	Function of x	[138]
$^{57}\text{Fe(I)}$	Fe_3Si	85	320	[143]
		Room	305 ± 10	[143]
$^{57}\text{Fe(II)}$	Fe_3Si	85	205	[143]
		Room	195 ± 10	[143]
$^{57}\text{Fe(I)}$	Fe_5Si_3	85	230	[143]
$^{57}\text{Fe(II)}$	Fe_5Si_3	85	130	[143]
^{57}Fe	FeSn	77	160	[122]
	Fe_2Ti	Room	5 ± 3	[121]
	Fe_2Ti	78	92	[148]
	Fe_2Zr	Room	190 ± 10	[121, 136, 148]
	γ-FeMn	Low temp.	20 − 40	[142]
^{57}Fe	$\text{Fe}_{0.5}\text{Rh}_{0.5}$	Room	267	[146]
$^{57}\text{Fe(I)}$	$\text{Fe}_{0.52}\text{Rh}_{0.48}$	Room	277	[147]
$^{57}\text{Fe(II)}$	$\text{Fe}_{0.52}\text{Rh}_{0.48}$	Room	384	[147]
^{57}Fe	$\text{Zr}_{0.9}\text{Ti}_{0.1}\text{Fe}_2$	300	194	[148]
	$\text{Zr}_{0.5}\text{Hf}_{0.5}\text{Fe}_2$	300	191	[148]
^{57}Fe	$\text{Zr}_{0.8}\text{U}_{0.2}\text{Fe}_2$	85	185 ± 4	[136]
	$\text{Zr}_{0.6}\text{U}_{0.4}\text{Fe}_2$	85	17 ± 4	[136]
	$\text{Zr}_{0.4}\text{U}_{0.6}\text{Fe}_2$	85	85 ± 4	[136]
	$\text{Zr}_{0.2}\text{U}_{0.8}\text{Fe}_2$	85	49 ± 4	[136]
	UFe_2	85	37 ± 4	[136]
$^{57}\text{Fe}^{3+}$ (tet)	YIG	300	392 ± 5	[106, 137]
		85	460 ± 15	[107]

TABLE V (continued)

Nucleus	Host	T (°K)	H_{eff} (kgausser)	Ref.
$^{57}Fe^{3+}$ (oct)	YIG	300	474 ± 7	[106, 137]
		300	485 ± 15	[107]
		85	535 ± 15	[107]
$^{57}Fe^{3+}$ (tet)	DyIG	300	395 ± 15	[107]
		85	460 ± 20	[107]
$^{57}Fe^{3+}$ (oct)	DyIG	300	485 ± 20	[107]
		85	540 ± 20	
^{57}Fe	$SmFe_2$, $DyFe_2$, $HoFe_2$, $ErFe_2$, $TmFe_2$	78	230 ± 5	[124]
	$CeFe_2$	78	212 ± 4	[124, 148]
			312 ± 4	[124, 148]
^{57}Fe	YFe_2	300	188	[148]
	$Y_xHo_{1-x}Fe_2$ $x = 0.25, 0.50, 0.75$	300	200	[148]
^{61}Ni	Ni	Room	-75 [b]	[113]
^{57}Co	Fe	4.5	$-300 ± 20$	[112, 128]
Pt	$Fe_{0.7}Pt_{0.3}$	4	4300	[150]
^{197}Au	Fe	—	$-1420 ± 180$	[129, 151]
	Ni	—	$-340 ± 60$	[129, 130, 151]
	$Co_{(cub)}$	—	$-980 ± 120$	[151]
	$Co_{(hex)}$	—	$-999 ± 120$	[129, 130, 151]
^{197}Au	$Fe_{0.995}Au_{0.005}$	—	1460 ± 160	[133, 134]
	$Co_{0.99}Au_{0.01}$	—	1180 ± 120	[134]
	$Ni_{0.99}Au_{0.01}$	—	530 ± 160	[134]
	$Ni_{0.99}Au_{0.01}$	—	420 ± 120	[134]
^{119}Sn	Mn_2Sn	0	-200	[115, 132, 153, 154]
	Mn_4Sn	0	-45	[115, 132, 153]
	$Fe_{0.99}Sn_{0.01}$	—	$-81 ± 4$	[114, 131]
	$Co_{0.99}Sn_{0.01}$	—	$-20 ± 1.5$	[131]
	$Ni_{0.99}Sn_{0.01}$	—	$+18.5 ± 1$	[131]
	Gd	80	<10	[152]
^{155}Gd	Gd	4	<1000	[120]
^{161}Dy	DyIG	85	3500 [c]	[116, 118]
		300	750 [c]	[116, 117, 118]
Eu	Eu	4.2	264 ± 8	[155, 156]
	EuO	—	296	[155]
	EuS	—	328	[155]
^{169}Tm	$TmFe_2$	0	7222 [d]	[157]

[a] The more precise NMR determinations and analyses of Boyd and Slonczewski [157a] are of considerable value here.

[b] This value is quoted on the basis of the moment of ^{61}Ni ($\mu = 0.7$ nm) determined by Locher and Geschwind [157b].

[c] Using $\mu(Dy^{161}) = ±0.3 ± 0.05$ nm.

[d] Using $\mu = 0.221$ nm determined by Ritter [157c].

information in addition to the hyperfine fields which are listed [158].*
Little of this can be reviewed here. Taken together with other data,
such as NMR and bulk magnetization measurements, they have greatly
aided our understanding of magnetic phenomena. As examples of this
we cite and call attention to the following:

(a) variation of the hyperfine field with temperature: The dependence
of the hyperfine field in a ferromagnet on temperature was determined
by Nagle et al. [104]. Figure 12 shows their results in which the relative
hyperfine field in Fe, $H(T)/H(297°K)$, is plotted vs. T/T_c, where T_c is
the Curie temperature. The important feature of the results is that the
hyperfine field is proportional to the saturation magnetization (shown as
the solid curve in Fig. 12). Note that the hyperfine field is zero above T_c.
(Similar results were obtained by these authors for a CoPd alloy.)
The proportionality of the hyperfine field and the magnetization have
since been assumed in many other cases—often without justification and
often incorrectly. Even in the simple case of iron, it has been found by

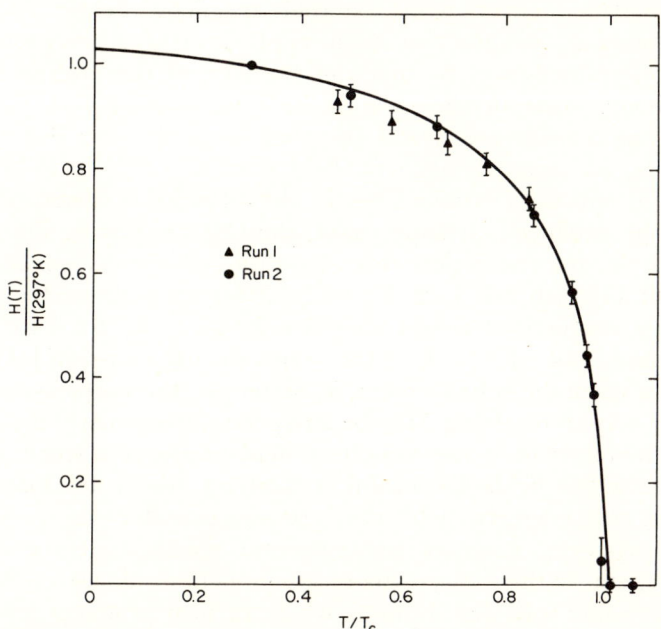

FIG. 12. Temperature dependence of the relative hyperfine field $H(T)/H$ (297°K).
The solid curve denotes the saturation magnetization. (After Nagle et al. [104].)

* In addition to the rare earth references given in Table V see also reference [159].

means of NMR studies that the proportionality constant, A, is temperature dependent [160-162]:

$$A = A_0(1-0.77 \times 10^{-7}T^2). \qquad (2.18)$$

Following this early work, many experiments have measured hyperfine interactions as a function of temperature to study the dependence of the magnetization of ferromagnetic (or antiferromagnetic) metals and alloys (and oxides). Mössbauer experiments (particularly in conjunction with external magnetic fields) are particularly useful for studying magnetic and crystallographic transitions, the paramagnetic region of these materials, and the variation of the magnetization near the Curie (or Néel) temperature. For the latter, different theories differ widely in their predictions as to the value of n in the $(T - T_c)^n$ power law giving the magnetization just below the Curie temperature, T_c. Mössbauer measurements could help in understanding this controversial and important subject.

(b) Mössbauer effect with external magnetic fields: Mössbauer measurements in conjunction with applied external magnetic fields allow a determination to be made of the sign of the internal field, as was originally done in the case of Fe^{57} in metallic iron [101, 102]. Such magnetic fields are useful in other ways as well. For example, by applying an external magnetic field parallel, or antiparallel, to the direction of emission various lines in the Mössbauer absorption spectrum can be enhanced, or suppressed, thereby simplifying the analysis. [For example, the theoretical intensity ratios of the six emission lines of Fe^{57} (of Fig. 10) are 3 : 2 : 1 : 1 : 2 : 3 for an unmagnetized source while for a magnetized source they are 3 : 0 : 1 : 1 : 0 : 3 for parallel emission and 3 : 4 : 1 : 1 : 4 : 3 for perpendicular emission.] There are many cases when the enhancement or, better yet, the suppression of lines is of great help in resolving Mössbauer spectra, particularly if numerous lines are involved or if the hyperfine field produces a small splitting.

While external fields are useful in systems which are magnetically ordered in zero magnetic field, they are indispensable when one wishes to study systems, such as paramagnetic metals, salts, and alloys, impurities, etc. In these cases, the magnetic field produces a polarization of the electronic spin distributions which in turn produces an effective magnetic field at the nucleus which in general is many orders larger than the applied field [48, 49, 163, 164]. Consider a paramagnetic substance. In the absence of an external field H_0, the thermal relaxation time of the paramagnetic ion is so fast ($\tau_R < 10^{-13}$ sec) relative to the γ-ray lifetime that the average value of the total electronic angular momentum in the time of measurement τ_M is zero (since $\tau_M \gg \tau_R$) and hence the hyperfine field is also observed to be zero. If, however, $H_0 \neq 0$, there is a net polarization of the paramagnetic ions at a temperature T; the thermal average of J, $\langle J \rangle_T$, is found, in the usual way, by averaging over the energy levels in the field H_0 with populations given by the Boltzmann distribution function.

Successful Mössbauer measurements in external fields have already been made with Fe^{57} in [163] iron, stainless steel, Cu, and MgO, and more recently [165] with Fe^{57} in Pd and other $3d$ and $4d$ transition metal alloys [166, 167]. In the former measurements the observations may be summarized in the form of a relation of the applied field H_0, the measured field at the nucleus H_n, and the field at the nucleus, H_{int}, when $H_0 = 0$:

$$H_n = H_{\text{int}} + H_0(1 + C) \tag{2.19}$$

where C is a constant for constant temperature. Of the cases cited, only in iron and chromium is H_{int} different from zero. In several of these measurements the field at the nucleus minus H_0 was found to have a Brillouin function dependence on temperature. [For example, for Fe in Pd the field corresponds to a large moment ($12.6\mu B$) and a large spin value ($S = 13/2$).] Interpretations of these observations have been given [164, 166, 167]; we shall discuss them later in Section IV.

Among the types of experiments possible with high magnetic fields is the investigation of the causes of line broadening in rare earth environments. For example, Mössbauer lines of Tm^{169} and Dy^{161} are observed to be 5 to 10 times broader than the natural line widths. The application of a very strong magnetic field can, in principle, resolve some of the possible broadening effects, especially if the line widths are measured as a function of the temperature [168].

Additional uses of high fields include the study of superconductors, the identification of ionicities of impurities, studies of deep-lying levels in semiconductors, the investigation of the unusual spin structures exhibited by rare earths arising from isotropic and anisotropic exchange interactions. In short, high fields extend the usefulness of the Mössbauer effect into many areas.

(2) *Quadrupole interactions.* Electric quadrupole interactions are observed in Mössbauer measurements when one has non-zero electric field gradients to interact with the nuclear quadrupole moment, Q. These electric field gradients are sensitive indicators of the electronic distributions and as such they may be used to gain information about these distributions in magnetic materials. To date, however, the most

useful information has been obtained from the temperature dependence of the quadrupole interactions and from the correlation of quadrupole coupling data with crystal-field and chemical bonding theories. Rare earths, which have particularly large quadrupole moments, may be studied by this means, and information such as symmetry and splittings of the crystal-field states may be obtained. From a study of electric quadrupole interactions, the nuclear quadrupole moment of excited states may be determined, as was demonstrated so strikingly for the excited state of Fe^{57} (the dominant Mössbauer nuclide). In addition, by combining the effects due to electric quadrupole and magnetic dipole interactions one may determine the sign of the nuclear quadrupole moment.

As mentioned earlier, since the nuclear ground states of Fe isotopes have spin $\frac{1}{2}$ no quadrupole splittings can be observed with conventional methods. In Mössbauer experiments, observations are made of the quadrupole splitting of the 14.4 kev excited state, $I = \frac{3}{2}$. Kistner and Sunyar [96] first measured this quadrupole splitting in the excited state of Fe^{57} in Fe_2O_3 and found that the spectrum could be fitted with an additional shift, ϵ, positive or negative, of the Zeeman substates of the excited level. In the case of axially symmetric field gradients, the shift is given by

$$\epsilon = \frac{e^2qQ}{4I(2I-1)}[3m^2 - I(I+1)] \quad (2.20)$$

where $eq = (\partial^2 V/\partial z^2)$. Kistner and Sunyar's experiments indicated that ϵ was <0 for the $m = \pm\frac{3}{2}$ substates and $\epsilon > 0$ for the $m = \pm\frac{1}{2}$ substates, giving a negative product for e^2q^Q and suggesting a small quadrupole moment for the excited state. As discussed in Section I, one must generally include an additional term if the asymmetry parameter $\eta \neq 0$ [cf. Eq. (1.24)], in which case the situation becomes much more complicated and the data must be interpreted with care. In later experiments proper account was taken of the direction of the magnetic hyperfine field with respect to the principal axis of the field gradient tensor [111, 119] and both the sign and magnitude of the quadrupole moment were deduced from calculated values of the electric field gradient [39, 99, 169-174].

The chief uncertainty in any such investigation is the value used for the electric field gradient at the nucleus [cf. Section I]. Considerable work has been done on estimating the divalent [39, 99, 169, 170, 172] and trivalent [171-174] ion field gradients in an effort to obtain Q for the excited state of Fe^{57}. When dealing with the Fe^{2+} ion, one normally ignores the effect of the crystalline environment and concentrates on the quadrupole interaction (and induced antishielding) due to the unfilled aspherical $(3d)^6$ shell. Estimates of the crystal-field gradient, and the antishielding factor (γ_∞) associated with it, are involved in the description of the half-closed shell $(3d)^5 Fe^{3+}$ ion. Originally the divalent and trivalent ion investigations yielded rather different predictions for Q but with increased efforts expended on estimating the various antishielding factors (and crystal-field gradient) the two types of investigations appear to have converged on a $Fe^{57m}Q$ value of $+0.18$ to $+0.28b$. This convergence may still prove to be fortuitous. As noted, there are a number of problems associated with the interpretation of electric quadrupole splittings, obtained from Mössbauer experiments. Ingalls [175] has recently discussed these for the ferrous compounds in terms of the energy splittings of the d states in crystalline fields and the covalency factor. The behavior of the quadrupole splitting as a function of axial and rhombic crystalline field strengths and temperature was studied. Table VI lists some materials in which quadrupole splittings have been studied and gives appropriate references [99, 176-195].*

The use of the Mössbauer effect to observe electric quadrupole as well as magnetic hyperfine effects has proven to be of particular interest for the rare earths. Recently such studies have dealt with quadrupole interactions of the form

$$e^2 Q[q_{\text{latt}}(1 - \gamma_\infty) + q_{4f}(1 - R_Q)] \tag{2.21}$$

involving both lattice and $4f$ shell field gradients (and the attendant antishielding factors) [168, 196]. Normally the valence electron gradient would completely dominate q_{latt} but large rare earth γ_∞ values of ~ -100 [40, 197, 198] enhance the lattice effects. The experiments (in particular their temperature dependence) also involve the $4f$ shell–crystal field interaction and the shielding of that interaction by the ion's closed shells (in much the same manner as the quadrupole antishielding). The crystal-field problem is of great interest both in itself and because it is fundamental to the observed temperature dependence of many of the magnetic properties of rare earth salts.

As mentioned in Section I, 2, recent Mössbauer measurements [38, 168] of electric quadrupole interactions in rare earths, when combined with Coulomb excitation data for Q, indicate that $R_q \approx 0.3$. While there are a number of uncertainties in such estimates, it is clear that these Sternheimer factors are real and experimentally significant and that one

---

* See Kerler and Neuwirth [176] for an extensive tabulation of quadrupole splittings in Fe compounds.

TABLE VI

LIST OF MATERIALS FOR WHICH QUADRUPOLE SPLITTINGS HAVE BEEN OBTAINED IN MÖSSBAUER EXPERIMENTS

Nucleus	Host	Ref.
Fe^{57}	YIG	[106, 107]
	DyIG	[107, 108]
	FeF_2	[99, 111, 137, 141]
	FeF_2	[123]
	$FeCl_3$	[99]
	$FeSO_4 \cdot 7 H_2O$	[99, 141, 179]
	$FeCl_2 \cdot 4 H_2O$	[99, 179]
	$Fe(NH_4)_2(SO_4)_2 \cdot 6 H_2O$	[99, 179]
	$FeSiF_6 \cdot 6 H_2O$	[170, 180]
	$Fe_2(SO_4)_3$	[99]
	Fe_2Ce	[148]
	Fe_2Ti	[148]
	$Fe_2Ti_xZr_{1-x}$	[148]
	$FeCoZr$	[148]
	FeS	[141, 168]
	Fe_3Al	[135, 149]
	Fe_5Si_3	[143]
	$(Fe_xMg_{1-x})O$	[109]
	$(Fe_xCr_{1-x})_2O_3$	[109]
	$(Fe_xV_{1-x})_2O_3$	[109]
	$(Fe_xAl_{1-x})_2O_3$	[109]
	$NiFe_2O_4$	[108]
	α-Fe_2O_3	[109, 119, 141]
	γ-Fe_2O_3	[108]
	Fe_2O_3	[96, 99, 181]
	Fe_2TiO_5	[109]
	$FeTi_2O_5$	[109]
	$FeTiO_3$	[109, 144]
	FeO	[109]
	Fe_3O_4	[140, 141]
	$(1-x)FeTiO_3 - x Fe_2O_3$	[144]
	Al_2O_3	[119]
	$BaTiO_3$	[186]
	$AlCl_3 \cdot 6 H_2O$	[186]
	TiO_2	[186]
	Ferrocene	[182]
	Ferricinium bromide	[182]
	$Co(C_5H_5)_2B(C_6H_5)_4$	[182]
	$NH_3CH_3Al(SO_4) \cdot 12 H_2O$	[186]
	$Y_3Ga_2(GaO_4)_3$	[186]
	Cu-Ni	[125]
	Pyrite	[145]

4. HYPERFINE INTERACTIONS IN MAGNETIC MATERIALS 211

TABLE VI (*continued*)

Nucleus	Host	Ref.
	Marcassite	[145]
	Ge	[185, 195]
	Si	[195]
Fe	Many compounds	[176, 183]
Au	AuCl	[189]
	AuBr	[189]
	AuI	[189]
	$AuCl_3$	[189]
	$KAuCl_4$	[189]
	$KAuBr_4$	[189]
Sn^{119}	Fe	[114]
	$Sn(\beta)$	[191]
	Mn_2Sn	[115, 154]
	$SnCdAs_2$	[191]
	SnF_2	[190, 192]
	$SnCl_2 \cdot 2 H_2O$	[190]
	SnS	[190]
	SnS_2	[190]
	SnO	[190, 191]
	$SnSO_4$	[190]
	SnO_2	[191]
Dy	DyIG	
Tm	(Graphs)	[178]
	Thulium ethyl sulfate	[193, 194]

is on the verge of obtaining reliable experimental estimates of these (sometimes) troublesome but interesting quantities.

(3) *Isomer shifts.* The information obtained from the isomer shift, δ, is twofold, consisting of, first, differences in the nuclear radius between ground and excited states, and second, changes in the total *s*-electron density at the nucleus. The second aspect is of importance for solid state effects and has been used to determine electron configurations in metals and alloys (ordered and disordered), to study ordering mechanisms in alloys, to distinguish valence states, to measure covalencies, etc.

Kistner and Sunyar [96] first reported an isomeric shift using Fe^{57}. Extensive applications have been presented by Walker *et al.* [98] and De Benedetti *et al.* [99] with Fe^{57} and by others* using Sn^{119}, Au^{197},

---

* See references in Tables V and VI for materials of interest.

Dy^{161}, and other nuclides. Walker et al. [98] reported a systematic variation for the isomer shift in divalent and trivalent compounds and in the 3d group metals, relative to stainless steel.

The restricted Hartree–Fock calculations of Watson [199] show significant differences in $|\psi_{3s}(0)|^2$ for different configurations, with a much smaller change in $|\psi_{2s}(0)|^2$ and $|\psi_{1s}(0)|^2$. The variation of the 3s density at the nucleus corresponds to different degrees of shielding by the different number of 3d electrons. Based upon several further assumptions, Walker and co-workers' [98] interpretation of the observed shifts is shown in Fig. 13 (taken from their paper). The total s-electron density is plotted as a function of the percentage of 4s character for various 3d electron configurations. (Another possible interpretation in the salts is to view the percentage of 4s character as indicating a measure of the degree of covalency in these environments.) Of particular interest are the results for some of the metals (approximately one "4s" electron for Fe, Ni, and Co), but it should be emphasized that such an assignment

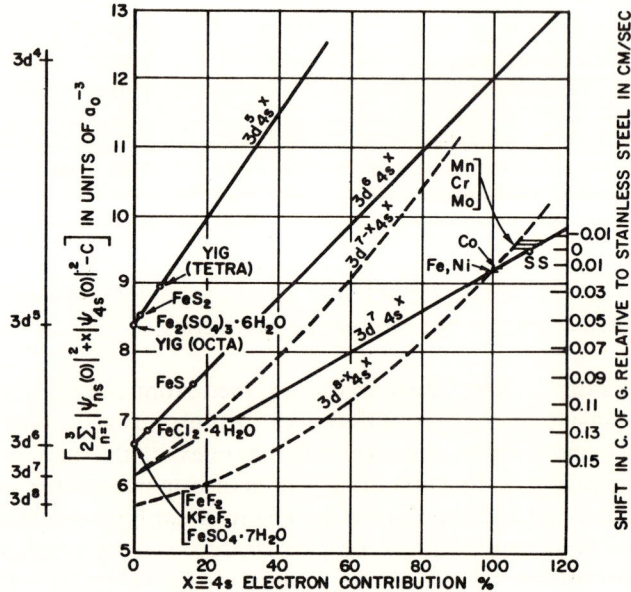

FIG. 13. The interpretation of the Fe^{57} Mössbauer isomer shifts in various solids. The total s-electron density is plotted as a function of the percentage of 4s character for various d-electron configurations. The reasons for placing the experimental data on given theoretical curves are discussed in reference [98]. The constant $C = 11873a^{-3}$. (C. of G. = center of gravity.) (After Walker et al. [98].)

depends very strongly on the use of free-atom results for solid state interpretations.

Isomer shifts have since been measured in various materials (cf. the references in Tables V and VI). In particular, a large amount of information has been obtained for metals and alloys (cf. the references in Tables V and VI). However, the importance of this data is not yet fully understood, mostly because the theory of metals and alloys has not been sufficiently developed. Still, a useful qualitative understanding has been obtained. In general, the simple ideas of the interpretation described above have been followed.

As an example of the value of isomer shift studies, consider the recent work of Roberts and Thompson [134] on dilute ferromagnetic gold alloys. For the Au-Ni alloys the observed isomer shift is linear over the whole range of composition in striking contrast to the observed magnetic behavior—the magnetization and the hyperfine field go to zero at 60% gold—and to theoretical descriptions [201] of the alloying process based on the rigid band model. See Fig. 14, which reproduces Roberts and Thompson's data. Most recently, Roberts *et al.* [200] have attempted to correlate these observed isomer shifts with the alloy residual resistivities using a Friedel-type [202, 203] phase shift analysis. They find the isomer shift to be proportional to $[Z(r_{Au}/r_h)^3 P_{av} - 1]$, where Z is the

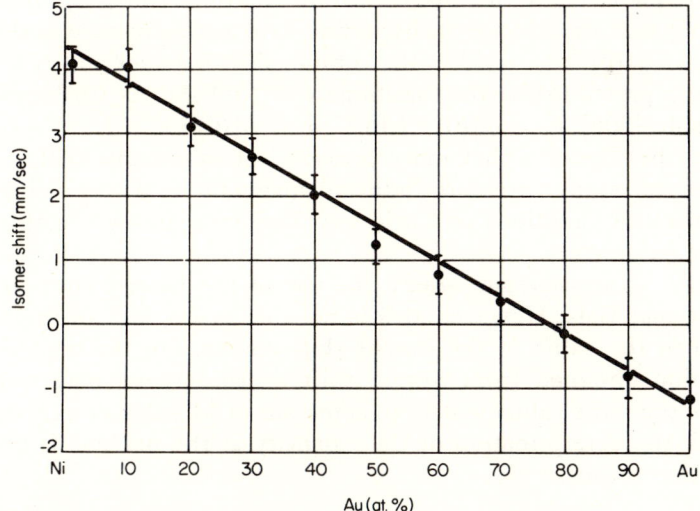

FIG. 14. The observed isomer shift as a function of composition for gold-nickel alloys. The solid line has been drawn to show the linearity of the data. (After Roberts and Thompson [134].)

effective number of s-band electrons per host atom, r_{Au} and r_h are radii of gold and the host atoms, and P_{av} is the average over the s band of the "charge polarization factor" $P(k)$ used at the Fermi surface in calculations of impurity Knight shifts. The calculated values are sensitive to the assumed s-band fillings (and other factors).

A somewhat unusual result has been obtained with ^{151}Eu in Eu metal and Eu^{2+} and Eu^{3+} salts [155, 156, 204]. The larger observed isomer shift for Eu^{2+} salts than that of Eu metal indicates that the s-electron density is higher in the salts. This is contrary to naïve expectations but once understood will help to correlate other results and further our understanding of Eu metal.*

An interesting example of the sort of unique information available by means of the Mössbauer effect has been given by Pound et al. [205]. These workers have measured the effect of hydrostatic pressure on the isomer shift of Fe^{57} in iron. Using free-ion wave functions [199] to estimate differences in s density, these authors find that for the total s density, $|\psi(0)|^2 = |\psi\uparrow(0)|^2 + |\psi\downarrow(0)|^2$ varies with volume as $\partial|\psi(0)|^2/\partial \ln V = -(2.7 \pm 0.4)$ in atomic units. Assuming that only $|\psi_{4s}(0)|^2$ is changed by compressions and that $|\psi_{4s}(r)|^2$ scales with volume, then with $|\psi_{4s}(0)|^2 \approx 3$, they find $\partial|\psi(0)|^2/\partial \ln V \approx -3$, in good agreement with observation. One other interesting result is obtained by combining these results with the pressure dependence of the Fe^{57} NMR frequency and the magnetization in iron [160-162]. Assuming [205, 206] that the change in hyperfine field with pressure is due to a change in $|\psi\uparrow(0)|^2 - |\psi\downarrow(0)|^2$, the combined Mössbauer and NMR data can be interpreted to give the variation in $|\psi\uparrow(0)|^2$ and $|\psi\downarrow(0)|^2$ separately. The result [206] is $\partial|\psi\uparrow(0)|^2/\partial \ln V \approx \partial|\psi\downarrow(0)|^2/\partial \ln V \simeq -1.3 au^{-3}$.

It is to be expected that more pressure measurements will be made in order to study pressure-induced magnetic and crystallographic transitions and variations of Curie and Néel temperatures with change of lattice parameters.

We have discussed but briefly the use of the isomer shift to study magnetic materials. However, this is not due to any lack of importance attached to the value of the isomer shift results. On the contrary, the information obtained from isomer shift measurements may well turn out to be the most valuable data to come out of Mössbauer experiments. After all, the determination of the s density at the nucleus is obtained by no other means.

---

* Brix et al. [204a] have just reported a systematic study of isomer shifts in europium compounds. By using optical data they report a scheme for analyzing isomer shifts of the rare-earths.

5. Angular Correlation of γ-Rays

a. Introduction. The probability $W(\theta)$ for the emission of a γ-ray is described as a function of angle θ between the emitted radiation and some fixed direction. If the nuclei have their spins randomly distributed in space, there is no preferred direction for the emission of the radiation, i.e., the unoriented nuclei give an isotropic angular distribution of γ-rays. In order to obtain *anisotropic angular distributions*, one must select or produce oriented nuclei. One way of doing this, nuclear orientation produced by low-temperature methods, was discussed earlier [Section II, 2].

Another method is to select nuclei with aligned spins, as in the case of angular correlation of successive nuclear radiations [45, 207-210]. Here nuclei which emit two γ-rays successively are used, and by choosing only those nuclei which emit the first γ-ray in a given direction, one has a means for selecting nuclei whose spins are partially aligned. In such a cascade, from an initial, to an intermediate, to a final state, the angular correlation will be attenuated if the nuclei in the intermediate state are subject to electric field gradients or magnetic fields. The interactions of these with Q or μ produce a precession of the nuclei about the symmetry axis and the changing nuclear orientation results in a changed angular correlation pattern. Aside from the case of static fields just described there are also time-dependent fields which produce a change in $W(\theta)$, i.e., the fluctuating electric field gradients induce transitions among the various nuclear states, E_m. Thus, the measurement of the change in degree of orientation (in the time between the orientation of the nucleus and the emission of the γ-ray) by electric and magnetic fields in the solid may be used to study hyperfine interactions.

For the successive emission of two γ-rays from unoriented nuclei with angle θ between the γ_1 and γ_2 directions, the angular correlation is written as

$$W(\theta) = \sum_{k} A_k P_k(\cos \theta) \tag{2.22}$$

where the $P_k(\cos \theta)$ are Legendre functions; and if only magnetic dipole or electric quadrupole radiations are involved, k can have only three values, 0, 2, 4. (For parity conservation, k must be even.) Since, in general, only even powers of $\cos \theta$ are involved, these experiments do not require a polarization of the nuclei but only an alignment. The coefficients A_k depend on the properties of the two transitions and, if pure multipoles are involved, they can be calculated for a number of different cases [45, 209]. If experiment disagrees with these predictions, then one knows that the correlation is not the originally supposed undisturbed one and the influence of external fields must be considered.

For oriented nuclei, observation is made of the angular distribution of the γ-rays emitted following (in most cases) beta decay. $W(\theta)$ is still given by Eq. (2.22), but A_k depends on some "orientation parameter" which describes the environmental effect in producing alignment and depends on temperature through the Boltzmann factor. These orientation parameters have been determined theoretically for most of the important nuclear orientation methods. As simple examples, consider the cases of dipole and quadrupole radiation [45]. For dipole radiation (nuclear spin change from I to $I - 1$) the directional distribution $W(\theta)$ is given by

$$W(\theta) = 2 + \frac{3I}{2I - 1} f_2 P_2(\cos \theta) \tag{2.23}$$

whereas for quadrupole radiation (nuclear spin change from I to $I - 2$)

$$W(\theta) = 2 - \frac{30}{7}\left(\frac{I}{2I-1}\right) f_2 P_2(\cos\theta) - 10 \frac{I^3}{(I-1)(2I-1)(2I-3)} f_4 P_4(\cos\theta) \tag{2.24}$$

We see that $W(\theta)$ depends only on even f_k and that, for 2^L pole radiation, the highest-order k which can occur is $k = 2L$. The f_k are the orientation parameters, discussed in Section II, 2.

Aside from the directional distribution of the emission pattern, valuable information is obtained from the polarization distribution of the γ-radiation [45, 208, 209]. For example, the electric and magnetic character of a transition can be distinguished by measuring the linear polarization of the emitted γ-rays. For cases of polarization ($f_1 \neq 0$) but no alignment ($f_2 = 0$) the γ-radiation pattern (for dipole radiation) is not anisotropic but the γ-rays are circularly polarized. From the sense of this circularly polarized radiation the sign of the nuclear magnetic moment can be determined.

In most experiments only the intensities at $\theta = 0$ and $\pi/2$ are observed. The anisotropy of γ-ray emission is then defined as

$$\epsilon = 1 - \frac{W(\theta)}{W(\pi/2)} \tag{2.25}$$

For example, in ferromagnets [55]

$$\epsilon = \frac{39}{14}\left(\frac{g_N \mu_N H}{kT}\right)^2 + 0\left(\frac{g_N \mu_N H}{kT}\right)^4 + \cdots \tag{2.26}$$

where $W(\theta)$ is the intensity of the γ-ray emission in a direction making an angle θ with the aligned spins.

The precession of the nuclear magnetic moment in a field **H** (applied perpendicular to the plane formed by the radioactive source and the two γ-ray detectors) has the Larmor frequency ω and a precession angle which is determined by ω and the time t that the nucleus spends in the intermediate state before emitting the second γ-ray. Hence, there is a rotation of the angular correlation pattern through an angle $\Delta\theta = \omega t$ and instead of Eq. (2.22) for $W(\theta)$ one must now write

$$W(\theta, t) = \sum_k A_k P_k[\cos(\theta + \omega t)] \qquad (2.27)$$

where $\omega = \mu_e H / I_e \hbar$. Here μ_e is the nuclear magnetic moment of the intermediate excited state and I_e is its spin.

In a differential angular correlation measurement, $W(\theta, t)$ is measured as a function of t. In these experiments, the resolving time of the apparatus is usually less than τ_e, the mean lifetime of the intermediate state. If the resolving time of the apparatus is large compared with τ_e, $W(\theta, t)$ must be averaged over the distribution function $\exp -(t/\tau_e)$ giving an integral correlation. Expressions for cases of physical interest to this chapter have been given by Abragam and Pound and more recently by Caspari *et al.* [212].

For the case where the hyperfine field can be considered static with respect to τ_e, the development indicated above is still valid for describing the rotation of the angular correlation pattern. However, in the above expressions H is replaced by H_{eff}, the effective magnetic field which described the hyperfine interaction and which in magnetic materials is much larger than the applied external field which is used to align the electronic moments. As has been emphasized by Caspari *et al.* [212], measuring hyperfine interactions by determining the rotation of the angular correlation pattern is a much more sensitive method than determining the anisotropy because the former depends on $\omega\tau_e$ whereas the latter varies as $(\omega\tau_e)^2$. In practice it is valuable to have both rotation and anisotropy measurements and to require consistency between them.

When the hyperfine field has a time-varying component, the angular correlation is determined by the average magnetic hyperfine field, acting at the nucleus in the direction of the applied field. The main effect of the time-dependent perturbation is to reduce both the amplitude and mean rotation. Details of a successful treatment of this case, for several circumstances, are given by Caspari *et al.* [212] based on the earlier work of Abragam and Pound [211] on time-dependent hyperfine interactions.

b. *Experimental results.* Among the interesting experimental investigations with this method are measurements of hyperfine interactions and relaxation times in magnetically ordered systems, paramagnetism, liquids, various hole effects, paramagnetic relaxation, static electric interactions, integral fields in this layers, time-dependent electric interactions, and combined electric and magnetic interactions.

A most extensively studied isotope is Co^{60}. Experiments were first made in 1951 in Oxford [213] and Leiden [214] with Co^{60} in a Tutton salt and an estimate of the nuclear magnetic moment (not then known) was obtained.

Cobalt was the first ferromagnet to be studied by this method in Oxford [215] and Moscow [216]. Cobalt^{-60} was again used as the radioactive source and the anisotropy ϵ [defined by Eq. (2.25)] was found (from the Oxford measurements) to be given by

$$\epsilon = 7.9 \times 10^{-5} T^{-2} \qquad (2.28)$$

down to $0.04°K$. From ϵ and the known magnetic moment the hyperfine field, H_{eff}, was determined to be 193 koe in reasonable agreement ($\sim 10\%$) with later specific heat (see Section II, 2), NMR, and Mössbauer (see Section II, 4) measurements. Aside from giving the correct magnitude of the hyperfine field, these measurements were the direct stimulus for a chain of measurements by specific heat [58–60] and NMR methods [217]. Various Co-Ni and Co-Fe ferromagnetic alloys have since been studied by the γ anisotropy method [218].

Using antiferromagnetic single crystals to produce nuclear alignment (the method of Gorter and Daunt) Daniels et al. [219] have recently done measurements with Co^{60} and Mn^{54} in $MnCl_2 \cdot 4\,H_2O(\epsilon = 9\%$ at $0.09°K$ for Mn^{54} and 0.7% at $0.1°K$ for Co^{60}), $MnBr_2 \cdot 4\,H_2O(\epsilon = 7\%$ at $0.09°K$ for Mn^{54} and 0% for Co^{60} at $0.1°K$), and $CoCl_2 \cdot 6\,H_2O$ (18% for Mn^{54} at $0.055°K$ and 1.5% for Co^{60} at $0.055°K$).

Anisotropic γ-ray emission measurements with other radioactive nuclei have also been made* [221], showing the feasibility of producing nuclear alignment (mostly in salts). Fortunately for the study of magnetism these nuclei are mainly from the $3d$ and $4f$ series. The first measurements on aligned rare earth nuclei were made by Ambler et al. [222], using radioactive Ce^{139}, Ce^{141}, and Nd^{147}, and estimates of the nuclear moments were obtained. Later measurements on paramagnetic rare earth ions in liquids have been used to give estimates of the nuclear moments of Nd^{150}, Sm^{152}, and Sm^{154} [223] and Dy^{160} and Eu^{152} ([224]; and the discussion in the review by Heer and Novey [209]) as well as the

---

* These include the use of Co^{58}, Co^{57}, Co^{56}, Mn^{54}, Mn^{52}, Ce^{139}, Ce^{141}, Nd^{147}, Yb^{169} and Yb^{175}. See Steenland and Tolhoek [220] for detailed references to this work.

effective field at the nucleus in these ions. The method used is similar to that described in Section II, 4.

Recently, Shirley et al. have studied Tb^{156} in single crystals of Nd ethyl sulfate and Y ethyl sulfate [225] and Pm^{144} in Nd ethyl sulfate [226]. Using the theory of Elliott and Stevens [227] and the usual spin Hamiltonian, they found the following: from $|A|/k = 0.120 \pm 0.025°K$ and $P/k = +0.006 \pm 0.002°K$ for Tb^{156}, values of 1.53 ± 0.30 nm and 1.8 ± 0.5 barns for the nuclear moment and quadrupole moment, respectively (assuming an $\langle r^{-3} \rangle$ value of 53.7×10^{-24} cm^3 [16]); and from $|A|/k = 0.0091 \pm 0.0003°K$ (for $I = 5$) for Pm^{144}, a value of 1.68 ± 14 nm (assuming $\sigma \langle r^{-3} \rangle$ of 36.8×10^{-24} cm^3). (The role of theoretical estimates $\langle r^{-3} \rangle$ will be elaborated upon in more detail in Section IV, 2.) Similarly, Postma and Huiskamp [228] have studied Ho^{166m} and Tb^{160} in Nd ethyl sulfate.

By measuring the anisotropy of γ-ray emission Samoilov et al. [229] and Kogan et al. [230] have measured the hyperfine field in nonmagnetic atoms which were added to a ferromagnet. In these measurements, a small percentage of the elements was added to iron, cooled to $10^{-2}°K$, and magnetized to saturation in order to polarize nuclei with respect to this field. The anisotropy parameter ϵ and H_{eff} are listed in Table VII.

TABLE VII

THE ANISOTROPY PARAMETER, ϵ, AND THE EFFECTIVE HYPERFINE FIELD, H_{eff}, AT VARIOUS NUCLEI IN Fe

	ϵ (%)	H_{eff}	T (°K)
^{198}Au	6	1×10^6	0.03
^{122}Sb	8	2.8×10^5	0.03
^{114m}In	28	2.5×10^5	0.03
^{46}Sc	1.8	$(100 \pm 30) \times 10^3$	0.05
^{60}Co	10	$(350 \pm 50) \times 10^3$	0.05
^{199}Au	3.0	$\geqslant 2000 \times 10^3$	0.05
^{192}Ir	10	$\sim 1000 \times 10^6$	0.05
^{186}Re	4	$\geqslant 200 \times 10^3$	0.05

The fields are seen to be very large, the uncertainty in the value of the nuclear moment making several of the fields (Ir^{192} and Au^{199}) listed in the table approximate. Also, for Re^{186} and Au^{199}, the influence of the β transition on γ anisotropy cannot be calculated accurately; hence the lower limit on H_{eff}. Experiments with Au-Ni, Au-Gd, and In-Ni alloys have found no γ-ray anisotropy within 0.5% accuracy [229].

From γ-anisotropy and β-symmetry studies Šott [231] has found the hyperfine fields at nuclei of Re, Ir, and Os (as dilute alloys) in iron to be negative in all these cases.

The sign of the magnetic hyperfine field at rare earth nuclei in rare earth-iron garnets has recently been determined by Caspari et al. [232]. In these experiments, the rotation of the angular correlation pattern, due to the precession of the nuclear magnetic moment of the intermediate excited state, is measured relative to a weak magnetizing field applied to the samples perpendicular to the plane of two γ-ray detectors. The sign of H_{eff} is found to be positive in Sm^{152} and Dy^{160} and of the order of a megagauss; for Gd^{154} and Eu^{154}, H_{eff} is negative and —50 kgausses.

The magnetic hyperfine interaction was measured for Sm^{152} and Gd^{154} in polycrystalline europium-iron garnet by Caspari et al. [212]. The data were analyzed using an extension of the theory of Abragam and Pound [211] to time-dependent magnetic hyperfine interactions in magnetic materials. Neglecting crystalline field and quadrupole effects, consistent results for the sign, magnitude, and temperature dependence of the average component of the effective hyperfine field were obtained by assuming that following K capture in Eu^{152} the electronic configuration is that for Sm^{3+}. The instantaneous hyperfine field was found to be 4.7×10^6 oe at room temperature and almost independent of temperature [212]. The situation for Gd^{154} is not as clear; the electronic configuration of Gd following beta decay in Eu^{154} was not determined. The magnetic hyperfine field at the Gd^{154} nucleus was found to be $8 \pm 2 \times 10^4$ oe at room temperature. Further, the hyperfine field of Sm^{3+} in SmIG is the same as that of Sm^{3+} in EuIG, whereas the experimentally determined sublattice magnetization of the Sm^{3+} ion sublattice in SmIG appears to be zero [233]. Cohen et al. [234] have used angular correlation techniques in a very recent interesting investigation of hyperfine fields at impurity nuclei in ferromagnets. Into ferromagnetic Gd they put dilute impurities of Lu^{3+} ions and measured both the sign and magnitude of the hyperfine field H_{eff} at the Lu nuclei (-2×10^5 oe at saturation) as a function of temperature. H_{eff} was found to follow the same temperature dependence as the Gd magnetization.

c. Comparison of angular correlation and Mössbauer methods. Angular correlation and Mössbauer scattering are, for the most part, complementary methods for studying hyperfine interactions. Although both may be used to study magnetic dipole and electric quadrupole interactions, only the Mössbauer method can be used for the study of isomer shifts [see Section II, 4]. A further limitation on angular correlation measurements is that they cannot easily distinguish magnetic from electric quadrupole effects (whereas Mössbauer measurements can) and so cannot be applied as simply to nuclides with large quadrupole moments. Measurements become difficult to perform with either method when the γ-ray lifetime, τ, is less than 10^{-9} sec.

What may be said about the relative sensitivity of the two methods? In the Mössbauer method one measures a line width, $\Delta E = 2\hbar/\tau$, where τ is the lifetime. In order for a splitting $\Delta E = \mu_m \cdot \mathbf{H}_m$, in a field H, to equal the line width it follows that $\tau = 2\hbar/\mu_m \cdot \mathbf{H}_m$. Now let us similarly consider an angular correlation experiment, using the same source. Here, $\mathbf{\mu} \cdot \mathbf{H} = h\nu$ and so the precession time is $P = h/\mu_m H_m = \pi\tau$. In the time τ, the precession angle is $\theta = \mu_m \cdot \mathbf{H}_m(\tau/\hbar)$ and for a splitting, $\Delta E = \mathbf{\mu} \cdot \mathbf{H}$, equal to the line width in the Mössbauer measurement the measured precession angle in the angular correlation method is $\theta = 2$ radians. Since an angular correlation precession measurement can easily observe a 5 degree change, it is therefore at least an order of magnitude more sensitive than the Mössbauer method. Hence, angular correlation is a good method for measuring small hyperfine interactions. Also, liquids may be studied by this method, but not by Mössbauer techniques.

An interesting use has been made of angular correlation measurements which show their applicability to solid state studies. It has been observed that in some Mössbauer experiments, e.g., Fe^{57} in stainless steel or Sn^{119} in iron, the line width is exceptionally large, i.e., ~ 2 times theoretical. Mössbauer methods cannot distinguish between quadrupole, magnetic dipole (due, say, to a distribution of H_{eff}), and isomeric sources of the broadening; hence this method cannot be used to determine the origin of the effect. Deutsch and associates [235] have, however, done this by studying the angular correlation pattern for $Sn^{119-120}$ diffused into iron. The lack of attenuation of the angular precession indicated that no major perturbation on the correlation due to internal quadrupole fields or nonunique magnetic fields are present; the widths of the Mössbauer lines (~ 1.5 times theoretical) must therefore be due to isomer shifts.

6. Interaction of Polarized Neutrons with Polarized Nuclei

The absorption cross section for polarized neutrons depends upon the polarization of the nuclei. If f_1^n denotes the fraction of neutrons which are polarized, f_1^N the fraction of polarized nuclei having spin I, and σ_0 the absorption cross section for zero polarization, the absorption cross sections for compound states having spins of $I + \tfrac{1}{2}$ and $I - \tfrac{1}{2}$ are

$$\sigma_+ = \sigma_0 \left(1 + \frac{I}{I+1} f_1^N f_1^n \right)$$

and

$$\sigma_- = \sigma_0(1 - f_1{}^n f_1{}^N) \quad (2.29)$$

respectively [236].* The parameters, $f_1{}^n$ and $f_1{}^N$ are given by Eq. (2.1). Measurements of the resonance absorption provide a straightforward way to determine the spin of the resonance level of the compound nucleus, provided the signs of $f_1{}^n$ and $f_1{}^N$ are known. Experiments have been done with Mn^{55}, In^{115}, Sm^{149}, Eu^{151}, Gd^{155}, Gd^{157}, and Ho^{165} resonances [238-240]; the first positive results were obtained in Oak Ridge [238] following unsuccessful attempts at Leiden [208].

Recently, transmission experiments with polarized monochromatic neutron beams and polarized nuclear samples at temperatures of $\sim 1°K$ at Brookhaven [240, 241] have investigated hyperfine interactions using resonances in Ho^{165}. For Ho^{3+} in holmium ethylsulfate the electronic ground state in a non-Kramers doublet. Using a term $As_z I_z$ in the Abraham and Pryce spin Hamiltonian to describe the hyperfine interaction [3], with $S = \frac{1}{2}$, Postma et al. [240] find that the hyperfine splitting constant $A/k = 0.580 \pm 0.045°K$. Assuming that the hyperfine splitting in holmium metal can be described by the same effective spin Hamiltonian, they determine $A/k = 0.61 \pm 0.06°K$. Within experimental errors the two results are seen to be equal. Specific heat measurements for Ho metal give [72] $A/k = 0.618$ and [78] $0.640°K$.

The field at the In nucleus in a ferromagnetic alloy of the Ho-In (92 atom. % Ho and 8 atom. % In) has been measured by Schermer et al. [241]. At the In nucleus H_{eff} is found to be ≈ 130 koe and directed antiparallel to an external field of 17.5 koe. This result is of interest for understanding the origin of hyperfine fields at nuclei of nonmagnetic atoms. More will be said on this subject later (cf. Section IV). An important study of gadolinium and europium metals has just been reported by Stolovy [242].† In Gd both the magnitude and sign (-324 koe) of the hyperfine field were measured. In Eu neither was found because the situation is much more complicated. These results are important because of the role they play in attempts at achieving an understanding of hyperfine fields in magnetic metals [cf. Section IV].

Details of experimental equipment, techniques, depolarization effects, etc. are given in reference [240].

---

* See de Klerk and Steenland [237] for a brief review of the early work with polarized neutrons.

† We are grateful to Dr. Stolovy for permitting us to quote his work prior to publication.

III. The Hartree–Fock Method in Its Conventional and "Unrestricted" Forms

We are interested in obtaining the stationary eigenfunctions of our many-electron Hamiltonian (time-dependent theory will not be considered here) in order to evaluate the expectation values of physical observables, such as the hyperfine interactions, of concern to us in this chapter. We need not consider the exact form of the Hamiltonian at this moment except to note that it consists of one- and two-electron operators (and perhaps of terms not involving the electrons at all).

Unfortunately, Schrödinger's equation has yet to be solved exactly for systems involving more than one electron.* One has, of necessity, resorted to the use of approximate wave functions which are often expanded in terms of some basis set of functions. Because of computational difficulties, an incomplete set which spans but a limited subspace of Hilbert space is chosen. The variation principle is then used to determine the "best" solution according to the criterion that the lower the total energy, the better the approximate wave function. In practice, the determination of the functions is limited by the fact that the energy is insensitive to further variation of the functional parameters. In addition, a small improvement in the solution, i.e., in its total energy, does not necessarily improve the prediction of some other physical observable.

Additional conditions and (preferably) the application of the variational principle to other observables would assist us in obtaining "good" wave functions; however, such efforts to improve the wave functions have not been very successful, despite progress in the ability to estimate bounds [243b] on the expectation values of observables [244][†] and in the ability to estimate more accurately some observable other than the energy with a given wave function [245]. Lowdin [243b] has discussed the convergence properties of the expectation value of the spin density operator with respect to the energy; the bounds of the two operators were shown to converge simultaneously.

We should note also that, while we are not able to obtain exact many-electron wave functions, we do not need all the information contained in them to predict physical observables. Because only one- and two-electron

---

* Pekeris [243] has obtained almost arbitrarily exact approximate wave functions for He-like systems, as shown in a recent study of the lower bounds of the wave function energies [243a].

† For recent investigations utilizing and comparing the lower energy bounds conditions of Temple, Weinstein, and Stevenson, see reference [244a]. See also reference [244b] and, in particular, Pekeris [243], who uses the Temple condition in his He investigations.

operators are associated with such observables, we actually need have knowledge of only the one- and two-electron reduced density matrices [246, 247].* Unfortunately, no one has yet developed a method whereby one deals solely with these, to the exclusion of the wave function, for the sort of system of interest to us here. Hence, in the discussion of this section we will consider wave functions and ignore density matrices.

We shall be mainly concerned with the one-electron approximation which represents one of the most successful schemes for approximating solutions of the many-electron Schrödinger equation. But basic to the scheme is the assumption that one has available one-electron wave functions, called orbitals, whose description depends only on the coordinates of a single particle. In many ways, the most accurate orbitals are those determined by means of the Hartree–Fock method. Let us consider it now.

1. The Hartree–Fock Formalism and the Fermi Exchange Hole

As is well known, the Hartree–Fock (H-N) [248, 249] formalism consists of approximating a true many-electron wave function by a single (or on occasion a linear combination† of) Slater determinant. A Slater determinant for an n electron system takes the form

$$\Psi = (n!)^{-1/2} \begin{vmatrix} \psi_1(\mathbf{x}_1) & \psi_2(\mathbf{x}_1) & \cdots & \psi_n(\mathbf{x}_1) \\ \psi_1(\mathbf{x}_2) & \psi_2(\mathbf{x}_2) & \cdots & \\ \cdots\cdots\cdots\cdots\cdots\cdots \\ \cdots\cdots\cdots\cdots\cdots\cdots \\ \psi_1(\mathbf{x}_n) & \psi_2(\mathbf{x}_n) & \cdots & \psi_n(\mathbf{x}_n) \end{vmatrix} \tag{3.1}$$

The $\mathbf{x}_i$ variables denote electron space and spin coordinates, and $(n!)^{-1/2}$ is the normalization constant if the one-electron spin orbitals ψ_i are orthonormal. Since such a determinant is unaffected by replacement of ψ_i by $\psi_i + C\psi_k$ for any constant C, it is no restriction to assume the ψ_i orthogonal. The matter of orthogonality will be of considerable interest to us shortly.

In abandoning the true eigenfunction of our Hamiltonian and going to the Hartree–Fock description, we have replaced a function with specific

---

* For a recent review pertinent to the Hartree–Fock theory see McWeeny [247a] and for a recent general review see ter Haar [247b].

† Single configuration, conventional H.F functions are always, via Hund's rule, single determinants for the ground states of ions. We have, for simplicity, limited discussion to the single-determinant case. Much, but not all, of the discussion can be trivially extended to the multideterminant case.

interelectronic dependence by a product of one-electron spin orbitals (the Hartree product) which was antisymmetrized in order to obtain a function which obeys the Pauli exclusion principle. In going to a product function we let each electron see only the average field due to the other electrons. Except for the Fermi hole, which we shall discuss shortly, no detailed description of local interelectronic behavior is built into the H–F function.

For a Hamiltonian consisting of only kinetic and electrostatic interaction terms* the total energy for a single determinant function is

$$E = \sum_{i=1}^{n} \int \psi_i^*(\mathbf{x}) K_{op} \psi_i(\mathbf{x}) \, d\mathbf{x} + \sum_{i=1}^{n} \sum_{k<i} \iint |\psi_i(\mathbf{x}_1)|^2 \frac{1}{|\mathbf{r}_1 - \mathbf{r}_2|} |\psi_k(\mathbf{x}_2)|^2 \, d\mathbf{x}_1 \, d\mathbf{x}_2$$

$$- \sum_{i=1}^{n} \sum_{k<i} \iint \psi_i^*(\mathbf{x}_1) \psi_k^*(\mathbf{x}_2) \frac{1}{|\mathbf{r}_1 - \mathbf{r}_2|} \psi_i(\mathbf{x}_2) \psi_k(\mathbf{x}_1) \, d\mathbf{x}_1 \, d\mathbf{x}_2 \quad (3.2)$$

where K_{op} is the one-electron kinetic plus nuclear potential energy operator. The integrations are over space and spin coordinates. The final terms, called exchange terms, are, owing to spin orthogonality, non-zero only for (ψ_i, ψ_k) pairs of common spin [251]. These terms arise from the antisymmetric nature of the many-electron H–F function. In the expression above, we have assumed that the nuclei are fixed and have therefore omitted energy terms involving only the nuclei.

Application of the variational principle to the energy by varying an individual ψ, say ψ_j, leads to a Hartree–Fock equation for ψ_j of the Schrödinger form,

$$\mathscr{H}_{\text{eff}} \psi_j(\mathbf{x}_1) = \left\{ K_{op} + \sum_{i \neq j} \int \psi_i(\mathbf{x}_2) \frac{1 - P_{12}}{|\mathbf{r}_1 - \mathbf{r}_2|} \psi_i(\mathbf{x}_2) \, d\mathbf{x}_2 \right\} \psi_j(\mathbf{x}_1) = \epsilon_j \psi_j(\mathbf{x}_1).$$
(3.3)

Here P_{12} is the permutation operator which permutes coordinates 1 and 2, so that the first term under the integral sign gives rise to Coulomb interactions and the second term to exchange interactions; ϵ_j is the one-electron energy and is in fact a Lagrange multiplier, present for the purpose of obtaining a normalized ψ_j. In practice it is convenient to let the sum within the bracket span all occupied spin orbitals. This does not affect the sum because the self-Coulomb and self-exchange terms of spin orbital ψ_j with itself are self-canceling, i.e.,

$$\int \psi_j(\mathbf{x}_2) \frac{1 - P_{12}}{|\mathbf{r}_1 - \mathbf{r}_2|} \psi_j(\mathbf{x}_2) \psi_j(\mathbf{x}_1) \, d\mathbf{x}_2 = 0 \quad (3.4)$$

* See Löwdin [250] for a discussion of the H–F scheme in which spin-orbit terms are explicitly included.

It is now apparent that the $\mathscr{H}_{\text{eff}}$ obtained by applying the variational principle to a particular ψ_i is identical with those effective Hamiltonians appropriate for all the other ψ_i appearing in the determinantal function. This is of great importance and represents a profound difference between Hartree and H–F theories, for in the former case different $\mathscr{H}_{\text{eff}}$ occur for different ψ_i. The presence of a common $\mathscr{H}_{\text{eff}}$ greatly simplifies the processes actually used to solve the H–F equations and, more important, the set of H–F equations yields orthogonal ψ_i. This follows from the well-known fact that eigenfunctions associated with different eigenvalues of the same Hamiltonian are orthogonal. Unfortunately, this simple and convenient situation is considerably complicated by the restrictions superimposed on the H–F equations in the conventional H–F method. These matters will be discussed later.

Let us now consider the exchange terms of Eq. (3.3) and, following Slater [252],* assume that the space and spin parts of the ψ_i are separable (as we did previously when we associated a specific spin assignment with a specific ψ_i) and write

$$-\left[\sum_i \int \psi_i^*(\mathbf{x}_2) r_{12}^{-1} P_{12} \psi_i(\mathbf{x}_2)\, d\mathbf{x}_2\right] \psi_j(\mathbf{x}_1)$$
$$= -\left[\frac{\sum_i \delta(m_{s_i}, m_{s_j}) \int \psi_j^*(\mathbf{x}_1) \psi_i^*(\mathbf{x}_2) r_{12}^{-1} \psi_j(\mathbf{x}_2) \psi_i(\mathbf{x}_1)\, d\mathbf{x}_2}{\psi_j^*(\mathbf{x}_1)\psi_j(\mathbf{x}_i)}\right] \psi_j(\mathbf{x}_1) \quad (3.5)$$

where the delta function term reminds us that this term is zero unless both ψ_i and ψ_j have the same spin. Note that we have retained the $i = j$ term in this sum. The sum can be regarded as representing the potential energy, at the position $\mathbf{x}_1$ of the electron being discussed, due to a charge distribution at point $\mathbf{x}_2$ of magnitude

$$\sum_i \delta(m_{s_i}, m_{s_j}) \frac{\psi_j^*(\mathbf{x}_1)\psi_i^*(\mathbf{x}_2)\psi_i(\mathbf{x}_1)\psi_j(\mathbf{x}_2)}{\psi_j^*(\mathbf{x}_1)\psi_j(\mathbf{x}_1)} \quad (3.6)$$

This charge density is called the exchange charge density and has three properties: (1) Its total charge is that of one electron, (2) it has the same spin as the spin orbital ψ_j under consideration, and (3) when $\mathbf{x}_2 = \mathbf{x}_1$, it is equal to the total density of all electrons of the same spin as the jth.

The first property follows immediately if one integrates over the $\mathbf{x}_2$ coordinates, for the orthogonality of the ψ_i wipes out all but the $i = j$ term. If we excluded the $i = j$ term from the H–F exchange and direct interelectronic interaction sum, we would have a charge of zero associated

---

* For a comparison of various exchange potentials see Hartree [252a].

with the resultant exchange density. In other words, the charge of one electron comes from keeping the Coulomb self-energy term in the Coulomb potential, which is common practice in H–F and more advanced theories.

If we set $\mathbf{x}_2 = \mathbf{x}_1$, and inspect the third property of the exchange density, Eq. (3.6) becomes

$$\sum_i \delta(m_{s_i}, m_{s_j}) |\psi_i(x_2)|^2 \qquad (3.7)$$

or just the density of all electrons of spin in common with the *j*th. This exchange charge density exactly cancels the total (if you will, Coulomb charge) density of all electrons of this spin and thus one can think of the electron as carrying a hole, the Fermi hole, centered at its position, $\mathbf{x}_1$, as a result of the exclusion principle. This hole has a total charge of one associated with it if one works in a frame where electrostatic self-potential terms are included. The Fermi hole also appears if one inspects the determinantal function and sets $\mathbf{x}_1 = \mathbf{x}_2$ for a pair of spin orbitals (remembering that an $\mathbf{x}_i$ includes a spin coordinate), for then the determinant becomes zero valued.* This form of correlation between electrons of the same spin is, of course, missing from the Hartree theory, just as correlation between electrons of opposite spin is absent from the H–F method.

Several more matters should be mentioned before we leave the discussion of the general features of H–F theory and exchange. First, the fact that the Fermi hole has the charge of one (or zero) electron associated with it by no means implies that different electrons in a system undergo the same exchange effects. In a system with net spin, electrons of differing m_s do undergo different exchange effects; this would be quite apparent if one inspected their exchange charge densities (which involve the ψ_j explicitly) as a function of $(\mathbf{x}_1 - \mathbf{x}_2)$.

Second, the region (i.e., the $\mathbf{r}_1 - \mathbf{r}_2$ range) over which the Fermi hole has an appreciable influence is small compared with the region over which an atomic electron's charge density extends and by causing pairs of electrons of common spin to locally avoid each other (i.e., for small $\mathbf{r}_1 - \mathbf{r}_2$), the Fermi hole reduces the average repulsive Hartree potential due to one electron as seen by the other. Hence, by reducing the Hartree Coulomb repulsion, exchange acts, in one sense, like an attractive interaction. One effect of this is to always make Hartree–Fock charge distributions for atoms more contracted than their Hartree counterparts.

* For a discussion of the Fermi hole in terms of density matrices see, for example, Löwdin [253].

An illustration of this is given in Fig. 15. This exchange attraction also plays a role in the first Hund's rule for atoms, which states that the ground multiplet state of a given atomic configuration will be of maximum allowable spin. When viewed from H–F theory there are two contributing causes to this: (1) Maximum spin produces the maximum number of exchange interactions among electrons of the open shell (or shells); these exchange interactions then reduce the positive energy associated with the interelectronic Coulomb repulsions. (2) The process of putting electrons of common spin into the H–F determinant requires that they go into spatially different orbitals (e.g., by having different angular functions); by being spatially different their Coulomb repulsive energy is again reduced. Both factors contribute to the familiar Slater–Condon equations [254, 255] for the atomic multiplet spacings.

Finally, it may be of interest to the reader to note that Slater [256]* has taken an averaged form of Eq. (3.5), analyzed its properties for free electrons, and obtained the familiar $\rho^{1/3}$ (density to the one-third power) approximate exchange potential which is commonly used in energy band calculations and in Hartree–Fock–Slater atomic self-consistent field calculations [257–259].

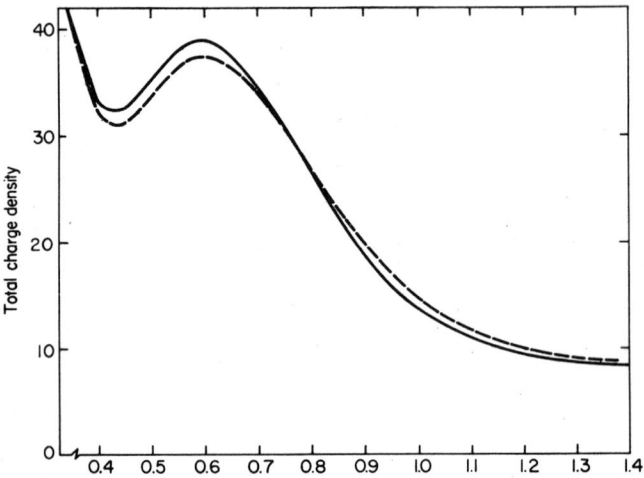

Fig. 15. Hartree [253a], versus Hartree–Fock charge density [253b], in the outer reaches of the Pr^{3+} ion.

* For another derivation of an effective exchange potential, yielding a result in somewhat different form, see Sternheimer [256a]. For a comparison of various exchange potentials see Hartree [256b]. For the inclusion of a screened Coulomb interaction (to introduce correlation effects) in the Slater potential see Robinson et al. [256c].

2. Correlation in Atoms

It has been known for some time that conventional Hartree–Fock wave functions are inadequate for treating various hyperfine effects in detail. Some indications of this were given in Section I. The inadequacies arise because, as has been noted, a H–F wave function differs from the exact eigenfunction of our many-electron Hamiltonian. Such differences are normally called correlation effects and these effects can be somewhat arbitrarily separated into two parts. First there are those effects which arise out of the restrictions associated with the conventional H–F formalism. Then there are the remaining correlation terms which, perhaps most appropriately, can be termed interelectronic correlation effects. There is little evidence as to the relative roles played by the two in hyperfine effects and what there is, is inconclusive, contradictory, and appropriate only to the Fermi contact term. (The authors believe that the evidence suggests the former to be the most, but by no means all, important, but not all workers will agree.) In any case, the former represents the more tractable problem to work on and as a result the existing investigations have dealt almost exclusively with it. For this reason it will be discussed in greater detail in this section. Although our discussion will center on the so-called "unrestricted" H–F (UHF) method it should be borne in mind that this is a simple but not entirely accurate approximation to a rigorous treatment of the matter. The UHF method has one important virtue. It can be applied to many-electron ions of the size of interest to us in this chapter and the results can be easily interpreted.

Before discussing the Hartree–Fock method as applied in practice to atomic or larger systems, we wish to consider a few features of the atomic correlation problem, as these will be pertinent to the discussion of the UHF method. Somewhat arbitrarily, we shall define correlation effects as the effects on the energy, wave functions, or expectation value of an observable on going from the conventional H–F wave function to the exact eigenfunction of our Hamiltonian.

The most obvious feature of the correlation problem is associated with the fact that while interelectron r_{ij} coordinates appear explicitly in our Hamiltonian, except for the Fermi hole, no interelectronic dependence appears in the H–F wave function. This is the classic correlation problem which has been studied so extensively for atomic helium [243, 243a] and for the electron gas.

Second, there are effects which will arise if any ψ_i in the H–F determinant is degenerate or nearly degenerate with any other unoccupied ψ. This is true, for example, in the four-electron beryllium-like ions [260], where we have a $1s^22s^2$ H–F ground state; due to a $2s - 2p$ near degeneracy, there is a strong $1s^22s^2 - 1s^22p^2$ configuration mixing which accounts for a large fraction of the correlation energy. In much of the discussion which follows, we shall quite arbitrarily refer to both this degeneracy effect and the first effect as "interelectronic correlation." Both contribute to a correlation hole which bears a strong resemblance to the Fermi hole already discussed. While interelectronic correlation corrections are most important between electrons of antiparallel spin, they also occur between electrons of parallel spin, i.e., H–F theory and its exchange interaction (or interactions) by no means supply the exact description of interelectronic behavior between such electron pairs.

Finally, there are the correlation effects associated with the restrictions of the conventional Hartree–Fock formalism, which we shall discuss in some detail later. Whether one calls these correlation effects or not is a matter of choice. We shall see that, when one attempts to compute this effect, he discovers it has strong similarities and striking differences with what he encounters when dealing with interelectronic correlation.

Let us now review two of the methods used for computing correlation effects in atoms. We shall not consider modern many-body theory [261], as this has only begun to be satisfactorily applied to the problems of interest to us here [261a].* We shall instead briefly review the Hylleraas method [263] and that of configuration mixing. These are both classical methods, in that they have been in use since the first days of quantum mechanics.

The Hylleraas method [263] introduces interelectronic r_{ij} coordinates explicitly into an analytic expression for the many-electron function. This, the obvious way to account for the fact that interelectronic coordinates appear in our Hamiltonian, has yielded highly accurate approximate wave functions for atomic He [243]. Helium is a case where the near-degeneracy and H–F restriction problems do not arise and where one is constructing a correlation hole which in some senses resembles the Fermi hole.

For many-electron systems, the Hylleraas approach is complicated by the fact that the number of interelectronic coordinates increases quadratically with the number of electrons and with problems associated with maintaining proper over-all wave function symmetry. These

---

* See Kelly [262], who has used Goldstone [262a] perturbation theory to estimate the correlation energy of Be. While most directly applicable to the correlation energy, this approach may prove useful in obtaining the expectation values of operators other than the energy. Also see reference [262b], which indicates the connection with the configuration interaction approach.

4. HYPERFINE INTERACTIONS IN MAGNETIC MATERIALS

symmetry problems bear some resemblance to the symmetry difficulties which we will see besetting "unrestricted" H–F theory.

Let us now turn to the configuration interaction method [264, 265],* in which one approximates the true eigenfunction by a sum of Slater determinants rather than with a polynomial in r_{ij} as in the Hylleraas approach. We shall limit discussion to the case where the determinants are constructed from an orthonormal set of ψ_i. While it is possible to do configuration interaction with nonorthogonal orbitals [266, 267], orthogonality follows logically from the Hartree–Fock formalism and is far simpler to discuss.

By "configuration" we mean the set of one-electron principal quantum numbers (for example, n and l in an atomic system) which specify the ψ_i making up a particular determinant. Configuration mixing consists of setting up and solving the secular equation of our Hamiltonian between a set of determinants or linear combination of determinants (which, for example, differ in m_l and m_s assignments of the ψ_i so that the linear combination is a proper eigenfunction of L or S). As the title suggests, each determinant or linear combination belongs to a different configuration from the others. This need not be so in practice, for situations occur where a single configuration yields more than one linearly independent function of appropriate symmetry. We shall see an example of this when we discuss lithium. Another example occurs when one investigates the partial breakdown of Russell–Saunders coupling and its effect[†] on atomic g-factors.

Let us take a H–F determinant as the "ground" determinant and inspect what matrix elements occur between it and other determinants. The rules are simple [254, 264, 265]. If the second determinant is constructed from a set of ψ_i which, with the exception of the jth, are common to the ground determinant, then the single substitution matrix element is

$$H_1 = \int \phi_j^*(\mathbf{x}) \mathcal{H}_{\text{eff}} \psi_j(\mathbf{x}) \, d\mathbf{x} \tag{3.8}$$

where ϕ_j denotes the substituted orbital, ψ_j the orbital associated with the ground determinant, the integration is over space and spin coordinates, and $\mathcal{H}_{\text{eff}}$ is the single-electron Hamiltonian appearing in Eq. (3.3). If the

---

* One should note that the superexchange problem, discussed by P. W. Anderson in Chapter 2 of Volume I of this treatise, has been traditionally a configuration interaction treatment of the coupling of magnetic ions.

† For example, see Judd and Lindgren [268], who used perturbation theory to estimate the configuration interaction effects.

excited determinant differs by two orbitals, ϕ_j and ϕ_k, from the ground determinant, the double substitution matrix element is

$$H_2 = \iint \phi_j^*(\mathbf{x}_1)\phi_k^*(\mathbf{x}_2) \frac{1 - P_{12}}{r_{12}} \psi_j(\mathbf{x}_1)\psi_k(\mathbf{x}_2) \, d\mathbf{x}_1 \, d\mathbf{x}_2 \qquad (3.9)$$

(where ψ_j has been replaced by ϕ_j and ψ_k by ϕ_k in the excited determinant). Orbital orthogonality leads to zero-valued contributions when more than two orbitals are substituted. This does not prevent determinants with more than two substitutions from appearing in the final many-determinant function but, if they do, their contribution will be from their interaction via intermediate determinants and not with the ground determinant directly.*

One more point must be made. If the ψ_i are eigenfunctions of H_{eff}, then the H_1 matrix elements have zero values. This follows from Brillouin's theorem [270]. However, as we shall see shortly, H–F orbitals are, in practice, often not eigenfunctions of Eq. (3.3), and so in configuration mixing with an H–F determinant, non-zero H_1 matrix elements do occur. Nesbet [264, 265] has investigated the symmetry properties of H–F orbitals and their role in these matters. The configuration interaction scheme indicates a natural division between those (single substitution) correlation effects arising from the failure to solve Eq. (3.3) and the interelectronic (double substitution) correlation effects. In practice this division will not be exactly defined, since there will in general be non-zero matrix elements connecting these two types of substituted determinants to one another [and to higher substituted configurations], and more importantly, proper maintenance of symmetry frequently requires their simultaneous presence.

Inspecting the correlation effects we wish to deal with suggests that the Hylleraas method is most appropriate for dealing with the r_{ij} singularity problem, while configuration interaction is better for orbital degeneracies and for any failure to solve the one-electron Schrödinger equations. Now, the r_{ij} problem tends to be the energetically most significant and for helium-like ions (where it is the only problem) the Hylleraas method converges far more satisfactorily.† Sinanoğlu [271, 272] has developed a version of double substitution configuration interaction where pair functions are used in the excited configurations. These pair functions can be of the Hylleraas type, and thereby give a marriage of the two

---

* It would be very convenient if one could limit one's investigation to singly and doubly substituted configurations. Unfortunately, more highly substituted configurations can be important (for example, see Watson [269]).

† For example, compare the results of Pekeris [243, 243a] and Weiss [266].

traditional approaches. Weiss and Martin [273] have introduced an alternate treatment of pair functions within particular shells for closed-shell ions. Their method can be extended and may prove more convenient in practice than Sinanoğlu's. The interested reader should also be aware of the relevant work of Szasz [274], Nesbet [264, 265], and Kelley [262]. Neither these nor the more traditional approaches have been used to obtain accurate solutions for systems of the size of interest to us here.

Before leaving the discussion of correlation effects we might note that, while both methods we have described are expansion methods, we have failed to mention such matters as the completeness of sets, linear dependence, or near-linear dependence. Despite their importance, we will not discuss them here. However, one should be aware that these matters, and serious problems associated with them, do exist.

3. Restrictions Associated with the Conventional Hartree–Fock Scheme

We have already indicated that the conventional H–F scheme is not as simple as what has been sketched above. Restrictions are introduced which affect the final form of the wave functions (and therefore of the matrix elements involving them) and which, in some senses, simplify the job of solving the H–F equations. We will discuss here the constraints for a case of particular interest: the atomic system. The extension of the restrictions to molecular and crystalline systems will then be apparent.*

Four restrictions are normally incorporated into conventional or restricted H–F (rhf) calculations:

(i) A one-electron spin orbital ψ_i is assumed, and thence constrained, to be separable into a product of spatial and spin functions, i.e.,

$$\psi_i(r, \theta, \varphi, \sigma) = \phi_i(r, \theta, \varphi)\mathscr{S}_i(\sigma), \tag{3.10}$$

where $\mathscr{S}_i(\sigma)$ is a spin function with a spin quantum number m_s equal to $\pm\frac{1}{2}$. Such separability was assumed in earlier discussions, but it should be noted that the general form of a spin orbital (since m_s takes on two values) is

$$\psi_i(r, \theta, \varphi, \sigma) = \phi_{i,+1/2}(r, \theta, \varphi)\mathscr{S}_{+1/2}(\sigma) + \phi_{i,-1/2}(r, \theta, \varphi)\mathscr{S}_{-1/2}(\sigma) \tag{3.11}$$

where the ϕ_{i,m_s} functions need not be at all similar; it is, in general, a restriction to require ψ_i to conform to Eq. (3.10). Symmetry difficulties,

---

* There is extensive literature dealing with molecular wave functions, the resulting magnetic hyperfine effects, and comparison with experiment (see reference [275]).

similar to those we shall encounter when dealing with other restrictions, can arise if we abandon Eq. (3.10). Overhauser [276], in his discussion of giant spin density waves, has relaxed this restraint for the free-electron gas problem. In general, it appears that the maintenance or relaxation of the restraint is most perinent to solids.* Since no applications have been made to hyperfine problems, we shall not refer to this matter again.

(ii) The spatial orbital ϕ_i is constrained to be separable into radial and angular functions, i.e.,

$$\phi_i(r, \theta, \varphi) = \frac{U_i(r)}{r} S_i(\theta, \varphi) \qquad (3.12)$$

For an atomic system, the $S_i(\theta, \varphi)$ are chosen to be spherical harmonics $Y_l^m(\theta, \varphi)$; in other words, ϕ_i is taken to be an eigenfunction of a spherical potential (for which the separability follows automatically). Now, the interelectronic Coulomb and exchange terms are exactly spherical only for S-state ions. This can be seen immediately if one inserts the self-energy terms into Eq. (3.3), for the Coulomb and exchange terms each become spherical. For other than S-state ions, the total potential is approximately, but not completely, spherical and the requirement of separability and of the use of spherical harmonics represents a real restraint in the formalism. The assumption of separability is often applied to cases involving nonspherical potentials, in which case the angular functions are no longer spherical harmonics. For example, the angular functions for a cubic potential would be cubic harmonics, which are linear combinations of spherical harmonics. The angular functions can be assigned prior to the application of the variational principle to Eq. (3.2), so that only the radial functions $U_i(r)$ are obtained variationally. In this way, the Hartree–Fock equations simplify from three-dimensional to one-dimensional equations. Of even greater importance is the fact that the separability restriction leads to the use of the one-electron quantum numbers n, l, and m_l for other than exactly spherical atoms. It must be emphasized that the use of n, l, and m_l for nonspherical atoms (which is necessary for the shell structure description of an atom) requires the introduction of this restriction into the H–F

---

* There is one important case in atoms where this restraint is modified, namely, during a change from the LS to the jj coupling schemes (when the one-electron m_l and m_s quantum numbers are replaced by j and m_j); in this case, one goes to a prescribed, i.e., not a variationally free, version of Eq. (3.11). In practice the nonrelativistic H–F scheme is inherently an LS scheme, the relativistic a jj scheme. Solids, of course, have translational symmetry which leads to an additional symmetry—the Bloch condition. Thompson [277] has discussed cases where one would relax this restraint.

formalism. This is Nesbet's [264, 265] "symmetry restriction" and its implications as well as those of the following restrictions (termed "equivalence restrictions" by him) on H–F and configuration interaction theory were first studied by him.

(iii) Assuming (ii), we constrain $U_i(r)$ to be independent to the m_l, value associated with ϕ_i. This is not a restriction for the case of a spherical atom.

(iv) Likewise, $U_i(r)$ is constrained to be independent of m_s also. This is not a constraint for ions where the total ion spin quantum number is a good quantum number and equals zero.

These last two restrictions imply a single $U_i(r)$ for any shell (i.e., n and l value) and thus a single H–F radial equation per shell rather than one per electron. The reduction in the number of simultaneous equations makes their solution easier, of course. In practice, the restricted H–F radial equation for a shell is an average of the H–F radial equations derived for the individual radial functions $U_i(r)$ of the non-S-state ions in that shell. For other than ^1S ions, this leads to a set of ψ_i functions which are not exact eigenfunctions of Eq. (3.3) and hence may not always be (automatically) orthogonal. The maintenance of orthogonality, then, requires the introduction of additional constraints, which can be off-diagonal Lagrange multipliers or take some other form. This matter is discussed in Appendix A.

Of greater importance than any reduction in the magnitude of a computation is the fact that the relaxation of any of the above restraints [excepting simple unitary transformations within a determinant, such as of pairs of spin orbitals from the form of Eq. (3.10) to that of Eq. (3.11)] leads to the collapse of the conventional atomic shell structure formalism. In other words, closed shells would cease to be "closed" in the sense of making inert ^1S contributions to an ion's spin and angular symmetry. This, in turn, implies a partial abandonment of a very successful description of atomic systems. We shall illustrate this with the example of atomic lithium later.

In many ways the H–F restrictions have but a small effect on the final wave function. This is because the unbalanced exchange terms and aspherical potential terms that arise are small compared with the spherical nuclear and interelectronic potential terms of the atom. The effect of maintaining or of relaxing these restraints has only a 10^{-3} to 10^{-4}% effect on the computed total energy of an atomic system. As we have already indicated, the conventional H–F method will sometimes yield nonorthogonal ψ_i functions unless additional constraints are applied. However, if one neglects to apply them, non-zero overlap integrals defined by

$$\Delta_{ij} \equiv \int \psi_i{}^*(\mathbf{x})\psi_j(\mathbf{x}) \, d\mathbf{x} \tag{3.13}$$

of only ≈ 0.0005 or less will occur. Thus, these nonorthogonality effects are also small. The effects associated with the conventional H–F restrictions, on the other hand, are not always small; for example, they affect calculations of hyperfine interactions in a nontrivial way.

Suppose we investigate the effect of the constraints in the restricted H–F (RHF) theory on the computation of observables. After first obtaining the RHF single determinant, we obtain the ψ_i functions for that determinant where the restrictions of interest have been relaxed. This results in an "unrestricted Hartree–Fock" (UHF) function.* Since we have relaxed a constraint, the resulting total energy is, of necessity, as low or lower than the RHF value. In addition, if sufficient restraints were relaxed, the ψ_i will be orthonormal (this may not automatically be the case for the RHF function) and they will be eigenfunctions of Eq. (3.3) defined for the resultant determinant. This implies that UHF theory is intimately associated with the single substitution problem [264, 265].

In addition to the H–F restrictions, we must also consider the interelectronic correlation effects. Whereas the UHF effects may be treated either by single substitution configuration interaction or within the H–F formalism, the interelectronic effects are handled by double substitution configuration interaction or some counterpart. The UHF problem has been found to be far more tractable than that of interelectronic correlation. This has led to a policy of studying the former in relatively great detail, of trying to understand and make accurate estimates of the effects associated with it, and of ascribing any remaining differences between theory and experiment to interelectronic correlation. There are obvious dangers in such a policy; but we believe it has considerable justification, particularly if one wishes to deal with systems involving more than a few electrons.

We have not indicated how one actually goes about solving such H–F equations and it is not our purpose to do this in this chapter. The interested reader will, however, find a few brief comments and sufficient references in Appendix A to learn about these matters in detail. We

---

* It should be noted that, in the published literature, the term "unrestricted Hartree–Fock" has been associated almost exclusively with the fourth (the m_s dependent) RHF restriction. The term should logically be associated with the relaxation of all four restraints.

should note that there are other problems in addition to the problem of actually solving the H–F equations, e.g., there is the question of the stability of the resulting H–F many-electron state [278–280]. This is of considerable interest when one is comparing RHF and assorted UHF [280], such as in the case of the Overhauser giant spin density waves [276].

4. Exchange Polarization

The phenomenon called "exchange polarization" was discussed by Slater [281, 252] in treating magnetic effects. For a system with a net unpaired spin, $S_z \neq 0$, Slater emphasized that electrons with spin parallel to the net spin would experience different exchange interactions from those of electrons having antiparallel spin. The result of this would be that electrons with different spins would have different wave functions. In view of what we have been discussing, exchange polarization is obviously associated with the so-called spin, or exchange, polarized H–F theory, i.e., the case where the m_s restriction has been relaxed. This will be reviewed in some detail as an example of the UHF theory in application; but before we do this, we will briefly note two other sorts of exchange polarization which are important for dealing with hyperfine effects in solids and which are not unrelated to the UHF case.

The first type is that which Zener [282] developed when investigating the sources of exchange interactions in ferromagnetic metals. It consists of the polarization of a conduction band by a localized spin via its exchange interaction. Here the polarization process simply depopulates the conduction band states of one spin and fills those of the other. The calculation of this polarization has been carried to a higher order in perturbation theory than was considered by Zener, for a conduction band described in the free electron approximation by Ruderman and Kittel [283], Kasuya [284], Yosida [285], and others [286]. The Ruderman-Kittel theory, as applied to s-d and s-f interactions in metals, is discussed in this treatise by Kasuya in Chapter 2 of Volume IIB. Contrary to the uniform polarization obtained with first-order theory, they find an oscillating spin density which can vary in sign and which appears to be pertinent [287] when one is discussing transferred hyperfine effects in magnetic alloys and intermetallic compounds. Obtained within the context of perturbation theory, these second order terms are *exactly* the UHF contribution to the conduction electron case and suffer from the same uncertainties associated with a lack of proper spin symmetry. It should be noted that a detailed treatment would involve a complicated mixture of the various types of polarization which we are discussing. The approach is that of an impurity problem for which perturbation theory is very convenient but, since it is essentially a treatment of single substitution configuration interaction effects, it could be approached alternatively with the exchange polarized H–F formalism. One promising area for this would be the polarization of conduction electrons in simple ferro- or antiferromagnetic metals. Callaway [288] has investigated this using the approximate Slater $\rho^{1/3}$ exchange potential.

The second type of polarization arises in the analysis [289–299, 299a] of transferred hyperfine effects in salts such as the iron series fluorides. It does not involve electrostatic exchange but is associated with the properties of an antisymmetric function which has been constructed from nonorthogonal basis functions. The nonorthogonality arises from adopting a tight binding approach to the solid. In a salt such as $KMnF_3$, for example, we are interested in the F^{19} hyperfine interactions due to the neighboring unfilled $3d$ shells of the Mn^{2+} ions. We shall sketch this matter but briefly here.

As is usual we shall limit discussion to a single $Mn^{2+} - F^-$ pair (since one may then sum over individual pair contributions after completing the computation). Adopting the tight binding method, we simply insert Mn^{2+} and F^- functions (e.g., free-ion H–F functions) into a determinant and evaluate the hyperfine field. Now the Pauli exclusion principle, .i.e., the antisymmetry of the determinant, has an automatic orthogonalization process built into it (the field obtained with nonorthogonal orbitals is identical with that obtained by orthogonalizing the orbitals first and inserting these into the determinant). This procedure leads, first, to an unpairing of the fluorine s-electron shells and hence to an isotropic hyperfine effect (from the Fermi contact term) and, second, to an unpairing of the $2p$ shell and, in turn, an anistropic interaction. For simplicity, let us consider a Mn–F pair and limit our attention to the F^- $2s$ shell. The effect of inserting the $2s$ orbitals and the unpaired $3d$ orbital nonorthogonal to them into the determinant would be identical to the following transformations:

$$\psi_{3d\uparrow}(\mathbf{x}) \to (1 - \Delta_{3d,2s}^2)^{-1/2} \{\psi_{3d,\uparrow}(\mathbf{x}) - \Delta_{3d,2s}\psi_{2s}(\mathbf{x})\}$$
$$\psi_{2s\uparrow}(\mathbf{x}) \to \psi_{2s\uparrow}(\mathbf{x}) \quad (3.14)$$
$$\psi_{2s\downarrow}(\mathbf{x}) \to \psi_{2s\downarrow}(\mathbf{x})$$

where $\Delta_{3d,2s}$ is the two-center overlap integral [Eq. (3.13)] and $\uparrow$ and $\downarrow$ denote m_s values of $+\frac{1}{2}$ and $-\frac{1}{2}$, respectively, for the spin orbitals. We see that there is now some unpaired $2s$ character associated with the orbitals. Neglecting ψ_{3d} contributions, the resulting Fermi contact term is

$$H_c = 4\pi \frac{\Delta_{3d,20}^2}{1 - \Delta_{3d,2s}^2} |\psi_{2s}(0)|^2 \quad (3.15)$$

where the 2s expectation value has been evaluated at the F^- nucleus. Although the above transformations are commonly used because of the ease with which covalent mixing can be included in the function, it may be of interest to inspect an exactly equivalent orthogonalizing transformation, namely,

$$\psi_{3d\uparrow}(\mathbf{x}) \to \psi_{3d\uparrow}(\mathbf{x})$$

$$\psi_{2s\downarrow}(\mathbf{x}) \to \psi_{2s\downarrow}(\mathbf{x}) \qquad (3.16)$$

$$\psi_{2s\uparrow}(\mathbf{x}) \to (1 - \Delta_{3d,2s}^2)^{-1/2}\{\psi_{2s\uparrow}(\mathbf{x}) - \Delta_{3d,2s}\psi_{3d\uparrow}(\mathbf{x})\}$$

which, of necessity, yields the H_c above. Here we see the unpairing effect written explicitly into the ψ_{2s} functions.

When dealing with transferred hyperfine effects for cases such as the iron series fluorides, one must [295, 296] include the unpairing effects on all ligand shells. In addition, one normally (for most recent discussion see [298, 299]) allows covalent mixing to occur between the outermost ligand orbitals and the iron series ion $3d$ orbitals, obtaining an H_c which is dependent on the extent of the mixing. Given theoretical one-electron orbitals and overlap integrals, one can analyze data for the iron series fluorides so as to determine the extent of this covalent mixing. Valuable information about the wave functions obtained this way has been utilized in further theoretical investigations (see references [298, 299]). The process is complicated [300] by a number of small contributions to the transferred hyperfine interaction whose total need not be negligible.

The unpairing effect is not unique to transferred hyperfine interactions; variants of it inevitably turn up when one deals with an antisymmetric function constructed from nonorthogonal orbitals. We shall see yet another example of it shortly. While it does not directly involve electrostatic exchange, the unpairing effect has the same source (the Pauli exclusion principle) and differentiates between electrons of differing spin in a similar way.

5. The Spin-, or Exchange-, Polarized Hartree–Fock Method

The spin-, or exchange-, polarized Hartree–Fock method is simply a treatment where the m_s restriction of the RHF formalism is relaxed and the single determinant is otherwise treated as before. This method has been discussed extensively [246, 264, 265, 281, 301–306] and many computations have been done with it [12, 258, 295, 307–313]. Calculations of the hyperfine interactions have also been done using perturbation

[18, 18a, b, 19, 314–316]* and configuration interaction [3, 20, 21, 22, 317–321] techniques; results with the former have, however, proven to be disappointing [313, 316].

The first spin-polarized H–F calculations where exchange was rigorously handled were made by Sachs [308] and Nesbet and Watson [309] for atomic lithium. While this atom is not of direct interest to us, in the rest of this chapter it provides us with a simple example of the exchange-polarized H–F formalism; in addition, much of the existing theoretical work has been concerned with it. Much of what we shall observe will also be pertinent to calcualtions which involve the relaxation of other RHF restraints. As we shall see, the situation is complicated even for this simplest system. In this section, we give some indication of the nature of these complications with Li as an example without either going into all the details or trying to resolve them; results of calculations for heavier ions are then presented and discussed.

a. **Methodology and complications.** Atomic Li as an example. Lithium in its ^2S ground state is described by a $(1s)^2 2s$ RHF single determinant. The RHF total energy is -7.4327 au (1 au $= 2$ ryd), compared with an experimental value [266] of -7.4781 au; but the wave function yields a Fermi contact term H_c, due entirely to the $2s$ spin orbital, whose value is only about 72% of the experimental value.

If we consider the Li state where $S = M_s = \frac{1}{2}$ (i.e., the $2s$ electron spin is "up"), derive the individual H–F equations [Eq. (3.3)] for the two $\psi_{1s}(\mathbf{x})$ spin orbitals, and integrate over spin coordinates, we obtain the following relationship between their spatial effective Hamiltonians $H_{\text{eff}}(\mathbf{r})$:

$$H_{1s}(\mathbf{r})_\uparrow = H_{1s}(\mathbf{r})_\downarrow - \int d\mathbf{r}_2 \phi_{2s}^*(\mathbf{r}_2) \frac{P_{12}}{r_{12}} \phi_{2s}(\mathbf{r}_2) \qquad (3.17)$$

As we might expect, the $1s\uparrow$ has an exchange interaction with the $2s$, whereas the $1s\downarrow$ does not, and thus the solution of their H–F equations will yield differing $\phi_{1s}(\mathbf{r})$ or $U_{1s}(r)$ spatial functions. The RHF requirement that there be one $U_i(r)$ per shell is normally† met by an averaging procedure which, for Li, yields

$$H_{1s,\text{RHF}}(\mathbf{r}) = H_{1s}(\mathbf{r}) - \tfrac{1}{2} \int d\mathbf{r}_2 \phi_{2s}^*(\mathbf{r}_2) \frac{P_{12}}{r_{12}} \phi_{2s}(\mathbf{r}_2) \qquad (3.18)$$

---

* Sternheimer [23] presented the first calculation which considered the effect of spin polarization of all s shells on the Fermi contact term.

† For further discussion and for the various RHF and UHF equations which are written out for Li, see Nesbet and Watson [309]. Note that the normalization constant appearing in their equation 19 is in error.

hence, neither resulting $\psi_{1s}(\mathbf{x})$ will be an eigenfunction of Eq. (3.3). It is these solutions which yield an H_c in 72% agreement with experiment.

Now let us lift the RHF m_s restriction and do an exchange-polarized calculation for Li. When this is done [308, 309, 313] one obtains a $\{\psi_{1s\uparrow} \psi_{1s'\downarrow} \psi_{2s\uparrow}\}$ determinant (we shall henceforth designate this by [$1s\uparrow\ 1s'\downarrow\ 2s\uparrow$]) whose ψ_i are orthonormal and are eigenfunctions of Eq. (3.3), defined for this determinant. The spatial parts $\phi_i(\mathbf{r})$ of the $1s$ and $1s'$ electrons differ (hence the prime), with $\langle \phi_{1s} | \phi_{1s'} \rangle \approx 0.999991$ and $\langle \phi_{2s} | \phi_{1s'} \rangle \approx 0.00167$. One can, of course, make linear combinations of ψ_{1s} and ψ_{2s} such that $\langle \phi_{2s} | \phi_{1s} \rangle = 0$ for the new ψ_{2s} without changing the character of [$1s\uparrow\ 1s'\downarrow\ 2s\uparrow$]. We would still have $\langle \phi_{1s} | \phi_{1s'} \rangle \neq 1$, i.e., the three space orbitals would remain linearly independent.

The exchange-polarized function yields a value for H_c which is in 97% agreement with experiment, although the improvement in the total energy is very small, as can be seen in Table VIII. While the $2s$ electron makes roughly the same contribution to H_c as in the RHF case (the difference is less than $\approx 1\%$), the opened-up $1s$ shell provides the improved agreement with experiment.

TABLE VIII

Comparison of Total Energy and H_c for Li as Predicted by Selected Wave Functions[a]

Wave function	Energy	H_c	% Exp. H_c	Ref.
RHF	−7.4327	2.095	72	[308]
UHF	−7.4327$_5$	2.825	97	[308]
Projected UHF	−7.4327$_5$	2.345	81	[308]
James and Coolidge correlated K shell times $2s$	−7.4748	2.648	91	[329]
Best James and Coolidge	−7.4763	2.883	99	[329]
Weiss configuration interaction correlated K shell times $2s$	−7.4740	3.989	137	[273]
Best Weiss function	−7.4771	2.595	89	[273]
Burke's Hylleraas function	−7.47797	2.8258	97	[333]
Experiment	−7.47807	2.906		

[a] All quantities are in atomic units.

Now, one can construct two spin doublets and one quartet from three linearly independent space orbitals. This immediately raises the question of whether the exchange-polarized function represents a pure doublet state; it does not. Sachs [308] has obtained a value of $\langle S^2 \rangle$ for Li of

0.750016, as compared with a pure doublet $S(S+1)$ value of $\frac{3}{4}$. In general, the $\langle S^2 \rangle$ values, computed for exchange-polarized H–F functions, deviate from the pure spin state values.* This means that they are not exact spin eigenfunctions, as they should be. Projection operators [246, 247, 247a, b, 322, 323] can be used to project out the function of proper symmetry from the exchange-polarized single determinant. If one does this for Li, he obtains

$$\Psi_1 \equiv \theta_s[1s\uparrow\ 1s'\downarrow\ 2s\uparrow] = N_1\{[1s\uparrow\ 1s'\downarrow\ 2s\uparrow] - \tfrac{1}{2}[1s\downarrow\ 1s'\uparrow\ 2s\uparrow] - \tfrac{1}{2}[1s\uparrow\ 1s'\uparrow\ 2s\downarrow]\} \tag{3.19}$$

where θ_s is the spin projection operator and N_1 a normalization constant. It must be emphasized that Ψ_1 is the doublet function appearing in $[1s\uparrow\ 1s'\downarrow\ 2s\uparrow]$ and that the determinants are not orthonormal because the $\phi_i(\mathbf{r})$ are not. Although Ψ_1 looks comparatively innocuous, its counterpart for exchange-polarized Mn^{2+}, for example, involves hundreds of thousands of determinants—a description which goes well beyond the simple one-electron orbital picture of single-determinant H–F theory. In this sense it can be argued that we are dealing with correlated functions, though of a rather special type.

There are a number of other properties of Ψ_1 and its constituent one-electron orbitals which are of varying interest and importance. We will consider but two of these here. First, we have observed that ϕ_{1s} and ϕ_{2s} could be transformed with no effect on $[1s\uparrow\ 1s'\downarrow\ 2s\uparrow]$, so that the only non-zero overlap integral was $\langle \phi_{1s} | \phi_{1s'} \rangle$. The same transformation can be made on Ψ_1 with no effect on it or on any expectation value computed with it. Löwdin [324] has shown that whenever one deals with the spin projection of a single determinant, one can transform the ϕ_i functions so that each ϕ_i is nonorthogonal to no more than one other orbital, say ϕ_i' (where the m_s value of ϕ_i' in the original single determinant differs from that of ϕ_i).† This "pairing" theorem offers tremendous simplifications when one is dealing with such functions, although complete orthogonality among the ϕ_i will not be obtained. This means that some determinants (in lithium, it is the third determinant in the expression for Ψ_1) will be constructed from nonorthogonal spin orbitals $\psi_i(\mathbf{x})$.

Second, the exchange-polarized H–F spin orbitals are not eigenfunctions of one-electron Hamiltonians defined for Ψ_1. In order to obtain

---

* See Watson and Freeman [12] and Appendix A therein for several $\langle S^2 \rangle$ values for exchange-polarized iron series functions.

† See the discussion in the vicinity of Löwdin's equation 85 for an indication of how to go about proving this.

such eigenfunctions, the proper procedure would be to apply the variational principle to the total energy of Ψ_i and obtain and solve what are now known as the "extended H–F equations" [246]. If one neglects to do this and inserts the exchange-polarized orbitals into Ψ_1, i.e., if the function is symmetrized after the orbitals are obtained, one still obtains a slightly better total energy and a value of H_c in 80%* agreement with experiment (see Table VIII). Note that H_c has been rather strongly affected by what has otherwise appeared to be but a small effect.

As previously stated, it is the extended H–F scheme which is really the appropriate way to deal with the problem; while it is quite feasible to solve the equations for Li, it has yet to be done rigorously. Two unpleasant features are associated with the straightforward application of the extended H–F scheme to larger multi-electron systems. First, one is solving equations for space orbitals $\phi_i(\mathbf{r})$ and not spin orbitals $\psi_i(\mathbf{x})$; hence, the orbitals will not have a common one-electron Hamiltonian. Because the resulting ϕ_i are not completely orthogonal, one will encounter a variety of new terms in the one-electron Hamiltonian unlike those appearing in the simple RHF or exchange-polarized H–F theories. These will prove somewhat inconvenient to deal with. Second, the very fact that the RHF formalism is a good approximation and that relaxing its restrictions has but a small effect on the orbitals leads to severe problems in maintaining computational accuracy. This can be seen for Li by inspecting the [$1s\uparrow$ $1s'\uparrow$ $2s\downarrow$] determinant, where the two $1s$ orbitals are of common spin. Now, the effect of inserting the two orbitals into a determinant is synonomous with orthogonalizing them, as we have just discussed with regard to transferred hyperfine effects. However, since they are almost identical, if we want to obtain a total wave function good to a certain number of digits, we must first obtain ϕ_{1s} and $\phi_{1s'}$ to a good number more. For Li, with $\langle \phi_{1s} | \phi_{1s'} \rangle \approx 0.999991$, we require five extra digits; the problem becomes even more severe with large systems. Some progress has been made in obtaining analytic expressions for extended H–F one- and two-electron density matrices [325]. Once these are obtained, then nonorthogonality problems will become less severe and also one will not have to deal with the array of determinants but with the simpler density matrix expressions instead. Unfortunately these results cannot be applied to the important class of multideterminant functions which are not projections of single determinants.

In the above discussion we ignored the fact that a second doublet function can be constructed from the three Li exchange-polarized orbitals $\phi_i(\mathbf{r})$. We should note that as soon as we abandon consideration

---

* Sach's [308] reported a value of 88 %, which is incorrect.

of Ψ_1 alone, we are abandoning the UHF procedure and its treatment of the RHF restrictions. Instead, we are considering a segment of the general correlation problem. Despite this, let us consider the matter of the two doublets briefly, as it may shed some light on the properties of Ψ_1. Instead of obtaining the doublet orthogonal to Ψ_1, it is perhaps more appropriate [326] to consider another pair of linearly independent doublet functions (from which Ψ_1 can be constructed), namely,

$$\Psi_2 \equiv N_2\{[1s\uparrow\ 1s'\downarrow\ 2s\uparrow] - [1s\downarrow\ 1s'\uparrow\ 2s\uparrow]\} \tag{3.20}$$

and

$$\Psi_3 \equiv N_3\{[1s\uparrow\ 1s'\uparrow\ 2s\downarrow] - \tfrac{1}{2}[1s\uparrow\ 1s'\downarrow\ 2s\uparrow] - \tfrac{1}{2}[1s\downarrow\ 1s'\uparrow\ 2s\uparrow]\}$$

$$= \theta[1s\uparrow\ 1s'\uparrow\ 2s\downarrow] \tag{3.21}$$

which are clearly orthogonal. Ψ_2 is a linear combination of the antisymmetrized product of $\psi_{2s\uparrow}(\mathbf{x})$ times $\{[1s\uparrow\ 1s'\downarrow] - [1s\downarrow\ 1s'\uparrow]\}$; the latter factor is of the form of the "open-shell" ^1S functions which describe very well [327, 328] radial correlation effects in helium-like ions. The $1s$ shell is treated symmetrically in such a function and therefore is "obviously" not associated with exchange polarization. (Note that this sort of functional character is also built into the projected spin-polarized function Ψ_1 to which we shall return shortly.) One unfortunate feature of Ψ_2 is that it is not projected from a single determinant and therefore one cannot invoke Löwdin's pairing theorem to simplify matters. Finally, we should note that the Fermi contact term H_c for Ψ_2 is not due entirely to the $2s$ electron, for there are additional small terms arising from orbital nonorthogonality. In other words, one has yet another case of the nonorthogonality polarization discussed earlier.

If the exchange polarization effects do not appear in Ψ_2, then it is "clear" that they are associated with Ψ_3. The function Ψ_3 has the property that if ϕ_{1s} and $\phi_{1s'}$ are made identical, Ψ_3 vanishes, while Ψ_1, Ψ_2 and $[1s\uparrow\ 1s'\downarrow\ 2s\uparrow]$ all go over into the RHF ^2S function. Because of this, one should not deal with Ψ_3 alone. In a function such as Ψ_1 (which is a linear combination of Ψ_2 and Ψ_3) the cross terms between Ψ_2 and Ψ_3 are significant and one cannot simply add their individual contributions to obtain the behavior of Ψ_1. This, in turn, means that Ψ_2, via the cross terms, plays an active role in the exchange-polarized behavior of Ψ_1.

As we have already noted, any combination of Ψ_2 and Ψ_3 other than Ψ_1 introduces interelectronic as well as exchange-polarization correlation effects (and it can be argued that such interelectronic effects occur in Ψ_1 as well). As these functions appear to be a poor way to deal with the interelectronic effects, we shall not discuss these matters further here. Various features of the properties of Ψ_2 and Ψ_3 have not been discussed, and in particular, the character of the $\phi_i(\mathbf{r})$'s appropriate to them (which will differ from one Ψ_i to another and with Ψ_1 or [1s↑ 1s↓ 2s↑]). For systems involving more electrons, the number of functions which are of the same symmetry as the extended H–F function (and which are constructed from the same set of determinants) increases rapidly. If one wished to deal with such an array of functions one would have a far less tractable situation than the already formidable extended H–F problem.

It has been noted that single substitution configuration interaction is associated with the failure of the RHF method to provide eigenfunctions of Eq. (3.3). Lefebvre [305] and Bessis et al. [320], using perturbation theory arguments, have shown that one-to-one relations exist between single substitution configuration interaction and the exchange-polarized and extended H–F methods. It has been single substitution configuration interaction, rather than the extended H–F method, which has been used to give a properly symmetric description of the effect on the lithium Fermi contact term of relaxing the RHF m_s restriction. Nesbet's [318] and Powell's [321] investigations were carried far enough to indicate that if one starts with an accurate RHF function and treats the single substitution problem in detail, one obtains a value of H_c which is in $\approx 95\%$ agreement with experiment and an energy improvement of $\sim 10^{-4}$ au over the RHF value (or two times that associated with the exchange-polarized function). If we had similar results for several other ions, we would be temped to argue that exchange effects, rather than those of interelectronic correlation, are largely (although by no means entirely) responsible for differences between RHF H_c values and experiment. Unfortunately, no such results exist. Bessis et al. [320] were unable to carry out their configuration interaction study of nitrogen far enough to supply such information unambiguously.

Recently, Berggren and Wood [329] arrived at conclusions which strongly contradict those inferred from the work of Nesbet [318] and Powell and Marshall [321]. They evaluated H_c for a number of the James and Coolidge [330, 331] Li functions. In particular, for the antisymmetrized product of a simple $2s$ orbital times a symmetrically treated correlated $1s^2$ core, they obtained an H_c value in $\approx 91\%$ agreement with experiment. Values of H_c obtained for functions where all three electrons were correlated (i.e., where r_{12}, r_{13}, and r_{23} coordinates were all present) tended to improve on this (see Table VIII). From the good values of the H_c's and of the wave function total energies, they concluded that they had converged on an accurate approximation to the true Li eigenfunction. Then, since exchange polarization was not explicitly introduced into the functions yielding a good H_c, they concluded that it plays little or no role in the Li hyperfine interaction. Martin and Weiss [273] have evaluated H_c for the Weiss nonorthogonal configuration interaction functions [266] for Li. For a correlated core plus simple 2s orbital, with an energy similar to its James and Coolidge counterpart, they obtained a 137% H_c value. The best Weiss function yields an 89% H_c and a total energy twice as good (i.e., half the unaccounted correlation energy) as the best James and Coolidge function (see Table VIII). Burke has obtained [332] a properly symmetrized Hylleraas function for Li whose energy in an order of magnitude superior to Weiss's and it yields [333] a 97% H_c. It remains to be seen if this finally represents "real" convergence onto true Li eigenfunction behavior. The above treatments of Li correlation used nonorthogonal expansions and did not use H-F functions as their first terms. This makes it almost impossible to casually inspect a particular function and uniquely analyze it from the H-F viewpoint presented in earlier subsections. Any analysis relying simply on such inspection is subject to challenge by alternate points of view. Nesbet has suggested [334] that the best basis for making comparisons between such functions would be to analyze them in terms of natural spin orbitals [246]. One advantage would be the natural connection which could then be made to H-F theory. Such analyses have been done [335, 336] for correlated two-electron functions but the method remains to be extended to the more difficult three-electron open-shell systems.

Before leaving this confused state, one final complication should be noted. The correlated function H_c's which agree numerically with experiment are, on naïve inspection of the matrix elements, closely associated with the open 2s-closed 1s shell nonorthogonality. The same nonorthogonal H_c terms would not occur for larger S-state ions, such as $N(2p^3)$, $P(3p^3)$ and $Mn^{2+}(3d^5)$, which have substantial H_c's despite the zero-valued prediction of RHF theory. If "pure" interelectronic correlation, rather than exchange polarization, should prove to determine H_c behavior, then it does this in a very different way for these ions than for Li.

We believe that there is considerable justification in trying to ascertain roughly the contributions of each of the various sorts of correlation effects to the expectation value of an observable such as H_c. At the moment, only the exchange-polarized H-F method can be carried through to reasonable completion for an ion the size of say, Mn. For this reason we shall return to this method and cite some results so obtained.

b. Computational results. In the absence of exchange-polarized calculations for magnetic solids one uses, of necessity, results which have been obtained for free ions and atoms and/or for ions in crudely approximated environments. We shall therefore present and discuss some of these results to show the sensitivity of the calculations to various factors and the sort of agreement with experiment one might expect. The values quoted here will be referred to later (Section IV) when experimental results in solids are discussed.

A convenient measure of the Fermi contact interaction arising from the difference between spin densities at the nucleus $|\psi\uparrow(0)|^2 - |\psi\downarrow(0)|^2$ for up ($\uparrow$) and down ($\downarrow$) spins was introduced by Abragam and Pryce [3] and is given by the quantity

$$\chi = (4\pi/2S) \sum_{ns \text{ shells}} \{\rho_{ns\uparrow}(0) - \rho_{ns\downarrow}(0)\} \tag{3.22}$$

Here $\rho = |\psi(0)|^2$ and χ is simply the effective field per unpaired electron; with χ in atomic units (au), the hyperfine field arising from the contact term H_c is found in gausses by using the conversion factor 1 au = 4.21×10^4 gausses.

(1) *Exchange-polarized results for some neutral atoms.* We list in Table IX some RHF, spin-polarized H–F, and experimental values of H_c for some low Z atoms and the percentage of these values with respect to experiment. The results are given in kilogausses to emphasize the large magnitude of the effective magnetic fields. Minus signs indicate effective fields due to an s-electron spin density whose spin direction is antiparallel to the spin of the ion. The exchange-polarized results are of the correct sign and magnitude; for the atoms Li, Na, and K, which have an unpaired outer s electron, they give values of H_c which are closer to experiment than are the RHF values. The poorest results occur for nitrogen and phosphorus. For nitrogen the exchange-polarized H–F result is ~ 2 times the experimental value. There is no obvious reason for this discrepancy in nitrogen. In the case of P, experiment and theory are found to yield results of the same magnitude but of differing sign. This result means that either an error in experimental sign has been made or that the first serious breakdown in the ability of the exchange polarization model to understand experimental fact has been found. Similar results have been obtained using configuration interaction [337]. Arsenic, the next ^4S p^3 ion after N and P has just been studied by Moser and co-workers who find agreement to $\sim 10\%$ between a spin polarized H–F calculation and experiment.

(2) *Exchange-polarized results for some iron series atoms.* Some results for H_c in neutral (but not necessarily ground state) iron series atoms are listed in Table IX. Here, the results, when compared with experiment,

TABLE IX

Fermi Contact Term Effective Fields (H_c) as Obtained by Restricted and Exchange-Polarized H–F Calculations and from Experiment for Various Atoms[a]

		H_c (RHF)	H_c (exchange-polarized H–F)	H_c (Exp.)
Li	$2s\ ^2S$	86 (71%)[b]	118 (97%)[b]	122[c]
Na	$3s\ ^2S$	297 (76%)[b]	339 (86%)[b]	394[c]
K	$4s\ ^2S$	360 (62%)[b]	447 (77%)[b]	581[c]
N	$2p^3\ ^4S$	0	92 (180%)	51
P	$3p^3\ ^4S$	0	−75 (−135%)	48
F	$2p^5\ ^2P\ (J=3/2)$	0	70 (126%)[b]	55[d]
Sc	$3d^1\ 4S^2\ ^2D$	0	−7[e]	0[f]
V	$3d^3\ 4S^2\ ^4F$	0	−57[e]	−315[f]
Mn	$3d^5\ 4S^2\ ^6S$	0	−114[e]	0[f]
Co	$3d^7\ 4S^2\ ^4F$	0	−76[e]	−210[f]

[a] All H_c's are in kilogausses.
[b] Goodings [313].
[c] Kusch and H. Taub [338].
[d] Radford *et al.* [339].
[e] Watson and Freeman [12].
[f] Abragam *et al.* [22].

are again not too satisfactory. The exchange-polarized Fermi contact terms for the neutral iron series atoms come about from a cancellation of the core contributions ($1s$, $2s$, and $3s$) with those of the $4s$ shell and, in essence, have zero values. The non-zero experimental H_c terms for V and Co appear to be due to low-lying configuration interaction [12] effects. In other words, we believe that the zero-valued experimental H_c terms of Sc and Mn are characteristic of the row, and that the exchange-polarized results crudely reproduce this behavior. The small magnitudes of the computed values arise from a strong cancellation between large positive $4s$ shell and large negative ($1s$, $2s$, and $3s$) ion core contributions to H_c. The latter are essentially identical with the results for the divalent ions, which we shall now see.

(3) *Exchange-polarized results for iron series ions.* The theoretical values of H_c and χ for a number of iron series ions [12] are listed in Table X. The calculated values of χ agree in both sign and magnitude with the experimental values reported by Abragam *et al.* [22] based on an analysis of experimental hyperfine data for these ions in hydrated salts, but are consistently more negative. As suggested by AHP we see that χ has a roughly constant value of ~ -3 au, equivalent to

TABLE X

Theoretical Contact-Term χ's, H_c's, and Individual s-Shell Contributions to χ as Obtained for Mn^{2+}, Fe^{3+}, Fe^{2+}, and Ni^{2+} [12]

Ion	$Mn^{2+}(3d^5)$	$Fe^{3+}(3d^5)$	$Fe^{2+}(3d^6)$	$Ni^{2+}(3d^8)$
χ (au)	−3.34	−3.00	−3.29	−3.94
1s-Shell contribution to χ	−0.16	−0.25	−0.21	−0.27
2s-Shell contribution to χ	−6.73	−8.51	−7.80	−9.62
3s-Shell contribution to χ	+3.55	+5.77	+4.72	+5.95
H_c (kgausses)	−700	−630	−550	−330

~ -125 kgausses. The negative sign indicates that the contact-term spin density is antiparallel to the net spin of the ion. The rough constancy of χ suggests that we are dealing with a polarizability which is linear in spin and this has important consequences in later discussions [cf. Section IV].

Inspection of the individual shell contributions to χ (results for Mn^{2+} are given in Table XI) shows that the resultant H_c arises from a competition of terms of opposing sign. The 1s and 2s shells, which lie inside the 3d, have their majority spin electrons "attracted" outward, leaving a region of negative spin density near the nucleus, which in turn gives a negative contribution to H_c. Owing to its close proximity to the 3d electrons one might expect the 3s-shell contribution to dominate, but this shell lies neither "inside" nor "outside" the 3d. The overlapping of the shells leads to competing tendencies and a contact contribution which is smaller than that of the 2s and is positive, i.e., the 3s shell acts as if it lies "outide" of the 3d insofar as χ is concerned.

In this discussion we have spoken of majority spin s electrons as

TABLE XI

Individual s-Electron Contributions (in kilgausses) to H_c as Computed for Mn^{2+} [a]

		$ns \uparrow + ns \downarrow$
$1s \uparrow$	2,502,840	~ -30
$1s \downarrow$	−2,502,870	
$2s \uparrow$	226,670	−1410
$2s \downarrow$	−228,080	
$3s \uparrow$	31,210	+740
$3s \downarrow$	−30,470	
Total		~ -700

[a] Arrow, ↑, denotes electrons with spin parallel (and ↓ antiparallel) to the 3d-shell spin.

being attracted toward the 3d shell. They are, of course, repelled via the Coulomb interaction, but due to exchange this repulsion is less than that suffered by the minority spin electrons (which are less affected by exchange). The resulting attraction is therefore relative to minority spin behavior. This picture can also be used for viewing the exchange polarization of ions such as Li, Na, and K [cf. Table IX] where an unpaired s electron is the source of the polarization. Here it appears that the innermost loop of the valence s electron density (i.e., that located at the nucleus) attracts inner closed-shell electrons of parallel spin into this region, thereby causing the closed shell contact term contributions to be of the same sign as that of the valence electron. In general, however, the noded character of atomic orbitals makes it difficult to carry out such arguments. Hence, one should not take the results of these arguments too seriously.

While the iron series χ's may show a rough constancy, the individual shell contributions show a definite tendency to increase in magnitude with increasing Z. The fact that the individual electron shell contributions are not constant suggests that it is an oversimplification to describe the s-electron polarizability as linear in the ion's spin. The make-up of the contact term is perhaps better seen by inspecting the individual s-electron contributions to H_c for Mn^{2+} also given in Table XI. We see that an H_c of ~ -700 kgausses arises from differences in individual terms each of which are many times larger. (Not surprisingly, the $1s$, $2s$, and $3s$ shells for the neutral atoms show almost exactly the same H_c contribution as is obtained for the divalent ions. For Mn, for example, the $4s$ shell, then, largely cancels this contribution, leaving a mere 100 kgausses H_c value [12].)

A spin-polarized H–F calculation has been reported [12] for Fe^{3+} in its $3d^5$, ^6S state. This gave a χ of -3.00 au or an H_c of -630 kgausses; the individual contributions from the $1s$, $2s$, and $3s$ shells being -0.25, -8.51, and $+5.77$ au, respectively. This is to be compared with the Mn^{2+} ion, to which it is isoelectronic. While the individual s-shell contributions are greater for Fe^{3+} than for Mn^{2+}, the calculated χ (-3.34 au) for Mn^{2+} is larger (in magnitude) than the χ for Fe^{3+}.

Spin-polarized H–F calculations have also been done [340] for a series of Mn ions in different stages of ionization in order to determine the effect of ionicity of the calculated Fermi contact terms. Results are given in Table XII and χ is seen to decrease with increasing ionization. Such a trend has been observed experimentally [340a].*

* This becomes apparent after subtracting orbital and spin dipolar contributions from experimental data such as that in [340a]. A detailed study of iron series orbital and spin dipolar hyperfine effects is being done by Geschwind and Blume [340b].

TABLE XII

VARIATION OF χ WITH DEGREE OF IONIZATION

Ion	χ
Mn^+	−3.55
Mn^{2+}	−3.10
Mn^{3+}	−2.91
Mn^{4+}	−2.34

Other questions such as the role of environment [310] and of computational accuracy [12] have been discussed elsewhere. Because of the cancellation of very large terms of opposing sign (and in view of the neglect of interelectronic correlation) it is surprising that the exchange-polarization method is successful in reproducing the observed sign of H_c; it is even more remarkable that the observed order of magnitude is also determined by the calculations.

(4) *Exchange polarization of rare earth ions.* Accurate spin-polarized H–F calculations have not been done for any rare earth ion. A spin-polarized Hartree–Fock calculation has been done [341] for Gd^{3+}, but quite aside from deficiencies in the formalism it is an inadequate source of information concerning the ion's contact-term interaction because the calculation is nonrelativistic and poorly describes electron behavior at the nucleus. Quite surprisingly, and clearly by accident, the calculated hyperfine field from core polarization agrees in sign and magnitude with the observed value ($\sim$−200 to −300 kgausses). The $1s$ contribution is $\sim$0, and the $2s$ and $3s$ contributions are negative, whereas the $4s$ and $5s$ contributions are positive. We shall return to the role of core polarization effects on the Fermi contact term in rare earth ions in Section IV.

We have so far been concerned only with one aspect of the closed-shell spin densities resulting from exchange polarization, i.e., the behavior at the nucleus. As an illustration of the over-all behavior of the spin density throughout the ion, we show in Fig. 16 the one obtained in the calculation for Gd^{3+} described above [341]. For comparison the $4f$ density is also graphed. Note the three regions of spin density, negative at both very small and large distances from the nucleus and positive in the vicinity of the $4f$ maximum. (The closed-shell spin density must, of course, integrate to zero.) This behavior may be understood in terms of the exchange attraction discussed earlier.

c. Discussion. Given the spin-polarized $\phi_i(r)$ functions, we must then ask whether it is more appropriate to reinsert them into the improperly

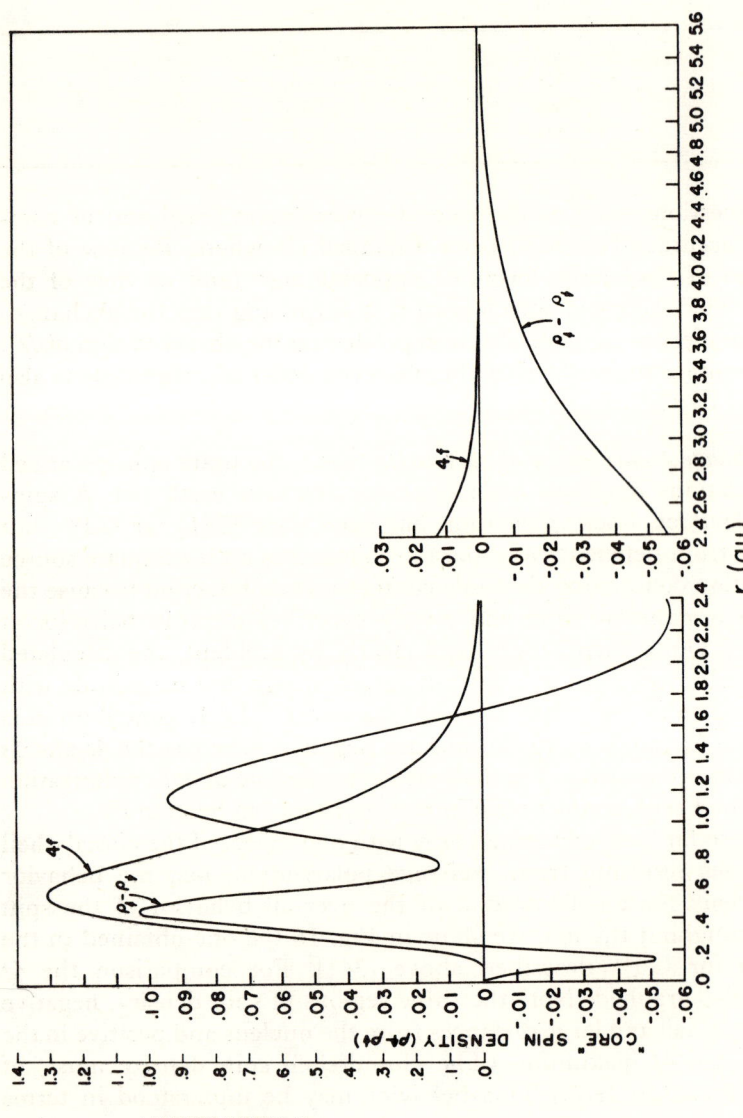

Fig. 16. The computed [341] "core" electron spin density $(\rho_\uparrow - \rho_\downarrow)$ for all the electrons other than the $4f$ shell and, for comparison, the $4f$ density as well. Note the change of scales at $r = 2.4$ au to a common scale for both $(\rho_\uparrow - \rho_\downarrow)$ and the $4f$ spin density.

symmetrized single determinant or into its multideterminant projection. On one hand, it is internally consistent to insert the $\phi_i(r)$'s into the function for which they were obtained; on the other, one favors dealing with functions of proper symmetry. Marshall [306] and Heine [342] have given quantitative arguments for Li indicating that the single determinant yields a more accurate contact term. This conclusion is based on comparisons with what are, roughly speaking, optimum combinations of Ψ_2 and Ψ_3. Bessis et al. [320] have presented a general argument based on perturbation theory which shows, to lowest order, that the single-determinant UHF function is closer than the projected function to the extended H–F function. This immediately implies some sort of cancellation between the deficiencies of the projected function and the error introduced by the components of improper symmetry appearing in the single determinant. Some workers find this an uncomfortable cancellation to rely on. Bessis et al., when dealing [41, 42] with O and F (ions discussed in the following subsection), have chosen to ignore their earlier argument and deal with the projected function, primarily because this policy yields better agreement with experiment. They conclude that the higher-order terms, omitted in their argument, are important.

Given the $\phi_i(r)$ functions, there is a third possible choice for their use, namely, to obtain the optimum linear combination (e.g., of ψ_2 and ψ_3 for Li) of the properly symmetric multideterminant functions which can be constructed from the $\phi_i(r)$. Ignoring computational matters, this choice gives us the best many-electron function available from $\phi_i(r)$ space, but it represents an inefficient and incomplete start on the full correlation problem. Also, the UHF $\phi_i(r)$'s are not very appropriate for the purpose. Computationally it does not let us exploit the simplifications available, or hopefully about to be,* for the projected single determinant, and without these the treatment of, say, Mn^{2+} is currently out of the question.

Quite aside from any matter of principle when we choose to concentrate on either the single determinant or its projection, uncertainties in the whole formalism make it most desirable to inspect the results associated with both whenever practicable.

If we accept the exchange-polarized results as meaningful, we have a simple picture of closed shells polarized by exchange effects. This picture is completely justified within the H–F framework and we believe it has considerable physical justification as well, although it must be remembered that it is a simplified version of a very complicated situation.

* For example, with further extensions of the work of Harriman [325].

We must not expect detailed quantitative agreement with experiment but, if we accept the qualitative (physical) meaning of such results, all sorts of other magnetically interesting effects appear [257]. For example, computations have been done investigating the effect of exchange polarization on neutron magnetic scattering [310], positive, and negative, Knight shifts [314, 343] in alkali and transition metals, and magnetic interactions and transferred hyperfine effects associated with rare earth ions [341]. The potentialities of doing spin-polarized Hartree–Fock–Slater [256] energy band calculations have been investigated [344].

6. Aspherical Distortions and Hyperfine Interactions

We have seen that the exchange-polarization effects associated with the relaxation of the m_s restriction caused closed shells to lose their singlet character and that the resulting spin densities appear to play a prominent role in understanding magnetic hyperfine interactions. Now, as was discussed by Nesbet [264, 265], a similar situation occurs if the Hartree–Fock potential is not spherical and we relax the m_l and spatial separability restrictions, for then a closed shell no longer retains its S-like character. It is, therefore, tempting to ask if, in such a case, there are repercussions on the wave functions which would cause them to make contributions to hyperfine interactions-contributions which are absent from the RHF description. Investigation [33] has shown that the relaxation of these constraints leads to a new way of calculating electric quadrupole polarizabilities and Sternheimer antishielding factors. In addition, the combined effect of relaxing both the m_s and these restrictions is to give rise to magnetic hyperfine interaction terms. The relaxation of the m_l restriction yields the Sternheimer radial polarizabilities (discussed in Section I, 3), while spatial separability is related to his angular polarizations. However, since such calculations are carried out within the framework of the Hartree–Fock method, certain problems of orthogonality, exchange, and self-consistency [cf. Section I, 3], which have complicated applications of the perturbation method, are easily resolved by this newer approach. The distortions induced in the inner closed shells by the distorted outer closed shells are thus included in a natural way and, by comparison with the results of the perturbation method (which normally does not take these into consideration), are found to be significant for large ions.

Self-consistent field calculations of γ_∞ have been done for Cu^+ and Cl^- [33], Ce^{3+} [40], and a series of ions isoelectronic with Br^- and with I^- [345]. Calculations have also been done for the $\langle r^{-3} \rangle_{\text{eff}}$ values of [42] O and [41] F, or [39] Fe^{2+} and Cl, and of [40] Ce^{3+}. In all cases only the radial polarizabilities were obtained, as the m_l relaxation can be dealt with straightforwardly with existing computational machinery. The angular polarizabilities require going to nonseparable $\phi(\mathbf{r})$'s which can be expanded as $\Sigma_l\, U_{lm}(r) Y_l{}^m(\theta\varphi)$. In general the sum will consist of but a few terms of common m value. Analytic H–F methods (see Appendix A) will deal with such a case; in practice, all existing computer programs are of a simplified version which yields $\phi(\mathbf{r})$'s of specific l (and m_l).*

When the source of the perturbing aspherical potential is other than the ion itself, we need not require the ion's shells to retain their original spherical symmetry. They will, of course, take on components with the symmetry of the "field." In such a case we are not dealing with correlation in the sense discussed earlier but are using the unrestricted H–F approach to replace standard perturbation theory methods for dealing with such a perturbing potential. This is particularly important for γ_∞ calculations if one of the outer shells makes a large contribution which, in turn, has repercussions on the other shells. The outer p shells of Cl^-, Br^-, and I^- are examples of this (see Table I, where the self-consistent field value of γ_∞ is listed for Cl^-).

The unrestricted H–F approach is also pertinent to any distortion of an ion due to a crystalline "field." For example, it can be used to gain information concerning crystal-field shielding in rare earths or crystal-field repercussions on an open shell and in turn on any of its electric or magnetic hyperfine interactions [346]. In all such investigations to date and (by definition) in all γ_∞ estimates, a most serious shortcoming has arisen in the unrealistic electrostatic "fields" which are used (see also Section I, 3).

When the source of the aspherical potential is an unclosed shell of the ion itself, one encounters symmetry difficulties similar to those met in the spin-polarized scheme. In this case the wave function is no longer a proper eigenfunction of L^2 when either (or both) restriction i or iii is relaxed, and much less of S^2 if the m_s restriction is relaxed as well. The perturbation calculations also yield functions of improper symmetry. However, little is known about such functions in either the projected or properly symmetrized "extended" forms. Further, the effects on the expectation values of operators, such as those of interest to us here, with respect to symmetrized wave functions vs.unsymmetrized ones, are far from understood. The few existing (unsymmetrized) calculations yield values of $\langle r^{-3}\rangle_{\text{eff}}$ which differ from RHF values in a way similar to the perturbation theory results of Sternheimer [19] and Ingalls [34]. The

---

* Nesbet's [264, 265] discussion of symmetry and equivalence restrictions is particularly pertinent to these matters.

variations between computed $\langle r^{-3}\rangle$ values are as much as 10%, strongly supporting the argument that the practice of equating hyperfine $\langle r^{-3}\rangle_{\text{eff}}$ values for magnetic and electric quadrupole interactions will at best yield experimental Q values with an uncertainty of 10%. We believe that considerably more theoretical work is required before this situation will be satisfactorily improved.

If the open shell is other than singly occupied, the UHF calculations have shown that the distortions [away from a common $U(r)$] within the unfilled shell make significant contributions to $\langle r^{-3}\rangle_{\text{eff}}$. These occur in addition to the closed-shell contributions. As an example of an ion with both effects, let us consider the various UHF radial distortion contributions [39] to the $\langle r^{-3}\rangle_{\text{eff}}$ values for $Fe^{2+}(^4D, 3d^6)$ which are listed in Table XIII. (Inspecting the table, one should note that the inclusion

TABLE XIII

THE RADIAL DISTORTION CONTRIBUTIONS TO THE HYPERFINE $\langle r^{-3}\rangle_{\text{eff}}$ VALUES AS OBTAINED BY A $m_l + m_s$ UHF CALCULATION FOR Fe^{2+} ^5D ($M_s = 2$, $M_L = +2$) COMPARED WITH THE RHF PREDICTION[a]

	$\langle r^{-3}\rangle_{\text{eff}}$ values for		
	Electric quadrupole interaction	Spin[b] dipolar interaction	L I interaction
2p Shell	0.3966	−1.1400	−0.5392
3p Shell	−0.0798	0.7539	0.3480
3$d_\uparrow$ ^{5} Subshell	−0.1600	0.1600	0
3$d_\downarrow$	4.7765	4.7765	4.7765
UHF total	4.9333	4.5504	4.5853
RHF	5.0802	5.0802	5.0802

[a] Units are a_0^{-3}.
[b] See reference [39].

of angular distortions would tend to reduce the magnitude of the total $\langle r^{-3}\rangle_{\text{eff}}$ values which are listed.) This calculation was m_s as well as m_l unrestricted (otherwise the closed shells would have made no contributions to the spin dipolar $\langle r^{-3}\rangle_{\text{eff}}$) and $\langle r^{-3}\rangle_{\text{eff}}$ contributions from the filled $3d^5$ (half-closed) subshell are listed separately from those of the single $3d$ orbital of minority spin. We see that the distortions of the $3d$ shell contribute substantially to the deviation of $\langle r^{-3}\rangle_{\text{eff}}$ from the RHF $\langle r^{-3}\rangle$ value; the distortion within the subshell and the variation of the sixth, distorting, $3d$ electron are both important. All electrons contribute to the quadrupolar and spin dipolar $\langle r^{-3}\rangle_{\text{eff}}$ values. The quadrupolar contributions are a simple sum of the distortions associated with the occupied subshells while the spin dipolar terms, depending on the m_s value of a particular subshell, arise from the difference in subshell distortions for any given shell.

The orbital $\mathbf{L} \cdot \mathbf{I}$ terms come only from subshells with spin parallel to the odd $3d$ electron. As mentioned in Section I, Coulomb interactions do not differentiate between orbitals of $+m_l$ and $-m_l$ while exchange interactions do. Since

$$\langle r^{-3}\rangle_{\text{eff, L·I}} = \frac{1}{m_{l,d}} \sum_i m_{l,i} \langle r^{-3}\rangle_i \qquad (3.23)$$

where $\mathbf{I}$ defines the z-direction, and $m_{l,d}$ is the m_l value of the odd $3d$ electron. Only subshells in which $\langle r^{-3}\rangle_{+m_l} \neq \langle r^{-3}\rangle_{-m_l}$ contribute (these have spin in common with the spin of the sixth $3d$ electron). We see that the resulting radial $\langle r^{-3}\rangle_{\text{eff}}$'s show the small variations discussed above; the extent of their agreement is somewhat surprising in view of their different origins.

In general, an unclosed aspherical shell has a quadrupolar field associated with it which is far larger than the electric field gradients in an ionic crystal, even if the latter is amplifed by a large γ_∞. There is one important exception to this rule: rare earth ions whose unfilled $4f$ shells are imbedded inside closed $5s$ and $5p$ shells. The $5p$ shells have a γ_∞ value of almost -100 associated with them [35, 40, 198] and these are large enough to compete with the $4f$ shells (cf. Sections II, 3 and II, 4).

7. Extended Hartree–Fock Functions for Closed-Shell Systems

For closed-shell atoms the RHF orbitals are internally consistent eigenfunctions of the H–F equations. To illustrate this, let us consider the He ^1S ground state. The RHF $\psi_{1s}(\mathbf{x})$'s are eigenfunctions of Eq. (3.3) for a potential derived from the $1s^2$ determinant. If we define a H–F many-electron function to be any single-determinant or multideterminant function obtained by applying symmetry projection operators to a single determinant, then the RHF $1s^2$ function is not the lowest energy function, within the H–F formalism, approximating the ^1S ground state. An "open-shell" H–F function* obtained by applying the singlet projection operator on a ϕ_{1s}, $\uparrow(\mathbf{x})\phi_{1s}$, $\downarrow(\mathbf{x})$ determinant (and variationally obtaining the two spatially different orbitals for the resulting two-determinant function) has a lower total energy than that of the RHF

---
* For additional work on the (analytic) open-shell approach to atoms see references [347-351].

function [327, 328]. The open-shell function is, in turn, not the lowest extended H–F eigenfunction for He. Abandoning the requirement of pure s-like orbital character, yet maintaining proper total wave function symmetry, will lead to a function with a lower total energy [although s-like character will predominate in the resulting $\phi(\mathbf{r})$]. While the RHF function does not have the lowest energy in H–F function space (as we have chosen to define it), Löwdin and Shull [327, 328] have observed an important property associated with it, i.e., that it may be a locally stable function in that space. For example, if one starts with a good approximation to the RHF function and iterates self-consistently for a better function, allowing it to go over to an open-shell function, it will converge on the RHF and not the open-shell result.

The above observations imply that the RHF formalism is a restricted formalism for any many-electron system. Further, since the extended H–F equations have yet to be solved for any many-electron system, the full features of H–F theory have yet to be exploited in any calculation. While extended H–F equations have yet to be solved, some information is obtainable concerning the resulting functions. This can be done by analyzing the accurate correlated wave functions which are available for a few systems. Davidson and Jones [335, 336] have done this with the Kolos and Roothaan [352] correlated wave function for H_2 and concluded that there exists an extended H–F function which contains 90% of the correlation energy. This is a surprisingly large improvement; if the extended H–F method did as well for appreciably larger systems it would be of great interest for many-electron applications. We doubt this to be the case because of the constraints introduced by simultaneously obtaining more orthogonal orbitals; besides, computational considerations make the approach unpromising for ions of the size of interest to us here. For this reason we will limit discussion to two further observations. First, extended H–F functions, such as the ones indicated here, are associated with double-substitution configuration interaction, i.e., Brillouin's theorem [270] holds, of course, for the closed-shell RHF function. Finally, these observations, which were made for closed-shell systems, are also pertinent to open-shell cases where the situation is considerably more complicated. These complications will not be discussed here, but some features of the problem should be apparent from the earlier discussion of exchange polarization.

8. Concluding Comments

In this section we have variously explored, but have by no means exhausted, the H–F formalism and certain related topics. We have seen that the UHF formalism predicts significantly different hyperfine effects arising in significantly different ways from those indicated by its RHF counterpart. In general the UHF predictions are in better, but not completely satisfactory, agreement with experiment. The roles of symmetry and correlation effects provide the greatest uncertainties in the UHF predictions. These and many other matters are far from understood and their resolution has far wider implications than we have indicated here. As an example of this, let us note a few features of the symmetry matter. We have discussed what little is known of its role in atomic hyperfine interactions as predicted by UHF theory. The same polarizations and the same questions of symmetry affect an ion's magnetic (and electric) interactions with its external environment, whether it be neutron scattering or a neighboring magnetic ion. Almost nothing is known about symmetry repercussions on such cases. More important is the fact that the symmetry problem arises (at least until proved benign) almost every time we deal with a magnetic polarization effect in a solid, molecule, or atom. The Ruderman–Kittel–Kasuya–Yosida polarization [283–285] is a case in point. The final function is not a spin eigenfunction of the solid and the implications of this fact are not understood.

In short, much nontrivial work remains to be done in these areas.

IV. Interpretation of Measured Hyperfine Interactions

It is apparent from Section II that there is a vast body of data currently available on hyperfine interactions in magnetic materials. In this section we discuss some possible interpretations of parts of these data with a view toward understanding the origin of the observed fields and the information one thus learns about electron distributions. Of necessity, part of this will consist of speculation rather than established fact because of the present inadequate state of theory. The field is by no means a closed one; much more theoretical work must be done before one can claim to have an understanding of the subject.

There is a natural division to the discussion: ionic crystals and metals. We shall observe such a division in what follows. In addition, since the subject is vast we shall have to be selective. Discussion will be almost exclusively restricted to the iron transition and rare earth series elements, i.e., to representatives of d- and f-electron behavior.

1. Hyperfine Fields in Iron Series Salts

There exists a massive array of experimental hyperfine data for iron series salts and we will, with few exceptions, neither go into a discussion of details nor attempt a complete tabulation of the data. We will instead review the trends which have emerged and some of the problems which remain.

While the field has grown immensely since the discovery of paramagnetic resonance [353] and the early work of the Oxford group, one still relies heavily on the theoretical work [3, 22, 44, 354–358] of that period. Let us briefly review a few of Abragam and Pryce's findings for a transition metal ion in a predominantly cubic environment with small tetragonal or tetrahedral components. They obtained the familiar spin Hamiltonian [cf. Eq. (2.3)] with a magnetic hyperfine term of the form

$$AS_zI_z + B(S_xI_x + S_yI_y) \tag{4.1}$$

A and B will, in general, be made up of contributions from the orbital, spin dipolar, and Fermi contact terms of Eq. (1.1). The detailed structure of A and B depends, of course, on the ion in question and in particular on whether the ion's ground state is or is not an orbital singlet.

For the case of an orbital singlet, which includes the d^3, d^4, d^5, d^8, and d^9 (e.g., V^{2+}, Cr^{2+}, Mn^{2+}, Ni^{2+}, and Cu^{2+}) configurations, the orbital momentum is "quenched" (i.e., the expectation value of **L** is zero) and the ground state hyperfine interaction is almost entirely due to the contact and spin dipolar terms. Spin-orbit coupling (either by itself or in conjunction with low-symmetry crystal-field matrix elements) mixes in excited states, thereby reintroducing an orbital contribution into A and B. The same excited state mixing causes the electron g-factor to deviate from 2 and one can often write the orbital hyperfine terms as

$$A_{\text{orb}} = 2g_I\mu_0\mu_N \Delta g_\| \langle r^{-3} \rangle \tag{4.2}$$

and

$$B_{\text{orb}} = 2g_I\mu_0\mu_N \Delta g_\perp \langle r^{-3} \rangle \tag{4.3}$$

where $\Delta g_\|$ and $\Delta g_\perp$ are the parallel and perpendicular electronic g shifts. We shall shortly see that these orbital terms need not be negligible. From symmetry it follows that the d^5, d^3, and d^8 configurations have zero-valued spin dipolar terms (as will be seen, a tetragonal or trigonal field can reintroduce a small term for the latter two configurations) but substantial fields may occur for the d^4 and d^9 configurations. For example, spin dipolar contributions to A of the order of 200 and 500 kilogausses exist* for Cr^{2+} and Cu^{2+}, respectively, when these ions are in a field which has a component of tetragonal symmetry. These configurations are orbital

---

* These estimates assume the presence of a tetragonal field (which is the source of an orbital singlet ground state) as Abragam and Pryce assumed [3].

doublets in a pure cubic field and a tetragonal field splits these states, yielding orbital singlets which lead to these large spin dipolar terms. The larger spin dipolar term for Cu^{2+} is due to its larger $\langle r^{-3} \rangle$ value. A trigonal field, by itself, does not split these levels but in conjunction with spin-orbit coupling this splitting does occur, normally leading to states with negligible spin dipolar terms outside of the tetragonal field case. The contact term dominates for the d^3, d^5, and d^8 configurations. The d^4 and d^9 configurations, which are not singlets in a purely cubic field, are perhaps more appropriately classed along with the orbitally degenerate configurations which we shall now consider.

The d^1, d^2, d^6, and d^7 (e.g., Sc^{2+}, Ti^{2+}, Fe^{2+}, Co^{2+}) configurations have degenerate cubic crystal-field ground states and their magnetic behavior is sensitive to spin-orbit coupling (which splits the ground degeneracy) and to trigonal or tetragonal fields. Orbital effects are far from quenched and in fact usually dominate in the hyperfine interaction; of these Co^{2+} is the classic example.* Abragam and Pryce [3] showed that it is convenient to use a spin Hamiltonian involving fictitious s and ℓ operators, but the extraction of information concerning orbital contributions to A and B from experimental g-factors then ceases to be trivial. Abragam and Pryce have shown how this is done in their treatment of the hydrated Co^{2+} ion [360].† However, we know of no similar investigation of the Fe^{2+} ion which, with the advent of Mössbauer data, would now be of great interest.

While care is required in the orbitally degenerate case, some slight caution is even desirable in applying Eqs. (4.2) and (4.3) to an orbital singlet. These equations will usually (but not inevitably) supply accurate estimates of A_{orb} and B_{orb} for such a case, but nonorbital g-shift contributions could arise from electron transfer or high-order spin effects.‡ These contributions are probably always negligible in their influence on an estimated orbital hyperfine term for a singlet.

Abragam et al. [22] reported values of the contact-term contribution (H_c) to the hyperfine field of several divalent ions. These were obtained by an analysis of experimental hyperfine data for the ions in hydrated salts, and involved some of the factors just discussed. The results of their, pioneering investigation, expressed as H_c per unpaired electron (i.e.,

---

* For example, see Kanamori [359], where the role played by the residual momentum in the magnetic susceptibilities and magnetic anisotropic energies of CoO and FeO is investigated.

† For a treatment of Co^{2+} in an almost exactly cubic environment (i.e., where the g- and A-tensors are almost isotropic) see Culvahouse et al. [361].

‡ Examples *not* appropriate to the orbital singlet appear in Clogston [362, equation 11] and Ham et al. [363].

$H_c/2S$ where S is the spin of the ion), are given in Table XIV. We see that a roughly constant value of -125 kgausses occurs, the negative sign indicating that the net s-electron spin causing H_c is antiparallel to the spin of the ion. The divalent ion spin-polarized H–F calculations discussed in the preceding section yielded $H_c/2S$ values which showed a similar crude constancy with an average value of ~ -150 kgausses. The agreement between experiment and theory is remarkable in view of the deficiencies in the spin-polarized formalism (and the fact that the theoretical results arise from the differences of far larger quantities) and tends to encourage the rash assumption that the calculations give a qualitatively correct picture of what is going on. This behavior of $H_c/2S$ is not the only constancy which we shall observe in hyperfine effects for iron series salt.

TABLE XIV

Values of Experimental Hyperfine Interaction χ, H_e, and $H_c/2S$ for the Divalent Iron Series Ions in Hydrated Salts[a]

Ion	$V^{2+}(3d^3)$	$Mn^{2+}(3d^5)$	$Co^{2+}(3d^7)$	$Cu^{2+}(3d^9)$
χ (au)	-2.8	-3.1	-2.5	-2.9
H_c/S (kgausses)	-118	-130	-105	-122
H_c (kgausses)	-354	-650	-315	-122

[a] Obtained from χ's reported by Abragam et al. [22].

While $H_c/2S$ is roughly constant for ions of common valency in common environments, it is by no means constant for differing environments. Van Wierengen first showed [364] this for the hyperfine field of Mn^{2+} (being a d^5, ^6S, ion, $H_e = H_c$ to a first approximation); his results are reproduced in Table XV. Low [44], for example, has listed some additional data including Mn^{2+} in Ge, where the field is but -318 kgausses or less than half that found for highly ionic environments. The more covalent the environment, the smaller H_e becomes,* and there are a number of potential contributing factors to this relationship.

In some of the hosts the ion is not in a pure ^6S state and additional small (positive) orbital and spin dipolar terms could arise in this way. Van Wierengen showed that exchange interactions between Mn^{2+} ions would also reduce H_e. These factors are, we suspect, minor compared

* For an extension of VanWierengen's hyperfine field observations see Matumura [365] and for an allied discussion of the variation of other Mn^{2+} spin Hamiltonian parameters with covalency see Title [366].

TABLE XV

Hyperfine Fields, H_e (in kgausses), for Mn^{2+}, for Various Ligand Neigbors[a]

Ligand neighbors:	H_2O	F^-	CO^{2-}	O^{2-}	S^{2-}	Se^{2-}	Te^{2-}
H_e	695	695	665	570–640	490	460	420

[a] After Van Wierengen [364].

with effects directly associated with H_c. First, covalent mixing involving the 3d orbitals tends to delocalize the ion's spin moment, therby reducing an effective **S** associated with (or rather localized at) the Mn ion site. With a constant $H_c/2S$ this would tend to reduce H_e. Second, $H_c/2S$ could be affected by covalent repercussions on either the s shells or the 3d radial spin distribution itself.* The crystal-field investigation of Sugano and Shulman [298] suggests that, at least in the case of the more ionic (e.g., fluoride and oxide) environments, the reduction in **S** is the most important source of H_c variation. If this is the case, these covalency effects could be characterized by a reduction factor associated with **S** of a form not dissimilar to the orbital reduction factor, k, used to account for the effects of covalency on orbital hyperfine and orbital Zeeman interactions [356].[†] Finally, g shifts of $+0.01$, and occasionally significantly greater ones, occur for Fe^{3+} and Mn^{2+} in highly covalent environments. Equation (4.2) predicts an orbital hyperfine field of $\sim+25$ kgausses for a Δg of 0.01 for these ions. H_e, being negative, will be reduced by this contribution, a contribution which is liable to be overlooked since these are S-state ions[‡] with g shifts, which one normally thinks of as zero valued. The ways in which covalency affects 3d ion hyperfine interactions have yet to be understood in detail.

Magnetic hyperfine data for Fe^{3+} and Fe^{2+} ions in different hosts were listed in Table V. (These have been obtained by Mössbauer experiments in recent years and are therefore absent from standard lists.) Notice that Fe^{3+} is isoelectronic with Mn^{2+} which we have just discussed and its hyperfine data are very similar but approximately 5% smaller in magnitude (for any given environment). The spin-polarized calculation

---

* Spin-polarized calculations suggest that H_c is sensitive to the 3d radial distribution, e.g., see figures 2 and 3 of reference [12]. The ordinate is mislabeled in figure 3 and should be read as 2χ rather than χ.

† For a discussion of covalency and its role in orbital reduction see reference [289, 358, 367].

‡ We wish to acknowledge conversations with S. Geschwind and J. Hensel concerning he potential importance of such orbital terms.

[12] for Fe^{3+} (see Table X) yielded an H_c some 10% smaller than that obtained for Mn^{2+}. Such a qualitative agreement between theory and experiment is very probably fortuitous.

Rather fewer data are available for Fe^{2+} and what there are, are smaller in magnitude than what one would expect from the contact term alone (e.g., $4 \times \langle H_c/2S \rangle_{avg} \sim -500$ kgausses; spin-polarized Fe^{2+} calculations yield ~ -550 kgausses). The smaller magnitude and more violent variation in H_e are typical of orbital effects. Unfortunately, the signs of these H_e's are not known. The known [368, 369] g shifts of Fe^{2+} in FeF_2 ($\Delta g \sim 0.24$) and Fe_3O_4 ($\Delta g \sim 0.06$) suggest that the H_e's are negative, i.e., that H_c dominates. Naïve, and not strictly appropriate, application of Eq. (4.2) yields orbital hyperfine estimates of $\sim +160$ and $\sim +40$ kgausses for the two salts, thus crudely accounting for the differences between observed H_e's (assuming them negative) and the anticipated H_c's.

Recently, hyperfine fields for sequences of isoelectronic ions in common environments have been studied and, as Geschwind [370] has emphasized, H_c (and at first glance H_e) is remarkably constant. Table XVI (which has been kindly supplied by him) reports several sets of hyperfine fields in terms of $H_e/2S$. The $3d^3$ and $3d^5$ sequences are strikingly constant. The Mn^{2+}-Fe^{3+} variations lie outside uncertainties in the values of the nuclear moments. The $3d^8$ sequences also do not show the strict constancy.

The observed g shifts and, we must emphasize, RHF $\langle r^{-3} \rangle$ values [200] were used by Geschwind [370] to estimate the orbital hyperfine terms for the $3d^3$ ions and, by Locher and Geschwind [157b], for the $3d^8$ cases; these results are also given in Table XVI. Equation (4.2) is expected to work quite well for these configurations. (The small $3d^3 H_{orb}$ terms show greater constancy than one might expect from the g-factors. While the Δg's decrease with increasing ionization, the $\langle r^{-3} \rangle$'s have compensated for this by increasing in magnitude.) The orbital effects are quite important for the $3d^8$ configuration and cause $H_e/2S$ to vary. Utilizing the observed $H_e/2S$ and the computed $H_{orb}/2S$ values, and neglecting minor spin dipolar terms, one obtains $H_c/2S$ values of roughly -95, -110, and -130 kgausses for the $3d^3$, $3d^5$, and $3d^8$ configurations (in sixfold oxygen coordination), respectively. The most striking feature is the constant behavior along isoelectronic sequences to within a few percent). The rougher constancy first observed by Abragam et al. [22] is seen, but with a slight increase as we go from the $3d^3$ to the $3d^8$ configuration. A similar trend in $H_c/2S$ is seen in the iron series spin-polarized calculations [12]. This trend is best seen in the calculations for the neutral $3d^n 4s^2$ atoms (for which more calculations were done than for the divalent ions).

TABLE XVI

H_c Values as Estimated[a] for Various Iron Series Ions in Oxygen, Fluorine, and Sulfur Coordinations[b]

Ion	Compound	Observed Δg	Observed H_e	Computed $H_{orb}(H_{SD})$	Resulting H_c
		$3d^3$ (Sixfold oxygen coordination)			
V^{2+}	MgO	-0.0223	-99 ± 1	-4	-95
Cr^{3+}	MgO	-0.0226	-101	-6	-95
Mn^{4+}	Al_2O_3	-0.0086	-100 ± 1	-3	-97
		$3d^5$ (Sixfold oxygen coordination)			
Mn^{2+}	MgO	—	$(-)115$	—	-115
Fe^{3+}	MgO	—	$(-)113 \pm 5$	—	-113
		$3d^8$ (Sixfold oxygen coordination)			
Ni^{2+}	MgO	0.2122	$(-)31.\pm 2$	$+94$	-125
	Al_2O_3	0.1925	-48.4^c	$+85(-6)^c$	-127
Co^+	MgO	0.1705	$(-)80.$	$+57$	-137
Cu^{3+}	Al_2O_3	0.0765	-85.5^c	$+43(-4)^c$	-125
		$3d^5$ (Fourfold sulfur coordination)			
Cr^+	ZnS	—	$(-)83 \pm 1$	—	-83
Mn^{2+}	ZnS	—	$(-)91$	—	-91
Fe^{3+}	ZnS	—	$(-)80 \pm 5$	—	-80
		$3d^5$ (Sixfold fluorine coordination)			
Mn^{2+}	$KMgF_3$	—	-130	—	-130
Fe^{3+}	FeF_3	—	$(-)124 \pm 1$	—	-124

[a] By Geschwind [157b, 370].
[b] All fields in kilogausses per $2S$.
[c] The experimental H_e and computed spin dipolar terms are parallel field (A) values.

$H_c/2S$ associated with the $1s$, $2s$ plus $3s$ shells varies from -119 kgausses for V to -143 for Mn and -158 for Cu.

Geschwind has also observed a second trend: Given a particular type and number of ligand neighbors, H_c is strikingly independent of the metal-ligand internuclear distance. Since one expects an individual contribution (e.g., overlap) to the local spin density, and hence to the spin polarization, to vary with the internuclear distance, such an observation is again highly remarkable.

Geschwind and Locher found minor discrepancies [i.e., the H_c obtained with experimental values of A and $g_\parallel$ and Eq. (4.2) differed from that of B, $g_\perp$ and Eq. (4.3)] of the order of 10 kgausses for the $3d^8$ Al_2O_3 sequence. This suggests the presence of spin dipolar effects which are anisotropic but which are not linearly related to H_{orb}. The non-linearity follows from the fact that the H_{SD} "configuration" mixing involves different excited crystal-field states, hence different off-diagonal matrix elements, than do those entering H_{orb}. In addition, there is no simple counterpart of Eqs. (4.2) and (4.3) relating H_{SD} linearly to other experimental parameters. One must thus compute it directly. Geschwind has investigated this mixing, which involves trigonal field matrix elements,* and found it to be of the correct magnitude and very likely the correct sign to account for the discrepancies.† While minor effects (e.g., varying $3d$ behavior in different crystal-field levels) may eventually have to be introduced in order to arrive at a detailed quantitative understanding of these hyperfine fields, they currently appear well understood in terms of contact and orbital terms plus minor spin dipolar contributions.

The most remarkable, and thus far not understood, observation in this section is the constancy of H_c. One normally expects the character of an ion's interaction with its environment to change rather severely as one passes through several stages of ionization, with the more heavily ionized ions being the more covalent. Such changes are reflected in the g shifts of Table XVI, but not in H_c (although Van Wierengen's experience with covalent effects would have us expect severe H_c variations). A similar situation holds with variations of other parameters but not with H_c for varying metal-ligand internuclear distances. We have yet to discover whether there is some profound physical reason for these regularities or whether they arise from an accidental cancellation of effects. We suspect the latter, but whatever the case, these results are most curious and are worthy of further investigation.

2. Hyperfine Fields in Rare Earth Salts

In contrast with the open d shells of transition elements, the $4f$ shell is imbedded in the interior of a rare earth ion. Figure 17 shows the

---

* For an example dealing with trigonal field effects in Al_2O_3, see McClure [371].

† The result depends on the sign of an off-diagonal trigonal field matrix element. The trigonal field is made up of $Y_2^m(\theta, \varphi)$ and $Y_4^m(\theta, \varphi)$ elements (see reference [371] and, if the former dominates, as one would expect, the sign of H_{SD} accounts for the discrepancies in the $3d^8$ ions.

4. HYPERFINE INTERACTIONS IN MAGNETIC MATERIALS

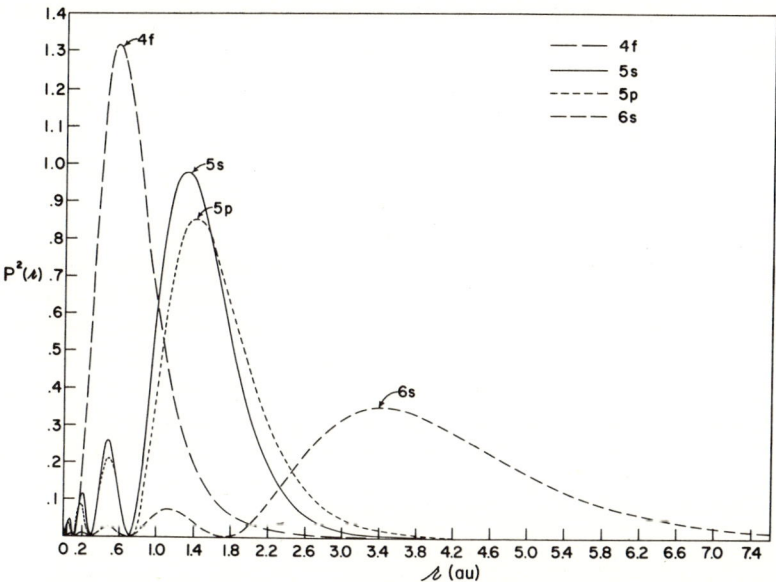

FIG. 17. Radial charge densities for the 4f, 5s, 5p, and 6s electrons of Gd$^+$ [253b].

position of the atomic 4f density relative to the "outer" 5s, 5p, and 6s electrons [253b]. This physical isolation causes the 4f shell to maintain much of its free ion character whatever the ion's environment, a fact which has proved most useful in understanding magnetic properties of rare earth ions in the past [372]. This is particularly so for hyperfine interactions in rare earth salts where the environment's major effect is to determine the occupancy of 4f crystal-field levels,* levels in which the 4f shell largely maintains its free ion character. Orbital and spin dipolar hyperfine fields are unquenched and together are of the order of 10^6 to 10^7 gausses. The contact term is least important, being of the order of 10^5 gausses or less [374, 375].

Rare earth hyperfine interactions have yet to be used extensively as a probe of rare earth behavior in various environments. This is not so much due to the physical isolation of the 4f shell (and the resulting insensitivity of the hyperfine field to environment) as it has been to a lack of accurately known nuclear moments. Much effort has been expended [227, 376–378]

---

* We shall not attempt a review of the crystal-field behavior of rare earth ions but instead refer the reader to the classic papers of Elliott and Stevens [373] and for the case of S-state ions to Baker et al. [373a]. In view of the rapid increase in interest in the rare earth ions in cubic environments, the reader may find the work of Ebina and Isuya [373b] on the rare earth cubic eigenfunctions of use.

in taking an experimental hyperfine interaction to estimate the hyperfine field and, from this, obtain the nuclear moment. In these estimates of the hyperfine field, the lack of precise $4f$ wave functions led to the practice [376–378] of relating an observed spin-orbit coupling constant to the $\langle r^{-3}\rangle_{4f}$ appropriate to the hyperfine field by assuming (1) the classical $1/r\,dV/dr$ version of spin-orbit coupling and (2) that the $4f$ orbitals were of hydrogenic (or modified hydrogenic) [377, 378] form. Nonrelativistic free-ion RHF functions are now available [372] and they yield $\langle r^{-3}\rangle_{4f}$ values which are 10 (for Yb^{3+}) to 25 (Ce^{3+}) percent larger than the Hartree parameterized* values by Lindgren [378]. The values of μ which have been observed [81] tend to fall between the values predicted by these two sets of $\langle r^{-3}\rangle_{4f}$'s. Exact agreement with the RHF (or Hartree) values is not to be expected, for correlation, relativistic, and Sternheimer antishielding (see Section I) effects remain to be accounted for. A naïve guess would say that the inclusion of correlation effects would lead to a small enhancement of the hyperfine interaction. Comparison [379] of relativistic and nonrelativistic Hartree functions suggests that relativistic repercussions on a $4f$ shell will tend to reduce its $\langle r^{-3}\rangle$ value [380].† Little is known about internal antishielding effects in the rare earths. A calculation has been done [40] for the electric quadrupole antishielding of Ce^{3+}, which yields an enhancement of the free-ion interaction. It is not at all obvious that the magnetic interactions would be enhanced as well; a reduction is not inconceivable. The relativistic and antishielding effects can be of the order of the deviations between RHF (or Lindgren's Hartree) predictions and experiment, and we suspect that the fair agreement between theory and experiment relies on the partial cancellation of such terms.

Quite recently, a number of moments have been obtained by more direct means‡ [384, 385], letting us now utilize hyperfine investigations to inspect hyperfine field behavior. Unfortunately, the above-mentioned complications will keep us from anything like a complete understanding

---

* Lindgren's $\langle r^{-3}\rangle$'s rely heavily on matching Ridley's [253a] Hartree functions for Pr^{3+} and Tm^{3+} and therefore can be looked upon as Hartree (as against H–F) $\langle r^{-3}\rangle_{4f}$ predictions.

† The results of Boyd et al. [380] suggest that the inclusion of relativistic effects will in general tend to enhance the hyperfine interaction of an open s or p shell but reduce that of an open d or f shell. These reductions will be small relative to the s and p shell enhancements.

‡ For Nd see reference [381]; for Er, [382]; for Yb, [383]; for Eu, [10]; and for Tm [157c]. The nuclear Zeeman effect is observed in these experiments and in order to obtain the nuclear moment one must estimate any shielding effects. For all cases but Yb, atomic beam techniques were used; Yb involved NMR in the metal. The Yb moment results are the least susceptible to error associated with shielding or other effects.

4. HYPERFINE INTERACTIONS IN MAGNETIC MATERIALS 269

of what is going on. The study of rare earth salt hyperfine data promises to be somewhat more difficult and, at the same time, less interesting than it was for transition ions because of the smaller role played by the environment in 4f shell behavior. Despite this, environmental effects are to be seen.

While an environment can completely change the magnitude and character of iron series hyperfine interactions, one is dealing with one or a few percent variations in the rare earths (ignoring their temperature dependence). These seem to be largely associated with 4f shell covalency. It has been suggested [386, 387] that covalency plays a significant role in rare earth crystal-field behavior. Such covalency is, of course, small, but so are the crystal-field effects to be accounted for. Recent studies of rare earth g-factor and hyperfine field behavior for the ions in oxide [388] and fluoride [389] environments indicate the presence of covalent orbital reduction effects [356], effects which cause a reduction of the order of 1% in the hyperfine interactions. The reduction is greater for the more covalent oxide as is to be expected (cf. the iron series salt discussion in Section IV, 1).

There are other environmental effects of the order of 1% which must be considered. Bleaney has concluded [389] that mixing from excited crystal-field levels also has roughly a 1% effect on the hyperfine field. In addition, the environment will have small direct repercussions on the spatial behavior of the ground state function. It will induce small radial (and angular) changes [40, 346] in the 4f orbitals, hence in $\langle r^{-3} \rangle_{4f}$, and it will induce aspherical distortions in the closed (e.g., 5s and 5p) shells. These distortions, by being coupled to the 4f shell, can also contribute [264, 265] to the ion's magnetic hyperfine interaction. These 4f and closed-shell distortion terms may make a contribution of the order of 1% but they will generally tend to be less important than the orbital reduction and crystal-field level mixing effects.

When dealing with 1% effects, the contact term becomes of interest. This term will be relatively insensitive to environment and can be readily observed by inspection of experimental hyperfine data for the S-state ions Eu^{2+} and Gd^{3+}. Estimates may be made [253b, 389] for the other ions if one assumes the contact term to be roughly proportional to the ion's spin (i.e., assume H_c/S is constant) as was seen to be the case (see Section IV, 1) for iron series ions. Using the observed [374, 375] hyperfine interaction for Eu^{2+} in CaF_2 and the Eu nuclear moment value obtained by Pichanick et al. [384] we would estimate

$$H_c \sim (-90 \text{ kgausses}) \times (g_J - 1)J \qquad (4.4)$$

for either divalent or trivalent ions. This is an exceedingly crude estimate and for this reason we have not attempted to differentiate between the divalent and trivalent cases. Geschwind's observations (see Section IV, 1) for the iron series ions suggest that H_c should be considered essentially free of valency. In addition, his observations suggest that H_c/S is not strictly a constant and, if we extrapolate from the iron series to the rare earths, that it increases slightly with increasing Z. We do not know enough to put a sensible estimate of this trend into Eq. (4.4). We would estimate this equation to be at best accurate to 20% for ions such as Ce^{3+} and Yb^{3+} which lie to either end of the rare earth sequence.

3. Hyperfine Fields in Ferromagnetic Metals

Much attention has been paid to the origin of the large observed hyperfine fields in ferromagnetic metals, in part because such an understanding would help in elucidating the nature of ferromagnetic exchange and in part because the negative fields were contrary to original expectations [55]. We shall see that this understanding is still incomplete.

For ferromagnetic metals, such as Fe, Co, or Ni, there are a number of terms, in addition to core polarization and the orbital contribution from any unquenched angular momentum, which must be considered as responsible for the observed hyperfine field, H_e. Marshall [55] discussed these terms in an investigation of the problem of nuclear alignment. Aside from the local magnetic field (which is composed of the external plus demagnetizing fields and the Lorentz field) these terms include contributions from the outer electrons as follows: (1) the field from the contact interaction with the "4s" conduction electrons, polarized by the 3d's, via the Zener–Ruderman–Kittel–Kasuya–Yosida mechanism [283-285]; (2) the field from the contact interaction with the s-like character of the "3d" bands; (3) a contribution from the dipolar field of the 3d electrons (zero for cubic symmetry). (As we shall see, there may be additional contributions to the one just listed.) In discussing the field in Co, Marshall estimated these various contributions and assuming the measured field to be $+219$ he determined term (2) to be $+137$ kgausses. We now know that this choice of sign was incorrect. Recent Mössbauer experiments by Hanna et al. [101, 102] showed that the effective field at the nucleus in Fe^{57} was in fact negative, i.e., directed opposite to the direction of magnetization. These workers [101, 102] concluded that the dominant contribution to H_e must come from core polarization since the other sources are (or were generally presumed to be) positive. From the discussion of Section IV, 1, the core polarization contribution in Fe with a spin of $2.2\mu_B$ could have been expected to be [12, 22] about -275 kgausses too small to overcome the anticipated positive contributions and yield greement with experiment (-340 kgausses). This suggested that the core polarization appropriate to the metal was intrinsically larger than that appropriate to the salts and several attempts were made [12, 312, 313] to augment the core polarization by means of expanded $3d$ wave functions, expansion suggested by energy band calculations [390-392], and in this way account for the large difference between core polarizations in metals and ions. It was found [12, 312] that expansions or contractions of the $3d$ spin density could lead to computed core polarization terms which were more than sufficient to "explain" the experimentally observed H_e, but it was noted [12] that such changes in spin density were incompatible with results of neutron diffraction determinations of the spin density. (Further spin-polarized calculations [12] with expanded charge densities but unchanged spin densities did result in a larger (in magnitude) core-polarized hyperfine field (~ -350 kgausses) which was, however, still too small to overcome the remaining, positive, contributions as estimated using results of energy band calculations for both the "$4s$" and "$3d$" bands.)

Anderson and Clogston [393] and Carr [394] pointed out that there exists yet another negative conduction electron contribution to H_e which arises from covalent mixing between the "s"- and "d"-band functions (single substitution $s \leftrightarrow d$ configuration mixing [395]). The negative character occurs because the covalent mixing is greater among electrons of minority spin.* Anderson and Clogston suggested, on the basis of perturbation theory, that this would cancel term (2) and could even dominate, causing a net negative conduction (or s) electron polarization.†
We will assume a net negative effect in subsequent discussions.

Hyperfine fields in ferromagnetic metals are remarkable in one striking respect. From the Mössbauer data for H_e listed in Table V, we see that for Fe^{57} as an impurity in CoPd H_e equals 330 kgausses in precise agreement with the Fe^{57}-in-Fe value, if we assume the sign to be negative. This and the data for Fe^{57} in Co and Ni (which show a small decrease in H_e) for and Fe^{57} as impurity in the Cu-Ni alloy system indicate that the field at the iron nucleus is predominantly due to its own electrons and depends only to a small degree on the magnetization of the host. Indeed one is tempted to make much of the observation (by way of a semi-empirical rule of thumb) that if one wants an estimate of H_e he

---

* To contribute to a physical effect, covalent mixing must occur between a pair of orbitals one of which is occupied and one of which is not (e.g., see reference [299]). We are interested in the effect of covalent mixing in lowering the occupied "$4s$" band energy levels (hence affecting the band's occupancy) and in the fact that the minority spin $3d$ band being less occupied leads to greater minority spin mixing.

† For a discussion of the various ways this will affect H_e, see the latter part of Section VA of Watson and Freeman [12].

should calculate H_c and neglect all the remaining terms. We have seen that this "works" for metallic cobalt; using a χ of -3.0 and 1.7 unpaired spins one obtains an H_c of -215 kgausses in remarkable agreement with experiment (see Table V). For ferromagnetic Ni with 0.6 unpaired spins and a χ of -3.0, this simple procedure predicts an H_c of -75 kgausses and was previously deemed a failure [12] because the observed field was taken as -185 kgausses. The recent measurements of Locher and Geschwind [157b] showed that the nuclear moment of Ni^{61} was 0.746 nm and not 0.30 nm as thought previously. With this new moment H_e becomes -75 kgausses in striking and undoubtably fortuitous agreement with the above-mentioned "rule of thumb" estimate.

An observed $3d$ metal H_e must, of course, arise from a detailed balance between conduction electron, $3d$ electron, and core-polarization contributions. Unfortunately, the present state of theory is such that it is currently impossible to obtain a detailed breakdown of their roles. One cannot even guarantee that all experimentally significant (say to ± 10 kgausses) terms have been enumerated. Phenomenological fits such as the above crude rule of thumb or the recent effort of Streever [396]* can be made but it is both difficult and dangerous to assign specific physical meaning to a particular resulting parameter. One fact of interest does emerge as we inspect Table V. If we assume for core polarization a $\chi \approx -3$ au, then the remaining contribution to H_e (from all sources) is negative and ranges from roughly 0 to -100 kgausses. This supports a picture of negative conduction electron polarization.

The relative importance of core polarization to iron series metal H_e behavior is largely due to the fact that $3d$ shell orbital and spin dipolar effects are almost entirely quenched. This is not the case for the rare earth metals where, as for the rare earth salts, orbital hyperfine contributions of millions of gausses occur. Exceptions arise for the Eu and Gd metals [397]† and intermetallics with their S-state $4f^7$ shells. The close agreement between the values of H_e, -354 koe in Gd metal [242] and -370 in GdN [397], and the corresponding values in the salts [374, 375] indicates that the crude rule of thumb of considering an H_c-like term to the exclusion of all others works remarkably well. In view of the importance of conduction electrons for the magnetic properties observed for the rare earth metals and the observations of hyperfine effects seen at nonmagnetic ion sites in these metals, the apparent lack of conduction electron contributions to H_e is striking and not well understood.

---

* A term proportional to the local moment plus a term appropriate to the host lattice is involved in this fit.

† Cf. the data referred to in Section IV, 2 (e.g., references [314, 375]).

Quadrupole interactions have been studied in a number of rare earth metals and intermetallics. Depending on crystal symmetry, some investigations [157, 168] yield information concerning the internal shielding factor R_q, others [196] concerning R_q, γ_∞, and also the shielding factor appropriate to the interaction between the 4f shell and the crystal-field component of $Y_0^2(\theta, \varphi)$ symmetry [346]. All the experimental analyses have relied on theoretical nuclear calculations (with an uncertainty $>10\%$) and/or experimental Coulomb excitation data (with an $\sim 10\%$ uncertainty) for Q values. If one utilizes nonrelativistic Hartree–Fock $\langle r^{-3}\rangle$ values (which are tabulated in Appendix B) one obtains $+0.3 \leqslant R_q \leqslant +0.4$ for the ions which have been investigated. If, on the other hand, one compares with an experimental magnetic $\langle r^{-3}\rangle_\text{eff}$ parameter, one obtains [157] an R of $\sim +0.2$ appropriate to

$$\langle r^{-3}\rangle_{\text{eff},q} = \langle r^{-3}\rangle_{\text{eff},m}(1-R) \tag{4.5}$$

The parameter R is then an experimental measure of the difference between magnetic and quadrupolar Sternheimer internal shielding. Although one might be tempted to interpret the difference between R_q and R behavior as an indication of closed-shell magnetic shielding, this does not seem to be the case. Comparison between nonrelativistic H–F estimates of rare earth fine and hyperfine parameters [398] suggests that the theoretical $\langle r^{-3}\rangle$ integrals are approximately 10% too high, because of the omission of such factors as relativistic effects in the H–F calculations. Further, this comparison suggests that genuine Sternheimer shielding has R_mag values of ± 0.05 associated with it. We have not attempted to observe trends in R_i behavior (though such a parameter need not be constant across the rare earths) because of the small number of the results and the uncertainties which must be attached to them. These uncertainties, which are of the order of the R_i values themselves, arise from a number of sources. Among these, in estimating an $\langle r^{-3}\rangle_\text{eff}$ from hyperfine field data one must make assumptions concerning (a) 4f shell occupancy and the associated angular and spin behavior, and (b) conduction electron contributions to the hyperfine fields [Eq. (4.4) should satisfactorily estimate the core polarization, H_c, contribution to H_e]. Despite such factors, there is the strong suggestion that internal Sternheimer antishielding effects and $R_q - R_\text{mag}$ differences are of experimental significance for the rare earths.

In this subsection we have reviewed some of the evidence indicating that conduction electron effects play but a minor role in ferromagnetic metal H_e behavior. While this does not mean that the contributions are negligible—they may at times be of the order 100 kgausses—it does imply that it is most difficult to deduce much about conduction electron behavior from such experiments. We return to this matter in the section which follows.

4. Hyperfine Fields in Nonmagnetic Ions in Magnetic Materials

H_e has been observed at the nuclei of nonmagnetic ions occurring in paramagnetic salts (e.g., F^- in MnF_2) and as impurities (or alloyed) in ferromagnetic metals (e.g., Sn in ferromagnetic Fe or Co). Let us briefly consider such cases here.

a. Salts. H_e normally consists of isotropic and anisotropic terms considered with respect to the relative orientation of an externally applied magnetic field and the crystal axes. The magnetic field is used to orient the paramagnetic ions and (with no local magnetic ordering of these ions) H_e is proportional to the salt's magnetic susceptibility. An isotropic contribution of 25 kgausses per near-neighboring magnetic ion (with $\langle S_z \rangle = S$) is typical for the Mn and Ni fluorides.

One frequent important source of the anisotropic H_e is the classical dipolar potential arising from the lattice of partially aligned magnetic ion moments. The other important source for both isotropic and anisotropic effects is the overlap (plus covalent mixing) polarization discussed in Section III, 4. These two sources appear to predominate in determining the qualitative nature of such transferred hyperfine effects. There are, of course, other possible contributions to these effects, one being the exchange polarization, in the UHF sense, of the ligand ion due to the unfilled d (or f) shells of its neighbors. Unfortunately, even if one accepts the exchange-polarized HF formalism, estimates of this are at best uncertain due to one's inability to adequately describe multicenter exchange effects.

In the case of iron series fluorides we come close to having a good quantitative understanding of the observed transferred hyperfine fields in terms of the overlap (plus covalent mixing) polarization, a situation which is due to the large body of work [289–299] which has accumulated over the years. Most of these data are for the closed cubic subshell configurations d^3, d^5, and d^8 which are relatively easier to treat. Work on the d^6 and d^7 configurations has also been reported [299a]. Details, such as the relative roles of any deviation(s) of the ion orbitals [most appropriate to Eqs. (3.14) and (3.16) away from free-ion H-F spatial behavior and of covalent mixing, have yet to be resolved [299].

The iron series fluoride observations are most useful when correlated [296–299] with observed neutron diffraction, optical spectra, and orbital reduction effects (see Section IV, 1). Once such details, as the above matter of orbital spatial behavior, are better resolved, it is hoped that one will be in a position to put the theory of superexchange [399, 400] on some sort of quantitative basis.

Similar analyses of transferred hyperfine effects for systems involving larger ligand ions, e.g., Cl^- or S^{2-}, are complicated by the fact that there are a number of p (or s) ligand shells whose overlap polarization must be simultaneously accounted for.* Such analyses have yet to be attempted on any scale and, because of the above complication, they promise to be a much less useful source of information concerning wave function behavior than the case of the fluorides. Transferred hyperfine effects have also been observed [401–403] for several rare earth ion systems. The type of analysis appropriate to the iron series salts is inadequate, for it gives the wrong sign (and too small a magnitude). Analysis using the results [341] of a spin-polarized H–F calculation for Gd^{3+} (cf. Fig. 16) correctly gives the observed sign and magnitude of the isotropic term [403]. The agreement in sign arises from overlap effects associated with the $5s$ and $5p$ shell electrons which contribute the negative spin density [cf. Fig. 16] of the outer regions of the ion.

b. Metals. In a metal, the core of a nonmagnetic ion may very well undergo overlap and exchange-polarization effects similar to those discussed above but these are considered to be of negligible importance.† Instead, the observed behavior is commonly presumed to be primarily due to the Fermi contact interaction of conduction electrons which have been polarized by the local moments of the magnetic ions. This polarization may arise in several ways (cf. Sections IV, 3 and III, 4), aside from the UHF polarization just discussed.

First, there is the Ruderman–Kittel–Kasuya–Yosida (R–K–K–Y) polarization [283–285], which produces a spin density distribution with regions of alternating spin sign about a local magnetic moment, of the sort indicated schematically in Fig. 18. The average induced spin density (the

---

* For the role of the $1s$ shell in the fluorides, see reference [396] (cf. the data referred to in Section IV, 2).

† There appears to be an exception to this rule when one deals with Knight shifts in intermetallic compounds such as V_3X (where X = Ga, Si, Ge, etc.). These metals have high densities of states associated with conduction electrons which reside almost entirely on the V sites [403a]. This implies that the spin density, induced by the applied magnetic field and causing the Knight shifts, resides mostly at the V sites. In order to understand Knight shifts [417] at the X sites one must assume an appreciable contribution from a negative polarization of the X site core s electrons due to the spin density at the neighboring V sites.

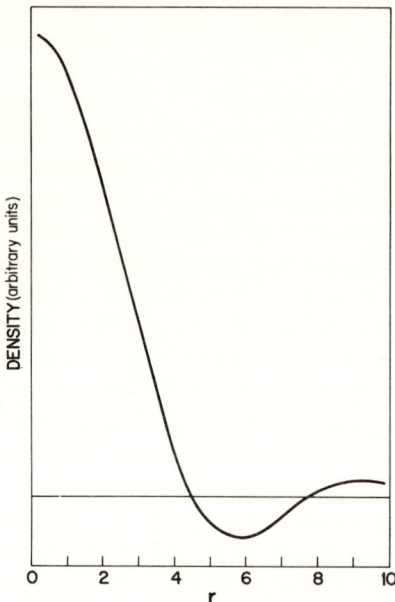

FIG. 18. Schematic representation of the Ruderman–Kittel–Yosida–Kasuya spin density oscillation about a spin impurity.

Zener term [282]) must be of spin parallel to that of the local moment.* The calculations which have been done to date have been in terms of free-electron conduction bands and a resulting spin density, such as that of Fig. 18, must be viewed with caution, particularly in discussions of the region about a nucleus. An orthogonalized plane wave (OPW) version of this exchange polarization has yet to be estimated but it is, we believe, a necessity in any serious attempt at a quantitative prediction of this polarization contribution to H_e.

The second conduction electron polarization contribution comes from the Anderson–Clogston $s \leftrightarrow d$ (or $p \leftrightarrow f$) mixing [393], which involves [395] single substitution configuration mixing between the conduction band and the d (or f) bands of more local character. This mixing makes a a contribution to the conduction electron spin density which is antiparallel to that of the more localized d (or f) moments. Recently, Peter and Koide [404] have considered the double-substitution $pf' \leftrightarrow ff''$ configuration mixing when dealing with electronic g shifts for rare earth

---

* This follows from the sign of an interelectronic exchange integral which, being the Coulomb interaction of an overlap charge density with itself, is of necessity always positive.

intermetallics. This mixing is of the same order as single-substitution $p \leftrightarrow f$ and the two must be considered simultaneously if one wishes to investigate the conduction electron polarization while preserving the spin (S^2) eigenvalue at a magnetic ion site (as one well might wish to do for a rare earth metal). Recent investigations [404a], assuming free-electronlike (or single orthogonalized plane wave) conduction electron character, indicate that it is the single substitution $f \leftrightarrow f'$ mixing (involving the f, not p, character of the conduction electron orbitals) involving important *spherical* matrix elements, which makes the most significant interband mixing contribution to the conduction electron polarization. The *net* spin polarization associated with this mixing will always be negative.

Experimentally, nonmagnetic ion H_e values, for ferromagnetic host metals, show a strong tendency to be negative, though exceptions, e.g., Sn in Ni (cf. Table V), occur. It is impossible, on the basis of this observation alone, to conclude anything concerning the relative responsibility of the two types of polarizarion, for one must perform a lattice sum over the various magnetic ion Kasuya–Yosida polarization contributions and the result can be either sign (within the free-electron approximation) despite the observation that the average polarization is positive.*

Other evidence may be considered concerning these matters. In the preceding subsection (IV, 3) we noted that conduction electrons play but a minor role (though not a negligible one by nonmagnetic ion standards) in magnetic ion H_e values and that there was inconclusive evidence to the effect that this contribution was negative. If one invokes the Kasuya–Yosida mechanism as being the predominant source of the nonmagnetic ion results, there would have to be a large positive contribution at the magnetic ion sites—a contribution which we believe to be incompatible with the above observation (whether accurate about the negative sign and magnitude or not). This suggests that this mechanism is not the dominant one (but it does not mean that it is experimentally insignificant). We may very well often be seeing a tendency for this and the Anderson–Clogston polarization to cancel at a magnetic ion site while, on occasion, enhancing one another at a nonmagnetic site.

More conclusive evidence appears in the work of Peter and co-workers [405, 406] where electronic g shifts, which provide information concerning the sign of average conduction electron polarization, have been obtained for rare earth intermetallics such as $GdAl_2$. Their results indicate a

---

* The computational result is very sensitive to such details as the position of the nonmagnetic ion with respect to the magnetic ions, particularly when considered in conjunction with the dependence of the exchange integral $J(\mathbf{q})$ on $\mathbf{q}$ (see Yosida [285]).

negative average polarization, in turn suggesting that $f \to f$ mixing is the dominant polarization term. The role of the two types of polarization in such g shifts has been studied by Kondo [407]. More recently Moore and Rodbell [408] and Gossard [409] have observed the g shift in metallic Gd and Eu, respectively, on going from the unmagnetized to the ferromagnetic state. While negative, the ferromagnetic state g shifts were found to be less negative than their unmagnetized counterparts. This suggests that the conduction electron polarization arising from the ferromagnetically ordered local f-electron moments may be positive, in contrast with the observations of Peter et al. for the intermetallics. Crangle [409a] has resolved this apparent discrepancy by reporting that while saturation magnetization results indicate a net positive conduction electron polarization in pure Gd [409b], a negative polarization occurs [409a] for Gd in Pd. In other words, for a given local moment, the average conduction electron polarization may be of either sign in different environments.

Before leaving this section, we would like to inspect briefly one other source of information concerning the polarization of electrons in ferromagnetic metals, i.e., the neutron diffraction studies of Shull and Yamada [410] and Moon [411] on ferromagnetic Fe and Co, respectively. These yield spin density plots such as is seen in Fig. 19. The feature of interest to us here is the region of negative spin density between the ion cores. If one ignores polarization mechanisms which yield a zero integrated density (e.g., exchange polarization in the UHF sense, or the second-order Kasuya–Yosida terms), one will conclude that the conduction electrons have undergone a substantial negative polarization, perhaps [311, 412] of the Anderson–Clogston type here too. On the other hand, one should remember that the d orbitals associated with the bottom of the d band (and hence are doubly occupied) are quite diffuse and overlap the interionic region heavily [390–392]. One should also note the occurrence (e.g., as seen in Fig. 16) of a negative spin density associated with paired orbitals in the outer reaches of an exchange polarized ion. Such a local polarization of paired "$3d$" and "$4s$" band orbitals may provide one source of the negative spin densities seen by Shull and Moon. An alternant and very crudely equivalent view of such a term is that the negative parts of the Ruderman–Kittel distribution (see Fig. 18) overpower the positive parts in the interionic region (when one sums over polarization contributions coming from all "local" magnetic ion cores in the lattice). The possibility of such terms makes it most difficult to derive any conclusions concerning the net conduction polarization from such experiments. In order to better understand some of the experiments cited above there has been recent theoretical activity

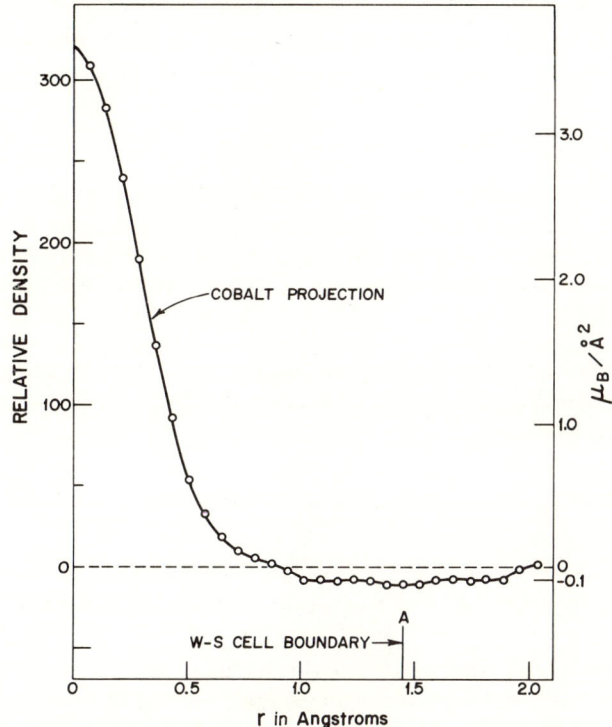

Fig. 19. Profile of projected density in hexagonal cobalt. [After Moon [411].

aimed at modifying the Kasuya–Yosida theory, which has largely concentrated [412a, 412b] on the "susceptibility" function. Common to these and earlier works have been certain assumptions (quite outside the basic assumption of free-electron conduction bands and orbital character) concerning the analytical character of the off-diagonal exchange integrals causing the redistribution of spin, away from a constant Zener term, seen in Fig. 18. A very recent investigation [412c] has shown these assumptions to be unrealistic and that a rigorous evaluation of the R–K–K–Y model may yield a spin density distribution which differs markedly from that seen in Fig. 18. For example, the spin density at the origin may be opposite in sign to the average polarization density. It may also be smaller in magnitude than the density at some radius corresponding to a neighbor ion position. In other words, contrary to the traditional view taken above, based on the spin distribution of Fig. 18, of the absence of a large conduction electron spin density at a magnetic ion nucleus *need not* imply an insignificant spin density away from that site. Finally, since the metals cited above are not describable strictly in terms of a free electron picture, one must view any results obtained by the use of this model with great caution.

5. Hyperfine Fields (Knight Shifts) in Metals

The application of a magnetic field H_0 to a metal produces a polarization of the electronic energy bands which contributes to the magnetic susceptibility and the hyperfine field. Ignoring diamagnetic effects we may describe these contributions as arising from the following three sources:

(1) The polarization of the conduction electrons by H_0 gives rise to a magnetic field at the nucleus through the Fermi contact term, which is usually written as

$$\Delta H_s = \frac{8\pi}{3}\chi_s H_0 \langle |\Psi_s(0)|^2 \rangle_F . \qquad (4.6)$$

Here, χ_s is the Pauli paramagnetic spin susceptibility of the conduction electrons and $\langle |\Psi_F(0)|^2 \rangle$ is the value of the probability density at the nucleus averaged over the electrons with energies near the Fermi level. The subscript s is derived from the usual observation that only those conduction electrons with nonvanishing density at the nucleus (for atoms these are the s electrons) contribute to the hyperfine field. Cohen *et al.* [314] pointed out that the conduction electrons which are polarized by the external field in turn produce a polarization of the core s electrons and hence an additional contribution to the hyperfine field. Since this additional contribution is proportional to the number of conduction electron spins which have been polarized by H_0, i.e., proportional to χ_s, Eq. (4.6) may be simply rewritten as

$$\Delta H_s = \frac{8\pi}{3}\chi_s H_0 \{\langle |\Psi(0)|^2 \rangle_F + |\Psi_s(0)|^2\}$$

$$= \frac{8\pi}{3}\chi_s H_0 \langle |\Psi(0)| \rangle_F^2 \{1 + R_s\} \qquad (4.7)$$

where $|\Psi_s(0)|^2$ is the spin density at the nucleus arising from the core s electrons and R_s is the ratio $|\Psi_s(0)|^2/\langle |\Psi(0)|^2 \rangle_F$.

(2) With the measurement of negative Knight shifts came the realization that another contribution to the hyperfine field was needed. Recognizing that core polarization effects for $3d$ ions give a negative hyperfine field, several authors [311, 412] suggested that such a term

should be included in interpreting Knight shift measurements in those metals and alloys which have unfilled d bands. Since the d electrons are polarized by H_0, the magnetic field produced at the nucleus by the d-electron polarization of the core s electrons may be written as

$$\Delta H_d = R_d \chi_d H_0 \qquad (4.8)$$

Here χ_d is the d-band susceptibility, which (unlike χ_s) is generally temperature dependent, and R_d is the core polarization field per unit magnetization of the $3d$ electrons.

(3) As emphasized very recently by Clogston et al. [413] in their explanation of Knight shifts in superconductors, one must also include a contribution due to the orbital moment induced by a Van Vleck [372, 414, 415] type of paramagnetism for metals [416]. This term is very difficult to calculate in an *ab initio* way but, using a simplified tight binding description, Clogston et al. used the form

$$\Delta H_{\text{orb}} = 2\chi_{\text{orb}} \xi \langle r^{-3} \rangle_{\text{atom}} H_0 \qquad (4.9)$$

to describe the additional hyperfine field. In Eq. (4.9), $\langle r^{-3} \rangle$ is the expectation value of the r^{-3} operator over the free atom $3d$ wave function and ξ is a reduction factor which is the ratio of $\langle r^{-3} \rangle$ for the metal to that for the atom.

Detailed quantitative *ab initio* calculations for each of these various contributions to the Knight shift have yet to be done. Experiment suggests that the several ΔH contributions play varying roles in different metals. Sorting out the various terms has been greatly assisted by graphical techniques [414, 415, 417] (where the temperature of ΔH is plotted vs. that of the magnetic susceptibility) and by systematic inspection of an alloy system as a function of concentration [418]. The original ΔH_s term dominates for free electron-like metals such as Na and K while the orbital term appears to be most important for systems such as [414, 415] V_3X (where $X = Si$, Ga, etc.) and [419] μ-Al_2. To date, ΔH_{orb} seems to play its greatest role in intermetallic and alloy systems; core polarization dominates in such metals as platinum [420].

The results of Cohen et al. [314] and unpublished computations of the present authors [343] for K, Cr, and Cu indicate that the core contribution tends to enhance the ΔH_s interaction by 0 to 30% (i.e., $0 \leqslant R_s \leqslant 0.3$). The exact result depends strongly on how much s-like character is associated with conduction electrons in the vicinity of the Fermi surface. This factor, of course, determines the magnitude of $\langle |\Psi(0)| \rangle_F^2$, but since both the s-like and the non-s-like parts of the polarized conduction electrons contribute (in differing ways) to the spin polarization of the core, the magnitude, and even the sign, of R_s depends on details of the relative s, p, d, etc., character of those conduction electrons. A negative R_s, reducing ΔH_s, could occur if there were relatively little s-like character present.

In order to facilitate estimates of ΔH_{orb} we list in Appendix B values of $\langle r^{-3} \rangle$, $\langle r^{-1} \rangle$, and $\langle r^2 \rangle$ which are involved in χ_{orb} and diamagnetic corrections, and other $\langle r^n \rangle$ values for various $3d$, $4d$, and $4f$ ions.

6. Mössbauer Effect in an External Magnetic Field

The use of external magnetic field in Mössbauer experiments to study hyperfine interactions was described but briefly in Section II, 4 where some results were also noted. We shall discuss here the interpretation of these measurements and describe the way in which information has been obtained about paramagnetic materials.

First consider those cases in which only the ion which is polarized by the field has a localized electronic moment, for example, for the measurements of Fe^{57} in stainless steel, Cu, and MgO [163] which were summarized in Section II, 4, and by Eq. (2.19).

Let us see how Eq. (2.19) comes about. For an array of paramagnetic atoms, each having an angular momentum J, thermal (Boltzmann) average J in a field H_0 is given by $\langle J \rangle_T = JB_J(x)$ where $B_J(x)$ is the usual Brillouin function with $x = g\mu_B H_0 kT$. For small x, $B_J(x) \approx (J+1)g\mu_B H_0 3kT$. Assuming that the hyperfine field due to the polarization of the electronic moments is proportional to $\langle J \rangle_T$, i.e.,

$$H_{\text{eff}} = A\langle J \rangle_T \qquad (4.10)$$

then we have

$$H_{\text{eff}} = AJB_J(x) \approx [AJ(J+1)g\mu_B/3kT]H_0 = C(T)H_0 \qquad (4.11)$$

Hence, since

$$H_n = H_0 + H_{\text{eff}} \qquad (4.12)$$

we have

$$H_n = H_0[1 + C(T)] \qquad (4.13)$$

which is the experimental result.

Consider now the measurements of dilute iron alloys. These Mössbauer measurements have been of interest because of their role in understanding the origin of the large magnetic moments which are associated with Fe atoms dissolved in nonmagnetic $3d$ and $4d$ metals. (For example, at the end of the series, the moments are found to be very large [421], i.e.,

~9–12μ_B.) These systems have been studied extensively by means of susceptibility measurements [421]; theoretical interpretations have emphasized the localized nature of these moments [395, 422, 423]. The Mössbauer measurements [165], which were on dilute solid solutions of Fe^{57} in Pd, found the field at the nucleus minus the externally applied field could be fitted, by a Brillouin-type function, over the entire range of measurement and an expression of the form

$$H_{\text{eff}}/H_{\text{sat}} = \frac{(S+1)}{3k} g\mu_B \frac{H_0}{T} \tag{4.14}$$

linear in the region; H_{sat} is the saturation value of H_{eff}. Using Eq. (4.14) Craig et al. [165] found a spin of 13/2 and a moment (assuming $g = 2$) of 12.6μ_B. These measurements confirmed the magnitude of the moment reported earlier by means of susceptibility measurements and that the moment associated with each Fe impurity acts, at low temperatures, like an isolated magnetic moment in an external magnetic field.

It has been shown [164] that measurements of the hyperfine field (e.g., by means of Mössbauer or NMR methods) can give information about the moment of the iron atoms themselves whereas susceptibility measurements cannot [421], and that the nature of the coupling between the spin on the Fe atom and the polarization of its surroundings can be inferred. We shall describe this work because it brings out, in a simple way, the explicit dependence of the hyperfine field H_{int} on the spin $\mathbf{S}'$ of the Fe atom and demonstrates why H_{int} vs. temperature follows a Brillouin function of the total localized spin $\mathbf{S}$ (with $\mathbf{S} \neq \mathbf{S}'$).

Consider now a spin $\mathbf{S}'$ rigidly coupled to its polarized surroundings, the system forming a resultant spin $\mathbf{S}$. Suppose we take a collection of such spins to form a paramagnetic system. In an external field, $\langle S \rangle_T$ is given by a Brillouin function for the spin S. Because of the rigid coupling, S' "follows" S, and so $\langle S' \rangle_T$ is now given by

$$\frac{\langle S' \rangle_T}{S'} = \frac{\langle S \rangle_T}{S} = B_B(x) \tag{4.15}$$

instead of by $\langle S' \rangle_T = S' B_{S'}(x)$, which holds only for a system of free spins S'. H_{eff} is again assumed to be proportional to the average value of the spin S' on the Fe atom:

$$H_{\text{eff}} = A \langle S' \rangle_T \tag{4.16}$$

This relationship is, of course, valid for the Fermi Hamiltonian for a free atom with $L = 0$, and appears to hold, at least qualitatively, for a number of cases in metals and alloys [424]. (The origin of H_{eff} in metals was discussed in detail earlier and here we need only point out that since contributions from the polarized host atoms are also expected to be proportional to S', these may also be included in A.) Substituting from Eq. (4.15) the value of $\langle S' \rangle_T$ gives

$$H_{\text{eff}} = \langle S' \rangle_T A = A S' B_S(x) \tag{4.17}$$

which, in the region of small x, is given by

$$H_{\text{eff}} = AS' \left[\frac{(S+1)}{3k} g\mu_B \frac{H_0}{T} \right] \tag{4.18}$$

Equations (4.17) and (4.18) display (1) the physical origin of H_{eff} through S' and A and (2) the temperature dependence of H_{eff} through the usual susceptibility factor (which is concerned with the total spin S). It follows from Eq. (4.17) that the saturation value of H_{eff} is given by

$$H_{\text{sat}} = AS' \tag{4.19}$$

and so we may therefore write

$$\frac{H_{\text{int}}}{H_{\text{sat}}} = \frac{(S+1)}{3k} g\mu_B \frac{H_0}{T} \tag{4.20}$$

Equation (4.20) is the phenomenological expression used to fit the data in the linear region [165].

This very simple formulation allows one also to determine S'. Taking $A \approx -300$ kgausses as a reasonable value for Fe in metallic iron and in several alloys [424] and using the observed H_{sat} value of -295 kgausses, we find that $S' \approx 1$ and $\mu \approx 2\mu_B$. This value seems quite reasonable and confirms the expectations of Clogston et al. [421].

We have discussed this work in detail because such studies in conjunction with those described in Sections IV, 3 and IV, 7, and with neutron magnetic scattering experiments, offer a means of understanding in detail the nature of localized magnetic states, the range of magnetic interactions, and the spatial distribution of the spin density. Perhaps the most interesting result from such combined inspection involves the spin distributions induced in Pd by magnetic impurities. Pd with its abnormally large magnetic susceptibility is, of course, of particular interest. Electron paramagnetic resonance [424a] on Gd impurities in Pd suggested that the first node in the induced conduction band spin density occurs at an appreciably greater radius than that appropriate to the Ruderman–Kittel theory (where the first node typically occurs at a nearest neighbor distance—see Fig. 18). Subsequent neutron diffraction [424b] and Mössbauer [424c] experiments for Fe in Pd suggest a node radius which is some 4 to 10 lattice spacings and give no evidence for oscillatory behavior outside this region. Pd is a transition metal which is perhaps appropriately described as an "incipient ferromagnet," and it should not come as a surprise that it does not conform to an exchange polarization scheme based on free-electron theory. Pd is not unique in this regard; in fact, energy band calculations for Gd [424d], as well as La specific heat results [424e], suggest that the rare earth metals, which have been traditionally treated as having free-electron conduction bands, do suggest that the bands have considerable 5d transition metal character. The implications of deviations from free-electron character on the conduction electron polarization for the rare–earths and other metals is, as yet, relatively unknown and requires considerable experimental and theoretical investigation.

7. The Mössbauer Effect and Hyperfine Fields in Iron Metal with Dilute Impurities

Considerable work [103, 138, 141, 425–429] has been done on the Mössbauer effect in a number of iron alloys with dilute impurities such as Si, Rh, V, Sn, Mn, or Co. A parameterization scheme has emerged which appears to provide successful fits of experiment and provides information about the range of the magnetic interaction and the spatial distribution of the spin density.

Consider an Fe atom with N_n impurities as nearest neighbors and N_{nn} next-nearest impurity neighbors. Let us write its hyperfine field as

$$H(N_n, N_{nn}) = H'\{1 + C_n N_n + C_{nn} N_{nn}\} \qquad (4.21)$$

where H', C_n, and C_{nn} are experimental parameters to be determined. If one has a crystal of given impurity concentration and if one assumes that the impurities have gone into the crystal in a statistically independent way, one can obtain the number of Fe atoms with particular values of N_n and N_{nn}. If we assume H', C_n, and C_{nn} to be independent of the N_i, a Mössbauer hyperfine spectrum can be computed and values for these parameters can be obtained by fitting experiment. In general, a non-zero C_{nn} term must be included in order to obtain an adequate fit. For a large number of impurities to the left and/or below Fe in the periodic table (e.g., Si, V, Cr, Mn, and Ru) the C_i's take on remarkably constant values of $C_n \sim -0.075$ (say to ± 0.02) and $C_{nn} \sim -0.006$, respectively. The negative signs, of course, indicate that the hyperfine field decreases with increasing numbers of impurity neighbors. Such a trend might be expected in terms of a rigid band model of the transition metals but details of Eq. (4.21) and the rough constancy of the C_i's are not entirely consistent with such a model. For impurities such as [429] Co and Ni it appears that C_i's of slightly smaller magnitude but positive sign occur.

The experimental fits involve the subtle matter of details in line shape and some uncertainty must be attached to the resulting experimentally derived parameters. These factors make it difficult to extend Eq. (4.21) to more distant neighbors. Stearns [429a] has extended the fitting scheme to include the six nearest shells of neighbors (i.e., through C_{6n}) when analyzing her Mössbauer results. From the resulting parameters, she and Overhauser [429b] have estimated the magnetic susceptibility behavior of metallic iron.

The parameter H' generally differs from the pure metal value by a few percent and is concentration dependent. This is not surprising since the various parameters of Eq. (4.21) must compensate for any shortcomings in that equation. H' increases linearly with increasing concentration for the dilute alloys which may be associated with a higher-field satellite (or a family of lines) which lies too close to the main hyperfine peak to be resolved by Mössbauer measurements [429c]. Whatever its source, the H' shift is small.

There of course exists a vast array of experimental alloy data which should not be ignored when one is contemplating the Eq. (4.21) results. The Mössbauer isomer shifts are one example of this. Our experience [98] with Fe^{57} isomer shifts indicates that they are sensitive to variations in ion configuration and bonding. The Fe alloy isomer shifts are almost identical, suggesting that the local chemical behavior at an Fe site is much the same for differing kinds and numbers of impurity neighbors.

Of the closely related non-Mössbauer work, perhaps the most interesting is that of Low and Collins. They have recently developed neutron diffraction techniques for studying the spatial extent of the magnetic disturbance associated with an impurity in an alloy. The method has been applied* to dilute Ni and Fe alloys and the results suggest that, in a number of the alloys studied by the Mössbauer effect, the disturbance extends considerably beyond next-nearest neighbors with a magnitude not appreciably less than that in the next-nearest neighbor sites. In addition, for V and Cr impurities in Fe, the disturbance spin distribution (i.e., the difference between the spin distribution in the vicinity of impurity and what would be there in pure Fe) has a spin direction of one sign at the impurity and nearest (and perhaps next

---

* See reference [430], in which the Ni alloy results appear; the Fe results remain to be published.

present)-neighbor sites and reverses sign, and is of smaller magnitude, outside this region. It is presumed that the disturbance spin direction at the impurity site is antiparallel to the net magnetization of the crystal. A sign reversal has already been seen in the Mössbauer data where the C_i's are negative and H' is greater than the pure metal value. This pattern is somewhat destroyed by the third Fe alloy for which Low and Collins have results, Mn in Fe. While the observed Mössbauer behavior appears to be quite similar to that for Cr and V impurities in Fe, the neutron measurements for Mn indicate that the magnetic disturbance is strongly localized on the impurity site. Wertheim and Jaccarino [430a] have suggested that the nearest neighbor coefficients, i.e., C_n, reflect chemical as well as magnetic influences and by ignoring the C_n values one obtains crude consistency between the Mössbauer and neutron diffraction data [430a].

From the discussions given earlier one might hope that, aside from conduction electron effects and any (presumably) small orbital contributions, one would observe a hyperfine field which is proportional to the local magnetic moment at the iron site. This also has yet to be proven or disproven.

Any final conclusion will require more data and inspection of yet other experimental information, e.g., alloy magnetization curves and Fe^{57} hyperfine results, where iron is the dilute impurity in an alloy (discussed elsewhere in this section). The difficulties encountered above between neutron and Mössbauer results suggest that the task may not prove simple.

Appendix A. Comments on the Solution of Conventional Hartree–Fock Radial Equations for Atoms

The conventional H–F equations are derived by applying the variational principle to the total energy of the system, subject to the restrictions discussed in Section III. Hartree discusses these matters in some detail in his book [251]. Some features of the averaging process associated with going from one H–F equation per electron to one radial equation per shell have been discussed elsewhere [309, 431].

The H–F equations we wish to solve are a set of simultaneous one-dimensional quartic equations (since the one-electron orbitals appear in the potential energy operator of the one-electron Hamiltonian). These equations have yet to be solved directly although work is under way [432] to investigate the possibilities of doing this. Instead, one transforms these into a set of quadratic equations by the judicious insertion of a set of trial ψ_i [or rather $U_i(r)$] functions into the original equations. One then solves these quadratic equations, obtaining a new set of $U_i(r)$ functions, which are then inserted into the quartic equations; the process is repeated until convergence is obtained. This is known as the self-consistent field (SCF) method and has associated with it convergence problems both in principle and in practice. One rarely uses a straight interative procedure, but even so, despite precautions, it is a rare practitioner who has not encountered difficulties on some occasion.

The traditional method for solving the H–F quadratic equations is by direct numerical integration. This method was developed to a high state by the Hartrees and is fully described in Hartree's book [251].

An alternate, analytic approach, where the $U_i(r)$ are expanded in terms of analytic basis sets, has also been used. Coulson originally developed [433] this approach for dealing with molecular problems and Roothaan [434] put it in a particularly useful form for computation. In this approach the space orbitals $\psi_i(r)$ are expanded so that

$$\psi_i(r) = \sum_j C_{ij}\chi_j(r) \tag{A-1}$$

where the χ's are not necessarily an orthogonal basis set. In the RHF formalism for atoms they are normally a product of a spherical harmonic times a radial factor $r^{n_j} \exp(-Z_j r)$ with a specific spherical harmonic associated with all the χ_j's of a specific ψ_i. All necessary matrix elements can then be obtained analytically; the SCF process is simply one of matrix addition, multiplication, and diagonalization. Nesbet [435] and Roothaan and Bagus [436] have discussed this approach in detail.

Each method has advantages and disadvantages, of course. The analytic method is limited by the finiteness of the basis set which is employed; the numerical approach, on the other hand, suffers from inaccuracies associated with the process of numerical integration. To date, the two methods have succeeded in producing results of similar accuracy for ions with atomic numbers up to 30. For appreciably larger ions, where relativistic effects cannot be ignored, the analytic approach is beset by the difficulty of maintaining, on the one hand, sufficient variational freedom in the basis set, while avoiding, on the other hand, the build-up of approximate linear dependencies (which will cause computational errors) within the basis set. Convergence difficulties have a greater tendency to appear, whatever the method, (1) the greater the atomic number and (2) the less positively charged the ion. In principle, the numerical method is superior, since it avoids the question of basis sets, but, in practice, there are some great virtues attached to the analytic approach. First, it is easier to obtain matrix elements for analytic functions, since one can then, in general, avoid numerical integration. Second, it is easier with the analytic approach to deal with the full UHF treatment or with antishielding factors, crystal-field effects, and the like. Finally, and perhaps most important, the analytic approach is, of necessity, used in most molecular and energy band investigations. Hence, an atomic method which connects logically with these approaches has obvious advantages.

We have indicated in the body of this chapter that maintaining the H–F restrictions while obtaining orthogonal ψ_i functions occasionally requires additional constraints on the solutions. This occurs when one has two differently occupied (not counting empty) shells of the same l value, for, as we have stated, the conventional H–F radial equations are obtained by averaging over the individually occupied electron equations for a shell. For two differently occupied shells of the same l value, this leads to differing radial H_{eff}. Hence, since they are eigenfunctions of different Hamiltonians, the $U_i(r)$ for the two shells will not be orthogonal.

Three methods, in general, are used to deal with this. First, there is the process of using off-diagonal Lagrange multipliers. Hartree [251] described this approach for the numerical method and Roothaan [437] and Huzinaga [438] have devloped the means for doing this in analytic calculations. Second, one can apply (e.g., see references [309, 431]) some form of Nesbet's symmetry and equivalence restrictions [261]. Third, one can simply ignore the matter and allow the final set of ψ_i to be slightly nonorthogonal. As often as not, the Hartrees used this last approach with satisfactory results in the few calculations of theirs where these matters arose. Finally, we should note that there is a fourth way to deal with this, namely, to abandon the H–F restriction which caused the nonorthogonality in the first place. This yields the many-electron function of lowest total energy and leads us to the matters discussed in the body of this chapter.

The above discussion is entirely for a nonrelativistic H–F theory. Unfortunately, just as we do not have exact solutions of the two-electron Schrödinger equation, neither do we have a complete relativistic theory for many-electron systems [439]. A relativistic Hartree theory has been available for some time as have wave functions obtained with it. A completely acceptable relativistic H–F theory encounters greater difficulties, but despite this Swirles [440, 441] has developed a relativistic H–F method which consists of inserting Dirac two(space)-component orbitals (with n, l, j, and m_j as good quantum numbers) into the determinantal function. Relativistic H–F functions have yet to be obtained, but the necessary computer programs are being developed in several places (i.e., by D. F. Mayers at Oxford and C. C. J. Roothaan at Chicago) utilizing numerical methods; results should be forthcoming shortly.

In the brief review given above, we have attempted, first, to supply the reader with a slight indication of what is involved in conventional H–F calculations and, second, to give sufficient references.* Once the reader has absorbed the more important material contained in these references, it should be obvious how one goes about setting up and solving the UHF equations. Deriving the UHF equations consists simply of following the same process as is used in the RHF formalism but omitting the RHF restrictions; the procedure is simpler and yields equations which are easier to solve than the RHF equations, although there are, of course, more equations to solve simultaneously. In other words, if one has a sufficiently large computer, a UHF calculation is inherently easier to deal with than its RHF counterpart. As should be clear from the discussion in the body of the chapter, this statement does not hold for the "extended" (properly symmetric) UHF formalism.

Appendix B. Hartree–Fock (r^n) Values for 3d, 4d, and 4f Ions

We list here in Tables XVII, XVIII, and XIX some values of $\langle r^n \rangle$ for 3d, 4d, and 4f ions, respectively, for which Hartree–Fock free-ion wave functions have been obtained. As mentioned in various parts of the text, these data provide a good zeroth-order approximation to those quantities of interest to the various hyperfine interactions, diamagnetic shielding, and crystal-field effects.

TABLE XVII

Some Hartree–Fock [199] $\langle r^n \rangle$ Values for Transition Metal Ions

	Sc($3d^3$)	Sc^{2+}($3d^1$)	Ti($3d^4$)	Ti^{2+}($3d^2$)	V($3d^5$)	V$^+$($3d^4$)	V^{2+}($3d^3$)	V^{4+}($3d^1$)
$\langle r^{-3} \rangle$	0.7647	1.5736	1.2658	2.1332	1.8355	2.2894	2.7476	3.6840
$\langle r^{-1} \rangle$	0.5763	0.8404	0.7097	0.9346	0.8231	0.9298	1.0192	1.1670
$\langle r^2 \rangle$	9.9462	3.0010	6.3481	2.4472	4.5167	2.8186	2.0702	1.3768
$\langle r^4 \rangle$	281.93	19.213	125.03	13.171	65.293	20.7145	9.6047	3.5926

* In addition to the references of this appendix, the reader may find the following useful: Lefebvre [442]; Pople and Nesbet [442a]; and, for the concept of the average of a configuration (which may prove useful for relativistic calculations), Shortly [442b].

TABLE XVII (continued)

	Cr(3d^6)	Cr^{2+}(3d^4)	Cr^{3+}(3d^3)	Cr^{4+}(3d^2)	Mn (3d^7)	Mn$^+$ (3d^6)	Mn^{2+} (3d^5)	Mn^{3+} (3d^4)
$\langle r^{-3} \rangle$	2.4255	3.4508	3.9588	4.4839	3.1215	3.6825	4.2499	4.7897
$\langle r^{-1} \rangle$	0.9102	1.1015	1.1759	1.2443	0.9982	1.0968	1.1828	1.2524
$\langle r^2 \rangle$	3.6643	1.7808	1.4470	1.2266	2.9992	2.0257	1.5482	1.2864
$\langle r^4 \rangle$	43.1345	7.2112	4.2967	2.9055	28.931	10.8718	5.5127	3.4457

	Fe(3d^8)	Fe^{2+}(3d^6)	Fe^{3+}(3d^5)	Fe^{4+}(3d^4)	Co(3d^9)	Co^{2+}(3d^7)	Co^{3+}(3d^6)	Co^{4+}(3d^5)
$\langle r^{-3} \rangle$	3.8822	5.0809	5.7242	6.3319	4.7877	6.0354	6.6988	7.4212
$\langle r^{-1} \rangle$	1.0792	1.2531	1.3290	1.3921	1.1647	1.3264	1.3976	1.4665
$\langle r^2 \rangle$	2.5370	1.3926	1.1500	0.9997	2.1450	1.2509	1.0492	0.9080
$\langle r^4 \rangle$	20.617	4.4958	2.7894	1.9861	14.704	3.6552	2.3423	1.6592

	Ni(3d^{10})	Ni^{2+}(3d^8)	Ni^{3+}(3d^7)	Ni^{4+}(3d^6)	Cu$^+$(3d^{10})	Cu^{2+}(3d^9)	Cu^{3+}(3d^8)	Cu^{4+}(3d^7)
$\langle r^{-3} \rangle$	5.7251	7.0941	7.7897	8.5516	7.5286	8.2515	9.0180	9.8137
$\langle r^{-1} \rangle$	1.2414	1.3994	1.4677	1.5340	1.3987	1.4711	1.5397	1.6035
$\langle r^2 \rangle$	1.8490	1.1300	0.9582	0.8371	1.2562	1.0275	0.8763	0.7719
$\langle r^4 \rangle$	10.640	3.0034	1.9708	1.4230	4.2389	2.4978	1.6623	1.2209

TABLE XVIII
Hartree–Fock[a] $\langle r^n \rangle$ Values for Some 4d Ions

	Y$^+$	Y^{2+}	Zr^{2+}	Zr^{3+}	Nb^{2+}	Nb^{3+}	Mo$^+$	Mo^{2+}
$\langle r^{-3} \rangle$	1.5898	2.0336	2.7057	3.1600	3.4141	3.9134	3.6623	4.1745
$\langle r^{-1} \rangle$	0.5322	0.5932	0.6607	0.7061	0.7211	0.7652	0.7318	0.7774
$\langle r^2 \rangle$	7.6007	5.5883	4.5260	3.8567	3.8291	3.3077	3.9538	3.3185
$\langle r^4 \rangle$	133.9293	58.9978	37.8608	25.3281	26.9820	18.6001	32.9845	20.2190

	Mo^{3+}	Tc^{2+}	Ru^{2+}	Ru^{3+}	Rh^{2+}	Rh^{3+}	Pd^{2+}	Pd^{3+}
$\langle r^{-3} \rangle$	4.7066	5.0149	5.8582	6.4961	6.8040	7.4467	7.8144	8.4871
$\langle r^{-1} \rangle$	0.8198	0.8329	0.8811	0.9242	0.9309	0.9708	0.9794	1.0180
$\langle r^2 \rangle$	2.9052	2.9033	2.6276	2.3132	2.3736	2.1171	2.1584	1.9390
$\langle r^4 \rangle$	14.3861	15.4096	12.8668	9.1689	10.5999	7.7851	8.8319	6.5906

	Ag$^+$	Ag^{2+}	Ag^{3+}	Cd^{2+}	Cd^{3+}
$\langle r^{-3} \rangle$	8.2232	8.9054	9.6110	10.1039	10.8265
$\langle r^{-1} \rangle$	0.9878	1.0274	1.0650	1.0763	1.1122
$\langle r^2 \rangle$	2.2379	1.9717	1.7820	1.8031	1.6424
$\langle r^4 \rangle$	10.7143	7.4098	5.6099	6.2249	4.8032

[a] Unpublished calculations of the authors.

TABLE XIX

Hartree–Fock [253b] $\langle r^n \rangle$ Values for Some Rare Earth Ions

	Ce^{3+}	Pr^{3+}	Nd^{3+}	Sm^{3+}	Eu^{2+}	Eu^{3+}	Gd^{3+}
$\langle r^{-3} \rangle$	4.72	5.37	6.03	7.364	7.53	8.081	8.84
$\langle r^{-1} \rangle$	1.301	1.361	1.417	1.514	1.510	1.561	1.609
$\langle r^{2} \rangle$	1.200	1.086	1.001	0.883	0.938	0.832	0.785
$\langle r^{4} \rangle$	3.455	2.822	2.401	1.897	2.273	1.697	1.515
$\langle r^{6} \rangle$	21.226	15.726	12.396	8.775	11.670	7.442	6.281

	Tb^{3+}	Dy^{3+}	Ho^{3+}	Er^{3+}	Tm^{2+}	Tm^{3+}	Yb^{3+}
$\langle r^{-3} \rangle$	9.577	10.34	11.158	12.01	12.220	12.857	13.83
$\langle r^{-1} \rangle$	1.650	1.690	1.732	1.773	1.764	1.810	1.856
$\langle r^{2} \rangle$	0.755	0.726	0.695	0.666	0.728	0.646	0.613
$\langle r^{4} \rangle$	1.419	1.322	1.219	1.126	1.552	1.067	0.960
$\langle r^{6} \rangle$	5.688	5.102	4.502	3.978	7.510	3.647	3.104

Acknowledgments

We are indebted to S. Geschwind, L. Grodzins, P. O. Löwdin, W. Marshall, and R. M. Sternheimer for informative discussions, and to many of the authors quoted in the text for valuable communications regarding their work. We are grateful to Miss J. Condon, Miss M. DeSesa, Miss A. Julian, and Mr. A. Tarmy for help with the manuscript. One of us (R. E. Watson) gratefully acknowledges the support of a National Science Foundation Postdoctoral Fellowship while he was at the University of Uppsala.

References

1. W. Pauli, *Naturwissenschaften* **12**, 741 (1924).
2. E. Fermi, *Z. Physik* **60**, 320 (1930).
3. A. Abragam and M. H. L. Pryce, *Proc. Roy. Soc.* **A205**, 135 (1951); **A206**, 164, 173 (1951).
4. R. Aronowitt, *Phys. Rev.* **92**, 1002 (1953).
5. G. Breit, *Phys. Rev.* **74**, 1278 (1948); F. Low and E. E. Salpeter, *ibid.* **83**, 478 (1951); N. M. Kroll and F. Pollack, *ibid.* **84**, 594 (1951); R. Karplus and A. Klein, *ibid.* **85**, 972 (1952); **87**, 848 (1952).
6. H. Kopfermann, "Nuclear Moments." Academic Press, New York, 1958.
7. C. Schwartz, *Phys. Rev.* **97**, 380 (1955).
8. R. E. Trees, *Phys. Rev.* **92**, 308 (1953).
9. J. C. Slater, "Quantum Theory of Atomic Structure," Vol. II. McGraw-Hill, New York, 1960.
10. H. H. Stroke, R. J. Blin-Stoyle, and V. Jaccarino, *Phys. Rev.* **123**, 1326 (1961).
11. W. D. Knight, *Solid State Phys.* **2**, 93 (1956).
12. R. E. Watson and A. J. Freeman, *Phys. Rev.* **123**, 2027 (1961).

13. H. B. G. Casimir, *Arch. Musee Teyler, Ser. III* **8**, 201 (1936).
14. R. V. Pound, *Phys. Rev.* **79**, 685 (1950).
15. T. P. Das and E. L. Hahn, *Solid State Phys.* Suppl. I (1958).
16. M. H. Cohen and F. Reif, *Solid State Phys.* **5**, 321 (1957).
17. A. Abragam, "The Principles of Nuclear Magnetism." Oxford Univ. Press, (Clarendon), London and New York, 1961.
18. R. M. Sternheimer and H. M. Foley, *Phys. Rev.* **92**, 1460 (1953); H. M. Foley, R. M. Sternheimer, and D. Tycko, *ibid.* **93**, 734 (1954); R. M. Sternheimer, *ibid.* **96**, 951 (1954).
18a. R. M. Sternheimer and H. M. Foley, *Phys. Rev.* **102**, 731 (1956).
18b. R. M. Sternheimer and H. M. Foley, *Phys. Rev.* **132**, 1637 (1963).
19. R. M. Sternheimer, *Phys. Rev.* **80**, 102 (1950); **84**, 244 (1951); **86**, 316 (1952); **95**, 736 (1954); **105**, 158 (1957).
20. E. Fermi and E. Segrè, *Rend. Accad. Nazl. Lincei* 4, 18 (1933); *Z. Physik* 82, 729, (1933).
21. G. F. Koster, *Phys. Rev.* **86**, 148 (1952).
22. A. Abragam, J. Horowitz, and M. H. L. Pryce, *Proc. Roy. Soc.* **A230**, 169 (1955).
23. R. M. Sternheimer, *Phys. Rev.* **86**, 316 (1952).
24. A. Dalgarno, *Advan. Phys.* (*Phil. Mag. Suppl.*) **11**, 281 (1962).
25. T. P. Das and R. Bersohn, *Phys. Rev.* **102**, 833 (1956).
26. E. G. Wikner and T. P. Das, *Phys. Rev.* **109**, 360 (1958).
27. L. C. Allen, *Phys. Rev.* **118**, 167 (1960).
28. S. Kaneko, *J. Phys. Soc. Japan* **11**, 1600 (1959).
29. A. Dalgarno, *Proc. Roy. Soc.* **A251**, 282 (1959), and references listed in ref. [24].
30. G. Burns, *Phys. Rev.* **115**, 357 (1959).
31. M. Sternheimer, *Phys. Rev.* **115**, 1198 (1959).
31a. R. M. Sternheimer, private communication, 1962.
32. P. G. Khubchandani, R. R. Sharma, and T. P. Das, *Phys. Rev.* **126**, 594 (1962).
33. R. E. Watson and A. J. Freeman, *Phys. Rev.* **131**, 250 (1963).
34. R. Ingalls, *Phys. Rev.* **128**, 1155 (1962).
35. G. Burns and E. G. Wikner, *Phys. Rev.* **121**, 155 (1961).
36. S. Sugano and R. G. Shulman, *Phys. Rev.* **130**, 517 (1963).
37. R. E. Watson and A. J. Freeman, *Phys. Rev.* **134**, A1526 (1964).
38. Proceedings of the Third International Conference on the Mössbauer Effect, *Rev. Mod. Phys.* **36**, No. 1, Part II (1964).*
39. A. J. Freeman and R. E. Watson, *Phys. Rev.* **131**, 2566 (1963).
40. A. J. Freeman and R. E. Watson, *Phys. Rev.* **132**, 706 (1963).
41. N. Bessis, H. Lefebvre-Brion, and C. M. Moser, *Phys. Rev.* **130**, 1441 (1963).
42. N. Bessis, H. Lefebvre-Brion, and C. M. Moser, *Phys. Rev.* **128**, 213 (1962).
43. J. S. M. Harvey, Ph.D. thesis, Oxford University, 1962 (unpublished); and to be published.
44. W. Low, *Solid State Phys.* Suppl. 2 (1960); W. Low (ed.), "Paramagnetic Resonance," Vols. 1 & 2. Academic Press, New York, 1963.
45. S. R. de Groot and H. A. Tolhoek, *in* "Beta- and Gamma-Ray Spectroscopy" (K. Siegbahn, ed.), p. 613. North-Holland Publ., Amsterdam, 1955; M. J. Steenland and H. A. Tolhoek, in "Progress in Low Temperature Physics" (C. J. Gorter, ed.), Vol. 2, Chapt. 10. North-Holland Publ., Amsterdam, 1957 (which contains complete references to theoretical results, methods for orienting nuclei, and description of experiments).

* References to this issue will be cited hereafter as *3rd Mössbauer Conf.*

46. C. J. Gorter, *Physik. Z.* **35**, 923 (1934).
47. N. Kurti and F. E. Simon, *Proc. Roy. Soc.* **A149**, 152 (1935).
48. C. J. Gorter, *Physica* **14**, 504 (1948).
49. M. E. Rose, *Phys. Rev.* **75**, 213 (1949).
50. B. Bleaney, *Proc. Phys. Soc.* **A64**, 315 (1951); *Phil. Mag.* [7] **42**, 441 (1951).
51. R. V. Pound, *Phys. Rev.* **76**, 1410 (1949).
52. E. Ambler, *Progr. Cryogenics* **2**, 235 (1960).
52a. C. D. Jeffries, *Progr. Cryogenics* **3**, 131 (1961).
53. B. Bleaney, *J. Phys. Soc. Japan* **17**, Suppl. B-1, 435 (1962).
54. B. Bleaney and R. W. Hill, *Proc. Phys. Soc.* **78**, 313 (1961).
55. W. Marshall, *Phys. Rev.* **110**, 1280 (1958).
56. B. Bleaney, *Phys. Rev.* **78**, 214 (1950).
57. R. J. Benzie and A. H. Cooke, *Nature* **164**, 837 (1949).
58. C. V. Heer and R. A. Erickson, *Bull. Am. Phys. Soc.* [2] **1**, 217 (1956).
59. C. V. Heer and R. A. Erickson, *Phys. Rev.* **108**, 896 (1957).
60. V. Arp, N. Kurti, and R. Petersen, *Bull. Am. Phys. Soc.* [2] **2**, 388 (1957).
61. V. Arp, D. Edmunds, and R. Petersen, *Phys. Rev.Letters* **3**, 212 (1959).
62. P. A. Beck, *3rd Mösbauer Conf.* p. 394.
63. C. H. Cheng, C. T. Wei, P. A. Beck, *Phys. Rev.* **120**, 426 (1960).
64. C. T. Wei, C. H. Cheng, P. A. Beck, *Phys. Rev.* **122**, 1129 (1961).
65. O. V. Lounasmaa, C. H. Cheng, and P. A. Beck, *Phys. Rev.* **128**, 2153 (1961).
66. N. Kurti and R. S. Safrata, *Phil. Mag.* [8] **3**, 780 (1958).
67. J. G. Dash, R. D. Taylor and P. P. Craig, *Proc. 7th Intern. Conf. Low Temp. Phys., Toronto, 1960* p. 705 (1961).
68. B. Dreyfus, B. B. Goodman, G. Trolliet, and L. Weil, *Compt. Rend.* **252**, 1743 (1961); **253**, 1085 (1961).
69. B. Dreyfus, B. B. Goodman, A. Lacaze, and G. Trolliet, *Compt. Rend.* **253**, 1764 (1961).
70. E. C. Heltemes and C. A. Swenson, *J. Chem. Phys.* **35**, 1264 (1961).
71. R. D. Parks, *Proc. 2nd Rare Earth Conf., Glenwood Springs, 1961* (to be published).
72. J. E. Gordon, C. W. Dempesy, and T. Soller, *Phys. Rev.* **124**, 724 (1961).
73. C. W. Dempesy, J. E. Gordon, and T. Soller, *Bull. Am. Phys. Soc.* [2] **7**, 309 (1962).
74. O. V. Lounasmaa, *Phys. Rev.* **126**, 1352 (1962); **133**, A211 (1964).
75. O. V. Lounasmaa and R. A. Guenther, *Phys. Rev.* **126**, 1357 (1962).
76. O. V. Lounasmaa and P. R. Roach, *Phys. Rev.* **128**, 622 (1963).
77. O. V. Lounasmaa, *Phys. Rev.* **119**, 2460 (1963).
78. O. V. Lounasmaa, *Phys. Rev.* **128**, 1136 (1963); **134**, A1620 (1964).
79. O. V. Lounasmaa, *Proc. Intern. Conf. Low Temp. Phys., London, 1962* p. 223 (1964). Butterworth, London.
80. O. V. Lounasmaa, *Proc. 3rd Rare Earth Res. Conf., Clearwater, 1964* (to be published).
81. B. Bleaney, *J. Appl. Phys.* **34**, 1024 (1963).
81a. R. M. Stanton, L. D. Jennings, and F. H. Spedding, *J. Chem. Phys.* **32**, 630 (1960).
81b. C. A. Bailey, unpublished material; also see Ref. 76.
82. B. C. Gerstein, F. J. Jelinek, and F. H. Spedding, *Phys. Rev.Letters* **8**, 425 (1962).
83. O. V. Lounasmaa, *Phys. Rev.* **133**, A502 (1964).
84. O. V. Lounasmaa, *Phys. Rev.* **133**, A219 (1964).
85. F. W. de Wette, *Phys. Rev.* **123**, 103 (1961).
86. R. L. Mössbauer, *Z. Physik* **151**, 124 (1958); *Naturwissenschaften* **22**, 538 (1958).
87. R. L. Mössbauer, *Z. Naturforsch.* **14a**, 211 (1959).

4. HYPERFINE INTERACTIONS IN MAGNETIC MATERIALS

88. H. Frauenfelder (ed.), "The Mössbauer Effect." W. A. Benjamin, New York, 1962.
88a. G. K. Wertheim, "Mössbauer Effect: Principles and Applications." Academic Press, New York, 1964.
89. R. V. Pound and G. A. Rebka, Jr., *Phys. Rev.Letters* **4**, 337 (1960).
89a. R. V. Pound and G. A. Rebka, Jr., *Phys. Rev. Letters* **3**, 554 (1959); G. DePasquali, H. Frauenfelder, S. Margulies, and R. N. Peacock, *ibid.* **4**, 71 (1960); S. S. Hanna, J. Heberle, C. Littlejohn, G. J. Perlow, R. S. Preston, and D. H. Vincent, *ibid.* p. 177.
89b. H. J. Hay *et al.*, *Phys. Rev.Letters* **4**, 165 (1960).
90. *3rd Mössbauer Conf.* [see Ref. 38].
91. W. Heitler, "Quantum Theory of Radiation." Oxford Univ. Press, (Clarendon), London and New York, 1949.
92. F. R. Metzger, *Nucl. Phys.* **7**, 964 (1959).
93. W. E. Lamb, *Phys. Rev.* **55**, 190 (1930).
94. W. M. Visscher, *Ann. Phys. (N.Y.)* **9**, 194 (1960).
95. K. S. Singvi and A. Sjölander, *Phys. Rev.* **120**, 1093 (1960).
96. O. C. Kistner and A. W. Sunyar, *Phys. Rev. Letters* **4**, 412 (1960).
97. I. Solomon, *Compt. Rend.* **250**, 3828 (1960).
98. L. R. Walker, G. K. Wertheim, and V. Jaccarino, *Phys. Rev. Letters* **6**, 98 (1961).
99. S. de Benedetti, G. Lang, and R. Ingalls, *Phys. Rev.Letters* **6**, 60 (1961).
100. D. M. J. Compton and A. Schoen (eds.), "The Mössbauer Effect," *Proc. 2nd Intern. Conf., 1961*. Wiley, New York, 1962.*
101. S. S. Hanna, J. Heberle, C. Littlejohn, G. J. Perlow, R. S. Preston, and D. H. Vincent, *Phys. Rev.Letters* **4**, 177 (1960).
102. S. S. Hanna, J. Heberle, G. J. Perlow, R. S. Preston, and D. H. Vincent, *Phys. Rev. Letters* **4**, 513 (1960); S. S. Hanna, R. S. Preston, and J. Heberle, *2nd Mössbauer Conf.* p. 85.
103. C. E. Johnson, M. S. Ridout, and T. E. Cranshaw, *2nd Mössbauer Conf.* p. 142.
104. D. Nagle, H. Frauenfelder, R. D. Taylor, D. R. F. Cochran, and B. T. Matthias, *Phys. Rev. Letters* **5**, 364 (1960).
105. D. Nagle, P. P. Craig, P. Barrett, D. R. F. Cochran, C. E. Olsen, and R. D. Taylor, *Phys. Rev.* **125**, 490 (1962).
106. C. Alff and G. K. Wertheim, *Phys. Rev.* **122**, 1414 (1961).
107. R. Bauminger, S. G. Cohen, A. Marinov, and S. Ofer, *Phys. Rev.* **122**, 743 (1961).
108. R. Bauminger, S. G. Cohen, A. Marinov, S. Ofer, and E. Segal, *Phys. Rev.* **122**, 1447 (1961).
109. G. Shirane, D. E. Cox, and S. L. Ruby, *Phys. Rev.* **125**, 1158 (1962).
110. W. H. Kelly, V. J. Folen, M. Hass, W. N. Schreiner, and G. B. Beards, *Phys. Rev.* **124**, 80 (1961).
111. G. K. Wertheim, *Phys. Rev.* **121**, 63 (1961).
112. J. G. Dash, R. D. Taylor, D. E. Nagle, P. P. Craig, and W. M. Visscher, *Phys. Rev.* **122**, 1116 (1961).
113. F. E. Obenshain and H. H. F. Wegener, *Phys. Rev.* **121**, 1344 (1961); *Z. Physik* **163**, 17 (1962).
114. O. C. Kistner, A. W. Sunyar, and J. B. Swan, *Phys. Rev.* **123**, 179 (1961).
115. L. Meyer-Schützmeister, R. S. Preston, and S. S. Hanna, *Phys. Rev.* **122**, 1717 (1961).
116. R. Bauminger, S. G. Cohen, A. Marinov, and S. Ofer, *Phys. Rev. Letters* **6**, 467 (1961).
117. S. Ofer, P. Avivi, R. Bauminger, A. Marinov, and S. G. Cohen, *Phys. Rev.* **120**, 406 (1960).

* References to this volume will be cited hereafter as *2nd Mössbauer Conf.*

118. R. Bauminger, S. G. Cohen, A. Marinov, and S. Ofer, *2nd Mössbauer Conf.* p. 177.
119. G. K. Wertheim, *Phys. Rev.* **124**, 764 (1961); G. K. Wertheim and D. N. E. Buchanan, *2nd Mössbauer Conf.* p. 130.
120. C. Littlejohn Herzenberg, L. Meyer-Schutzmeister, L. L. Lee, Jr., and S. S. Hanna, *Bull. Am. Phys. Soc.* [2] **7**, 39 (1962).
121. C. W. Kocher and P. J. Brown, *J. Appl. Phys.* [2] **33**, Suppl., 1091 (1962).
122. E. A. Friedman and D. Nicholson, *Bull. Am. Phys. Soc.* [2] **7**, 402 (1962).
123. D. N. E. Buchanan and G. K. Wertheim, *Bull. Am. Phys. Soc.* [2] **7**, 227 (1962)
124. G. K. Wertheim and J. H. Wernick, *Phys. Rev.* **125**, 1937 (1962).
125. G. K. Wertheim and J. H. Wernick, *Phys. Rev.* **123**, 755 (1961).
126. G. K. Wertheim, *Phys. Rev. Letters* **4**, 403 (1960).
127. G. Shirane, W. J. Takei, and S. L. Ruby, *Phys. Rev.* **126**, 49 (1962).
128. J. G. Dash, R. D. Taylor, P. P. Craig, D. E. Nagle, D. R. F. Cochran, and W. E. Keller, *Phys. Rev. Letters* **5**, 152 (1960).
129. D. A. Shirley, M. Kaplan, and P. Axel, *Phys. Rev.* **123**, 816 (1961).
130. L. D. Roberts and J. O. Thompson, *Bull. Am. Phys. Soc.* [2] **6**, 230 (1961).
131. A. J. F. Boyle, D. St. Bunbury, and C. Edwards, *Phys. Rev. Letters* **5**, 553 (1960).
132. S. S. Hanna, L. Meyer-Shutzmeister, R. S. Preston, and D. H. Vincent, *Phys. Rev.* **120**, 2211 (1960).
133. L. D. Roberts and J. O. Thompson, *Bull. Am. Phys. Soc.* **7** 350, 351 (1962).
134. L. D. Roberts and J. O. Thompson, *Phys. Rev.* **129**, 664 (1963).
135. O. Kazus, Y. Ishikawa, and A. Ito, *J. Phys. Soc. Japan* **17**, 1747 (1962).
136. S. Komura and N. Shikazono, *J. Phys. Soc. Japan* **18**, 323 (1963).
137. G. K. Wertheim, *J. Appl. Phys.* **32**, 110S (1961).
138. M. B. Stearns, *Phys. Rev.* **129**, 1136 (1963) for FeSi series.
139. R. S. Preston, S. S. Hanna, and J. Heberle, *Phys. Rev.* **128**, 2207 (1962).
140. A. Ito, K. Ono, and Y. Ishikawa, *J. Phys. Soc. Japan* **18**, 1465 (1963).
141. K. Ono, Y. Ishikawa, and A. Ito, *J. Phys. Soc. Japan* **17**, 125S (1962).
142. C. Kimball, W. D. Gerber, and A. Arrott, *Bull. Am. Phys. Soc.* [2] **7**, 278 (1962).
143. S. Teruya, Y. Naomoto, and N. Shikazono, *J. Phys. Soc. Japan* **18**, 797 (1963).
144. G. Shirane, D. E. Cox, W. J. Takei, and S. L. Ruby, *J. Phys. Soc. Japan* **17**, 1598 (1962).
145. I. Solomon, *Compt. Rend.* **251**, 2675 (1960).
146. F. E. Obenshain, L. D. Roberts, H. H. F. Wegener, *Bull. Am. Phys. Soc.* [2] **8**, 43 (1963).
147. G. Shirane, C. W. Chen, P. A. Elinn, and R. Nathans, *Phys. Rev.* **131**, 183 (1963).
148. W. E. Wallace and L. M. Epstein, *J. Chem. Phys.* **35**, 2238 (1961).
149. C. E. Johnson, M. S. Ridout, and T. E. Cranshaw, *Proc. Phys. Soc.* **81**, 1079 (1963).
150. G. W. Rothberg, N. Benczer-Koller and J. K. Harris, *3rd Mössbauer Conf.* p. 357.
151. R. W. Grant, M. Kaplan, D. A. Keller, and D. A. Shirley, *Bull. Am. Phys. Soc.* [2] **7**, 601 (1962); and *3rd Mössbauer Conf.* p. 352.
152. L. Meyer-Schutzmeister, *2nd Mössbauer Conf.* p. 190.
153. S. S. Hanna, J. Heberle, J. Diaz, and R. W. Reno, *3rd Mössbauer Conf.* p. 407.
154. A. J. Boyle, D. St. Bunbury, and C. Edwards, *Proc. Phys. Soc.* **17**, 1062 (1901).
155. D. A. Shirley, R. B. Frankel, and H. H. Wickman, *3rd Mössbauer Conf.* p. 392.
156. P. H. Barrett and D. A. Shirley, *Phys. Rev.* **131**, 123 (1963).
157. R. L. Cohen, *3rd Mössbauer Conf.* p. 393.
157a. E. L. Boyd and J. C. Slonczewski, *J. Appl. Phys.* **33**, 1077S (1962).
157b. P. R. Locher and S. Geschwind, *Phys. Rev. Letters* **11**, 333 (1963).
157c. G. J. Ritter, *Phys. Rev.* **128**, 2238 (1963).

158. C. E. Violet, R. Both, and R. J. Borg, *Phys. Rev. Letters* **11**, 464 (1963); V. A. Bryukhanov, N. N. Delyagin, and V. S. Shpinel, *Zh. Eksperim. i Teor. Fiz.* **42**, 1183 (1962); V. S. Shpinel, "Proceedings of Soviet Conference on the Mössbauer Effect, Dubna" (Translation), p. 16-1. Consultants Bureau, New York; V. A. Bucharev, *ibid.* p. 6-1; V. I. Gol'Danskii, G. M. Gorodinskii, L. A. Korytko, L. M. Krizhanskii, E. F. Makarov, I. P. Suzdalev, and V. V. Khrapou, *ibid.* p. 23-1; P. P. Craig, D. E. Nagle, and D. R. F. Cochran, *Phys. Rev. Letters* **4**, 561 (1960); M. de Coster, H. Pollak, and S. Amelinckx, *Phys. Stat. Solidi* **3**, 283 (1963); V. A. Lyubimov and A. I. Alikhanov, *Soviet Phys. JETP* (*English Transl.*) **11**, 1375 (1960); A. J. F. Boyle, D. St. Bunbury, C. Edwards, and H. E. Hall, *2nd Mössbauer Conf.* p. 182; S. L. Ruby and G. Shirane, *Phys. Rev.* **123**, 1239 (1961); P. A. Flinn and S. L. Ruby, *ibid.* **124**, 34 (1961).
159. V. P. Alfimenko, Yu. M. Ostanevich, T. Rukov, A. V. Strelhov, F. L. Shapiro, and Yen Wu-Kuang, *Soviet Phys. JETP* (English Transl.) **15**, 718 (1962); M. Kalvius, P. Kienle, K. Bockmann, and H. Eicher, *Z. Physik* **163**, 87 (1961); F. W. Stanek, *ibid.* **166**, 6 (1962); F. E. Wagner, F. W. Stanek, P. Kienle, and H. Eicher, *ibid.* p. 1; *2nd Mössbauer Conf.* p. 1; F. W. Stanek, F. E. Wagner, H. Eicher, and W. Wiedemann, *ibid.* p. 185; S. G. Cohen, *ibid.* p. 73; R. L. Mössbauer, *3rd Mössbauer Conf.* p. 362; P. Kienle, *ibid.* p. 372; S. G. Cohen, I. Nowick, and S. Ofer, *ibid.* p. 378.
160. C. Robert and J. M. Winter, *Compt. Rend.* **250**, 3831 (1960).
161. J. S. Kouvel and R. H. Wilson, *J. Appl. Phys.* **32**, 435 (1961).
162. F. Galperin, S. Larin, and A. Shishkov, *Dokl. Akad. Nauk SSSR* **89**, 419 (1953).
163. N. Blum and L. Grodzins, *Bull. Am. Phys. Soc.* [2] **7**, 39 (1962).
164. A. J. Freeman, *Phys. Rev.* **130**, 888 (1963).
165. P. P. Craig, D. E. Nagle, W. A. Steyert, and R. D. Taylor, *Phys. Rev. Letters* **9**, 12 (1962).
166. N. Blum, A. J. Freeman, and L. Grodzins, *3rd Mössbauer Conf.* p. 406.
167. R. D. Taylor, W. A. Steyert, and D. E. Nagle, *3rd Mössbauer Conf.* p. 406.
168. R. Bauminger, L. Grodzins, and A. J. Freeman, *Phys. Rev.* to be published (1965).
169. A. Abragam and F. Boutron, *Compt. Rend.* **252**, 2404 (1961).
170. C. E. Johnson, W. Marshall, and G. J. Perlow, *Phys. Rev.* **126**, 1503 (1962).
171. G. Burns, *Phys. Rev.* **124**, 524 (1961).
172. R. Ingalls, *Phys. Rev.* **128**, 1155 (1962).
173. R. M. Sternheimer, *Phys. Rev.*, **130**, 1423 (1963).
174. G. Burns, unpublished material (1963).
175. R. Ingalls, *3rd Mössbauer Conf.* p. 351.
176. W. Kerler and W. Neuwirth, *Z. Physik* **167**, 176 (1962).
177. N. N. Delyagin, V. S. Shpinel', V. A. Bryukhanov, and B. Zvenglinskii, *Soviet Phys. JETP* (*English Transl.*) **12**, 159 (1961).
178. R. Cohen, U. Hauser, and R. L. Mössbauer, *2nd Mössbauer Conf.* p. 172.
179. L. G. Lang, S. DeBenedetti, and R. I. Ingalls, *2nd Mössbauer Conf.* p. 168.
180. C. E. Johnson, W. Marshall, and G. J. Perlow, *2nd Mössbauer Conf.* p. 163.
181. J. Gastebois, *2nd Mössbauer Conf.* p. 160.
182. R. H. Herber and G. K. Wertheim, *2nd Mössbauer Conf.* p. 160.
183. W. Kerler and W. Neuwirth, *2nd Mössbauer Conf.* p. 90.
184. J. Gastebois and J. Quidort, *Compt. Rend.* **253**, 1257 (1961).
185. P. C. Norem and G. K. Wertheim, *Bull. Am. Phys. Soc.* [2] **7**, 79 (1962).
186. W. J. Nicholson and G. Burns, *Bull. Am. Phys. Soc.* [2] **7**, 85 (1962).
187. K. Ôno, A. Ito, and E. Hirahara, *J. Phys. Soc. Japan* **17**, 1675 (1962).

188. N. L. Costa, J. Danon, and R. M. Xavier, *Phys. Chem. Solids* **23**, 1783 (1962).
189. D. A. Shirley, R. W. Grant, D. A. Keller, *3rd Mössbauer Conf.* p. 352.
190. D. C. Kistner, V. Jaccarino, and L. R. Walker, *2nd Mössbauer Conf.* p. 264.
191. G. M. Gorodinskii, L. M. Krizhanskii, and E. M. Kruglov, *Zh. Eksperim. i Teor. Fiz.* **43**, 2050 (1962).
192. V. S. Shpinel', V. A. Bryukhanov, and N. N. Delyagin, *Soviet Phys. JETP (English Transl.)* **14**, 1256 (1962).
193. R. L. Mössbauer, R. G. Barnes, E. Kankeleit, and J. Poindexter, *Bull. Am. Phys. Soc.* **8**, 470 (1963).
194. R. L. Mössbauer, R. G. Barnes, E. Kankeleit, and J. Poindexter, *Phys. Rev. Letters* **11**, 253 (1963).
195. P. C. Norem and G. K. Wertheim, *Phys. Chem. Solids* **23**, 1111 (1962).
196. e.g., see S. Hüfner, M. Kalvius, P. Kienle, W. Wiedemann, and H. Eicher, *Z. Physik* **175**, 416 (1963); R. G. Barnes, E. Kankeleit, R. L. Mössbauer, and J. M. Poindexter, *Phys. Rev. Letters* **11**, 253 (1963); R. L. Cohen, *Phys. Letters* **5**, 177 (1963).
197. E. G. Wikner and G. Burns, *Phys. Letters* **2**, 225 (1962).
198. R. M. Sternheimer, *Phys. Rev.* **132**, 1637 (1963).
199. R. E. Watson, Tech. Rept. No. 12, Solid State and Molecular Theory Group, M.I.T., June 15, 1959 (unpublished).
200. L. D. Roberts, R. L. Becker, and F. E. Obenshain, ref. [90], p. 408.
201. E. P. Wohlfarth, *Proc. Roy. Soc.* **A195**, 434 (1948).
202. A. Blandin, E. Daniel, and J. Friedel, *Phil. Mag.* [8] **4**, 180 (1959).
203. A. Blandin and E. Daniel, *Phys. Chem. Solids* **10**, 126 (1959).
204. I. Nowick and S. Ofer, *3rd Mössbauer Conf.* p. 392.
204a. P. Brix, S. Hüfner, P. Kienle, and D. Quitmann, *Phys. Letters* **13**, 140 (1964).
205. R. V. Pound, G. B. Benedek, and R. Drever, *Phys. Rev. Letters* **7**, 405 (1961).
206. G. B. Benedek, "Magnetic Resonance at High Pressure," p. 47. Wiley (Interscience), New York, 1963.
207. R. M. Steffen, *Advan. Phys. (Phil. Mag. Suppl.)* **4**, 293 (1955).
208. H. Frauenfelder, in "Beta- and Gamma-Ray Spectroscopy" (K. Siegbahn, ed.), Chapt. XIX. North-Holland Publ., Amsterdam, 1955.
209. E. Heer and T. B. Novey, *Solid State Phys.* **9**, 000 (1959).
210. S. Devons and L. J. B. Goldfarb, in "Handbuch der Physik" (S. Flügge, ed.), Vol. 42, p. 362. Springer, Berlin, 1957.
211. A. Abragam and R. V. Pound, *Phys. Rev.* **92**, 943 (1953).
212. M. E. Caspari, S. Frankel, and M. A. Gilleo, *J. Appl. Phys.* **31**, 320S (1960); M. E. Caspari, S. Frankel, and G. T. Wood, *Phys. Rev.* **127**, 1519 (1962).
213. J. M. Daniels, M. A. Grace, and F. N. H. Robinson, *Nature* **168**, 780 (1951).
214. C. J. Gorter, O. J. Poppema, M. J. Steenland, and J. A. Beun, *Physica* **17**, 1050 (1951).
215. M. A. Grace, C. E. Johnson, N. Kurti, R. G. Scurlock, and R. T. Taylor, *Commun. Conf. Basses Temp., Paris, 1955* p. 263 (1955).*Bull. Am. Phys. Soc.* [2] **2**, 136 (1957); B. Bleaney, J. M. Daniels, M. A. Grace, H. Halban, N. Kurti, F. N. H. Robinson, and F. E. Simon, *Proc. Roy. Soc.* **A221**, 170 (1954).
216. G. R. Khutsishvili, *Zh. Eksperim. i Teor. Fiz.* **29**, 894 (1955).
217. A. M. Portis and A. C. Gossard, *J. Appl. Phys.* **31**, 205S (1960).
218. B. N. Samoilov, V. V. Sklyarevskii, and E. Stepanov, *Zh. Eksperim. i Teor. Fiz.* **36**, 972 (1959); W. J. Huiskamp and H. A. Tolhoek, in "Progress in Low Temperature Physics" (C. J. Gorter, ed.), Vol. 3, North-Holland Publ., Amsterdam, 1961.
219. J. M. Daniels, J. C. Giles, and M. A. R. Le Blanc, *Can. J. Phys.* **39**, 53 (1961).

4. HYPERFINE INTERACTIONS IN MAGNETIC MATERIALS

220. M. J. Steenland and H. A. Tolhoek, *in* "Progress in Low Temperature Physics" (C. J. Gorter, ed.), Vol. 2, p. 292. North-Holland Publ., Amsterdam, 1957.
221. *Proc. 7th Intern. Conf. Low Temp. Phys., Toronto, 1960* (1961). G. M. Graham and A. C. Hollis Hallett (eds.) Univ. of Toronto Press, Toronto.
222. E. Ambler, R. P. Hudson, and G. M. Temmer, *Phys. Rev.* **97**, 1212 (1955); **101**, 196 (1956); see also C. F. M. Cacho, M. A. Grace, C. E. Johnson, A. C. Knipper, R. G. Scurlock, and R. T. Taylor, *Phil. Mag.* [7] **46**, 1287 (1955).
223. G. Golding and P. P. Scharenberg, *Phys. Rev.* **110**, 701 (1958).
224. P. Debrunner, W. Kündig, J. Sunier, and P. Scherrer, *Helv. Phys. Acta* **31**, 326 (1958).
225. C. A. Lovejoy and D. A. Shirley, *Proc. 7th Intern. Conf. Low Temp. Phys., Toronto, 1960* p. 164 (1961). Univ. of Toronto Press, Toronto.
226. J. F. Schooley, D. A. Shirley, and J. O. Rasmussen, *Proc. 7th Intern. Conf. Low Temp. Phys., Toronto, 1960* p. 188 (1961). Univ. of Toronto Press, Toronto.
227. R. J. Elliott and K. W. H. Stevens, *Proc. Roy. Soc.* **A218**, 553 (1953).
228. H. Postma and W. J. Huiskamp, *Proc. 7th Intern. Conf. Low Temp. Phys., Toronto, 1960* p. 180 (1961). Univ. of Toronto Press, Toronto.
229. B. N. Samoilov, V. V. Sklyarevsky, and E. P. Stepanov, *Zh. Eksperim. i Teor. Fiz.* **36**, 644, 1944 (1959); **38**, 359 (1960).
230. A. V. Kogan, V. D. Kulkov, L. P. Nikitin N., M. Reinov, I. A. Sokolov and M. F. Stelmach, *Proc. 7th Intern. Conf. Low Temp. Phys., Toronto, 1960* p. 193 (1961). Univ. of Toronto Press, Toronto.
231. I. Sött, Private communication, 1963.
232. M. E. Caspari, S. Frankel, D. Ray, and G. T. Wood, *Phys. Rev. Letters* **6**, 345 (1961).
233. R. Pauthenet, *Ann. Phys. (Paris)* **3**, 424 (1958).
234. S. G. Cohen, N. Kaplan, S. Ofer, and H. Zmora, *Phys. Letters* **7**, 91 (1963).
235. M. Deutsch, A. Buryn, and L. Grodzins, unpublished material, 1963.
236. M. E. Rose, *Nucleonics* **3**(6), 23 (1948); *Phys. Rev.* **75**, 213 (1949).
237. D. de Klerk and M. J. Steenland, *in* "Progress in Low Temperature Physics" (C. J. Gorter, ed.), Vol. 1. North-Holland Publ., Amsterdam, 1957.
238. S. Berstein, L. D. Roberts, C. P. Stanford, J. W. T. Dabbs, and T. E. Stephenson, *Phys. Rev.* **94**, 1243 (1954); L. D. Roberts, S. Bernstein, J. W. T. Dabbs, and C. P. Stanford, *ibid.* **95**, 105 (1954); J. W. T. Dabbs, L. D. Roberts, and S. Bernstein, *ibid.* **98**, 1512 (1955).
239. A. Stolovy, *Phys. Rev.* **118**, 211 (1960); *Bull. Am. Phys. Soc.* [2] **5**, 294 (1960); *ibid.* [2] **6**, 275 (1960).
240. H. Postma, H. Marshak, V. L. Sailor, F. J. Shore, and C. A. Reynolds, *Phys. Rev.* **126**, 979 (1962).
241. R. Schermer, V. L. Sailor, C. A. Reynolds, F. J. Shore, H. Marshak, and H. Postma, *Bull. Am. Phys. Soc.* [2] **6**, 502 (1961); *Phys. Rev.* **127**. 1124 (1963).
242. A. Stolovy, *Phys. Rev.* **134**, B68 (1964).
243. C. L. Pekeris, *Phys. Rev.* **112**, 1649 (1958).
243a. C. L. Pekeris, *Phys. Rev.* **126**, 1470 (1962).
243b. P. O. Löwdin, *Rev. Mod. Phys.* **35**, 538 (1963).
244. N. W. Bazley and D. W. Fox, *J. Math. Phys.* **3**, 469 (1962); L. Redei, *Phys. Rev.* **130**, 420 (1963).
244a. G. Temple, *Proc. Roy. Soc.* **A119**, 276 (1928); A. Weinstein, *Proc. Natl. Acad. Sci. U.S.* **20**, 529 (1934); A. F. Stevenson, *Phys. Rev.* **53**, 199 (1938); A. F. Stevenson and M. F. Crawford, *ibid.* **54**, 375 (1938).
244b. A. Fröman and G. G. Hall, *J. Mol. Spectry.* **7**, 410 (1961); G. L. Caldow and C. A. Coulson, *Proc. Cambridge Phil. Soc.* **57**, 341 (1961).

245. C. Schwartz, *Ann. Phys. (N.Y.)* [7] **2**, 170 (1959).
246. P. O. Löwdin, *Phys. Rev.* **97**, 1474, 1490, 1509 (1955).
247. A. J. Coleman, *Rev. Mod. Phys.* **35**, 668 (1963); Tech. Note II 80, Quantum Chemistry Group, University of Uppsala, Uppsala, 1962 (unpublished); F. Sasaki, Tech. Note No. 77, Quantum Chemistry Group, University of Uppsala, Uppsala, 1962 (unpublished).
247a. R. McWeeny, *Rev. Mod. Phys.* **32**, 335 (1960).
247b. D. Ter Haar, *Rept. Progr. Phys.* **24**, 304 (1961).
248. D. R. Hartree, *Proc. Cambridge Phil. Soc.* **24**, 89 (1928); V. Fock, *Z. Physik* **61**, 126 (1930); J. C. Slater, *Phys. Rev.* **35**, 210 (1930).
249. P. A. M. Dirac, *Proc. Cambridge Phil. Soc.* **26**, 376 (1930); **27**, 240 (1931).
250. P. O. Löwdin, Tech. Note No. 27, Quantum Chemistry Group, University of Uppsala, Uppsala, 1959 (unpublished).
251. D. R. Hartree, "The Calculations of Atomic Structure." Wiley, New York, 1957.
252. J. C. Slater, *Rev. Mod. Phys.* **6**, 209 (1934); J. C. Slater, *Phys. Rev.* **81**, 385 (1951).
252a. D. R. Hartree, *Phys. Rev.* **109**, 840 (1958).
253. P. O. Löwdin, *Phys. Rev.* **97**, 1475 (1955).
253a. E. C. Ridley, *Proc. Cambridge Phil. Soc.* **56**, 41 (1960).
253b. A. J. Freeman and R. E. Watson, *Phys. Rev.* **127**, 2058 (1962).
254. J. C. Slater, "Quantum Theory of Atomic Structure," Vol. I. McGraw-Hill, New York, 1960.
255. E. U. Condon and G. H. Shortley, "The Theory of Atomic Spectra." Cambridge Univ. Press, London and New York, 1953.
256. J. C. Slater, *Phys. Rev.* **81**, 385 (1951).
256a. R. M. Sternheimer, *Phys. Rev.* **78**, 235 (1950).
256b. D. R. Hartree, *Phys. Rev.* **109**, 840 (1958).
256c. J. E. Robinson, F. Bassani, R. S. Know, and J. R. Schrieffer, *Phys. Rev. Letters* **9**, 215 (1962).
257. G. W. Pratt, Jr., *Phys. Rev.* **88**, 1217 (1952).
258. J. H. Wood and G. W. Pratt, Jr., *Phys. Rev.* **107**, 995 (1957).
259. F. Herman and S. Skillman, "Atomic Structure Calculations." Prentice-Hall, Englewood Cliffs, New Jersey, 1963.
260. J. Linderberg and H. Shull, *J. Mol. Spectry.* **5**, 1 (1960); R. E. Watson, *Ann. Phys.(N.Y.)* **13**, 250 (1961).
261. D. Pines (ed.), "The Many-Body Problem." W. A. Benjamin, New York, 1961.
261a. P. L. Altick and A. E. Glassgold, *Phys. Rev.* **133**, A632 (1964).
262. H. P. Kelly, *Phys. Rev.* **131**, 684 (1963).
262a. J. Goldstone, *Proc. Roy. Soc.* **A234**, 267 (1957).
262b. H. P. Kelley and A. M. Sessler, *Phys. Rev.* **132**, 2091 (1963).
263. E. A. Hylleraas, *Z. Physik* **54**, 347 (1929); **65**, 209 (1930); E. A. Hylleraas and J. Midtal, *Phys. Rev.* **103**, 829 (1956); **109**, 1013 (1958).
264. R. K. Nesbet, *Proc. Roy. Soc.* **A230**, 312 (1955).
265. R. K. Nesbet, *Rev. Mod. Phys.* **33**, 28 (1961).
266. A. W. Weiss, *Phys. Rev.* **122**, 1826 (1961).
267. S. F. Boys, *Proc. Roy. Soc.* **A201**, 125 (1950); **A217**, 136, 235 (1953).
268. B. R. Judd and I. Lindgren, *Phys. Rev.* **122**, 1802 (1961).
269. R. E. Watson, *Phys. Rev.* **119**, 170 (1960).
270. L. Brillouin, *Actualités Sci. Ind.* **71** (1953); **159** (1934).
271. O. Sinanoğlu, *Proc. Roy. Soc.* **A260**, 376 (1961); *Phys. Rev.* **122**, 491, 493 (1961).
272. O. Sinanoğlu, *J. Chem. Phys.* **33**, 1212 (1960); **34**, 1078, 1237 (1961); **36**, 706 (1962).

273. A. W. Weiss and J. B. Martin, unpublished material (1964).
274. L. Szasz, *Z. Naturforsch.* **15a**, 909 (1960); *J. Chem. Phys.* **35**, 1072 (1961); *Phys. Rev.* **126**, 169 (1962).
275. H. M. McConnell, *J. Chem. Phys.* **28**, 1188 (1958); H. M. McConnell and J. Strathdee, *Mol. Phys.* **2**, 129 (1959); A. D. McLachlan, *ibid.* p. 271;**3**, 233 (1960); R. Lefebvre, H. H. Dearman, and H. M. McConnell, *J. Chem. Phys.* **32**, 176 (1960); T. H. Brown, D. H. Anderson, and H. S. Gutovsky, *ibid.* **33**, 720 (1960); A. T. Amos and G. G. Hall, *Proc. Roy. Soc.* **A263**, 483 (1961); H. Lefebvre-Brion and C. Moser, *Phys. Rev.* **118**, 675 (1960); C. C. Lin, K. Hijikata, and M. Sakamoto, *J. Chem. Phys.* **33**, 878 (1960); G. J. Hoijtink, J. Townsend, and S. I. Weissman, *ibid.* **34**, 507 (1961); G. Vincow and G. K. Fraenkel, *ibid.* p. 1333; M. Yamazaki, M. Sakamoto, K. Hijikata, and C. C. Lin, *ibid.* p. 1926; D. R. Eaton, A. D. Josey, W. D. Phillips, and R. E. Benson, *J. Chem. Phys.* **37**, 347 (1962).
276. A. W. Overhauser, *Phys. Rev.Letters* **4**, 415, 462 (1960); W. Kohn and S. J. Nettel, *ibid.* **5**, 8 (1960); K. Sawada and N. Fukuda, *Progr. Theoret. Phys. (Kyoto)* **25**, 653 (1961); A. W. Overhauser, *Phys. Rev.* **128**, 1437 (1961).
277. E. Thompson, *Ann. Phys. (N.Y.)* **22**, 309 (1963).
278. D. J. Thouless, *Nucl. Phys.* **21**, 225 (1960).
279. D. J. Thouless, "The Quantum Mechanics of Many-Body Systems." Academic Press, New York, 1961.
280. W. H. Adams, *Phys. Rev.* **127**, 1650 (1962).
281. J. C. Slater, *Phys. Rev.* **82**, 538 (1951).
282. C. Zener, *Phys. Rev.* **81**, 440 (1951); **83**, 299 (1951).
283. M. A. Ruderman and C. Kittel, *Phys. Rev.* **96**, 99 (1954).
284. T. Kasuya, *Progr. Theoret. Phys. (Kyoto)* **16**, 45 (1956).
285. K. Yosida, *Phys. Rev.* **106**, 893 (1957).
286. A. H. Mitchell, *Phys. Rev.* **105**, 143a (1957); A. Paskin, *J. Appl. Phys.* **31**, 318S (1960).
287. V. Jaccarino, B. T. Matthias, M. Peter, H. Suhl, and J. H. Wernick, *Phys. Rev. Letters* **5**, 251 (1960).
288. J. Callaway, *Nuovo Cimento* **26**, 625 (1962).
289. M. Tinkham, *Proc. Roy. Soc.* **A236**, 535, 549 (1956).
290. R. G. Shulman and V. Jaccarino, *Phys. Rev.* **103**, 1126 (1956); **108**, 1219 (1957).
291. F. Keffer, T. Oguchi, W. O'Sullivan, and J. Yamashita, *Phys. Rev.* **115**, 1553 (1959).
292. A. Mukherji and T. P. Das, *Phys. Rev.* **111**, 1479 (1958); G. Benedek and T. Kushida, *ibid.* **118**, 46 (1960).
293. R. G. Shulman and K. Knox, *Phys. Rev.Letters* **4**, 603 (1960); *Phys. Rev.* **119**, 94 (1960); R. G. Shulman, *ibid.* **121**, 125 (1961).
294. A. M. Clogston, J. P. Gordon, V. Jaccarino, M. Peter, and L. R. Walker, *Phys. Rev.* **117**, 1222 (1960).
295. A. J. Freeman, and R. E. Watson, *Phys. Rev. Letters* **6**, 343 (1961).
296. W. Marshall and R. Stuart, *Phys. Rev.* **123**, 2048 (1961).
297. R. G. Shulman and S. Sugano, *Phys. Rev.* **130**, 506 (1963).
298. S. Sugano and R. G. Shulman, *Phys. Rev.* **130**, 517 (1963).
299. R. E. Watson and A. J. Freeman, *Phys. Rev.* **134**, A1526 (1964).
299a. T. P. Hall, W. Hayes, R. W. H. Stevenson, and J. Wilkens, *J. Chem. Phys.* **39**, 35 (1963).
300. W. Marshall, *in* "Paramagnetic Resonance" (W. Low, ed.), Vol. 1, p. 347. Academic Press, New York, 1963.
301. G. W. Pratt, Jr., *Phys. Rev.* **102**, 1303 (1956).

302. J. A. Pople and R. K. Nesbet, *J. Chem. Phys.* **22**, 571 (1954).
303. G. Berthier, *J. Chem. Phys.* **51**, 363 (1954).
304. P. O. Löwdin, *Ann. Acad. Reg. Sci. Upsalien.* **2**, 127 (1958).
305. R. Lefebvre, *Cahiers Phys.* **381**, 1 (1959).
306. W. Marshall, *Proc. Phys. Soc.* **78**, 113 (1961).
307. V. Heine, *Phys. Rev.* **107**, 1002 (1957).
308. L. M. Sachs, *Phys. Rev.* **117**, 1504 (1960).
309. R. K. Nesbet and R. E. Watson, *Ann. Phys.* (*N.Y.*) **9**, 260 (1960).
310. R. E. Watson and A. J. Freeman, *Phys. Rev.* **120**, 1125, 1134 (1960).
311. D. A. Goodings and V. Heine, *Phys. Rev. Letters* **5**, 370 (1960).
312. A. J. Freeman and R. E. Watson, *Phys. Rev. Letters* **5**, 498 (1960).
313. D. A. Goodings, *Phys. Rev.* **123**, 1706 (1961).
314. M. H. Cohen, D. A. Goodings, and V. Heine, *Proc. Phys. Soc.* **73**, 811 (1959).
315. T. P. Das and A. Mukherjee, *J. Chem. Phys.* **33**, 1808 (1960).
316. S. M. Blinder, *Bull. Am. Phys. Soc.* [2] **5**, 14 (1960).
317. A. Abragam, *Phys. Rev.* **79**, 534 (1950).
318. R. K. Nesbet, *Phys. Rev.* **118**, 681 (1960).
319. N. Bessis and H. Lefebvre-Brion, *Compt. Rend.* **251**, 648 (1960).
320. N. Bessis, H. Lefebvre-Brion, and C. M. Moser, *Phys. Rev.* **124**, 1124 (1961).
321. M. J. D. Powell and W. Marshall, unpublished material (1961); W. Marshall, *J. Phys. Soc. Japan* **17**, 20BI (1962).
322. R. K. Nesbet, *Ann. Phys.* (*NY*) **3**, 397 (1958).
323. R. K. Nesbet, *J. Math. Phys.* **2**, 701 (1961).
324. P. O. Löwdin, *J. Appl. Phys.* **33**, 251 (1962).
325. J. E. Harriman, Tech. Note No. 97, Quantum Chemistry Group, University of Uppsala, Uppsala, unpublished (1963).
326. V. Heine, private discussions (1962).
327. H. Shull and P. O. Löwdin, *J. Chem. Phys.* **25**, 1035 (1956).
328. P. O. Löwdin and H. Shull, *Phys. Rev.* **101**, 1730 (1956).
329. K. F. Berggren and R. F. Wood, *Phys. Rev.* **130**, 198 (1963).
330. H. M. James and A. S. Coolidge, *Phys. Rev.* **47**, 700 (1935).
331. H. M. James and A. S. Coolidge, *Phys. Rev.* **49**, 688 (1936).
332. E. A. Burke, *Phys. Rev.* **130**, 1871 (1963).
333. E. A. Burke, *Phys. Rev.* **135**, A621 (1964).
334. R. K. Nesbet, private communication, 1962.
335. E R. Davidson, *J. Chem. Phys.* **37**, 577 (1962).
336. E. R. Davidson and L. L. Jones, *J. Chem. Phys.* **37**, 2966 (1962).
337. N. Bessis, H. Lefebvre-Brion, C. M. Moser, A. J. Freeman, R. K. Nesbet, and R. E. Watson, *Phys. Rev.* **135**, A588 (1964).
338. P. Kusch and H. Taub, *Phys. Rev.* **75**, 1477 (1949).
339. H. E. Radford, V. W. Hughes, and V. Bettran-Lopez, *Bull. Am. Phys. Soc.* [2] **5**, 272 (1960).
340. R. E. Watson and A. J. Freeman, unpublished, 1962.
340a. S. Geschwind, P. Kisliuk, M. P. Klein, J. P. Remeika, and D. L. Wood, *Phys. Rev.* **126**, 1684 (1962); N. Laurance and J. Lambe, *Phys. Rev.* **132**, 1029 (1963).
340b. S. Geschwind and M. Blume, to be published.
341. R. E. Watson and A. J. Freeman, *Phys. Rev. Letters* **6**, 277, 388E (1961).
342. V. Heine, *Czech. J. Phys.* **13**, 619 (1963).
343. A. J. Freeman and R. E. Watson, *Bull. Am. Phys. Soc.* [2] **6**, 105 (1961).
344. L. F. Matthiess, unpublished material, 1963.

345. A. J. Freeman and R. E. Watson, *Phys. Rev.* **135**, A1209 (1964).
346. R. E. Watson and A. J. Freeman, *Phys. Rev.* **133**, A1571 (1964).
347. E. A. Hylleraas, *Z. Physik* **48**, 469 (1928); **54**, 347 (1929).
348. G. H. Brigman and F. A. Matsen, *J. Chem. Phys.* **27**, 829 (1957).
349. R. P. Hurst, J. D. Gray, G. H. Brigman, and F. A. Matsen, *Mol. Phys.* **1**, 189 (1958).
350. E. A. Burke and J. F. Mulligan, *J. Chem. Phys.* **28**, 995 (1958).
351. G. H. Brigman, R. P. Hurst, J. D. Gray, and F. A. Matsen, *J. Chem. Phys.* **29**, 251 (1958).
352. W. Kolos and C. C. J. Roothaan, *Rev. Mod. Phys.* **32**, 219 (1960).
353. E. K. Zavoisky, *J. Phys. Chem. USSR* **9**, 245, 447 (1945).
354. M. H. L. Pryce, *Proc. Phys. Soc.* **A63**, 25 (1950).
355. A. Abragam and M. H. L. Pryce, *Proc. Phys. Soc.* **A63**, 409 (1950).
356. K. W. H. Stevens, *Proc. Roy. Soc.* **A219**, 542 (1953).
357. B. Bleaney and K. W. H. Stevens, *Rept. Progr. Phys.* **16**, 108 (1953).
358. J. S. Griffith, "The Theory of Transition Metal Ions." Cambridge Univ. Press, London and New York, 1961.
359. J. Kanamori, *Progr. Theoret. Phys. (Kyoto)* **17**, 177, 197 (1957).
360. W. Low, *Phys. Rev.* **109**, 256 (1958).
361. J. W. Culvahouse, U. P. Unruh, and D. K. Brice, *Phys. Rev.* **129**, 2430 (1963).
362. A. M. Clogston, *Phys. Rev.* **118**, 1229 (1960).
363. F. S. Ham, G. W. Ludwig, G. D. Watkins, and H. H. Woodbury, *Phys. Rev. Letters* **5**, 468 (1960).
364. J. S. Van Wierengen, *Discussions Faraday Soc.* **19**, 118 (1955).
365. O. Matumura, *J. Phys. Soc. Japan* **14**, 108 (1958).
366. R. S. Title, *Phys. Rev.* **130**, 17 (1963).
367. J. Kondo, *Progr. Theoret. Phys. (Kyoto)* **23**, 106 (1960).
368. R. C. Ohlmann and M. Tinkham, *Bull. Am. Phys. Soc.* [2] **3**, 416 (1958).
369. J. Smit and H. P. J. Wijn, "Ferrites." Wiley, New York, 1959.
370. S. Geschwind, *Bull. Am. Phys. Soc.* [2] **8**, 212 (1963).
371. D. S. McClure, *J. Chem. Phys.* **36**, 2757 (1962).
372. J. H. Van Vleck, "Electric and Magnetic Susceptibilities." Oxford Univ. Press, London and New York, 1932.
373. K. W. H. Stevens, *Proc. Phys. Soc.* **A65**, 209 (1952); R. J. Elliott and K. W. H. Stevens, *Proc. Roy. Soc.* **A215**, 437 (1952); **A218**, 553 (1953); **A219**, 387 (1953).
373a. J. M. Baker, B. Bleaney, and W. Hayes, *Proc. Roy. Soc.* **A247**, 141 (1958).
373b. Y. Ebina and N. Tsuya, *Sci. Rept. Res. Inst. Tohoku Univ.* **B13**, 1, 25, 43 (1960).
374. W. E. Blumberg and J. Eisinger, *Bull. Am. Phys. Soc.* [2] **6**, 141 (1961).
375. J. M. Baker and F. I. B. Williams, *Proc. Roy. Soc.* **A267**, 283 (1962).
376. B. Bleaney, *Proc. Phys. Soc.* **A68**, 937 (1955).
377. B. R. Judd and I. Lindgren, *Phys. Rev.* **122**, 1802 (1961).
378. I. Lindgren, *Nucl. Phys.* **32**, 151 (1961).
379. D. F. Mayers, *Proc. Roy. Soc.* **A240**, 93 (1957).
380. R. G. Boyd, A. C. Larson, and J. T. Waber, *Phys. Rev.* **129**, 1629 (1963).
381. D. Halford, *Phys. Rev.* **127**, 1940 (1962).
382. W. M. Doyle and R. Marrus, *Phys. Rev.* **131**, 1586 (1963).
383. A. C. Gossard, V. Jaccarino, and J. H. Wernick, *Bull. Am. Phys. Soc.* [2], **7** 482 (1962); *Phys. Rev.* **133**, A881 (1964).
384. F. M. Pichanick, P. G. H. Sandars, and G. K. Woodgate, *Proc. Roy. Soc.* **A257**, 777 (1960).
385. J. B. Baker and F. I. B. Williams, *Proc. Roy. Soc.* **A267**, 283 (1962).

386. C. K. Jørgensen, private communication, 1962; C. K. Jørgensen, R. Pappalardo, and H. H. Schmidtke, *J. Chem. Phys.* **39**, 1422 (1963).
387. W. Marshall, private communication, 1963; also see M. J. M. Leask, R. Orbach, M. J. D. Powell, and W. P. Wolf, *Proc. Roy. Soc.* **A272**, 371 (1963).
388. W. Low and R. S. Rubens, unpublished material (1964).
389. B. Bleaney, unpublished material (1964).
390. J. H. Wood, *Phys. Rev.* **117**, 714 (1960); Quarterly Progress Report, Solid State and Molecular Theory Group, p. 79, Massachusetts Institute of Technology, January 15, 1961 (unpublished).
391. J. W. Wood, private communications, 1962.
392. F. Stern, *Phys. Rev.* **116**, 1399 (1959).
393. P. W. Anderson and A. M. Clogston, *Bull. Am. Phys. Soc.* [2] **6**, 124 (1961).
394. W. J. Carr, Jr., Winter Institute in Quantum Chemistry and Solid State Physics, Sanibel Island, Florida, January 1-13, 1961 (unpublished).
395. P. W. Anderson, *Phys. Rev.* **124**, 41 (1961).
396. R. L. Streever, unpublished material (1964).
397. E. L. Boyd and R. J. Gambino, *Phys. Rev. Letters* **12**, 20 (1964).
398. M. Blume, A. J. Freeman, and R. E. Watson, *Phys. Rev.* **134**, A320 (1964).
399. P. W. Anderson, *Solid State Phys.* **14**, 99 (1963).
400. P. W. Anderson, "Magnetism," (G. T. Rado and H. Suhl, eds.), Vol. 1, p. 25. Academic Press, New York, 1963.
401. R. G. Shulman and B. J. Wyluda, *J. Chem. Phys.* **30**, 335 (1959).
402. W. B. Lewis, J. A. Jackson, J. F. Lemons, and H. Taub, *J. Chem. Phys.* **36**, 694 (1962).
403. J. M. Baker and J. P. Hurrell, *Proc. Phys. Soc.* **82**, 742 (1963).
403a. L. F. Matthiess, *Phys. Rev.*, to be published (1965).
404. M. Peter and S. Koide, *Rev. Mod. Phys.* **36**, 160 (1964).
404a. R. E. Watson, S. Koide, and M. Peter, to be published (1965).
405. M. Peter, D. Shaltiel, J. H., Wernick, H. J. Williams, J. B. Mock, and R. C. Sherwood, *Phys. Rev.* **126**, 1395 (1962).
406. M. Peter et al., *Phys. Rev. Letters* **9**, 50 (1962).
407. J. Kondo, *Progr. Theoret. Phys.* (*Kyoto*) **28**, 846 (1962).
408. T. W. Moore and D. S. Rodbell, *J. Appl. Phys.* **55**, 906S (1964).
409. A. C. Gossard, unpublished material (1964).
409a. J. Crangle, Discussions at the International Conference on Magnetism, Nottingham, England, 1964, *Phys. Rev. Letters* **13**, 569 (1964).
409b. H. E. Nigh, S. Legvold, and F. H. Spedding, *Phys. Rev.* **132**, 1092 (1963).
410. C. G. Shull and Y. Yamada, *J. Phys. Soc. Japan* **17**, Suppl. B-III (1962).
411. R. M. Moon, *Phys. Rev.* **136**, A195 (1964).
412. A. M. Clogston and V. Jaccarino, *Bull. Am. Phys. Soc.* [2] **5**, 430 (1960).
412a. B. Giovannini, M. Peter, and J. R. Schrieffer, *Phys. Rev. Letters* **12**, 736 (1964).
412b. A. W. Overhauser and M. B. Stearns, *Phys. Rev. Letters* **13**, 316 (1964).
412c. R. E. Watson, A. J. Freeman, unpublished material (1964).
413. A. M. Clogston, A. C. Gossard, V. Jaccarino, and Y. Yafet, *Phys. Rev. Letters* **9**, 262 (1962).
414. R. J. Elliott, *Proc. Phys. Soc.* **B70**, 119 (1957); J. M. Baker and B. Bleaney, *Proc. Roy. Soc.* **A245**, 156 (1958).
415. M. H. L. Pryce, *Proc. Phys. Soc.* **A63**, 25 (1950); D. J. I. Fry, P. M. Llewelyn, and M. H. L. Pryce, *Proc. Roy. Soc.* **A266**, 84 (1962); W. E. Blumberg, J. Eisinger, and S. Geschwind, *Phys. Rev.* **130**, 900 (1963).

416. R. Kubo and Y. Obata, *J. Phys. Soc. Japan* **11**, 547 (1956).
417. A. M. Clogston and V. Jaccarino, *Phys. Rev.* **121**, 1357 (1961).
418. E. g., D. M. Lam, D. O. Van Ostenburg, M. V. Nevitt, H. D. Trapp, and D. W. Pracht, *Phys. Rev.* **131**, 1428 (1963).
419. A. C. Gossard, V. Jaccarino, and J. H. Wernick, *Phys. Rev.* **128**, 1038 (1962).
420. A. M. Clogston, V. Jaccarino, and Y. Yafet, *Phys. Rev.* **134**, A1650 (1964).
421. A. M. Clogston, B. T. Matthias, M. Peter, H. J. Williams, E. Corenzwit, and R. C. Sherwood, *Phys. Rev.* **125**, 541 (1962).
422. P. A. Wolff, *Phys. Rev.* **124**, 1030 (1961).
423. A. M. Clogston, *Phys. Rev.* **125**, 439 (1962).
424. D. E. Nagle, P. P. Craig, P. Barrett, D. R. F. Cochran, C. E. Olsen, and R. D. Taylor, *Phys. Rev.* **125**, 490 (1962).
424a. D. Shaltiel, J. H. Wernick, H. J. Williams, and M. Peter, *Phys. Rev.* **135**, A1346 (1964).
424b. G. Low, *Intern. Conf. Magnetism, Nottingham, England, 1964*, to be published; W. Phillips, *Phys. Rev.*, to be published (1965).
424c. R. Segnan and P. Craig, *Phys. Rev.*, to be published (1965).
424d. J. O. Dimmock and A. J. Freeman, *Phys. Rev. Letters*, **13**, 750 (1964).
424e. A. Berman, M. W. Zemansky, and H. A. Boorse, *Phys. Rev.* **109**, 70 (1958).
425. C. E. Johnson, M. S. Ridout, T. E. Cranshaw, and P. E. Madsen, *Phys. Rev. Letters* **6**, 450 (1961).
426. P. A. Flinn and S. L. Ruby, *Phys. Rev.* **124**, 34 (1961).
427. E. A. Friedman and W. J. Nicholson, *J. Appl. Phys.* **335**, 1048 (1963).
428. R. Segnan, B. Mozer, and R. Nathans, *Bull. Am. Phys. Soc.* [2] **8**, 250 (1963).
429. G. K. Wertheim, V. Jaccarino, J. H. Wernick, and D. N. E. Buchanan, *Phys. Rev. Letters* **12**, 24 (1964).
429a. M. B. Stearns, *Bull. Am. Phys. Soc.* **11**, 276 (1964); M. B. Stearns and S. S. Wilson, *Phys. Rev. Letters* **13**, 313 (1964).
429b. A. W. Overhauser and M. B. Stearns, *Phys. Rev. Letters* **13**, 316 (1964).
429c. J. I. Budnick, private communication (1964).
430. G. E. Low and M. F. Collins, *J. Appl. Phys.* **345**, 1195 (1963).
430a. G. K. Wertheim, *Bull. Am. Phys. Soc.* **11**, 276 (1964); G. K. Wertheim and V. Jaccarino, unpublished material (1964).
431. R. E. Watson and A. J. Freeman, *Phys. Rev.* **123**, 521 (1961).
432. M. J. D. Powell, unpublished material (1964).
433. C. A. Coulson, *Proc. Cambridge Phil. Soc.* **34**, 204 (1938).
434. C. C. J. Roothaan, *Rev. Mod. Phys.* **23**, 69 (1951).
435. R. K. Nesbet, *Rev. Mod. Phys.* **35**, 552 (1963).
436. C. C. J. Roothaan and P. S. Bagus, *Methods Computational Phys.* **2**, 47 (1963).
437. C. C. J. Roothaan, *Rev. Mod. Phys.* **32**, 179 (1960).
438. S. Huzinaga, *Phys. Rev.* **120**, 866 (1960); **122**, 131 (1961).
439. H. A. Bethe and E. E. Salpeter, "Quantum Mechanics of One- and Two-Electron Atoms." Springer, Berlin, 1957.
440. B. Swirles, *Proc. Roy. Soc.* **A152**, 625 (1935).
441. I. P. Grant, *Proc. Roy. Soc.* **A262**, 555 (1961).
442. R. Lefebvre, *Cahiers Phys.* **110**, 1 (1959).
442a. J. A. Pople and R. K. Nesbet, *J. Chem. Phys.* **22**, 571 (1954).
442b. G. H. Shortley, *Phys. Rev.* **50**, 1072 (1936).

5. Nuclear Resonance in Antiferromagnets

V. Jaccarino

Bell Telephone Laboratories,
Murray Hill, New Jersey

I. Introduction	307
II. Some General Considerations on Nuclear Resonance	308
1. The Basic Resonance Relations	308
2. Criteria for Observing Nuclear Resonance in Magnetic Materials	309
III. Nuclear Hamiltonian: Origins and Magnitudes of the Internal Fields	312
1. Nonmagnetic Solid	312
2. Magnetic Solid	312
3. Magnitudes of the Internal Fields	319
IV. Temperature Dependence of the Sublattice Magnetization from the Time-Averaged Field	319
1. The Magnetization of an Antiferromagnet at Low Temperatures	319
2. Effect of a Large Energy Gap on $M(T)$	327
3. $CuCl_2 \cdot 2H_2O$	327
4. Temperature Dependence of the Sublattice Magnetization Near T_N	330
V. Relaxation and Line Widths	333
1. Fluctuating Local Field	333
2. Indirect Nuclear Spin-Spin Interaction	344
VI. Summary	347
Appendix A. Spin Waves on a Two-Sublattice Antiferromagnet	348
Appendix B. The Isotropic Hyperfine Interaction Expressed in Terms of the Spin Wave Operators	351
Appendix C. Importance of Spin Wave-Phonon Interactions	352
References	353

I. Introduction

The techniques of nuclear magnetic resonance (NMR) and nuclear quadrupole resonance (NQR) have been extensively applied to the study of the electronic properties of solids. Both the experimental and theoretical aspects of nuclear resonance methods in nonmagnetic materials have been the subject of several recent books [1, 2, 3] to which the reader is referred for detailed background material.

In this chapter we confine ourselves to the studies of nuclear resonance in those nonmetallic magnetic materials that exhibit spontaneous antiferromagnetic ordering of the electronic spins at low temperatures. Considerations of space necessitate that we treat only a few select topics. The virtue of the nuclear resonance method is that it provides us with a weakly interacting probe, allowing a precise determination to be made of the internal fields that are present in magnetic solids. More specifically we will see that nuclear resonance is sensitive to the time-averaged magnetic field as well as to certain spectral components of the fluctuating local fields.

II. Some General Considerations on Nuclear Resonance

1. The Basic Resonance Relations

Among the intrinsic properties of nuclei are the total angular momentum $\mathbf{I}\hbar$ and the magnetic dipole moment μ. The magnetic interaction energy E_m of a nuclear moment and an external field $\mathbf{H}_0$, $E_m = -\mu \cdot \mathbf{H}_0$, results in an equal splitting of the $2I + 1$ energy levels. The energy separation between adjacent levels is

$$\Delta E = \gamma \hbar H_0 \qquad (2.1)$$

where γ is a scalar quantity called the "gyromagnetic ratio" defined by the relation $\mu = \gamma \mathbf{I}\hbar$. (The vectors μ and $\mathbf{I}$ are collinear but not necessarily parallel.) Since the quantum-mechanical selection rules for magnetic dipole transitions allow for transitions to be induced only between adjacent levels, a nuclear spin system will undergo resonance absorption if a perturbing oscillatory magnetic field H_1, of frequency ω, is applied perpendicular to H_0 provided that $\omega = \Delta E/\hbar$, or

$$\omega = \gamma H_0 \qquad (2.2)$$

The dynamic aspects of a nuclear spin system, subject to interactions between the individual spins and between the spins and their environment (lattice), while undergoing resonance absorption, may be obtained from a modified form of the classical equations of motion for a magnetic moment $\mathbf{M}$ in a magnetic field; $\dot{\mathbf{M}} = \gamma \mathbf{M} \times \mathbf{H}$. (In our case $\mathbf{M}$ would represent the magnetization density of nuclear moments.) If the steady magnetic field H_0 is along the z-direction and M_0 is the equilibrium magnetization, then the forced precession of the magnetization by the perturbing oscillation field H_1 causes the z-component of the magnetization M_z to decrease. We assume that, when H_1 is reduced to zero, M_z will exponentially decay to its equilibrium value. The equation of motion for $\dot{M}_z$ is then

$$\dot{M}_z = \gamma(\mathbf{M} \times \mathbf{H})_z + (1/T_1)(M_0 - M_z) \quad (2.3)$$

T_1 is the spin-lattice relaxation time and characterizes the rate at which energy is transferred from the nuclear spin system to its environment. Similarly, the fact that the instantaneous local field H_{loc} at each nucleus may differ (e.g., due to nuclear spin-spin interactions) causes the precession rate to vary, thereby randomizing the transverse component of the magnetization $M_\perp$. $M_\perp$ will decay with a characteristic time $T_2 \simeq (\gamma H_{\text{loc}})^{-1}$ and the equation of motion for $\dot{M}_\perp$, corresponding to Eq. (2.3) for $\dot{M}_z$, is

$$\dot{M}_\perp = \gamma(\mathbf{M} \times \mathbf{H})_\perp - (M_\perp/T_2) \quad (2.4)$$

We will consider T_1 and T_2 in greater detail later on as regards nuclei embedded in magnetic solids.

Nuclei for which $I \geqslant \hbar$ may have an aspherical charge distribution and therefore an electrical quadrupole moment. If $\mathbf{Q}$ is the tensor defining the nuclear charge distribution then the interaction $\mathcal{H}_Q$ of the quadrupole moment with an electrical field gradient tensor ∇E is $\mathcal{H}_Q = \mathbf{Q} \cdot \nabla E$. For an axially ($z$-directed) symmetrical electrical field gradient eq the eigenvalues of $\mathcal{H}_Q$, in a representation which is diagonal in I_z, are

$$E_Q(I_z) = \frac{e^2 q Q}{4I(2I-1)}[3I_z^2 - I(I+1)] \quad (2.5)$$

where Q is the scalar quadrupole moment [3]. When $E_Q(I_z) \ll \gamma \hbar H_0$, a small deviation from the equal spacing of the $2I+1$ levels results and resonance absorption may take place at more than one frequency. In Fig. 1 we schematize the two cases we have discussed, for $I = \frac{3}{2}\hbar$: (a) $e^2 qQ = 0$, and (b) $e^2 qQ \neq 0$, with the external magnetic field and electrical field gradient collinear.

2. Criteria for Observing Nuclear Resonance in Magnetic Materials

The elementary considerations given above apply to nuclei in nonmagnetic materials. We know in magnetic materials the electronic

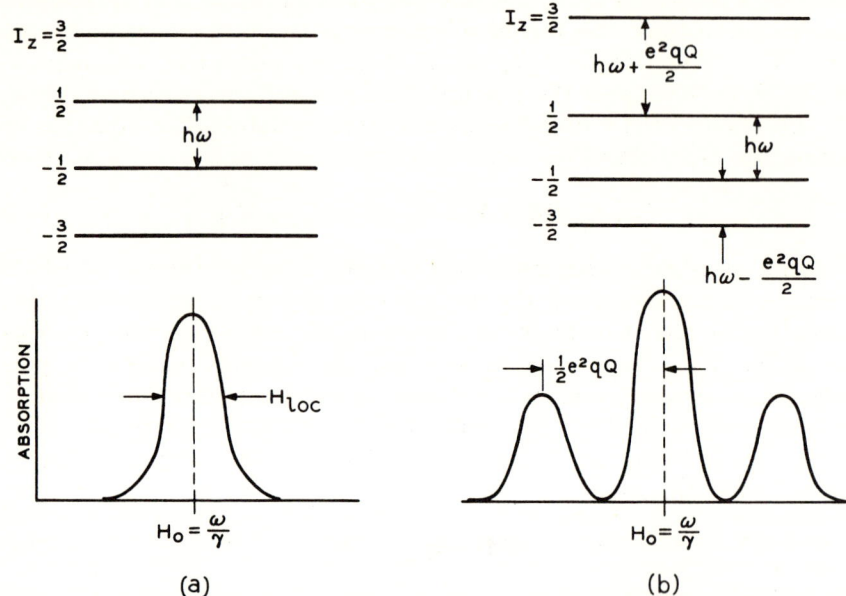

FIG. 1. The energy levels of a nuclear spin system in a magnetic field H_0, for the case $I = \frac{3}{2}\hbar$, with (a) no quadrupole interaction and (b) an axial quadrupole interaction with H_0 parallel to the symmetry axis. The corresponding resonance absorption is schematized in both cases as a function of magnetic field at a fixed frequency ω; in (a) the absorption is maximum at $H_0 = \omega/\gamma$ while in (b) three maxima occur at $H_0 = \omega/\gamma$ and $(\omega/\gamma) \pm (e^2qQ/2)$.

dipolar fields are some 10^3 times larger than the corresponding nuclear dipole fields, and the atomic hyperfine fields of magnetic ions possibly 10^6 times larger. Obviously if H_{loc} scaled proportionately nuclear resonance would be unobservable in magnetic materials. What fields *do* the nuclei "see"? To best answer this question we have to know what is the observation period in the resonance process. In an NMR experiment it is the Larmor period which, from Eq. (2.2), is

$$T_L = (\gamma H_{\text{eff}})^{-1} \qquad (2.6)$$

where H_{eff} is the time-averaged field—the sum of the external field and the constant components of the fields produced by the electronic spins, whether of dipolar or hyperfine origin (or both). If, during T_L, the local fields fluctuate very rapidly, then their effects will average to zero. The transverse components of the fluctuating field at the Larmor frequency will induce transitions between the magnetic substates (see

Fig. 1) and will contribute to T_1 and to T_2 inasmuch as each substate is lifetime broadened. The longitudinal components of the fluctuating field at frequencies much less than the Larmor frequency contribute to T_2 since, in effect, they make T_L vary from one Larmor period to the next and when averaged over many periods a distribution in T_L will exist. (A detailed formulation of this problem is given in the section on linewidths and relaxation times.)

It is important then to know what is the frequency distribution of the local field spectra as seen by the nuclei. At temperatures much above T_N (the antiferromagnetic transition temperature) the strong exchange interaction between electronic spins produces a rapid reorientation of the spin moments. The frequency dependence of the amplitude of the local field spectra $|H(\omega)| \propto S(\omega)$ is schematized in Fig. 2, where we

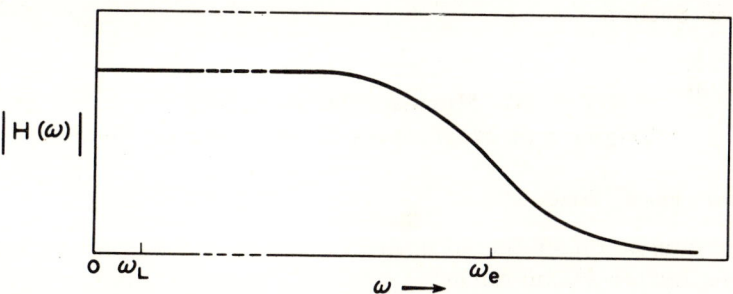

FIG. 2. A schematic representation of the frequency dependence of the amplitude of the local field at a nucleus in a magnetic material for $T \gg T_N$. The ratio ω_e/ω_L is usually in excess of 10^5.

give $|H(\omega)|$ vs. ω, with ω_e the exchange frequency. The amplitude of $|H(\omega)|$ will vary, depending on whether the field arises from hyperfine or dipolar coupling, but the frequency distribution will not. Most important, however, is that $|H(\omega \simeq 0)|$ and $|H(\omega_L)|$ will be much less than estimates based on the static hyperfine or dipolar field H_{int} would yield. This immediately follows from the fact that the integrated spectral density has to remain constant and since the effect of exchange is to distribute the local field spectra over a very large frequency range the low-frequency components are severely attenuated.

Since many factors enter into observability of any given NMR (e.g., γ, density of the particular nuclei being studied, etc.) it is difficult to give other than what amounts to a crude necessary condition for observing resonance in a magnetic solid. Let $T_{2-\min}$ (or $T_{1-\min}$) be the smallest value of T_2 (or T_1) for which resonance is still observable and let H_{int} be the static value of the perturbing field at the nucleus in question in the magnetic material. Then it is required that ω_e be sufficiently large for the inequality

$$\frac{1}{T_{1,2-\min}} > \frac{(\gamma H_{\text{int}})^2}{\omega_e} \tag{2.7}$$

to be satisfied.

Considerations of the local field spectra (anisotropy, temperature dependence, etc.) in the ordered state $T < T_N$ are sufficiently more complex so that we must refrain from attempting to adhere to any simple criterion such as Eq. (2.7). For the present let it suffice to remark that the condition (2.7) qualitatively applies to the ordered state with regard to $1/T_2$ although the reasons for this being so are remarkably different in origin. (See the discussion on line widths and relaxation times in Section V.)

III. Nuclear Hamiltonian: Origins and Magnitudes of the Internal Fields

1. Nonmagnetic Solid

The interaction of the nuclear moment with the external field H_0 is given by the Hamiltonian

$$\mathscr{H} = -\gamma \hbar \mathbf{I} \cdot \mathbf{H}_0 \tag{3.1}$$

The difference between adjacent eigenvalues of Eq. (3.1) yields (2.1). The nucleus experiences only the external field in a nonmagnetic solid, apart from relatively small corrections [1] to be associated with the dipolar fields of other nuclei, atomic diamagnetism and chemical shifts.

2. Magnetic Solid

The rapid fluctuation of the electronic spins resulting from exchange, dipolar, and spin lattice interactions causes the nuclei to see a static field which is proportional to the time average of the individual electronic spin moment $\langle \mathbf{S} \rangle$. In the paramagnetic state, $\langle \mathbf{S} \rangle$ is nonvanishing only in a finite applied field; it is related* to the uniform magnetization per unit volume $\mathbf{M}_0$ by

$$\mathbf{M}_0 = N g \beta \langle \mathbf{S} \rangle \tag{3.2}$$

---

* If the medium is magnetically anisotropic then g is a tensor and $\langle \mathbf{S} \rangle$ and $\mathbf{H}_0$ need not be parallel.

Here N is the volume density of spins, β the Bohr magneton, g the electronic gyromagnetic ratio, and $\mathbf{M}_0$ and $\langle \mathbf{S} \rangle$ are functions of temperature and applied field. In the ordered state a spontaneous magnetization persists even in the absence of magnetic field. The internal fields at the nucleus produced by the electronic spin moments fall into three broad classes depending on the nature of the atom in the magnetic solid.

a. Nuclei of nonmagnetic atoms (e.g., H^1 in $CuCl_2 \cdot 2H_2O$). In this case the interaction Hamiltonian for the kth atom in the unit cell is

$$\mathcal{H}_k = -\gamma \hbar \mathbf{I}_k \cdot \left[\mathbf{H}_0 - g\beta \sum_n r_n^{-3} \left(\langle \mathbf{S}_n \rangle - \frac{3\mathbf{r}_n(\mathbf{r}_n \cdot \langle \mathbf{S}_n \rangle)}{r^2} \right) \right] \quad (3.3)$$

where the second term in the square brackets is the dipolar field $\mathbf{H}_D$ produced by all of the electronic spins in the solid; the summation extends over spins $\mathbf{S}_n$ which are at a distance r_n from the kth atom. It may be that there is more than one physically equivalent site for like atoms in the unit cell so that, say, the kth and lth atoms have different dipolar fields. For crystals of low symmetry and crystals that have a number of inequivalent like atoms per unit cell the diagonalization of Eq. (3.3) will result in a number of different allowed transitions which vary in frequency as a function of the angle between $\mathbf{H}_0$ and the crystal axes. If the electronic g tensor is anisotropic then $\langle \mathbf{S}_n \rangle$ will be orientation dependent as well as field and temperature dependent. In any case $\mathbf{H}_0$ and $\mathbf{H}_D$ need not be parallel, especially at temperatures below T_N.

The behavior of the proton NMR in the orthorhombic crystal $CuCl_2 \cdot 2H_2O$ ($T_N = 4.336°K$) exemplifies some of the conditions discussed above [4-12]. Four distinct proton resonances (doublets) are seen corresponding to the four physically inequivalent water molecules in the unit cell. The angular dependence of the frequencies for resonance [5], at a fixed $\mathbf{H}_0$ lying in the ab-plane, is shown in Fig. 3. The observed angular dependence is different above and below T_N and it was, in fact, from such studies that antiferromagnetism was first found in this crystal.

b. Nuclei of magnetic atoms (e.g., Co^{59} in CoF_2). The nuclei of the paramagnetic atoms will be subjected to the intense hyperfine fields produced by the electronic spin and orbital moments. (The origins of these fields are discussed in detail in a companion article [13].) If we represent the electron-nuclear hyperfine interaction by a tensor $\tilde{A}$, the complete nuclear Hamiltonian has the form

$$\mathcal{H}_k = -\gamma \hbar \mathbf{I}_k \cdot [\mathbf{H}_0 + \mathbf{H}_{kD}] + \mathbf{I}_k \cdot \tilde{A} \cdot \langle \mathbf{S}_k \rangle \quad (3.4)$$

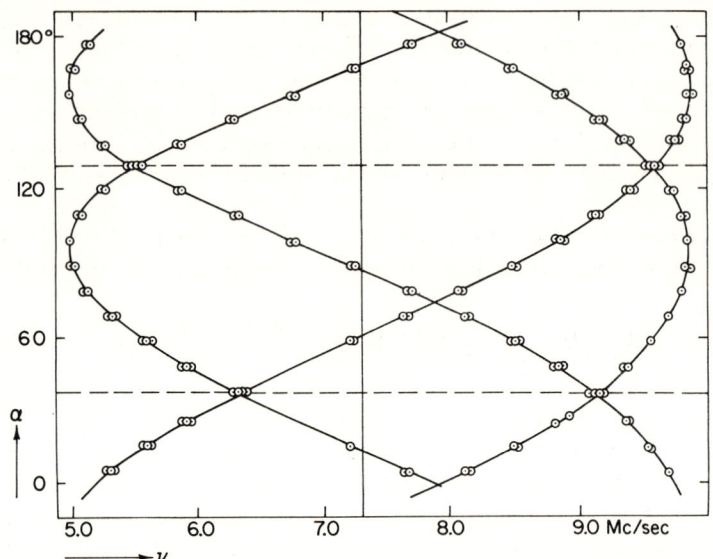

FIG. 3. The proton NMR in $CuCl_2 \cdot 2H_2O$ in the antiferromagnetic state ($T = 3.03°K$); the frequency for resonance for a constant external field ($H_0 = 1705$ oe) rotated in the ab-plane. The dashed lines are the $a(\alpha_a = 125°)$- and $b(\alpha_b = 35°)$-axis positions. The vertical line is the frequency of the proton resonance in water. (From Poulis and Hardeman [5].)

where $\mathbf{H}_{kD}$ is the dipolar field at the kth ion from all other paramagnetic spins but the kth one. In effect the hyperfine coupling introduces an additional field $\mathbf{H}_{HF} = -(\gamma\hbar)^{-1}\tilde{A} \cdot \langle \mathbf{S} \rangle$. The frequency for resonance obtained from the diagonalization of Eq. (3.4) is

$$\omega = \gamma \left[\sum_i^{x,y,z} (H_0^i + H_D^i + H_{HF}^i)^2 \right]^{1/2} \quad (3.5)$$

in which we have assumed that the principal axes (x, y, and z) of the electronic $\tilde{g}$ and $\tilde{A}$ tensors coincide.

As an example of NMR of the nucleus of a paramagnetic atom in the antiferromagnetic state we consider Co^{59} ($I = \frac{7}{2}\hbar$) in CoF_2. The difluorides of Mn, Fe, and Co have the rutile-type structure. The magnetic ions form a body-centered tetragonal lattice with the spin moments of the corner and body-centered magnetic ions aligned antiparallel, with respect to each other, along the unique (c) axis [001] of the crystal (see Fig. 4). The point symmetry of the Co site is less than cubic, allowing for an electric quadrupole interaction in addition to

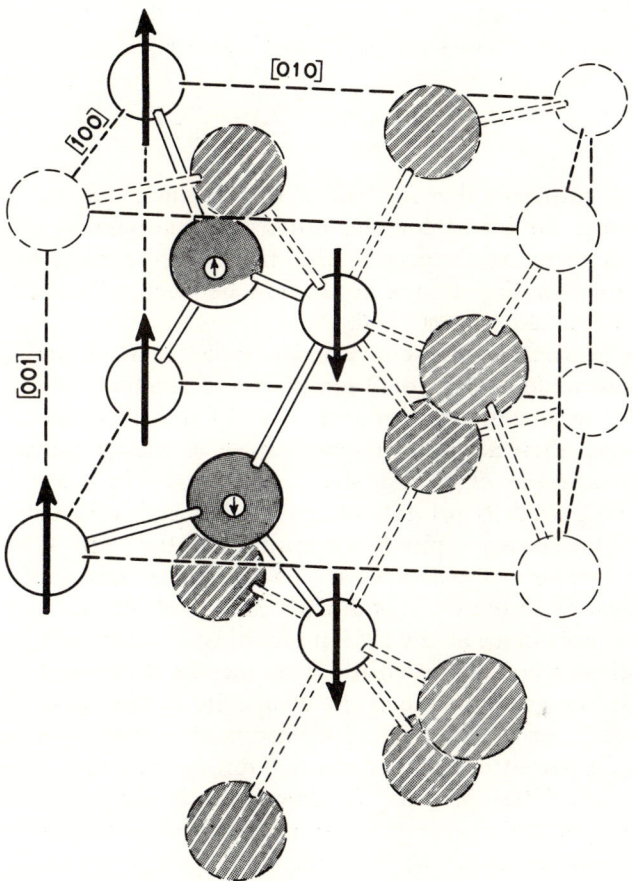

Fig. 4. The body-centered tetragonal lattice of antiferromagnetic MnF_2, FeF_2, and CoF_2 with the unique (c) axis indicated as [001]. The corner and body-centered positions (large open circles) are occupied by the metal ions and the ordering is schematized by the heavy black arrows. The two physically inequivalent F ions (large shaded circles) have three nearest-neighbor metal ions, two of which are on one sublattice and the remaining one on the other sublattice. The directions of the net spin imbalance at two representative inequivalent sites are shown by the small arrows in the small circles. This spin density on the nominally nonmagnetic ions, a direct result of overlap and charge transfer effects, provides the means for investigating the temperature dependence of the sublattice magnetization as is discussed in Section IV.

the magnetic terms in Eq. (2.4). Let $+z$ be one of the [001] directions and assume for simplicity that the quadrupole tensor is axially symmetric about z. With H_0 parallel to the $+z$-direction the frequencies of the allowed transitions, from Eqs. (2.5) and (3.5), are

$$\omega_{\pm}(I_z \leftrightarrow I_z - 1) = \hbar^{-1} A_z \langle S_z \rangle - \gamma(H_D \pm H_0) + \frac{3e^2 qQ\hbar^{-1}}{4I(2I-1)}(2I_z - 1) \quad (3.6)$$

where the $\pm$ indicates that for half of the Co nuclei H_0 will be parallel to the internal field and for the other half antiparallel. As a result $2 \times 2I$ resonances are expected and the corresponding observations [14] are shown in Fig. 5; the separation between adjacent transitions when $H_0 = 0$ is $\Delta\omega = e^2qQ/14\hbar$.

c. *Nuclei of partially magnetic atoms* (e.g., F^{19} in MnF_2 and CoF_2). If there is appreciable overlap between the wave functions of the electrons of nominally nonmagnetic ions and those of the electrons of the paramagnetic ions, then a spatial redistribution of the spin magnetization occurs. An example of this is the F^- ion and Mn^{2+} ion in MnF_2. Consider the paired $2s$ orbitals of the F^- ion and a $3d$ orbital on the Mn^{2+} ion. The "down" spin F^- $2s$ orbital and the "up" spin Mn^{2+} $3d$ orbital are orthogonal whereas the ones of like spin are not. The orthogonalization of the F^- $2s$ and Mn^{2+} $3d$ orbitals of like spin produces a net $2s$ spin imbalance at the F^- ion resulting in a hyperfine interaction characteristic of the $2s$ spin density. This may be thought of as a process that exists in the ground state of the composite system. A second contribution arises from the transfer of electrons of a given spin orientation from the spin-paired orbitals on the F^- ion to unpaired $3d$ orbitals on the Mn^{2+} ion. This process, fundamental to the theory of superexchange, is primarily an interaction that results from consideration of the allowed excited states of the composite system. Again in this case the preferential transfer of F^- electrons of a given spin orientation relative to that of the Mn^{2+} $3d$ spin results in a spin imbalance on the F^- ion with an attendant hyperfine interaction. Alternative descriptions of these processes have been given in terms of covalent bonding involving augmentation of ligand wave functions of appropriate symmetry to the metal ion wave functions. These points are discussed in the companion article on the origins of hyperfine fields [13].

The nuclear Hamiltonian in the case of "transferred" hyperfine interactions is, for the nucleus of the kth ion in the unit cell,

$$\mathcal{H}_k = -\gamma\hbar \mathbf{I} \cdot [\mathbf{H}_0 + \mathbf{H}_{kD}] + \sum_n^{(nn)} \mathbf{I}_k \cdot \tilde{A}_{kn} \cdot \mathbf{S}_n \quad (3.7)$$

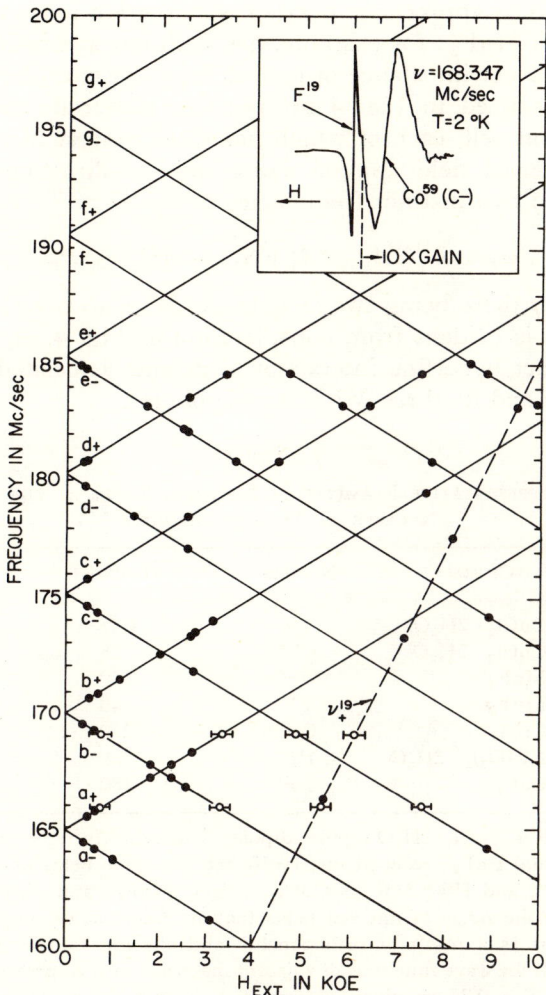

FIG. 5. The NMR of Co^{59} in antiferromagnetic CoF$_2$. The observed frequency for resonance as a function of external field applied parallel to [001] at $T = 1.3°K$ is shown by the solid dots and the solid line is the best fit to the data using Eq. (3.6). The F^{19} NMR, indicated by ν_+^{19}, is the frequency-increasing branch predicted by Eq. (3.8). Recorder traces of the derivatives of the two resonances are shown in the upper part of the figure. (From Jaccarino [14, 14a].)

where the first term has its usual meaning and the sum in the second term is to extend over all nearest-neighbor (nn) magnetic ions of spin **S** for which there is appreciable overlap with the kth ion wave functions. The F$^-$ site in MnF$_2$, for example, has such low symmetry that in the ordered state a net imbalance of spin density exists at the F^{19} nucleus. This is schematized in Fig. 4 for the two physically inequivalent F sites in the unit cell, each of which has three nearest Mn^{2+} neighbors. With the external field applied along the $+z$-direction there result from Eq. (3.7) two distinct frequencies below T_N:

$$\omega_{\pm} = \hbar^{-1}(2A_\mathrm{I}^z - A_\mathrm{II}^z)\langle S_z \rangle - \gamma[H_D^z \pm H_0] \qquad (3.8)$$

The reason for there being two frequencies is identical to that given for Eq. (3.6) and is evident from consideration of Fig. 4. A_I^z and A_II^z are the two distinct hyperfine interaction constants associated with a particular F$^-$ ion and its three Mn near neighbors.

TABLE I

REPRESENTATIVE EXAMPLES OF INTERNAL FIELDS AT THE NUCLEUS IN ANTIFERROMAGNETS

Reference[b]	Crystal	Nucleus	H_int(koe)	$\nu(0)$(Mc/sec)
1	CuCl$_2 \cdot$ 2H$_2$O	H^1	0.76	3.5
2	CuCl$_2 \cdot$ 2H$_2$O	Cl^{35}	28	10, 19, 28
3	MnF$_2$	Mn^{55}	650	680
4	MnF$_2$	F^{19}	40	160
5	CrCl$_3$	Cr^{53}	260	63
6	Ni(IO$_3$)$_2 \cdot$ 2H$_2$O	I^{127}	3.4	144, 154, 297
7	CoF$_2$	Co^{59}	180	180

[a] The H^1 NMR in CuCl$_2 \cdot$ 2H$_2$O typifies dipolar fields; the Mn^{55}, Cr^{53} and Co^{59} NMR in MnF$_2$, CrCl$_3$, and CoF$_2$, respectively, typify the hyperfine fields of magnetic atoms; and the Cl^{35}, F^{19}, and I^{127} NMR in CuCl$_2 \cdot$ 2H$_2$O, MnF$_2$, and Ni(IO$_3$)$_2 \cdot$ 2H$_2$O are representative of the range of internal fields that are found at the nuclei of nominally nonmagnetic ions as a result of orthogonality and electron transfer effects resulting from overlap with the wave functions of adjacent magnetic ions. Approximate frequencies for resonance at $T = 0°$K are given with multiple listings if the quadrupole interaction is large compared to the magnetic interaction.

[b] References:
1. Poulis et al. [12].
2. O'Sullivan et al. [15].
3. Jefferts and Jones [15a].
4. Jaccarino and Shulman [16].
5. Narath [16a].
6. Burgiel et al. [16b].
7. Jaccarino [14].

3. Magnitudes of the Internal Fields

The three kinds of internal fields are reasonably distinguishable by their magnitudes. Assuming that well below $T_N \langle S_z \rangle \simeq S$, we would expect the orders of magnitude of the fields at the nucleus to be at the

(a) nonmagnetic ions (dipolar fields ~ 0.5–7×10^3 oe);

(b) magnetic ions (magnetic hfs fields ~ 0.5–7×10^5 oe);

(c) partially magnetic ions (transferred hfs fields ~ 0.5–7×10^4 oe).

Although it is not intended to attach any particular significance to the bounds of the above estimates, it should be apparent that they are at least distinct in magnitude. Representative examples of the internal fields corresponding to the different cases are given in Table I along with pertinent data on the approximate frequencies for resonance at $T = 0°$K. (Note $\nu(0) = \omega(0)/2\pi$).

IV. Temperature Dependence of the Sublattice Magnetization from the Time-Averaged Field

By definition the true antiferromagnet exhibits no macroscopic spontaneous moment because the spin moments on both sublattices, though equal in magnitude, are oppositely directed. However, within the unit cell the magnetization density $\mathbf{M}(r)$ will be finite and vary from point to point in magnitude and direction. Observation of the magnitude and direction of the time-averaged field at a certain select point (or points) in the unit cell via nuclear resonance should allow, in principle, for a determination of $\mathbf{M}(r)$. Of particular interest is the temperature dependence of the sublattice magnetization which is directly obtained by measurement of the temperature dependence of the time-averaged field at any point in the unit cell.

1. The Magnetization of an Antiferromagnet at Low Temperatures

Both the ferromagnet and antiferromagnet have in common that the exchange interaction between electrons on neighboring atoms leads to an almost complete ordering of the spin system at low temperatures. At 0 °K the simple ferromagnet, without dipolar interactions, has as its ground state the fully aligned state, i.e., every spin is in the state of maximum $S_z = S$. For the antiferromagnet this is not the case, since if (see Appendix A for more complete Hamiltonian)

$$\mathcal{H} = -2J \sum_{i>j}^{nn} \mathbf{S}_i \cdot \mathbf{S}_j \tag{4.1}$$

is the Heisenberg exchange Hamiltonian then it is readily shown that the state with $S^z = S$ for spins on the "up" sublattice and $S^z = -S$ on the "down" sublattice is not an eigenfunction of Eq. (4.1) [17]. Though the true ground state (which is not known) is not far removed in energy from that state in which the spins are fully aligned on each sublattice, it has as one of its properties that the magnetization of each of the sublattices should be less than $Ng\beta S$ at $0°K$.

Nevertheless, let us take the fully aligned state as a pseudo-ground state. At temperatures above absolute zero the thermal excitation of the spin system will cause the spins to deviate from their value $|S^z| = S$. For any individual spin $\mathbf{S}_i$ to have its S_i^z changed by $\hbar$ would require an amount of energy equal to the exchange energy, by Eq. (4.1), so that the first excited state would be removed in energy by J from the ground state. Suppose, however, that this unit of spin deviation $S - S^z$ could be shared by many spins as would be the case if long-wavelength sinusoidal variations of the amplitudes of S^z were possible. Then, because the spin deviation could be spread out in space, any individual spin would have as a near neighbor a spin which was almost antiparallel* and the requisite energy to produce such a disturbance could be much less than J. Such disturbances, or spin waves, because of the small possible excitation energies, form a natural starting point for describing the thermodynamics of the ferromagnet and antiferromagnet at low temperatures [17–20]. (An outline of the antiferromagnetic spin wave operator formalism is given in Appendix A.)

a. **Noninteracting spin waves.** For the ferromagnet the operators that create and destroy spin waves are the Fourier transforms of the spin deviation operators, from which it follows that the excitation of any spin wave, of arbitrary wave vector $\mathbf{k}$, corresponds to decreasing the total S^z by $\hbar$. To compute the decrease in the magnetization at a temperature T it is only necessary to know the total number of spin waves that have been excited; i.e., for the ferromagnet

$$M(0) - M(T) = g\beta \sum_k n_k = g\beta \sum_k \left(\exp\frac{\hbar\omega_k}{kT} - 1\right)^{-1} \quad (4.2)$$

where we have utilized the fact that spin waves obey Bose statistics and n_k and $\hbar\omega_k$ are the number and energy, respectively, of spin waves of wave number $\mathbf{k}$ and the sum extends over the first Brillouin zone in $\mathbf{k}$ space. If the dispersion relation for spin waves ($\mathbf{k}$ dependence of $\hbar\omega_k$) in a given crystal is known then $M(0) - M(T)$ may be determined

---

* Parallel for the ferromagnet.

as a function of temperature. For a ferromagnet with no anisotropy or applied field the long-wavelength (small $\mathbf{k}$) approximation for E_k, i.e., $\hbar\omega_k \sim k^2$, leads to the well-known $T^{3/2}$ law for the initial decrease of the magnetization with temperature.

By contrast, for the antiferromagnet, the amount of spin deviation produced for a given spin wave excitation is a function of $\mathbf{k}$ because the amplitudes of the transformation coefficients, u_k^2 and v_k^2, are $\mathbf{k}$ dependent [see Eq. (A-10)]. The decrease in the sublattice magnetization in the case of the antiferromagnet is given by (see Appendix A for definition of symbols)

$$M(0) - M(T) = g\beta \sum_k n_k \left(\frac{\omega_E + \omega_A}{\omega_k}\right) \tag{4.3}$$

When $\omega_A = 0$ (no anisotropy), the long-wavelength (small k) approximation to the spin wave dispersion relation, i.e., $\hbar\omega_k \sim |\mathbf{k}|$, leads to an initial decrease of the sublattice magnetization with temperature that is proportional to T^2.

Anisotropy has a more profound effect on the thermodynamic properties of an antiferromagnet than it does on a ferromagnet. In the former case a given amount of anisotropy introduces a larger gap E_G in the spin wave spectrum at small $\mathbf{k}$ because of the interplay of anisotropy and exchange energies, whereas in the ferromagnet there is a uniform raising of all E_k by just the anisotropy energy. This is schematized in Fig. 6.

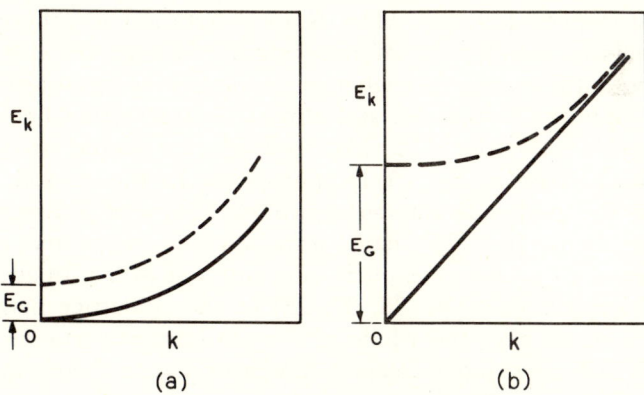

FIG. 6. The modification of the long-wavelength spin wave dispersion relation introduced by anisotropy, for spin waves in (a) a ferromagnet and (b) an antiferromagnet. For the ferromagnet the gap energy $E_G = \hbar\omega_A$ whereas in the antiferromagnet $E_G = \hbar(2\omega_A\omega_E)^{\frac{1}{2}}$. The solid line in each case represents E_k for $\omega_A = 0$ and the dashed line that of finite ω_A.

In Table II we have summarized the low-temperature behavior of the magnetization, with and without anisotropy, for the ferromagnet and antiferromagnet in the long-wavelength (small **k**) approximation.

TABLE II

Long-Wavelength (Small-k) Dependence of the Spin Wave Energy E_k and Initial Temperature Dependence of Magnetization[a]

	Ferromagnet		Antiferromagnet	
	E_k	$M(0) - M(T)$	E_k	$M(0) - M(T)$
No anisotropy	ak^2	$\sim T^{3/2}$	$b\|k\|$	$\sim T^2$
With anisotropy	$E_G + ak^2$	$\sim \exp(-E_G/kT)T^{3/2}$	$(E_G^2 + b^2k^2)^{1/2}$	$\sim \exp(-E_G/kT)T^{3/2}$

[a] For the ferromagnet $E_G = \hbar\omega_A$, whereas for the antiferromagnet $E_G = \hbar(2\omega_A\omega_E)^{1/2}$. Since in general $\omega_E \gg \omega_A$, E_G(ferro-) $\ll E_G$(antiferro-) for the same ω_A.

That the energy gap at small **k** may be a sizeable fraction of the width of the spin wave band, i.e., $0.1 < kT_{AE}/kT_N < 1$, strongly modifies the usual criterion of the region of validity of the long-wavelength approximation; to wit, that temperature region for which $M(0) - M(T)$ is less than, say, $0.1\,M(0)$. This can be most clearly seen by a detailed consideration of one antiferromagnet, MnF_2. MnF_2 appears to be ideal in several respects [19, 21]. Mn^{2+} is an S-state ion and therefore spin-orbit effects as manifest in an anisotropic exchange are absent. There is a strong antiferromagnet exchange interaction between near-neighbor spins on opposite sublattices and almost no interaction between spins on the same sublattice [22, 23], making the simple, two-sublattice Hamiltonian given in Eq. (A-1) appropriate to this crystal. Measurements of the perpendicular susceptibility [24] and antiferromagnetic resonance frequency [25] have provided the necessary parameters (J, $\hbar\omega_A$, and kT_{AE}) for an explicit calculation of $M(T)$. In addition very precise measurements [16, 26, 27, 28] of the sublattice magnetization have been made over the whole of the antiferromagnetic region using the F^{19} NMR, allowing a detailed comparison with theory to be made.

In Fig. 7 we show the dispersion relation for spin waves in a particular symmetry direction (perpendicular to the unique axis) in the first Brillouin zone in MnF_2 [19]. Without anisotropy and with the discreteness of the lattice neglected, $\hbar\omega_k$ would be proportional to $|\mathbf{k}|$ as is shown by the dashed line. The discreteness of the lattice introduces "zone boundary" effects since it is required from Eq. (A-16) that

5. NUCLEAR RESONANCE IN ANTIFERROMAGNETS

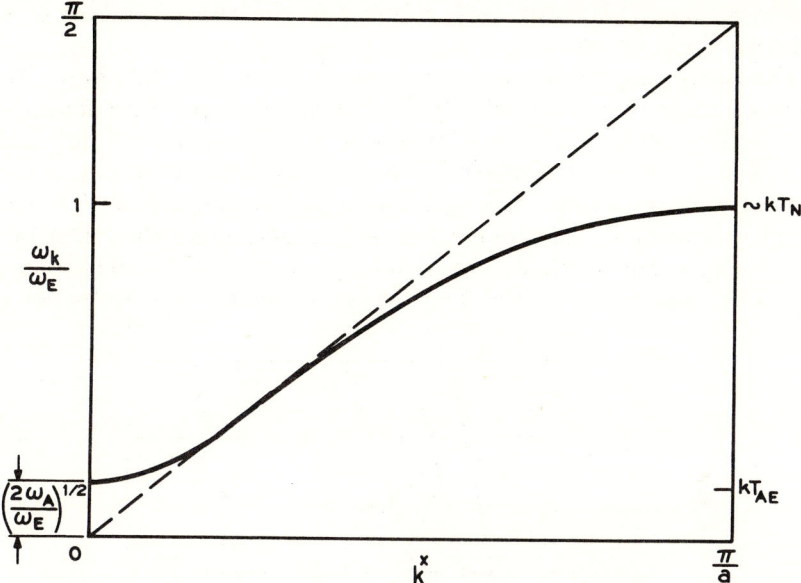

FIG. 7. The dispersion relation for spin waves along a particular symmetry direction, perpendicular to the c-axis, in the first Brillouin zone in MnF_2. The dashed line is the long-wavelength approximation to E_k without anisotropy and disregards zone boundary effects. The solid line is the true dispersion relation (for noninteracting spin waves) with the anisotropy energy assumed not to be $\mathbf{k}$ dependent. (From Keffer [19].)

$(\partial E_k/\partial \mathbf{k})_{k-\mathrm{max}} = 0$. The combined effects of anisotropy and the discreteness of the lattice are shown by the solid line and represent the true dispersion relation for noninteracting spin waves, assuming that the anisotropy energy is independent of $\mathbf{k}$. (We will return to the latter two points shortly.) It is clear that the gap suppresses the excitation of the small $\mathbf{k}$ spin waves which would be important for $T < T_{AE}$. Most significant is that when the temperature exceeds T_{AE}, in addition to the small $\mathbf{k}$ excitations, spin waves of relatively large $\mathbf{k}$ are excited whose energy is poorly approximated by the long-wavelength dispersion relation given in Table II. Account may be taken of this by expanding $\hbar\omega_k$ [see Eq. (A-16)] in a power series in $\mathbf{k}$

$$\hbar\omega_k = \left[(\hbar\omega_0)^2 + \sum_{n=1} a_n \mathbf{k}^{2n}\right]^{1/2} \tag{4.4}$$

where ω_0 is the frequency of the uniform mode and the a_n are distinct functions of the exchange and anisotropy energies, and by evaluating

Eq. (4.3) at a sufficiently large number of the allowed values of **k** in the first Brillouin zone.

The low-temperature measurements of the F^{19} NMR in MnF_2 illustrate these points very clearly. In Fig. 8 the temperature dependence of the F^{19} NMR frequency as given by Eq. (3.8), in the region $T < T_{AE} \simeq 13\,°K$, is displayed. The resonance frequency $\nu_T = \omega_T/2\pi$ is strictly proportional to $\langle S_z \rangle_T$ if the hyperfine and dipolar interaction coupling constants are independent of temperature. The effect of the energy gap is quite marked since $[d \ln (\nu_0 - \nu_T)]/(d \ln T)$ increases steadily as $T \to 0$ in contrast to the T^2 behavior expected in the absence of a

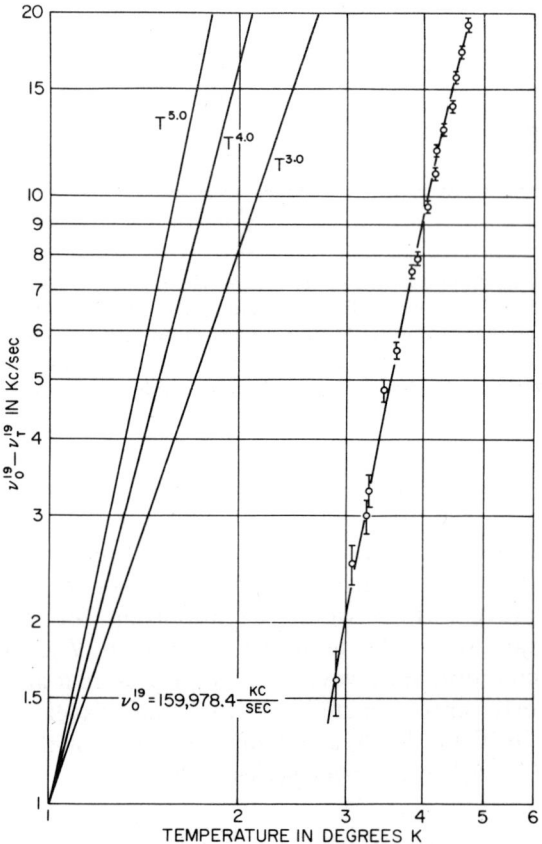

FIG. 8. A log-log plot of the temperature dependence of the F^{19} NMR in antiferromagnetic MnF_2 for T much less than T_{AE}. The pronounced effect of the energy gap kT_{AE} on spin wave excitations is manifest in the steadily increasing value of $[d \ln (\nu_0 - \nu_T)]/(d \ln T)$ as the temperature is decreased. (From Jaccarino and Walker[26].)

gap. It is to be noted that an accuracy in ν_T of a few parts per million is achieved in these measurements. At temperatures in the vicinity of T_{AE} no reasonable fit to experiment may be made using the long-wavelength approximation for $\hbar\omega_k$, as is evident from Fig. 9 where the fractional change in frequency divided by T^2 is plotted vs. T on a doubly logarithmic scale. (The solid line represents the experimental data with $\Delta\nu/\nu_0 = [M(0) - M(T)]/M(0)$). The reason for dividing the fractional change in the magnetization by T^2 is that in the small-**k** limit a change in kT_{AE}, if J is kept constant, roughly corresponds to a horizontal translation of the predicted curve. Thus the two calculated dashed curves for two slightly different values of T_{AE} are seen to be almost parallel [27]. However, despite the fact that the magnetization has decreased by no more than 1% at $T = T_{AE}$, one is obviously outside the region where the small-**k** approximation is valid.

The open circles represent the results [29] of a computer calculation of the integral representation of Eq. (4.3) in which explicit account is taken of zone-boundary effects. Up to 25 °K ($\sim 2T_{AE}$) theory and experiment agree to within $\pm 3\%$ for $[M(0) - M(T)]/M(0)$. The values of the parameters that provide the best fit are given in the figure legend. Despite the fact that most of the anisotropy energy E_A for MnF_2 is of dipolar origin, no appreciable improvement in the fit is made by making E_A to be dependent on **k** in proper fashion.

b. *Interacting spin waves.* The results outlined above are obtained from a noninteracting spin wave theory in which it is assumed that many spin waves of a given **k** may be excited without modifying the energy E_k of any one. This is the assumption of linear superposition of spin waves. However, it clearly is easier to reverse a given spin if the neighboring spins are already partially reversed. Account may be taken of this by considering the dynamical interaction [30] between spin waves. This results in a renormalization of the energy of any given spin wave which leads to a changed population of that spin wave mode at a given tempearture. Since the energy is lowered for any given spin wave the population of that mode at a given temperature is increased and a corresponding decrease in the magnetization, over and above that predicted by the noninteracting spin wave theory, is to be expected.

Again for MnF_2, just such a calculation has been made [21] by computing the energy renormalization resulting from the lowest-order (two spin wave) dynamical interactions. It is found that striking agreement between experiment and theory is now extended to $T = \frac{3}{4}T_N$. Thus, as regards the thermodynamics of the magnetization in this simple

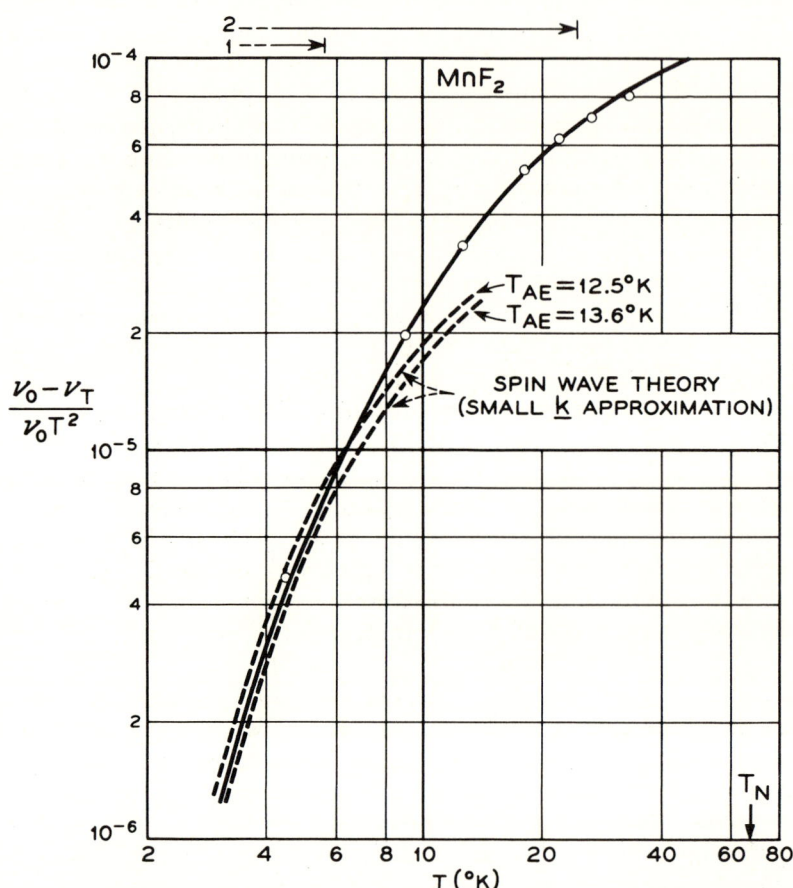

FIG. 9. A comparison of experiment with the predictions of noninteracting spin wave theory for the temperature dependence of the sublattice magnetization in antiferromagnetic MnF_2. The solid line is the experimental data obtained from the F^{19} NMR; the two dashed lines are the predicted behaviors for $[M(0) - M(T)]/M(0)T^2$ in the small-$\mathbf{k}$ approximation for two slightly different values of T_{AE} (from Jaccarino and Walker [27]). The open circles are the results obtained from properly including zone-boundary effects on the energy spectrum for noninteracting spin waves J for next-nearest neighbors $= 1.79$ °K and $H_A = 8800$ oe (from Lines [29]). It can be seen that the small-$\mathbf{k}$ approximation is inadequate beyond $T/T_N = 0.1$, where $\Delta M/M \ll 1\%$, as indicated in region 1 on top of figure, although noninteracting spin wave theory is in very good agreement with experiment to temperatures as high as $T/T_N = \frac{1}{3}$, indicated by region 2.

antiferromagnet, one is led to believe that a most satisfactory description of the elementary excitations is achieved by using spin wave theory.

2. Effect of a Large Energy Gap on M(T)

For antiferromagnets with non-S-state magnetic ions occupying positions of low point symmetry the single-ion crystalline anisotropy energy may be as large as the exchange energy. This will necessarily lead to a case where the gap energy kT_{AE} and the transition temperature T_N are comparable in magnitude: FeF_2 exemplifies this situation. The free Fe^{2+} ion ($3d^6$) has a ^5D ground state. The combined effects of spin-orbit coupling and the orthorhombic point symmetry of the Fe^{2+} site produce a large magnetocrystalline anisotropy energy [31], one consequence of which is the large energy gap at 0 °K in the spin wave spectrum. Antiferromagnetic resonance measurements [32] in the far infrared show this to be $T_{AE} = 76$ °K.

The crystal and magnetic spin structures of FeF_2 and MnF_2 are isomorphic and their Néel temperatures are approximately the same: $T_N = 78.4°$ and 66.9 °K, respectively. A comparison of their low-temperature magnetization curves is therefore of some interest since there is such a large disparity in the initial gap energies for the two: $T_{AE} = 76°$ and 12.55 °K, respectively.

Measurements [27] of the temperature dependence of the fractional change in the NMR frequency in FeF_2 and MnF_2 are displayed in Fig. 10. The almost complete suppression of spin wave excitations at low temperatures in FeF_2 is manifest in the extraordinarily small decrease in the sublattice magnetization at low temperatures; $\Delta M/M \sim 1 \times 10^{-4}$ at a temperature equal to the T_{AE} of MnF_2.

Agreement of experiment with long-wavelength spin wave theory is still less satisfactory for FeF_2 than it is for MnF_2. It is to be presumed that in addition to zone-boundary effects any attempt to understand the thermodynamic properties of FeF_2 at low temperatures should allow for the possibility of an anisotropic exchange interaction and should include terms in $(S_i^x)^2 + (S_i^y)^2$ for the single-ion anisotropy (because of the orthorhombic point symmetry), neither of which is contained in the simple Hamiltonian (A-1).

3. $CuCl_2 \cdot 2H_2O$

One of the most well-studied antiferromagnets from an experimental point of view is $CuCl_2 \cdot 2H_2O$ [4–12, 15]. Results on the temperature dependence of the sublattice magnetization in $CuCl_2 \cdot 2H_2O$ ($T_N =$

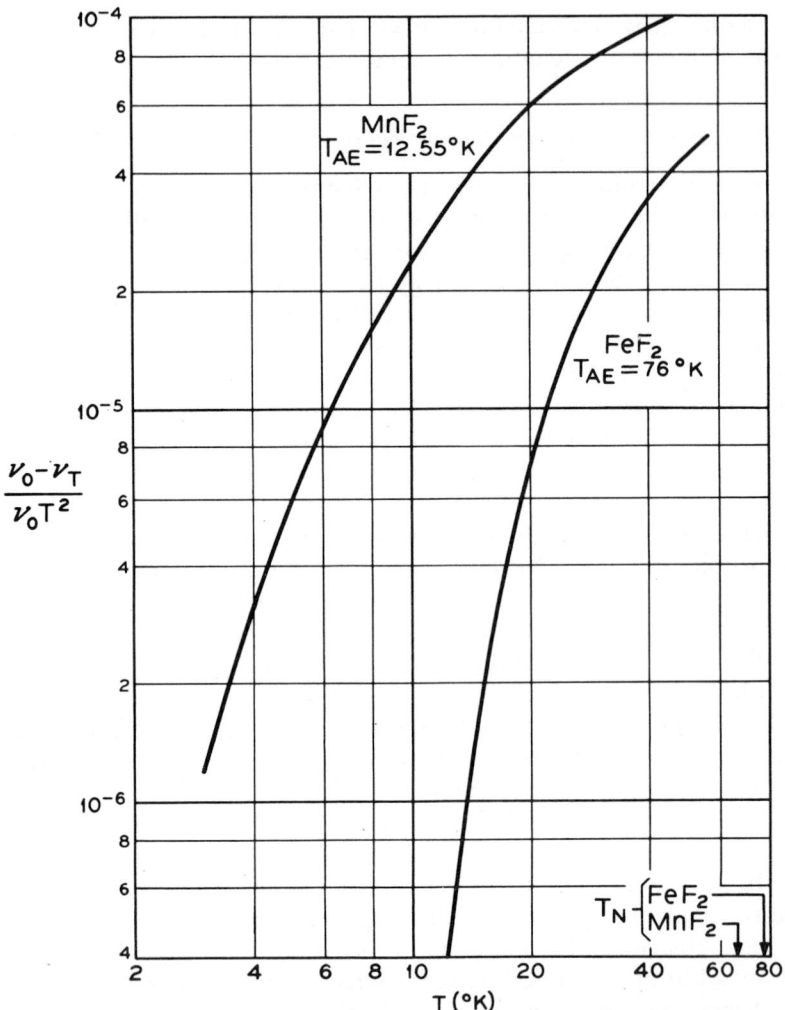

FIG. 10. A comparison of the temperature dependence of the F^{19} NMR (and thus the sublattice magnetizations) in antiferromagnetic FeF_2 and MnF_2. It is seen that the large value of T_{AE} in FeF_2 results in extremely small changes in the magnetization at temperatures comparable to the initial energy gap in MnF_2; $T_{AE} = 12.55°K$.

4.32 °K) as determined [12, 15] from H^1 and Cl^{35} NMR are shown in Fig. 11. Most peculiar is the fact that the decrease in the sublattice magnetization seems to follow a T^4 law over a rather extended temperature region, $1.0 < T < 3.5$ °K. Because the crystal has orthorhombic symmetry there are two energy gaps $T_{AE}(1) \simeq 1$ °K and $T_{AE}(2) \simeq 2$ °K, resulting in a splitting of the spin wave modes. The temperature depen-

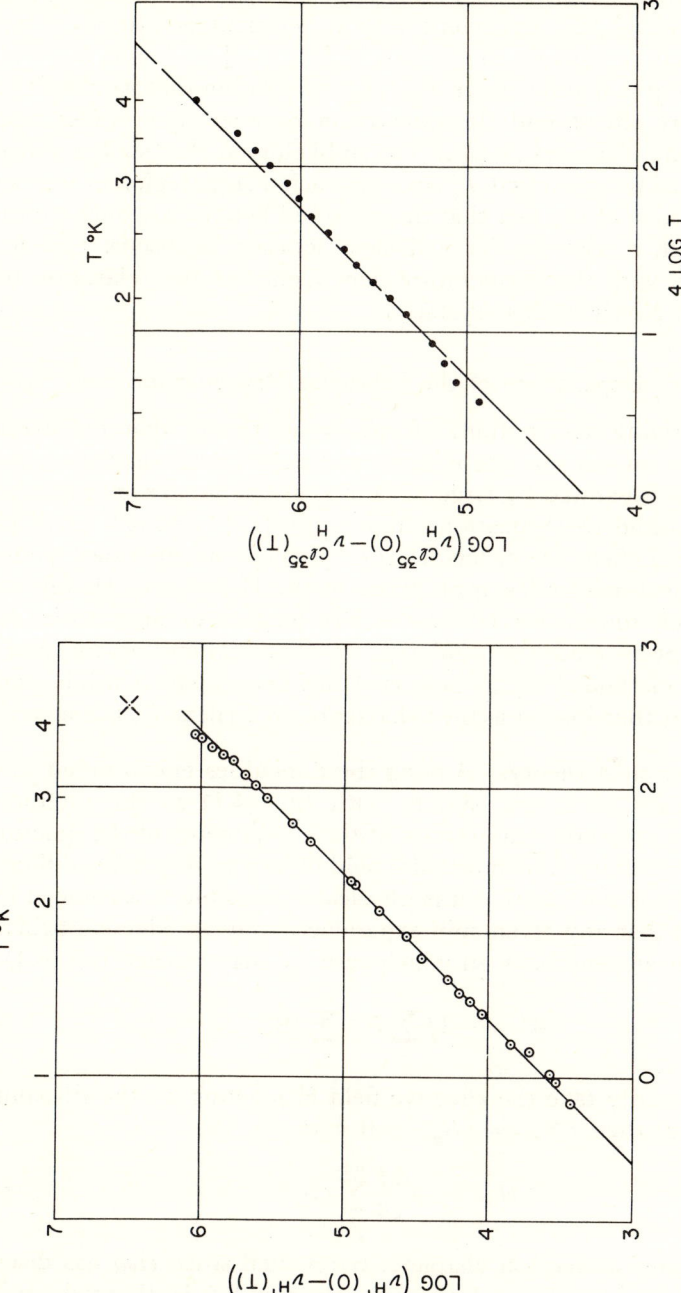

Fig. 11. A log-log plot of the temperature dependence of the (a) proton and (b) Cl^{35} nuclear resonance in antiferromagnetic $CuCl_2 \cdot 2H_2O$. In the region indicated it appears that $\Delta M \sim T^4$. (From (a) Poulis and Hardeman [8] and (b) O'Sullivan et al. [15].)

dence of the magnetization has been calculated in the small-**k** limit for this case [33] and the agreement with experiment is extremely poor. It is reasonable to assume that, since $T_{AE}/T_N \sim \frac{1}{4} - \frac{1}{3}$ and the region of measurement is mainly above $T_{AE}(1)$, a calculation that would include zone-boundary effects and the interaction between spin waves would account for the observed results. The unlikelihood that such a calculation would lead to an explicit T^4 law over such a large region of temperature is suggestive of the fact that the observed behavior is only approximated by such a relation. We will have occasion to discuss this again in connection with the temperature dependence of the relaxation time of the proton NMR in this crystal.

4. Temperature Dependence of the Sublattice Magnetization Near T_N

The underlying assumptions of spin wave theory that the average spin deviation on any one site is small and that strong correlation exists between spins separated by large distances, are no longer appropriate to the region around the transition temperature [17]. Though much progress has been made in order-disorder phenomena no exact solution to the statistical mechanics appropriate to the Heisenberg Hamiltonian (4.1) has been found, for either the ferromagnet or antiferromagnet. Resort has been made to studying simplified Hamiltonians and to approximate methods [34], the results of only two of which will be considered and only then insofar as the behavior of $M(T)$ near T_N is concerned.

   a. *Molecular field theory.* Among the difficulties encountered in the thermodynamics of a spin system with Eq. (4.1) as the interaction Hamiltonian is the noncommutivity of the components of the quantum-mechanical operators. The molecular field theory is free of this difficulty. It has as its basis the fact that it is physically plausible to assume at high temperatures that any given spin experiences only an effective field due to all of its near neighbors, in which case we may replace Eq. (4.1) by

$$\mathcal{H} = -2J \sum_i \mathbf{S}_i \cdot \sum_j^{\text{nn}} \langle \mathbf{S}_j \rangle \tag{4.5}$$

If we arbitrarily take the effective field $\mathbf{H}_{\text{eff}}$ acting on the ith spin to be z directed, then $\langle S_x \rangle = \langle S_y \rangle = 0$ with

$$H_{\text{eff}} = -\frac{2J}{g\beta} \sum_j^{\text{nn}} \langle S_j^z \rangle \tag{4.6}$$

Some quantum-mechanical character is retained since H_{eff} has discrete values because of the $2S + 1$ allowed values of S_j^z. In thermal equilibrium the average spin moments on the two sublattices must be equal in magnitude and oppositely directed. The temperature dependence of the sublattice magnetization derived from these postulates is the familiar molecular-field curve: a comparison of the predictions of the theory for $S = 5/2$ and the observed result derived from the F^{19} NMR [27, 28] in MnF_2 is shown in Fig. 12. Of most interest to us here is the behavior

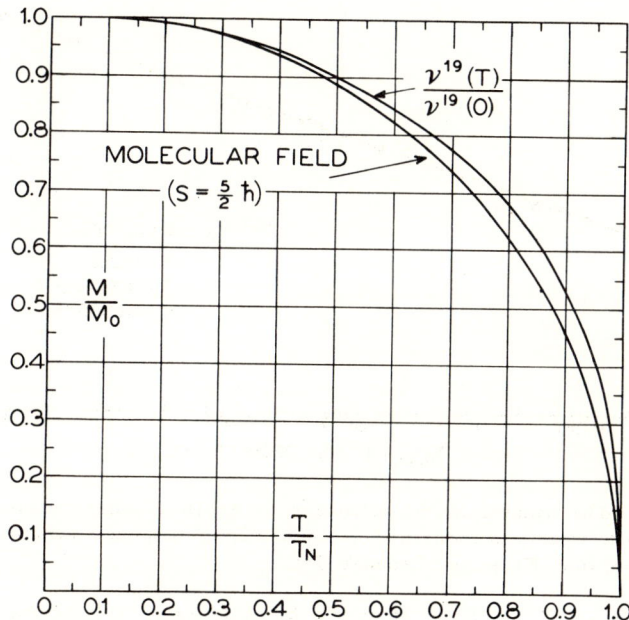

FIG. 12. A comparison of the temperature dependence of the sublattice magnetization from molecular field theory, for $S = \frac{5}{2}\hbar$ and that obtained from the F^{19} NMR in MnF_2.

near the critical point, where molecular field theory predicts a $(T_N - T)^{1/2}$ temperature dependence for $M(T)$.

b. Ising model. Approximation methods have been applied to the statistical mechanics of the Ising Hamiltonian;

$$\mathcal{H} = 2J \sum_{i>j} S_i^z S_j^z \qquad (4.7)$$

For a two-dimensional lattice an exact solution has been obtained [35] for the dependence of $M(T)$ on T near T_N; $M(T) \sim (T_N - T)^{1/8}$. The method of Padé approximants has been applied to a power series analysis of the three-dimensional lattice and it appears that $M(T) \sim (T_N - T)^{5/16}$ in this case [36].

c. Comparison with experiment. Extremely precise measurements [28] of the F^{19} NMR in MnF_2 have been made in the immediate vicinity of T_N. The results are shown in Fig. 13 and are suggestive of the fact

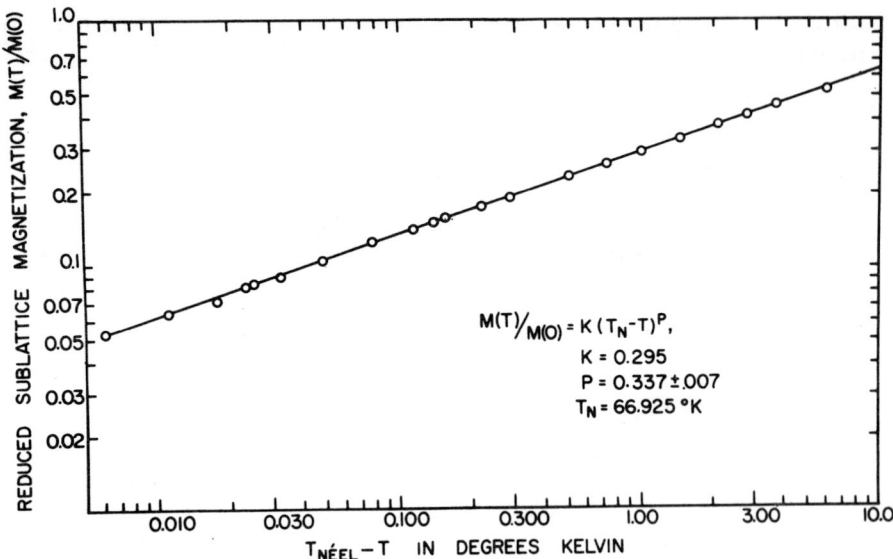

FIG. 13. The temperature dependence of the sublattice magnetization in MnF_2 in the immediate vicinity of the transition temperature. The precise measurements of the F^{19} NMR are from Heller and Benedek [28].

that $M(T) \sim (T_N - T)^{1/3}$. It would be instructive to have measurements of the temperature dependence for other antiferromagnets near T_N, particularly those with large anisotropy, such as FeF_2, and for layer-type structures which approximate the two-dimensional antiferromagnet.

The above results are summarized in Table III.

TABLE III

The Exponent n in the Theoretical Temperature Dependence $(T_N - T)^n$ of the Sublattice Magnetization Immediately below $T_N{}^a$

Dimensionality	Molecular field	Ising	Heisenberg	Experiment
2	1/2	1/8	—	—
3	1/2	5/16	—	1/3

[a] For two- and three-dimensional antiferromagnetic lattices. The experimental result for MnF_2 is that shown in Fig. 13.

V. Relaxation and Line Widths

1. Fluctuating Local Field

As we have noted earlier the observability of nuclear resonance in magnetic materials depends on the magnitude of certain spectral components of the fluctuating local field at the nucleus resulting from hyperfine and dipolar interactions. The amplitudes of the fluctuating fields vary with the degree of order in the electronic spin system and are therefore quite temperature dependent. We outline the theory [37–41] of nuclear spin relaxation and line widths for the three main regions of interest in an antiferromagnet: (1) $T \gg T_N$, where high-temperature approximations for the local field spectra exist; (2) $T \ll T_N$, where spin wave theory may be employed; and (3) the region immediately above T_N. For simplicity only the isotropic hyperfine interaction is considered but generalization of the method is alluded to when appropriate.

Let $\langle S \rangle$ be the thermal average of S and the fluctuations in S be defined by

$$\delta S = S - \langle S \rangle \tag{5.1}$$

Then a time-dependent perturbation $\mathcal{H}' = A\mathbf{I} \cdot \delta \mathbf{S}$ of the nuclear spin system will result from the isotropic hyperfine interaction. Let z be the direction of quantization for the nuclear and electronic spin systems. A formal prescription exists [42] for obtaining the NMR line profile $I(\omega)$ and the nuclear spin lattice relaxation time T_1 in terms of the time correlation functions $\langle \delta S_i(\tau) \delta S_i(0) \rangle^*$ of the spin fluctuation δS. If ω_0 is the nuclear Larmor frequency then

$$I(\omega) = \int_{-\infty}^{\infty} \exp\left[i\omega t - \Psi(t)\right] dt \tag{5.2}$$

with

$$\Psi(t) = \left(\frac{A}{\hbar}\right)^2 \int_0^t (t - \tau)[\langle\{\delta S^z(\tau)\delta S^z(0)\}\rangle \\ + \tfrac{1}{2} \exp(-i\omega_0 \tau) \langle\{\delta S^+(\tau)\delta S^-(0)\}\rangle]\, d\tau \tag{5.3}$$

---

* The operator product $S_i(\tau)S_i(0)$ is defined by

$$\Omega(\tau) \equiv S_i(\tau)S_i(0) = \exp(i\hbar^{-1}\tau\mathcal{H})\, S_i \exp(-i\hbar^{-1}\tau\mathcal{H})\, S_i$$

where $\mathcal{H}$ is the electronic Hamiltonian (4.1). The statistical average $\langle \Omega \rangle$ is given by

$$\langle \Omega \rangle = \frac{\text{Trace}\,[\exp(-\mathcal{H}/kT)\,\Omega]}{\text{Trace}\,[\exp(-\mathcal{H}/kT)]}$$

For an elementary discussion of correlation functions see reference [2].

where the { } indicates that symmetrized operator products are to be taken. T_1 is given by

$$\frac{1}{T_1} = \frac{1}{2}\left(\frac{A}{\hbar}\right)^2 \int_{-\infty}^{\infty} \cos \omega_0 t \, \langle\{\delta S^+(t)\delta S^-(0)\}\rangle \, dt \qquad (5.4)$$

which is just the Fourier component at the resonance frequency of those elements of the fluctuating field that are transverse to the quantization direction.

In principle, a knowledge of the correlation functions in the appropriate temperature region would provide an explicit solution to the problem but, in general, Eq. (5.2) is too unwieldy to be useful. Fortunately a considerable simplification of (5.3) may be had when the characteristic decay time τ_c of the correlation function is such that

$$\omega_0 \tau_c \ll 1 \qquad (5.5)$$

that is to say, all correlation in the electronic spin system persists for times short compared with the nuclear Larmor period. $\Psi(t)$ simplifies to

$$\Psi(t) = \left(\frac{A}{\hbar}\right)^2 |t| \int_0^\infty \langle\{\delta S^z(\tau)\delta S^z(0)\}\rangle + \tfrac{1}{2}\langle\{\delta S^+(\tau)\delta S^-(0)\}\rangle \, d\tau \qquad (5.6)$$

the Fourier transform of which is a Lorentzian-shaped line profile whose half-width $\Delta\omega_{1/2}$ may be expressed as

$$\Delta\omega_{1/2} \equiv \frac{1}{T_2} = \frac{1}{T_2'} + \frac{1}{T_1'} \qquad (5.7)$$

where

$$\frac{1}{T_2'} = \left(\frac{A}{\hbar}\right)^2 \int_0^\infty \langle\{\delta S^z(\tau)\delta S^z(0)\}\rangle \, d\tau \qquad (5.8)$$

and

$$\frac{1}{T_1'} = \frac{1}{2}\left(\frac{A}{\hbar}\right)^2 \int_0^\infty \langle\{\delta S^+(\tau)\delta S^-(0)\}\rangle \, d\tau \qquad (5.9)$$

are the secular and nonsecular contributions to the line width, respectively. In the limit set by Eq. (5.5) $1/T_1$ reduces to

$$\frac{1}{T_1} = \left(\frac{A}{\hbar}\right)^2 \int_0^\infty \langle\{\delta S^+(\tau)\delta S^-(0)\}\rangle \, d\tau \qquad (5.10)$$

so that

$$\frac{1}{T_1} = 2\left(\frac{1}{T_1'}\right) \qquad (5.11)$$

It remains then to find the explicit dependence of the correlation functions on temperature and the parameters that characterize the magnetic order.

a. $T \gg T_N$. At temperatures such that $T \gg T_N$ the local-field spectra may be assumed to have a Gaussian distribution centered about zero frequency: it is found that*

$$\langle\{\delta S^i(t)\delta S^i(0)\}\rangle = \frac{S(S+1)}{3} \exp\left(-\tfrac{1}{2}\omega_e^2 t^2\right) \qquad (5.12)$$

where $i = x, y, z$ and $\omega_e^2 = (J/\hbar)^2 ZS(S+1)$. Equations (5.7)–(5.11) then yield

$$\frac{1}{T_1} = \frac{1}{T_2} = (2\pi)^{1/2}\left(\frac{A}{\hbar}\right)^2 \frac{S(S+1)}{3} \frac{1}{\omega_e} \qquad (5.13)$$

The simplicity of Eq. (5.13) depends upon the isotropy of A. Otherwise $1/T_2 \neq 1/T_1$, except for possibly special orientations of the field relative to the axes of the hyperfine interaction tensor. In any case the magnitudes of $1/T_1$ and $1/T_2$ will not be altered appreciably. The effectiveness of exchange in lengthening T_2 is to be noted since in its absence $1/T_2 \approx AS/\hbar$, whereas with exchange $1/T_2 \approx (AS/\hbar)/(AS/\hbar\omega_e)$.

There is reasonable agreement with experiment in the few cases that have been studied, two examples of which are given in Table IV.

TABLE IV

Examples of the Hyperfine-Broadened Exchange-Narrowed NMR in Antiferromagnets with $T \gg T_N$[a]

	Crystal	$\omega_e(\text{sec}^{-1})$	Nucleus	$1/T_2(\text{sec}^{-1})$ Theory[b]	Exp.[c]
(a)	MnF_2	1.6×10^{12}	F^{19}	1.35×10^6	0.95×10^6
(b)	CoO	1.5×10^{13}	Co^{59}	3×10^6	0.86×10^6

[a] Since only the line width is measured in the experiments and not T_1 the theoretical results are given for $1/T_2$.
[b] Moriya [38].
[c] (a) Shulman and Jaccarino [42a]; (b) Shulman [42b].

* At high temperatures $\delta\mathbf{S} \simeq \mathbf{S}$. Note that as $t \to 0$ Eq. (5.12) reduces to the familiar result $\langle(S^i)^2\rangle = \tfrac{1}{3}S(S+1)$.

b. $T \ll T_N$. At those temperatures, in the ordered state, for which a spin wave description of the thermodynamic properties is adequate, the local-field spectra again may be calculated. A sketch of the theory will be given, emphasizing three points: (1) the strong angular dependence of $1/T_1$ and not of $1/T_2$, (2) the T^3 dependence of both T_1 and T_2 in the small **k** limit for $T_{AE} \ll T \ll T_N$, and (3) the marked decrease of $1/T_1$ and $1/T_2$ in the region $T < T_{AE}$ when anisotropy is present. Again we will consider only the isotropic hyperfine interaction.

Whereas, in the region $T \gg T_N$, the spectral density of the fluctuating local field is approximately independent of temperature for $\omega \lesssim \omega_e$, this is not the case well below T_N. Moreover, because of the directional character of the ordering, there is a pronounced anisotropy to the local field fluctuations.

Two types of processes will contribute to nuclear relaxation as a result of the perturbation $\mathscr{H}' = A\mathbf{I} \cdot \delta\mathbf{S}$: (1) direct processes in which a nuclear spin is flipped and a spin wave of energy equal to the nuclear Zeeman energy $\hbar\omega_0$ is absorbed or emitted; and (2) Raman processes in which the nuclear spin flip is accompanied by the absorption of a spin wave of energy $\hbar\omega_k$ and the simultaneous emission of another spin wave of energy $\hbar\omega_{k'}$, such that $|\omega_k - \omega_{k'}| = \omega_0$. Since only at temperatures $T = \hbar\omega_0/k \lesssim 0.01\ ^\circ K$ are there spin waves of energy equal to the nuclear Zeeman energy, and then only in the absence of any anisotropy, the direct processes are unimportant. However, for the Raman processes all possible spin wave modes, consistent with energy and momentum conservation, can participate.

The components of $\delta\mathbf{S}_i$ and $\mathbf{S}_i$ are related by $\delta S_i^\pm = S_i^\pm$ and $\delta S_i^z = S_i^z - \langle S_i^z \rangle \simeq S_i^z - S_i$. From expressions (A-20)–(A-22), which express the components of $\mathbf{S}_i$ in terms of the spin wave creation and destruction operators α_k, α_k^*, β_k, and β_k^*, we see that $\delta S_i^\pm$ are linear in the α_k, α_k^*, etc., and therefore would only contribute to the direct processes, in lowest order. Fortunately, δS_i^z is bilinear in the spin wave operators and can therefore induce Raman processes, provided that the direction of antiferromagnetic alignment $\pm z$ is not collinear with the direction of nuclear quantization ζ. The latter qualification is merely an expression of the fact that only the components of the fluctuating local field transverse to the nuclear quantization direction are effective in causing relaxation.

Let θ be the angle between the z- and ζ-directions. The portion of $\mathscr{H}'$ that induces nuclear spin transitions by Raman scattering of spin waves is, from Eq. (A-22),

$$\mathscr{H}_i' = \frac{A\sin\theta}{N}(I_i^+ + I_i^-)\sum_{k,k'}\exp[i(\mathbf{k}-\mathbf{k'})\cdot\mathbf{r}_i](u_k u_{k'}\alpha_k^*\alpha_{k'} + v_k v_{k'}\beta_k\beta_{k'}^*) \tag{5.14}$$

Using time-dependent perturbation theory to calculate the transition probability $W = 1/2T_1$ and taking the small-**k** limit for $\hbar\omega_k$ (see Section IV), we have for $1/T_1$ [37–40]

$$\frac{1}{T_1} = \sin^2\theta \left[\frac{(A\Omega)^2 \eta^4 (S+1)^4}{81\pi^3 \hbar b^3 k T_n}\right] \left(\frac{T}{T_N}\right)^3 \int_{T_{AE}/T}^{\infty} \frac{x\,dx}{e^x - 1} \qquad (5.15)$$

where Ω = atomic volume, $b = \frac{1}{3}r_{nn}^2$ for a cubic crystal, and $\eta = 3kT_N/2JZS(S+1)$ and is of order unity. The temperature dependence of $1/T_1$ from Eq. (5.15) is such that

$$\frac{1}{T_1} = C \sin^2\theta \left(\frac{T}{T_N}\right)^3 \quad \text{for} \quad T \gg T_{AE} \qquad (5.16)$$

and

$$\frac{1}{T_1} = C \sin^2\theta \left(\frac{T_{AE}}{T_N}\right)\left(\frac{T}{T_N}\right)^2 \exp(-T_{AE}/T) \quad \text{for} \quad T \ll T_{AE} \qquad (5.17)$$

where C is $\pi^2/6$ times the expression in the square brackets in Eq. (5.15). The pronounced effect of the energy gap kT_{AE} is manifest in Eq. (5.17).

Since the spectral density of the fluctuating field contributing to the Raman processes extends to frequencies $\omega \gg \omega_0$ the decay of the correlation functions is rapid and the condition (5.5) holds. From Eqs. (5.8), (5.9), and (5.11) and the knowledge that only the fluctuations in S_i^z induce Raman processes one finds, in the small-**k** approximation,

$$\frac{1}{T_2} = \left(\frac{1 + \cos^2\theta}{2}\right) C \int_{T_{AE}/T}^{\infty} \frac{x\,dx}{e^x - 1} \qquad (5.18)$$

Although $1/T_2$ and $1/T_1$ have the same temperature dependence, the former is only slightly dependent on θ in contrast to $1/T_1$.

For many antiferromagnets $\theta = 0$ and no contribution to $1/T_1$ arises from Raman process in lowest order. One may still wonder if any appreciable contribution to $1/T_2$ is to come from Eq. (5.18). Even for those metal nuclei with large hyperfine interactions (e.g., Mn^{2+} or Co^{2+}) Raman processes are not effective in contributing to the line width in that temperature range for which a spin wave calculation is appropriate. For example, in MnF_2 even at temperatures as high as $T \simeq 2T_{AE}$, which, as we have noted in Section IV, was beyond the temperature limit for noninteracting spin wave theory, and far exceeds the region in which the small-**k** approximation is adequate, Eq. (5.18) yields $1/T_2 \simeq 10^4$ sec^{-1}. This corresponds to a line width $\delta H \simeq 1$ oe and, as we shall see later, this is three orders of magnitude smaller than the line width resulting from the indirect spin-spin interaction.

However, in some low-symmetry crystals $\theta \neq 0$. Moreover, if the externally applied field and the local field are comparable in magnitude (and not collinear), then an appreciable variation in θ can be realized. An example of this is the proton resonance [5–12] in $CuCl_2 \cdot 2H_2O$. Here the local field is dipolar in origin but this essentially only changes the coupling constants in Eq. (5.15). Indeed, the correct $\sin^2 \theta$ angular dependence is observed for $1/T_1$. Although there is qualitative agreement for the magnitude of $1/T_1$ the temperature dependence is quite different from that predicted by Eq. (5.15). It is observed that $1/T_1 \propto T^7$ in the same temperature region in which $[M(0) - M(T)]/M(0) \propto T^4$ (see Section IV). In Fig. 14 the experimental results on T_1 [10] are shown and a comparison is made with the small-$\mathbf{k}$ limit theory with, and without, anisotropy [37]. A theoretical result which gives good agreement with experiment for $1/T_1$ (and supposedly for $[M(0) - M(T)]/M(0)$) is also included in the figure. This curve was obtained by an arbitrary modifi-

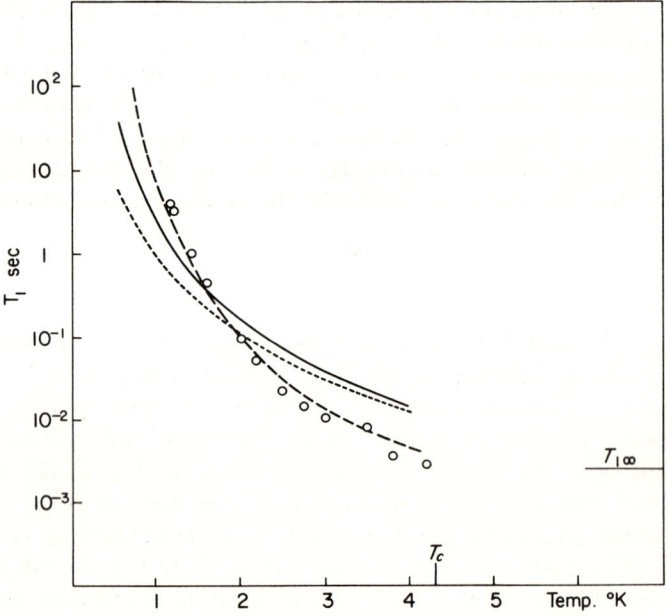

FIG. 14. The temperature dependence of T_1 for the proton NMR in $CuCl_2 \cdot 2H_2O$. The open circles are the experimental data (Hardeman *et al.* [10]). The solid and short-dash curves are those obtained from the long-wavelength spin wave theory [Eq. (5.15) with dipolar instead of hyperfine coupling constants] with and without anisotropy, respectively. The long-dash curve is that obtained from an arbitrary modification of the spin wave density of states (Moriya [37]).

cation of the spin wave spectrum [37]. At best, this approach is of heuristic value. It is to be emphasized again that since sizeable gaps are present in the spin wave spectrum of $CuCl_2 \cdot 2H_2O$ and the measurements of all experimental quantities have been made above (T_{AE})min the small-$\mathbf{k}$ limit of spin wave theory is inappropriate. More than likely good agreement will be found if zone-boundary effects and spin wave interactions are included.

c. Transition temperature region: $T - T_N \ll T_N$. As the transition temperature is approached from above, correlation in the electronic spin system increases, both in space and in time [41]. Spatial correlation may be expressed in terms of a range-dependent tensor susceptibility $\tilde{\chi}(\mathbf{r})$ or a wave number-dependent tensor susceptibility $\tilde{\chi}(\mathbf{k})$ related by the Fourier transform

$$\tilde{\chi}(\mathbf{k}) = \int \tilde{\chi}(\mathbf{r}) \exp(-i\mathbf{k} \cdot \mathbf{r}) \, d\mathbf{r} \tag{5.19}$$

For the antiferromagnet we are interested in the staggered susceptibility corresponding to values of $\mathbf{k} \geqslant \mathbf{K}_0$ where, for example, for a body-centered cubic lattice $\mathbf{K}_0$ is the reciprocal lattice vector $(2\pi/a, 0, 0)$. Thus, as in a ferromagnet where $\chi(0)$ diverges at $T = T_c$, so in an antiferromagnet $\chi(K_0)$ diverges at $T = T_N$. In what follows $\mathbf{k}$ will be measured from $\mathbf{K}_0$; i.e., $\mathbf{k} \to \mathbf{k} + \mathbf{K}_0$.

Let $\mathbf{H}_k$ be the kth Fourier component of the external field; then $\langle \mathbf{S}_k \rangle$ is defined by

$$\langle \mathbf{S}_k \rangle = (g\beta)^{-1} \tilde{\chi}(\mathbf{k}) \cdot \mathbf{H}_k \tag{5.20}$$

Time correlations in the fluctuations of the $\mathbf{S}_k$'s may be related to $\tilde{\chi}(\mathbf{k})$ by assuming a relaxation function

$$f_k(t) = \exp(-t/\tau_k) \tag{5.21}$$

that characterizes the stochastic motion in the electronic spin system. For example,

$$\langle \{\delta S_k^i(t) \delta S_{-k}^i(0)\} \rangle = \frac{kT}{g\beta^2} \chi^i(k) f_k(t) \tag{5.22}$$

where τ_k is the characteristic decay time of $\chi(k)$ and $i = x, y, z$. From Eqs. (5.12) and (5.22) we see that, at $t = 0$, $\chi^z(0) = (g^2\beta^2/3kT)S(S+1)$, which is the correct high-temperature limit for the uniform susceptibility. Though a molecular field treatment for $\chi(k)$ is adequate near T_N the calculation of τ_k is quite involved. We give certain approximate results for some special cases, again only considering the isotropic hyperfine interaction.

(1) *Cubic crystal; no magnetic field or anisotropy.* From Eqs. (5.4) and (5.22) (cubic symmetry, $\tilde{\chi}(k)$ a scalar quantity) one finds

$$\frac{1}{T_1} = \left(\frac{A}{\hbar}\right)^2 \left(\frac{kT}{g^2\beta^2}\right) \frac{1}{N} \sum_k \chi(k) \int_{-\infty}^{\infty} \cos(\omega_0 t) \exp(-|t|/\tau_k)\, dt \qquad (5.23)$$

which, in the extreme narrowing limit (5.5), reduces to

$$\frac{1}{T_1} = 2\left(\frac{A}{\hbar}\right)^2 \frac{kT}{Ng\beta^2} \sum_k \tau_k \chi(k) \qquad (5.24)$$

It is found [41] that $\sum_k \tau_k \chi(k) \sim (1/T)[T_N/(T-T_N)]^{1/2}$ as $T \to T_N$ and Eq. (5.22) becomes

$$\frac{1}{T_{1\infty}} = \frac{1}{T_{2\infty}} = C \frac{1}{T_{1\infty}} \left(\frac{T_N}{T-T_N}\right)^{1/2} \qquad (5.25)$$

where $1/T_{1\infty}$ is the high-temperature limit (5.13) and C is a lattice-dependent factor of order 10^{-1}. Equation (5.25) is valid in only a very limited temperature region,

$$\frac{\omega_0}{\omega_e} < \frac{T-T_N}{T_N} < 10^{-2} \qquad (5.26)$$

(2) *Cubic crystal; with magnetic field.* As is well known [17], the effect of a magnetic field H on an antiferromagnet in the temperature region about T_N is to reduce T_N by an amount proportional to H^2. The effect is quite anisotropic. We define two temperatures $T_\parallel$ and $T_\perp$ as those temperatures for which $\chi_\parallel(K_0)$ and $\chi_\perp(K_0)$, the susceptibilities parallel and perpendicular to H, respectively, diverge. In the molecular field approximation $T_\parallel$ and $T_\perp$ are related to T_N by

$$T_N - T_\parallel = 3(T_N - T_\perp) \qquad (5.27)$$

where the fractional change $[(T_N - T_\parallel)/T_N] \approx (H/H_E)^2$. The spin fluctuation correlation functions will no longer be the same parallel and perpendicular to H and therefore $T_1 \neq T_2$ at a given temperature. When proper account of this is taken one finds

$$\frac{1}{T_1} = C \frac{1}{T_{1\infty}} \left(\frac{T_\perp}{T-T_\perp}\right)^{1/2} \qquad (5.28)$$

and

$$\frac{1}{T_2} = \frac{C}{2} \frac{1}{T_{1\infty}} \left[\left(\frac{T_\|}{T - T_\|} \right)^{1/2} + \left(\frac{T_\perp}{T - T_\perp} \right)^{1/2} \right] \quad (5.29)$$

in the same region defined by Eq. (5.25). Note that this does not imply that either one is anisotropic.

(3) *Tetragonal crystal*; $H_A \gg H$. The anisotropy energy has a marked effect on the angular dependence of the spin fluctuation spectra in noncubic crystals. Consider the case of tetragonal symmetry with the unique (c) axis the easy axis of magnetization; e.g., MnF_2 and FeF_2. If the anisotropy energy is large compared to $g\beta H$ then $T_\| = T_N$ and $T_\perp < T_\|$. For the external field applied parallel and perpendicular to the c-axis, which determines the nuclear quantization direction as shown in Fig. 15, we find

$$\left(\frac{1}{T_2} \right)_\| = \frac{1}{2} \frac{C}{T_{1\infty}} \left[\left(\frac{T_N}{T - T_N} \right)^{1/2} + \left(\frac{T_\perp}{T - T_\perp} \right)^{1/2} \right] \quad (5.30)$$

and

$$\left(\frac{1}{T_2} \right)_\perp = \frac{1}{4} \frac{C}{T_{1\infty}} \left[\left(\frac{T_N}{T - T_N} \right)^{1/2} + 3 \left(\frac{T_\perp}{T - T_\perp} \right)^{1/2} \right] \quad (5.31)$$

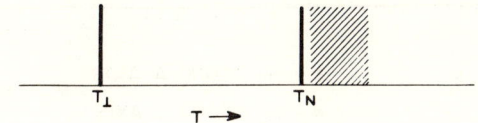

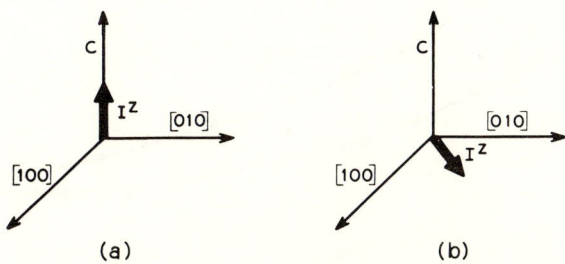

FIG. 15. The direction of nuclear quantization in a tetragonal crystal with H applied (a) parallel to the unique axis and (b) in the plane perpendicular to the unique axis. The temperature region of interest for measurements of the line width and relaxation time is indicated by the shaded area in the upper part of the figure.

all in the temperature region $T > T_N > T_\perp$. Now when $(T - T_N) \ll (T - T_\perp)$

$$\left(\frac{1}{T_2}\right)_\parallel = 2\left(\frac{1}{T_2}\right)_\perp \tag{5.32}$$

However, $1/T_1$ is considerably more anisotropic: we calculate

$$\left(\frac{1}{T_1}\right)_\parallel = \frac{C}{T_{1\infty}} \left(\frac{T_\perp}{T - T_N}\right)^{1/2} \tag{5.33}$$

and

$$\left(\frac{1}{T_1}\right)_\perp = \frac{1}{2} \frac{C}{T_{1\infty}} \left[\left(\frac{T_N}{T - T_N}\right)^{1/2} + \left(\frac{T_\perp}{T - T_\perp}\right)^{1/2}\right] \tag{5.34}$$

or

$$\left(\frac{1}{T_1}\right)_\perp = \left(\frac{1}{T_2}\right)_\parallel \tag{5.35}$$

Again in the region $(T - T_N) \ll (T - T_\perp)$

$$\left(\frac{1}{T_1}\right)_\perp \gg \left(\frac{1}{T_1}\right)_\parallel \tag{5.36}$$

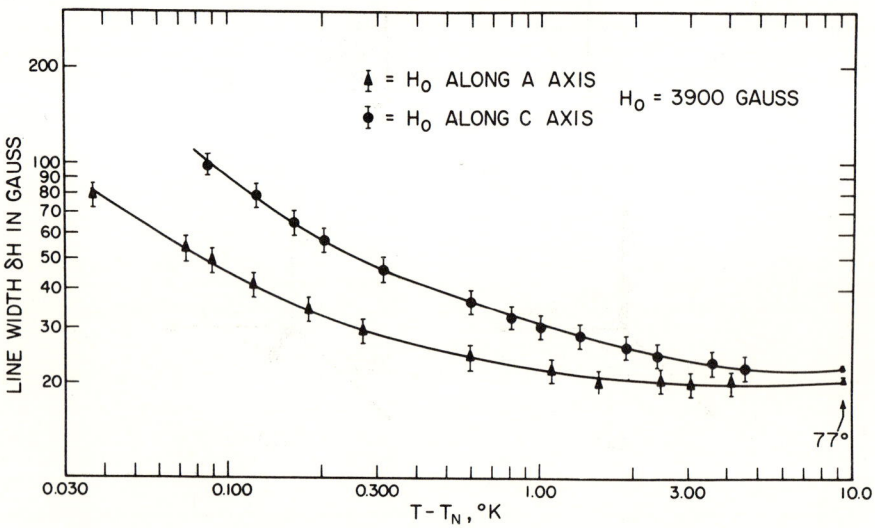

Fig. 16. A doubly logarithmic plot of the line width ($\sim 1/T_2$) of the F^{19} NMR in paramagnetic MnF_2 vs. temperature in the region close to the transition temperature. (From Heller and Benedek [28].)

All of the formulas given above are readily obtained from the basic expressions (5.7)–(5.11) and the assumption $\langle\{S^i(t)S^i(0)\}\rangle \sim [T_i/(T - T_i)]^{1/2}$ where T_i is the temperature associated with the singularity in $\chi^i(K_0)$.

The F^{19} NMR in MnF_2 has been studied in great detail in the region close to T_N. The experimental results with H parallel and perpendicular to the c-axis are shown in Fig. 16, where the logarithm of $\delta H \sim 1/T_2$ is plotted vs. the logarithm of $(T - T_N)$. Because of the anisotropy energy, $T_N - T_\perp = 1.36\ °K$, so that the temperature region $T - T_N \ll T_N - T_\perp$ may be explored with ease. As predicted by Eq. (5.32) it is

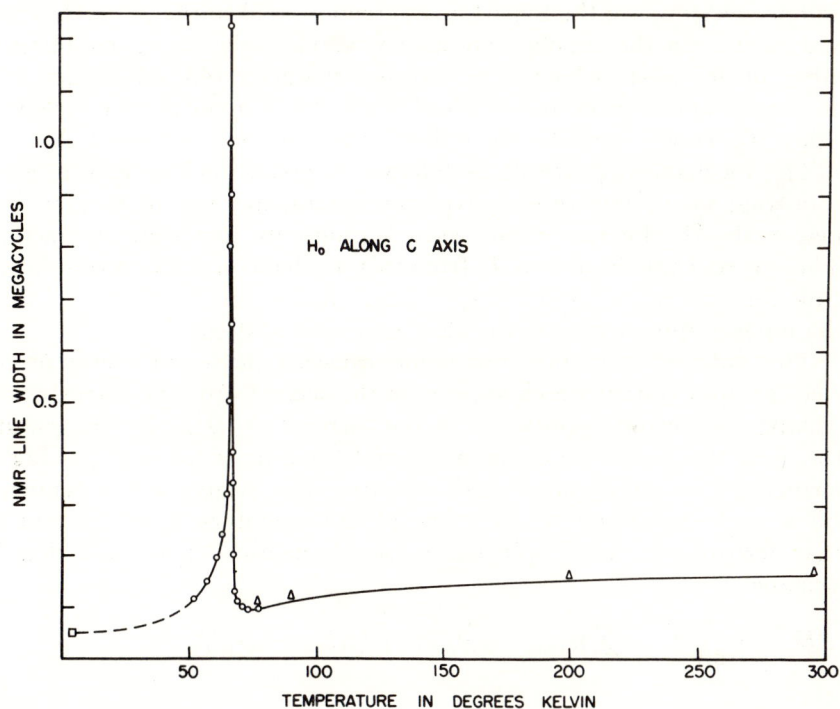

FIG. 17. The temperature dependence of the F^{19} NMR line width ($\sim 1/T_2$) in MnF_2 from well below to well above T_N. In the high-temperature region the line width is dominated by the exchange-narrowed direct hyperfine broadening while in the low-temperature region the indirect spin-spin interaction predominates. The effects of increased spatial and temporal correlation appear in the dramatic increase in the line width in the region immediate to the transition temperature, while it is believed that short-range order is responsible for the temperature dependence of δH between 77° and 300°K. The symbol ○ refers to the experimental results of Heller and Benedek [28]; △, Shulman and Jaccarino [42a]; and □, Jaccarino and Shulman [16].

seen then that $(1/T_2)_\parallel$ approaches $2(1/T_2)_\perp$ below $T - T_N \simeq 0.1\ °K$. However, only qualitative agreement with the magnitude and temperature dependence is found for $1/T_2$. It is not clear whether this is a result of the inadequacy of the assumptions as to the behavior of $\Sigma_k \chi(k)\tau_k$ near T_N or of the fact that the theory is inappropriate for the F^{19} NMR in MnF_2: the latter exhibits relatively large effects associated with short-range order which are not treated in the above theory [41] (see Fig. 17).

2. Indirect Nuclear Spin-Spin Interaction

In addition to the direct coupling of a nuclear moment to its own atomic electrons via the hyperfine interaction, and to the other electrons and nuclei via the dipolar interaction, a relatively strong coupling to other nuclei exists which is of indirect character [43, 44]. Since it is an important source of line width of the NMR in many antiferromagnets below T_N we will consider the indirect coupling in some detail.

The mechanism proceeds as follows: A given nucleus may excite a spin wave via its own atomic hyperfine interaction and another nucleus may reabsorb the same spin wave through its hyperfine interaction. There is no contribution to T_1 from this mechanism since in this virtual emission and reabsorption of spin waves no energy is transferred from the nuclear spin system to the electronic spin system.

The indirect spin-spin interaction causes a change in energy of the electron spin system which appears in the second order of perturbation. Taking the perturbation to be of the form $H' = A \sum_n^N \mathbf{I}_n \cdot \mathbf{S}_n$, noting that only the transverse components of H' will be effective in producing changes in the polarization in the electron spin system at low temperatures, and expressing H' in terms of the operators α, α^*, β, and β^* that destroy and create spin waves (see Appendix A), we find that H' becomes

$$H' = \left(\frac{S}{2N}\right)^{1/2} A \sum_k [(u_k I_{ik}^- + v_k I_{jk}^-)\alpha_k + (v_k I_{ik}^+ + u_k I_{jk}^+)\beta_k$$

$$+ (u_k I_{ik}^+ + v_k I_{jk}^+)\alpha_k^* + (v_k I_{ik}^- + u_k I_{jk}^-)\beta_k^*] \quad (5.37)$$

The change in energy, to second order, is

$$\Delta E = -\sum_k \frac{\langle 0 | H' | 1_k \rangle \langle 1_k | H' | 0 \rangle}{\hbar \omega_k} \quad (5.38)$$

where $|0\rangle$ is the ground state and $|1_k\rangle$ those spin wave states of wave number k removed in energy $\hbar\omega_k$ from $|0\rangle$ by excitation of a single

5. NUCLEAR RESONANCE IN ANTIFERROMAGNETS

spin wave. Since $\beta_k | 0 \rangle = \alpha_k | 0 \rangle = 0$ and $\beta_k^* | 0 \rangle = \alpha_k^* | 0 \rangle = | 1_k \rangle$ we find from Eqs. (5.37) and (5.38) that

$$\Delta E = -\tfrac{1}{2} A^2 S \frac{1}{N} \sum_k \frac{1}{\hbar\omega_k} [\tfrac{1}{2}(u_k^2 + v_k^2)(I_{ik}^+ I_{ik}^- + I_{ik}^- I_{ik}^+ + I_{jk}^+ I_{jk}^- + I_{jk}^- I_{jk}^+)$$
$$+ 2 u_k v_k (I_{ik}^+ I_{jk}^- + I_{ik}^- I_{jk}^+)] \qquad (5.39)$$

Omitted from Eq. (5.39) are terms, leading to a small shift in the resonance frequency, which arise from the virtual emission and reabsorption of spin waves by the same nucleus via the hyperfine coupling.

Only operators of the nuclear spin appear in Eq. (5.39) so that it is, in effect, a nuclear spin Hamiltonian $\mathscr{H}_N$ which can be utilized to calculate the various moments of the line profile [45]. For this purpose a further truncation of (5.39) may be made. The terms $I_{ik}^+ I_{jk}^-$ have a temporal dependence similar to that of $I_{ik}^+ I_{jk}^+$ and contribute only "satellite" lines at $\pm 2\omega_0$; that is to say, the local-field spectra that a nucleus on one sublattice produces at a nucleus on the other sublattice have mainly high-frequency components and therefore will not contribute to $1/T_2$. $\mathscr{H}_N$ simplifies then to

$$\mathscr{H}_N = -\tfrac{1}{2} A^2 S \sum_{j > j'} \tfrac{1}{2}(I_j^+ I_{j'}^- + I_j^- I_{j'}^+) \frac{1}{N} \sum_k \frac{u_k^2 + v_k^2}{\hbar\omega_k} \exp[i\mathbf{k} \cdot (\mathbf{r}_j - \mathbf{r}_{j'})]$$
$$(5.40)$$

where $\mathbf{r}_j - \mathbf{r}_{j'}$ is a lattice vector connecting two spins I_j and $I_{j'}$ on the same sublattice. A quadrupolar-like $[\sim(I_j^z)^2]$ term has been omitted from H_N which is important for $I > \tfrac{1}{2}$.

The rate at which the coupling between two spins a distance R apart diminishes is governed by the range function

$$F(R) = \frac{1}{N} \sum_k \frac{u_k^2 + v_k^2}{\hbar\omega_k} \exp(i\mathbf{k} \cdot \mathbf{R}) \qquad (5.41)$$

Because the contributions to Eq. (5.41) for large R (i.e., $R \gg a$, the lattice spacing) will come from long-wavelength spin waves the small-k approximation may be made for $\hbar\omega_k$ and the sum replaced by an integral. It is found then that

$$F(R) \sim \frac{a}{R} \exp\left[-\left(\frac{H_A}{H_E}\right)^{1/2} \frac{R}{a}\right] \qquad (5.42)$$

Thus for a typical antiferromagnet where $H_A/H_E \sim 0.1$ to 0.01, nuclei as far as 3 to 10 lattice spacings away will interact strongly through the indirect coupling.

Perhaps more interesting is the magnitude of the coupling. Since for small k, $u_k^2 + v_k^2 \approx (H_E/2H_A)^{1/2}$ and $\hbar\omega_k \approx g\beta(2H_AH_E)^{1/2}$, then from Eq. (5.40)

$$|\mathcal{H}_N| \approx \frac{A^2}{\hbar\omega_E} \qquad (5.43)$$

An evaluation of the second moment $\overline{(\Delta\omega)^2}$ of the Gaussian line profile yields

$$\hbar^2\overline{(\Delta\omega)^2} = \tfrac{1}{3}I(I+1)\left(\frac{A^2S}{\hbar\omega_E}\right)^2 \frac{1}{N}\sum_k\left(\frac{\omega_E}{\omega_k}\right)^4 \qquad (5.44)$$

For $\omega_A \ll \omega_E$, $(1/N)\sum_k (\omega_E/\omega_k)^4 \simeq (1/\pi)(\omega_E/2\omega_A)^{1/2}$; hence the frequency width is

$$[\overline{(\Delta\omega)^2}]^{1/2} \simeq \frac{1}{\hbar}\left[\frac{I(I+1)}{3\sqrt{2\pi}}\right]^{1/2}\left(\frac{A^2S}{\hbar\omega_E}\right)\left(\frac{\omega_E}{\omega_A}\right)^{1/4} \qquad (5.45)$$

It is interesting to note that apart from the factor $(\omega_E/\omega_A)^{1/4}$ the right-hand side of Eq. (5.45) is the magnitude of the line width in the high-temperature limit in the paramagnetic state (5.13). Certain qualifications need to be given concerning this remark. The indirect coupling is only effective between (1) nuclei with identical gyromagnetic ratios and (2) nuclei on the same sublattice. Thus if one could vary the isotopic abundance of the nuclei under consideration then the line width obtained from Eq. (5.44) would change whereas that obtained from (5.13) would not. It is also to be seen that a negligible temperature dependence is to be expected for the indirect spin-spin interaction for temperatures such that $[M(0) - M(T)]/M(0) < 0.1$ since the interaction takes place in the ground state of the antiferromagnet.

In the event that symmetry allows for an axial electric quadrupole interaction [see Eq. (2.5)] and

$$|E_Q(I_z+1) - E_Q(I_z)| \gg [\hbar^2\overline{(\Delta\omega)^2}]^{1/2} \qquad (5.46)$$

where the right-hand side of the inequality would be (5.44) in the absence of the quadrupole interaction, then the indirect spin-spin interaction will lead to a dependence of the line width on I_z. It is found [46] that

$$[\overline{(\Delta\omega)^2}]^{1/2}_{I_z \leftrightarrow I_z+1} \sim [(I-I_z)^2(I+I_z+1)^2 + 2(2I_z+1)^2] \qquad (5.47)$$

The line widths of the F^{19}, Mn^{55}, and Co^{59} NMR have been examined at low temperatures and a comparison between theory and experiment is given in Table V. The results encourage one to believe that indirect processes are the main source of line widths in these antiferromagnets.

TABLE V

THE LINE WIDTHS OF THE NMR RESULTING FROM THE INDIRECT SPIN-SPIN
INTERACTION IN ANTIFERROMAGNETIC MnF_2 AND CoF_2

		Line width (oe)	
Crystal	Nucleus[a]	Exp.	Theory
MnF_2	F^{19}	14[b]	13.6[c]
	Mn^{55}	~1000[d]	>600[c]
CoF_2	Co^{59}		
	$1/2 \leftrightarrow -1/2$	268[e]	268[f]
	$\left\{\begin{array}{l}3/2 \leftrightarrow 1/2\\-3/2 \leftrightarrow -1/2\end{array}\right\}$	258	244
	$5/2 \leftrightarrow 3/2$	222	185
	$7/2 \leftrightarrow 5/2$	165	127

[a] The theoretical results on the Co^{59} NMR are arbitrarily normalized to the $(\frac{1}{2} \leftrightarrow \frac{1}{2})$ line width observed and it is only the I_z dependence that is to be considered, although the order of magnitude of coupling constant is as predicted.
[b] Jaccarino and Shulman [16].
[c] Nakamura [44].
[d] Jones and Jefferts [15a].
[e] Jaccarino [14].
[f] Suhl [46].

VI. Summary

We have examined, in some detail, two of the many topics in antiferromagnetism that can be further elucidated via nuclear resonance studies. It appears that in the antiferromagnet, as in the ferromagnet, a complete spin wave theory including zone-boundary effects and spin wave interactions gives an excellent description of the low-temperature thermodynamic properties, in particular, the temperature dependence of the sublattice magnetization. The line width of the NMR is seen to be dominated, in many instances, by the indirect nuclear spin-spin interaction via virtual excitation and reabsorption of spin waves at low temperatures. The situation is less clear with respect to nuclear spin lattice relaxation and more work is needed in this field.

Only in MnF_2 has the region about the critical point been studied in great detail. Since this is a region of considerable theoretical interest and is one for which NMR techniques are eminently suitable, it would seem that further studies are desireable. Such studies would serve to complement neutron scattering and electron resonance measurements.

The behavior of line widths and relaxation times of the NMR at temperatures well above the transition temperature appear to be adequately described by a high-temperature expansion of the correlation function using a Gaussian model for the local-field spectra. There are indications that the effects of short-range order modify the spectra at somewhat lower temperatures and clearly more experimental and theoretical work is called for in this region.

Appendix A. Spin Waves on a Two-Sublattice Antiferromagnet

Since recourse is made more than once in the text to describing the individual electron spin operators in terms of the operators that annihilate and create spin waves, we give a brief treatment of the subject here [17–20], valid for temperatures much below T_N.

For simplicity we consider an antiferromagnet with uniaxial anisotropy which may be divided into two sublattices of "up" (i) and "down" (j) spins such that the Z nearest neighbors (nn) to a given spin on the ith sublattice are all spins on the jth sublattice (and conversely) and only nn exchange interactions (assumed to be equal) are nonvanishing. The Hamiltonian for this system is, in the absence of an applied field,

$$\mathscr{H} = -2J \sum_{i>j} \mathbf{S}_i \cdot \mathbf{S}_j - \tfrac{1}{2} K \sum_{i,j} \{(S_i^z)^2 + (S_j^z)^2\} \qquad \text{(A-1)}$$

where i and j are lattice points on the respective sublattices, K the anisotropy constant, and J the exchange integral between two spins. To cast Eq. (A-1) in spin wave form several steps need to be taken.

Since at low temperatures the spins on the two sublattices spend most of their time in the states $S_i^z = S$ and $S_j^z = -S$, respectively, it follows that the expectation values of the spin deviation operators, $S - S_i^z$ and $S_j^z - S$, of the two sublattices will be small. This fact is utilized in spin wave theory insofar as it is possible to show that the low-lying spin states are indistinguishable from those of a harmonic oscillator.

Defining the raising and lowering operators for the individual spins on the two sublattices in the usual fashion,

$$S_{i,j}^+ = S_{i,j}^x + iS_{i,j}^y; \qquad S_{i,j}^- = S_{i,j}^x - iS_{i,j}^y \qquad \text{(A-2)}$$

an exact representation of the spin operators in terms of Bose creation and destruction operators may be given:

$$\begin{aligned}
S_i^+ &= (2S - a_i^* a_i)^{1/2} a_i, & S_j^+ &= b_j^*(2S - b_j^* b_j)^{1/2} \\
S_i^- &= a_i^*(2S - a_i^* a_i)^{1/2}, & S_j^- &= (2S - b_j^* b_j)^{1/2} b_j \\
S_i^z &= S - a_i^* a_i, & S_j^z &= -S + b_j^* b_j
\end{aligned} \qquad \text{(A-3)}$$

where the a_i and $a_i{}^*$ are operators which create and destroy spin deviations at the ith spin on the "up" sublattice. The $a_i{}^*$, a_i, $b_j{}^*$, and b_j satisfy the bose commutation relations

$$a_i a_{i'}^* - a_{i'}^* a_i = \delta_{ii'}; \qquad a_i{}^* a_{i'}^* - a_{i'}^* a_i{}^* = 0 \qquad a_i a_{i'} - a_{i'} a_i = 0 \qquad \text{(A-4)}$$

and similarly for the $b_j{}^*$ and b_j operators. Of course, all a_i and all b_j operators commute.

From the last of the relations (A-3) it is seen that $\langle a_i{}^* a_i \rangle$ and $\langle b_j{}^* b_j \rangle$ are the expectation values of the spin deviation at the ith and jth sites on the two sublattices. The factors $(2S - a_i{}^* a_i)^{1/2}$ and $(2S - b_j{}^* b_j)^{1/2}$ occurring in (A-3) reflect the difference between a spin and a harmonic oscillator. The latter quantities vanish in the state for which $\langle a_i{}^* a_i \rangle = 2S$ or $S_i{}^z = -S$ and thus prevent the raising and lowering operators from going outside the manifold of $2S + 1$ spin states. The initial linearization of the spin operator problem comes about if we notice that when $\langle a_i{}^* a_i \rangle$ is small it is reasonable to make the approximation

$$(2S - a_i{}^* a_i)^{1/2} \to (2S)^{1/2} \qquad \text{(A-5)}$$

In this "harmonic oscillator" approximation the raising and lowering operators are then proportional to the Bose creation and destruction operators of the spin deviation since Eq. (A-4) reduces to

$$\begin{aligned} S_i{}^+ &= (2S)^{1/2} a_i, & S_j{}^+ &= (2S)^{1/2} b_j{}^* \\ S_i{}^- &= (2S^{1/2}) a_i{}^*, & S_j{}^- &= (2S)^{1/2} b_j \\ S_i{}^z &= S - a_i{}^* a_i, & S_j{}^z &= -S + b_j{}^* b_j \end{aligned} \qquad \text{(A-6)}$$

To completely linearize the problem it is required in substituting the operators (A-6) into the original Hamiltonian (A-1) that we neglect terms that involve products of Bose operators such as arise from considering $(S_i{}^z)^2$.

Next, to obtain spin wave operators, we introduce the Fourier transforms of the creation and destruction operators of spin deviations

$$\begin{aligned} a_k &= N^{-1/2} \sum_i a_i \exp(-i\mathbf{k} \cdot \mathbf{r}_i), & b_k &= N^{-1/2} \sum_j b_j \exp(i\mathbf{k} \cdot \mathbf{r}_j) \\ a_k{}^* &= N^{-1/2} \sum_i a_i{}^* \exp(i\mathbf{k} \cdot \mathbf{r}_i), & b_k{}^* &= N^{-1/2} \sum_j b_j{}^* \exp(-i\mathbf{k} \cdot \mathbf{r}_j) \end{aligned} \qquad \text{(A-7)}$$

where $\mathbf{r}_i$ and $\mathbf{r}_j$ are lattice vectors of the ith and jth site on the appropriate sublattices and the values of $\mathbf{k}$ are the allowed wave vectors in the first Brillouin zone of the reciprocal lattice obtained from considerations of the primitive cell containing two magnetic ions. Then using Eqs. (A-6) and (A-7) we may write (A-1) as

$$\mathcal{H}_0 = -2JZS \sum_k \left[\left(1 + \frac{K}{2JZ}\right)(a_k^* a_k + b_k^* b_k) + \gamma_k(a_k^* b_k^* + a_k b_k)\right] \tag{A-8}$$

where

$$\gamma_k = \frac{1}{Z} \sum_{nn}^{Z} \exp(i\mathbf{k} \cdot \mathbf{r}_{nn}) \tag{A-9}$$

and $\mathbf{r}_{nn}$ is the vector connecting a given spin with its nearest neighbor. Clearly $\mathcal{H}_0$ is not diagonal in the $a_k b_k$ representation.

The canonical transformation which diagonalizes $\mathcal{H}_0$ is

$$a_k = u_k \alpha_k + v_k \beta_k^*, \qquad b_k = v_k \alpha_k^* + u_k \beta_k$$
$$a_k^* = u_k \alpha_k^* + v_k \beta_k, \qquad b_k^* = v_k \alpha_k + u_k \beta_k^* \tag{A-10}$$

with $u_k = \cosh \theta_k/2$, $v_k = -\sinh \theta_k/2$, and θ_k defined by

$$\tanh \theta_k = -\frac{\gamma_k}{[1 + (K/2JZ)]} \tag{A-11}$$

and

$$u_k^2 - v_k^2 = 1 \tag{A-12}$$

In terms of the orthogonalized spin wave operators α_k, α_k^*, β_k, and β_k^*, Eq. (A-1) becomes, apart from a constant,

$$\mathcal{H}_0 = \sum_k \tfrac{1}{2}\hbar\omega_k(\alpha_k^*\alpha_k + \alpha_k\alpha_k^* + \beta_k^*\beta_k + \beta_k\beta_k^*) \tag{A-13}$$

Since the spin wave operators obey the Bose commutation relations $\alpha_k \alpha_k^* - \alpha_k^* \alpha_k = 1$, etc., we find that if we define $n_k' \equiv \beta_k^* \beta_k$ then

$$\mathcal{H}_0 = \sum_k \hbar\omega_k[(n_k' + \tfrac{1}{2}) + (n_k'' + \tfrac{1}{2})] \tag{A-14}$$

The zero-point energy E_{zero} arising from the noncommuting properties of the spin operators comes from the terms in Eq. (A-14) that are independent of n_k' and n_k''

$$E_{zero} = \sum_k \hbar\omega_k \tag{A-15}$$

For small anisotropy, i.e., $K \ll 2JZ$, the dispersion relation for spin waves (or "magnons") is given by

$$\hbar\omega_k = 2JZS\left(1 - \gamma_k^2 + \frac{K}{JZ}\right)^{1/2} \tag{A-16}$$

Since for $k = 0$, $\gamma_k^2 = 1$ the initial gap in the spin wave spectrum is

$$\hbar\omega_0 = 2S(JKZ)^{1/2} \equiv kT_{AE} \qquad (A\text{-}17)$$

or, expressed in terms of the effective exchange and anisotropy fields, H_E and H_A,

$$H_E = \frac{2ZJS}{g\beta}; \qquad H_A = \frac{KS}{g\beta} \qquad (A\text{-}18)$$

or

$$\hbar\omega_0 = g\beta(2H_A H_E)^{1/2}; \qquad \hbar\omega_A = g\beta H_A; \qquad \hbar\omega_E = g\beta H_E \qquad (A\text{-}19)$$

Finally from Eqs. (A-6), (A-7), and (A-10) we obtain

$$S_i^+ = 2\left(\frac{S}{N}\right)^{1/2} \sum_k \exp(-i\mathbf{k}\cdot\mathbf{r}_i)(u_k\alpha_k + v_k\beta_k^*) \qquad (A\text{-}20)$$

$$S_i^- = 2\left(\frac{S}{N}\right)^{1/2} \sum_k \exp(i\mathbf{k}\cdot\mathbf{r}_i)(u_k\alpha_k^* + v_k\beta_k) \qquad (A\text{-}21)$$

$$S_i^z = S - \frac{1}{N}\sum_k\sum_{k'} \exp[i(\mathbf{k}-\mathbf{k}')\cdot\mathbf{r}_i](u_k\alpha_k^* + v_k\beta_k)(u_{k'}\alpha_{k'} + v_{k'}\beta_{k'}^*) \qquad (A\text{-}22)$$

$$S_j^+ = 2\left(\frac{S}{N}\right)^{1/2} \sum_k \exp(i\mathbf{k}\cdot\mathbf{r}_j)(v_k\alpha_k + u_k\beta_k^*) \qquad (A\text{-}23)$$

$$S_j^- = 2\left(\frac{S}{N}\right)^{1/2} \sum_{kk} \exp(-i\mathbf{k}\cdot\mathbf{r}_j)(v_k\alpha_k^* + u_k\beta_k) \qquad (A\text{-}24)$$

$$S_j^z = -S + \frac{1}{N}\sum_k\sum_{k'} \exp[i(\mathbf{k}-\mathbf{k}')\cdot\mathbf{r}_j](v_k\alpha_k + u_k\beta_k^*)(v_{k'}\alpha_{k'}^* + u_{k'}\beta_{k'}) \qquad (A\text{-}25)$$

Appendix B. The Isotropic Hyperfine Interaction Expressed in Terms of the Spin Wave Operators

The hyperfine interaction Hamiltonian

$$\mathscr{H}_{\text{hf}} = A\mathbf{I}\cdot\mathbf{S} = AI_z S_z + \frac{A}{2}(I^+S^- + I^+S^-) \qquad (B\text{-}1)$$

may be cast in spin wave form using Eqs. (A-20)–(A-25). Let $I_{i,j}^{\pm j} = I_{i,j}^{x} \pm iI_{i,j}^{y}$ be the analogous raising and lowering nuclear spin operators for nuclei of magnetic ions on the $i(j)$ sublattice and

$$I_i^{\pm} = N^{-1/2} \sum_k \exp(-i\mathbf{k} \cdot \mathbf{r}_i) I_{k,i}$$

and $\quad I_j^{\pm} = N^{-1/2} \sum_k \exp(+i\mathbf{k} \cdot \mathbf{r}_j) I_{k,j} \quad$ (B-2)

be the same operators expressed in terms of their Fourier transforms. Then

$$\mathscr{H}_{\text{hf}} = AS(I_i^z - I_j^z) + \frac{A\sqrt{S}}{2} \sum_k [I_{k,i}^-(u_k\alpha_k + v_k\beta_k^*) + I_{k,i}^+(u_k\alpha_k^* + v_k\beta_k)$$

$$+ I_{k,j}^+(v_k\alpha_k + u_k\beta_k^*) + I_{k,j}^-(v_k\alpha_k^* + u_k\beta_k)] \quad \text{(B-3)}$$

where only terms that are linear in the spin wave operators have been retained.

Appendix C. Importance of Spin Wave-Phonon Interactions

We have seen that the existence of an energy gap kT_{AE} in the spin wave spectrum of an antiferromagnet severely attenuates all magnetic processes that depend on thermal excitations, for temperatures $T \ll T_{AE}$. Since the phonon dispersion relation has no gap in its spectrum, $E \sim |\mathbf{k}|$ for small $\mathbf{k}$, any appreciable coupling between spin wave and phonon modes introduces spin wave character into the phonons and thereby allows the phonons to induce magnetic processes directly [47]. This pseudo-spin wave character of the phonons will be particularly important for $T \ll T_{AE}$.

The interaction between spin waves and phonons arises from the magnetostrictive coupling constant G and may be expressed as

$$\mathscr{H}_{\text{sw-ph}} = G \sum_{i=1}^{N} (S_i^x S_i^z \epsilon_i^{xz} + S_i^y S_i^z \epsilon_i^{yz}) \quad \text{(C-1)}$$

where ϵ_i^{xz} and ϵ_i^{yz} are the components of the strain tensor. By expressing the spin operators and components of the strain tensor in terms of the operators that annihilate and create spin waves and phonons, respectively, a virtual spin wave character is impressed on the thermal phonons. From the thermodynamics of the mixed spin wave-phonon modes one obtains the following results for the fractional change in the magnetization and the direct and Raman nuclear relaxation rates, for $T \ll T_{AE}$:

$$\frac{M(0) - M(T)}{M(0)} = K_1 \left(\frac{G}{\omega_A}\right)^2 \left(\frac{T}{\theta_D}\right)^4 \quad \text{(C-2)}$$

$$\left(\frac{1}{T_1}\right)_{\text{dir.}} = K_2 \left(\frac{G}{\omega_A}\right)^2 A^2 \left(\frac{T}{\theta_D}\right) \quad \text{(C-3)}$$

and

$$\left(\frac{1}{T_1}\right)_{\text{Ram.}} = K_3 \left(\frac{G}{\omega_A}\right)^4 A^2 \left(\frac{T}{\theta_D}\right)^7 \quad \text{(C-4)}$$

where K_1, K_2, and K_3 are different constants dependent on the Debye temperature θ_D, S, the ionic mass, and the lattice constant.

Unfortunately, for the three antiferromagnets that have been investigated below T_{AE}, namely, MnF_2, FeF_2, and CoF_2, an exponential rather than a power law behavior has been found for $[M(0) - M(T)]/M(0)$. In addition, for MnF_2, an unreasonably large value of G must be assumed [47] to obtain qualitative agreement for the magnitude of the change in M in that region.

Although an apparent T^7 behavior for $1/T_1$ has been observed for the proton NMR in $CuCl_2 \cdot 2H_2O$ (see Section V) it is only so in the region $T > T_{AE}$ where presumably "pure" spin wave processes predominate over virtual spin wave–thermal magnon processes.

Acknowledgments

It is with great pleasure that we here acknowledge private communications and discussions with A. C. Gossard, M. E. Lines, P. Pincus, and H. Suhl. To L. R. Walker the author owes a great debt for holding his hand many times in some of the dark corridors of theory. L. N. Finnie has helped to edit the manuscript and proofs.

References

1. A. Abragam, "The Principles of Nuclear Magnetism." Oxford Univ. Press (Clarendon), London and New York, 1961.
2. C. P Slichter, "Principles of Magnetic Resonance." Harper & Row, New York, 1963.
3. T. P. Das and E. L. Hahn, "Nuclear Quadrupole Resonance Spectroscopy." Academic Press, New York, 1958.
4. N. J. Poulis, *Physica* **17**, 392 (1951).

5. N. J. Poulis and G. E. G. Hardeman, *Physica* **18**, 201 (1952).
6. N. J. Poulis and G. E. G. Hardeman, *Physica* **18**, 315 (1952).
7. N. J. Poulis, G. E. G. Hardeman, and B. Bolger, *Physica* **18**, 429 (1952).
8. N. J. Poulis and G. E. G. Hardeman, *Physica* **19**, 391 (1953).
9. N. J. Poulis and G. E. G. Hardeman, *Physica* **20**, 7 (1954).
10. G. E. G. Hardeman, N. J. Poulis, and W. van der Lugt, *Physica* **22**, 48 (1956).
11. G. E. G. Hardeman, N. J. Poulis, W. van der Lugt, and W. P. A. Haas, *Physica* **23**, 907 (1957).
12. N. J. Poulis, G. E. G. Hardeman, W. van der Lugt, and W. P. A. Haas, *Physica* **24**, 280 (1958).
13. A. J. Freeman and R. E. Watson, this volume, p. 167.
14. V. Jaccarino, *Phys. Rev. Letters* **2**, 163 (1959).
14a. V. Jaccarino, *J. Chem. Phys.* **30**, 1627 (1959).
15. W. J. O'Sullivan, W. W. Simmons, and W. A. Robinson, *Phys. Rev. Letters* **10**, 476 (1963).
15a. E. D. Jones, and K. Jefferts, *Phys. Rev.* **135**, A1277 (1964).
16. V. Jaccarino and R. G. Shulman, *Phys. Rev.* **107**, 1196 (1957).
16a. A. Narath, *Phys. Rev.* **131**, 1929 (1963).
16b. J. C. Burgiel, V. Jaccarino, and A. L. Schawlow, *Phys. Rev.* **122**, 429 (1961).
17. T. Nagamiya, K. Yosida, and R. Kubo, *Advan. Phys.* **4**, 1 (1955).
18. J. Van Kranendonk and J. H. Van Vleck, *Rev. Mod. Phys.* **30**, 1 (1958).
19. F. Keffer, *in* "Handbuch der Physik" (S. Flügge, ed.), Springer, Berlin (to be published).
20. L. R. Walker, *in* "Magnetism" (G. T. Rado and H. Suhl, eds.), Vol. 1, Chapt. 8. Academic Press, New York, 1963.
21. G. G. Low, *Proc. Phys. Soc. (London)* **82**, 992 (1963).
22. J. Owen, M. R. Brown, and B. A. Coles, *J. Phys. Soc. Japan* **17**, Suppl. BI, 428 (1962).
23. M. R. Brown, B. A. Coles, J. Owen, and R. W. H. Stevenson, *Phys. Rev. Letters* **7**, 246 (1961).
24. C. Trapp and J. W. Stout, *Phys. Rev. Letters* **10**, 157 (1963).
25. F. M. Johnson and A. H. Nethercott, *Phys. Rev.* **114**, 705 (1959).
26. V. Jaccarino and L. R. Walker, *J. Phys. Rad.* **20**, 341 (1959).
27. V. Jaccarino and L. R. Walker, unpublished data, 1959.
28. P. Heller and G. B. Benedek, *Phys. Rev. Letters* **8**, 428 (1962).
29. M. E. Lines, Private communication—to be published.
30. T. Oguchi, *Phys. Rev.* **117**, 117 (1960).
31. A. Honma, *J. Phys. Soc. Japan* **15**, 456 (1960). References are given in this article to earlier theoretical and experimental work on FeF_2.
32. R. C. Ohlman and M. Tinkham, *Phys. Rev.* **123**, 425 (1961).
33. J. A. Eisele and F. Keffer, *Phys. Rev.* **96**, 929 (1954).
34. C. Domb, *Advan. Phys.* **9**, 149, 245 (1960).
35. C. N. Yang, *Phys. Rev.* **85**, 808 (1952).
36. M. E. Fisher and R. J. Burford, to be published. Result quoted by M. E. Fisher at *Int. Conf. on Magnetism, Nottingham, England 1964*.
37. T. Moriya, *Progr. Theoret. Phys. (Kyoto)* **16**, 23 (1956).
38. T. Moriya, *Progr. Theoret. Phys. (Kyoto)* **16**, 641 (1956).
39. J. Van Kranendonk and M. Bloom, *Physica* **22**, 545 (1956).
40. A. H. Mitchell, *J. Chem. Phys.* **27**, 59 (1957).
41. T. Moriya, *Progr. Theor. et. Phys. (Kyoto)* **28**, 371 (1962).
42. R. Kubo and K. Tomita, *J. Phys. Soc. Japan* **6**, 888 (1954).

42a. R. G. Shulman and V. Jaccarino, *Phys. Rev.* **108**, 1219 (1957).
42b. R. G. Shulman, *Phys. Rev. Letters* **2**, 459 (1959).
43. H. Suhl, *Phys. Rev.* **109**, 606 (1958); *J. Phys. Radium* **20**, 333 (1959).
44. T. Nakamura, *Progr. Theoret. Phys.* (*Kyoto*) **20**, 542 (1958).
45. J. H. Van Vleck, *Phys. Rev.* **74**, 1168 (1948).
46. H. Suhl, Private communication, 1959.
47. P. Pincus and J. Winter, *Phys. Rev. Letters* **7**, 269 (1961); private communications with P. Pincus, 1964.

6. Nuclear Resonance in Ferromagnetic Materials

A. M. Portis

University of California,
Berkeley, California

R. H. Lindquist

California Research Corporation,
Richmond, California

I. Introduction	357
II. Radio-Frequency Excitation	358
1. Apparatus	359
2. Intensity of Absorption	360
3. Domain-Wall Displacement	362
III. Field Contributions	363
1. Theory	363
2. Experiments in Magnetic Insulators	364
3. Experiments in Metals	368
4. Experiments in Alloys and Intermetallic Compounds	373
IV. Nuclear Relaxation	376
1. Spin-Spin Interactions	377
2. Spin-Lattice Interactions	379
References	380

I. Introduction

Nuclear resonance in ferromagnetic materials elucidates the static and dynamic nuclear hyperfine coupling in ordered magnetic systems. The magnitude and direction of the nuclear hyperfine field, its temperature and pressure dependence, and the field inhomogeneities resulting from impurity substitution are examples of the static magnetic information gained from resonance. The dynamic interactions of the nuclear spins are observed by the characteristic relaxation times for spin-spin and spin-lattice coupling. Spin diffusion and spin wave processes may be observed as a function of excitation intensity.

Neutron scattering studies supplement the results of resonance studies. Chapter 3 in Volume III compares the two techniques; Chapters 4 and 5 of Volume III interpret the results of neutron scattering studies in ionic structures and metals. Another powerful method for determination of hyperfine interactions is the observation of recoil-free gamma radiation, the Mössbauer effect [1, 2]. See also Chapter 4 of this volume.

Nuclear resonance studies in ferromagnetic materials received their initial stimulus from N. Kurti's low-temperature nuclear orientation studies. Kurti and his collaborators first examined nuclear hyperfine coupling in ferromagnetic materials by studying nuclear orientation in a Co^{60} single crystal [3]. This work was followed by heat capacity measurements of cobalt metal below 4.2 °K [4]. Stimulated by Kurti's interest, a successful search was made for the nuclear magnetic resonance of Co^{59} in finely divided particles of fcc cobalt [5].

This chapter will summarize the main experimental and theoretical work done through 1963 on nuclear resonance in ferromagnetic materials. The nuclear resonance studies in antiferromagnetics are described in Chapter 5 of this volume; in Chapter 4 the detailed theory of the hyperfine field is explained. The emphasis in this chapter is on resonance excitation and nuclear relaxation in ferromagnetic materials.

II. Radio-Frequency Excitation

The presence of strong interactions among electrons in ordered magnetic materials permits a considerable simplification in the theoretical treatment of the nuclear-electron coupling. Since the energy associated with the nuclear-electron interaction is very much smaller than that associated with the electron-electron interaction, we may think of the nucleus as coupled to a time-average field produced by electron spin and orbital currents [6]. The nuclear energy may then be written as

$$E = m\mu_n H_n/I, \tag{2.1}$$

where m is the magnetic quantum number of the nucleus, μ_n is the nuclear moment, I is the nuclear spin, and H_n is the field established by the electron magnetization. We can expect to induce nuclear resonance when the quantum energy in the radio-frequency field $\hbar\omega$ is equal to the energy splitting, $\Delta E = \mu_n H_n/I$. The transition probability may be computed from conventional resonance theory [7], and it is found that for the Co^{59} resonance the fractional energy absorption should be about 10^{-5}. The first experiments [5] gave nearly unit absorption, almost five orders of magnitude greater than computed from conventional theory. The reason for the tremendous enhancement is that the resonance is not driven directly by the external rf field, but instead indirectly through the nuclear-electron hyperfine coupling.

In the limit that the electron and nuclear spins are decoupled, the effect of a transverse rf field is to turn the electron magnetization M through an angle H_x/H_a, thus establishing a transverse magnetization

$$M_x = M \frac{H_x}{H_a}, \qquad (2.2)$$

where H_a is the static anisotropy field acting on the electrons and H_x is the external rf field. The nuclei then see a transverse field

$$2H_1 = H_x + \left(\frac{M_x}{M}\right) H_n = \left(1 + \frac{H_n}{H_a}\right) H_x. \qquad (2.3)$$

The transition probability, and thus the absorption rate, will then be enhanced by the factor $(1 + H_n/H_a)^2$; for fcc cobalt metal H_n is 217 koe and H_a is 1 koe, giving an enhancement factor of 4.7×10^4.

The enhancement of a nuclear transition probability by a factor of this order of magnitude is also seen in the excitation of hyperfine transitions of free atoms. In molecular beam experiments [8], the absorption is enhanced by the ratio of the electronic to the nuclear moment squared, analogous to the ordered magnetic system expression.

1. Apparatus

The observation of nuclear resonance in zero applied field requires the use of swept-frequency spectrometers, rather than the swept field, fixed-frequency spectrometers used for nonferromagnetic materials. For continuous wave excitation, marginal oscillators of the type used for nuclear quadrupole resonance studies [9] have been adapted for use at elevated temperatures [10], low temperatures [11], and high pressures [12]. The sample to be observed is placed in a coil or cavity constituting part of the oscillator tank circuit. Detection of power absorption is by the conventional resonance technique of modulating the oscillator frequency and amplifying the modulation envelope via a lock-in amplifier to present the first derivative of the resonance line.

Pulsed spectrometers have proved useful for high-sensitivity and nuclear relaxation studies. Free-induction decay signals have been observed [13] and superregenerative techniques used [14] for weak resonance absorption. Spin-echo pulse techniques [15] provide a method for observing nuclear relaxation [16].

In the spin-echo technique, nuclear precession is established by a short intense rf pulse. After the nuclear spins have dephased a second rf pulse is applied to the system. The spins then come back into coherence and produce an echo. From a determination of the intensity of the echo, processes which destroy spin coherence (spin-spin) may be detected. The technique may also be used as a way of measuring the magnitude of the nuclear magnetization and thus studying longitudinal (spin-lattice) relaxation processes.

2. Intensity of Absorption

Nuclear resonance excitation in ferromagnetic materials takes place indirectly through the response of the sample magnetization rather than by the rf field acting directly on the nuclear moments. For this reason the character of the resonant response depends on the magnetization processes in the material. In order to obtain a high degree of penetration by the rf field, most of the resonance studies on metals have been done on small particles prepared either by reduction of the oxide powder or by grinding. Additional studies on films, whiskers, and fine chemisorbed metal particles will be described in the following sections.

Metal particles produced by grinding or by reduction of oxides are generally multidomain in character. Special precautions must be taken to prevent fine metal particles from sintering to multidomain size. For a metal particle to be single domain, it must be less than several hundred Angstroms in diameter, while particles of ferromagnetic insulators need only be less than several thousand angstroms in diameter [17] to be single domain.

In multidomain ferromagnetic insulators and single-domain metal particles, the principal magnetization process at radio frequencies is by domain rotation [14, 18, 19]. In metal multidomain particles, however, the principal magnetization process is the displacement of domain walls; the walls move until a reverse demagnetizing field is established which is equal to the applied field. Thus, only those nuclear spins actually within domain walls will be excited since the domain volume is shielded from rf excitation. Since the theory of the dynamic nuclear-electron coupling is simpler for domain rotation, single-domain particles will be discussed first.

a. *Single-domain particles.* For single-domain ferromagnetic particles or for larger particles magnetically saturated by an applied field, we can expect to describe the sample response in terms of a uniform magnetization **M**. We may then regard the nuclear magnetization as uniform and represent it by the vector **m**. The transverse parts of the nuclear and electronic magnetizations are given by the expressions

$$\mathbf{m}_t = \chi_n \cdot (\mathbf{H}_t + H_n \mathbf{M}_t/M), \qquad (2.4)$$

$$\mathbf{M}_t = \chi_e \cdot (\mathbf{H}_t + H_n \mathbf{m}_t/M), \qquad (2.5)$$

where the χ are susceptibility tensors and the subscript t indicates the transverse components. The rate of energy absorption is given by

$$P = \tfrac{1}{2} \operatorname{Re} \{i\omega \mathbf{H}_t \cdot (\mathbf{M}_t + \mathbf{m}_t)\}, \qquad (2.6)$$

where ω is close to the resonance frequency ω_0 defined by the relation $\hbar\omega_0 = \mu_n H_n/I$. By separating the electron and nuclear magnetizations and retaining only the leading terms, we obtain for the absorption rate

$$P = \tfrac{1}{2}\omega\chi_e''[1 + 2\chi_e'(H_n/M)^2\chi_n']H_x^2 + \tfrac{1}{2}\omega\chi_n''(\chi_e'H_n/M)^2 H_x^2. \qquad (2.7)$$

We have developed the above expression in order to demonstrate that the resonance losses involve not only the imaginary part of the nuclear susceptibility χ_n'' but also the real part χ_n' through a modulation of the electronic losses. Because of the enhancement of the driving field by the factor $\chi_e' H_n/M$, the nuclear absorption χ_n'' will become saturated at very low external rf fields in the 1/10-oersted range. At normal rf levels only the real part of the susceptibility will remain, and we may characterize the resonance signal by a fractional variation in the electronic losses

$$\Delta P/P = 2\chi_n'\chi_e'(H_n/M)^2. \qquad (2.8)$$

Not only does the enhancement tremendously increase the observed signal, but it also makes possible certain transient experiments with very much less power than would otherwise be required. Spin-echo studies [16] can be easily performed with quite short pulses. Another striking result has been the observation of magnetic resonance excitation in the excited nuclear state of Fe^{57} [20].

b. Multidomain particles. As discussed, the only nuclei of multi-domain particles that respond to an rf field are the nuclei in the domain walls. A calculation of domain-wall enhancement and fractional wall volume is necessary to determine the intensity of wall absorption.

Döring [21] has shown that the variation in magnetization through a moving domain wall, with damping neglected, is the same as for the wall at rest. This result makes it possible to characterize a wall by a single parameter, its position. For small-amplitude motion we make the assumption that damping will not seriously modify the magnetization within the wall so that we may write an equation of the form

$$\mu d^2z/dt^2 + \beta dz/dt + \alpha z = 2MH_x, \qquad (2.9)$$

where μ is called the wall mass, β is the damping constant, α is the stiffness constant, and z is the direction normal to wall. The field, H_x, is applied parallel to the wall. The term on the right side plays the role of a pressure on the wall. We may treat the nuclear interaction as an additional pressure on the wall. Carrying the leading terms, the expression for the rate of energy absorption is

$$P = \omega\beta(1 + 2w'/\alpha)(2\omega M^2/\alpha^2)H_x^2 + w''(2\omega M^2/\alpha^2)H_x^2, \qquad (2.10)$$

where w is given by:

$$w = 2H_n^2\chi_n \int (d\theta/dz)^2 \, dz. \qquad (2.11)$$

In order to compare the above result with that for single-domain particles, we imagine a spherical particle of diameter d split by a single-domain wall in which the magnetization turns uniformly through an angle of π radians in a distance δ—the wall thickness. Then the rf field seen by the nuclei in the wall, which is enhanced by a factor $2\sqrt{2}MH_n(d\theta/dz)/\alpha$, is about an order of magnitude larger than in single-domain particles. At moderately high rf fields, when the absorption is saturated, the fractional variation in losses is given by

$$\Delta P/P = 2\chi_e'\chi_n'(H_n/M)^2(\pi^2/3)d/\delta. \qquad (2.12)$$

The fractional losses are greater for wall motion than for domain rotation by a factor $(\pi^2/3)d/\delta$, which is of the order of 50.

3. Domain-Wall Displacement

a. *Static fields.* In a static external field the nuclear resonance signal decreases much more rapidly than does the permeability of the particles [5]. This drop has been attributed to the sweeping of domain walls out of the sample by the static field [23]. An interesting study has been made of the effect of a static field on the nuclear resonance intensity in oriented whiskers [24]. It is found that the resonance signal drops much more rapidly with the static field along the axis than perpendicular to the axis. This behavior has been interpreted in terms of the sweeping of walls out of the whiskers by an axial field.

b. Pulsed fields. A direct demonstration that the nuclear resonance in ferromagnetic metal particles arises from nuclei in walls comes from pulsed wall displacement spin-echo experiments [16, 25]. In a typical spin-echo experiment a pair of rf pulses separated by a time τ is applied, and an echo is observed at a time 2τ after the first pulse. Since the echo is produced by coherent interaction between the pulses, it is clear that both pulses must be applied to the same nuclei. It is found that if a dc field pulse is applied at the same time as the second rf pulse the echo is reduced in intensity. The rate of decrease is consistent with domain-wall displacement by the dc field accompanying the second pulse such that different nuclei are driven by the two pulses.

III. Field Contributions

1. Theory

The detailed discussion of hyperfine fields in metals is given in Chapter 4 of this volume; Watson and Freeman [26] have also summarized the situation in insulators. We give here a brief outline of the sources contributing to the hyperfine field in magnetic materials as a basis for the following discussion of resonance experiments. Three sources of hyperfine field are the local field, the core field, and the conduction electron field. Only the local and core fields are significant for magnetic insulators; all three sources are present in ferromagnetic metals.

a. Local field. We include in the local field those contributions to the nuclear hyperfine field which arise from moments outside the central atomic cell. These are the static field, the demagnetizing field, and the Lorentz field. In multidomain particles the first two contributions will generally cancel as a result of shielding. We will therefore not expect the resonance frequency of nuclei in walls to be shifted by an external field, although there may be some line broadening from the component of the field along the direction of the magnetization within the walls. For single-domain particles the frequency should shift linearly with applied field for fields in excess of the anisotropy field H_a. In the case of ferrimagnetic single-domain materials, the applied field will add to one magnetic sublattice, while subtracting from the other.

b. Core field. The dominant contributions to the hyperfine field in metals and insulators arise from core electrons. Three distinguishable core field contributions are:

(1) The field from polarized core s electrons. The core s electron wave functions are distorted by the exchange potential associated with their

interaction with the d electrons. Since the exchange potential is spin-dependent the up and down s electron spins will be distorted differently, resulting in a net spin density. The sign of the spin density at the nucleus is opposite to that of the d electrons, producing a negative hyperfine field. This interaction provides the dominant contribution to the hyperfine field in both metals and insulators.

(2) The field from the residual *orbital* moment associated with the d electrons. The magnitude of this field can be estimated from the known orbital contribution to the magnetic moment. The rare earth ions, except those with half-closed shells, and Fe^{2+} and Co^{2+}, have large unquenched orbital moments that contribute a positive field at the nucleus.

(3) The *dipolar* field resulting from a noncubic spin density of the d electrons. In effect, this term introduces the anisotropy of the site symmetry [27] and results in line broadening for noncubic magnetic insulators.

c. **Conduction electron field.** Four kinds of conduction electron contributions may be distinguished:

(1) Distortion of the s conduction electron wave functions. The exchange interaction with the core d electrons may produce a small and positive field at the nucleus.

(2) The exchange interaction between the s conduction electrons and the d core electrons should also result in a magnetization of the s band and a positive contact field at the nuclei in the metal.

(3) We can also expect some *admixture* of s electron wave function into the d band, providing an additional positive contribution.

(4) The covalent mixing of the $4s$-$3d$ electrons results in an *antiferromagnetic polarization* [28] which may nearly cancel the admixture contribution to the hyperfine field.

2. **Experiments in Magnetic Insulators**

The results of nuclear resonance studies in magnetic insulators are given in Table I, together with the derived information on hyperfine fields at the magnetic nuclei. For discussion purposes, these materials are divided into three categories: garnets and ferrites, iron oxides, and ferromagnetic insulators. Chapter 1 of Volume III describes the crystal structures necessary for background knowledge in the following discussion.

a. **Garnets and ferrites.** The sublattice magnetizations of the rare earth garnets have been extensively investigated by resonance techniques

TABLE I

NUCLEAR RESONANCE DATA IN MAGNETIC INSULATORS

Nucleus	Host	Temp. (°K)	Frequency (Mc/sec)	Field (| koe |)	Comment	Ref.
Cr^{53}	$CrCl_3$	0	63.246	262.87	Hexagonal layer compound, weak antiferromagnetic coupling between layers	[34]
	$CrBr_3$	0	58.096	241.46	Center of triplet, spacing 295 kc/sec	[42]
			56.9	237	Domain-wall nuclear signal	
	CrI_3	4	49.35	205.11	—	[47]
	CrO_2	0	36.65	152.3	$\Delta \nu$ 200 kc/sec, possible sublines	[46]
Mn^{55}	$MnFe_2O_4$	4.2	585.1	554.5	—	[35]
Fe^{57}	α-Fe_2O_3	273	71.5	525	Intensity drops rapidly below 263° K and disappears at 259° K (Morin transition)	[40] [41] [39]
	Fe_3O_4	300	67.621	491	Tetrahedral site	[32]
		300	63.549	462	Octahedral site	[36]
	$Eu_3Fe_5O_{12}$	303	54.76	398	Tetrahedral site	[32]
			67.84	493	Octahedral site	
	$Gd_3Fe_5O_{12}$	303	54.69	397	Tetrahedral site	[32]
	$Tb_3Fe_5O_{12}$	300	54.80	—	Tetrahedral site	[32]
	$Dy_3Fe_5O_{12}$	303	53.93	392	Tetrahedral site	[49]
			67.29	489	Octahedral site	
	$Ho_3Fe_5O_{12}$	303	53.74	390	Tetrahedral site	[32]
			67.15	488	Octahedral site	
	$Er_3Fe_5O_{12}$	303	53.56	389	Tetrahedral site	[32]
			66.94	486	Octahedral site	
	$Tm_3Fe_5O_{12}$	303	53.31	387	Tetrahedral site	[32]
			66.79	485	Octahedral site	
	$MnFe_2O_4$	296	54.18	329	—	[50]
			58.78	429	—	
	$NiFe_2O_4$	300	67.5	490	Tetrahedral site	[51]
			72.0	525	Octahedral site	
	$Li_{0.5}Fe_{2.5}O_4$	298	68.94	501	Tetrahedral site[a]	[48]
			70.77	514	Octahedral site[a]	[52]
	$Y_3Fe_5O_{12}$	77	64.47	468	Tetrahedral site	[32]
			75.71	550	Octahedral site	
		300	54.2	395	Tetrahedral site	[53]
			67.4	485	Octahedral site	

[a] Same positions in ordered and disordered state.

TABLE I (continued)

Nucleus	Host	Temp. (°K)	Frequency (Mc/sec)	Field (| koe |)	Comment	Ref.
Fe^{57}	$Sm_3Fe_5O_{12}$	303	54.98	399	Tetrahedral site	[32]
			67.97	494	Octahedral site	
	$Yb_3Fe_5O_{12}$	303	52.97	385	Tetrahedral site	[32]
			66.54	483	Octahedral site	
	$Lu_3Fe_5O_{12}$	303	52.82	384	Tetrahedral site	[32]
			66.43	483	Octahedral site	
Eu^{151}	EuS	0	340	342	Δf 1500 kc/sec	[18]
Eu^{153}	EuS	0	151	342	Five peaks, 450 kc apart, $\sim$6 Mc for quadrupole interaction	[18]

[29–32]. Two resonances are observed for Fe^{57} in yttrium-iron garnet, arising from Fe^{+3} ions in octahedral sites (a sites) with hyperfine field of 485 koe and in tetrahedral sites (d sites) with hyperfine field of 395 koe [53]. Superexchange interactions via the intervening oxygen atoms order the a and d sites antiparallel; thus the effect of an applied field is to add to the hyperfine field of one sublattice and decrease the hyperfine field of the other as found in single-crystal experiments [33]. Watson and Freeman [26] have shown that the observed hyperfine field at the a site agrees with their free ion calculation for the field produced by spin polarization of the core s electrons; the field at the d sites is reduced by covalent interaction with oxygen valence electrons.

The intensity of the resonance decreases as the square of the applied field strength, in agreement with Eq. (2.7) [30]. The temperature dependence of the nuclear resonance frequency follows the observed spontaneous magnetization [29].

We give two examples of nuclear resonance studies of the crystal-field interactions in garnets: (1) The temperature dependence of the sublattice magnetizations in yttrium-iron garnet and gadolinium-iron garnet may be compared by observing the shift in nuclear resonance frequency [27]. For the d sites the resonance frequency difference shows an increase with increasing temperature, while the reverse is observed for a sites. The effect is interpreted as indicating that Gd^{3+} ions interact much more strongly with d sites than with a sites. (2) The resonance frequency of Fe^{57} ions in the rare earth–substituted garnets increases linearly from Lu to Sm [31, 32], paralleling the increase in the crystal lattice parameter. From Lu to Sm, the d-site Fe^{57} frequency increases by 4% at room temperature; the a-site Fe^{57} frequency increases by 2%.

The smaller increase in the octahedral a-site resonance frequency may be due to its covalent character.

The nuclear resonance of $(Mn^{55})^{2+}$ ions has been observed in $MnFe_2O_4$ using a double resonance technique [35]. The nuclear resonance is observed by applying a modulated rf signal and detecting the electron ferromagnetic resonance. Since the hyperfine interaction with the polarized nuclear spins contributes to the local electronic field, saturation of the nuclear spins produces a shift in the electron resonance. A single line observed with a half-width of 1.7 Mc/sec was assigned to Mn^{2+} ions on a sites, since d-site Mn^{2+} ions should exhibit five lines due to quadrupole splitting with nearest-neighbor oxygen ions.

b. Iron oxides. Magnetite, Fe_3O_4, has been the object of several nuclear resonance studies to investigate the sublattice magnetization and test several hypotheses of electronic conduction. As in the case of the garnets and ferrites previously described, the Fe^{57} resonance absorption occurs in the domains rather than in domain walls. The Fe^{3+} ions in tetrahedral sites produce a single sharp line at 67.6 Mc/sec at room temperature [36]. The Verway model [37] explains the high conductivity of cubic magnetite by a rapid interchange of electrons among the octahedral sites which are occupied by an equal number of Fe^{2+} and Fe^{3+} ions. The temperature dependence of the nuclear resonance frequencies of the tetrahedral and octahedral sites fit the Verway model when an electronic temperature-independent g value is assumed [38]. The observed broadening of the resonance lines as the temperature is lowered through the cubic-to-orthorhombic phase transition at 118 °K is attributed to two factors: the electron interchange frequency approaching the nuclear resonance frequency and the cell distortion of the tetrahedral sites [32, 38].

The nuclear resonance behavior of $(Fe^{57})^{3+}$ ions in α-Fe_2O_3 has been followed through the Morin transition (-14 °C) [39]. The weak ferromagnetism due to slight canting of the spins in the basal plane disappears as the spins orient antiferromagnetically along the c-axis. The verification of domain-wall excitation is found in the disappearance of resonance when the sample temperature is lowered through the Morin transition [40] or a small external field is applied [39]. The enhancement factor has been calculated to be 2.5×10^3 [41], typical of domain-wall excitation.

c. Ferromagnetic insulators. Several ferromagnetic insulators have been examined by nuclear resonance techniques. Since the dominant interactions are ferromagnetic rather than antiferromagnetic as in the ferrites, these materials are of particular interest for the theory of ferromagnetism. The most extensive studies have been made of $CrBr_3$ [42, 43]. The hyperfine field is negative in $CrBr_3$, as shown by a shift to lower frequencies of the Cr^{53} nuclear resonance with applied field. Of the four absorption lines observed in the frequency range of 30 to 200 Mc/sec, three were assigned to nuclei in the bulk of domains and arise from a quadrupole split triplet; the fourth line was assigned to nuclei in domain walls. At 1.4 °K the center of the triplet occurs at 58 Mc/sec; the wall resonance is found at 46.6 Mc/sec. The nuclear field differences in domains and walls fit the relation

$$H_n(\text{domain}) - H_n(\text{wall}) = 4450 + 1200T$$

from 0° to 8 °K.

The difference between domain and domain-wall magnetization of 4450 oe at 0 °K is attributed to differences in dipolar fields for M parallel and perpendicular to the c-axis and to a substantial anisotropic orbital field. The temperature-dependent decrease in wall magnetization is thought to arise from wall excitation and spin wave scattering [44, 45].

The Cr^{53} nuclear resonance in CrO_2 has been observed [46] from 77° to 240 °K. The temperature dependence of the frequency follows a T^2 law, with an extrapolated resonance frequency at 0 °K of 36.65 Mc/sec corresponding to a hyperfine field of 152.3 koe. Future studies of Cr^{53} resonance in ferromagnetic oxides, halides, and sulfides appear promising.

The only ferromagnetic sulfide studied by nuclear resonance has been EuS [18]. The Eu^{153} nuclear resonance observed at 4.2 °K consists of one broad line with five sharp lines superimposed on it. The five sharp lines are attributed to nuclei in domains, with quadrupole splitting producing the structure. The one broad line is probably associated with domain walls, as in the case of $CrBr_3$.

3. Experiments in Metals

The results of nuclear resonance studies in ferromagnetic metals, alloys, and intermetallic compounds are given in Table II. This section will discuss the effects of variation in physical parameters on the hyperfine field in pure ferromagnetic metals.

Relaxation studies have demonstrated that the line width in the iron group metals is associated with a static variation in the hyperfine field from site to site, rather than a dynamical interaction [44]. As discussed in Section II, 2, low rf power levels are necessary to observe the absorp-

6. NUCLEAR RESONANCE IN FERROMAGNETIC MATERIALS

TABLE II
Nuclear Resonance Data in Metals and Alloys

Nucleus	Host	Temp. (°K)	Frequency (Mc/sec)	Field (| koe |)	Ref.
B^{11}	MnB	300	32.4	23.7	[54]
			31.2	22.8	[54]
P^{31}	MnP	80	74.0	42.9	[54]
			69.7	40.4	[54]
V^{51}	Fe	273	94.7	84.6	[55]
Mn^{55}	MnB	300	217.7	206	[54]
			203.6	193	[54]
	Mn_4N	282	117.9	111.7	[56]
	MnP	82	143.4	136	[54]
			116.4	110.8	[54]
	Ni	298	309	293	[57]
	MnAs	77	235.08	222.7	[55]
	Mn_5Ge_3	90	201.8	191.2	[56]
		1.6	196–216	—	[58]
	Mn_2Sn	82	~250	237	[54]
	MnSb	273	234.37	222	[59]
	Mn_2Sb	82	126.26	119.6	[54]
			143.7	137.1	[54]
	MnBi	77	234.4	222.1	[54]
		196	225.0	213.3	[78]
	Cu_2MnAl	0	227.5	215.6	[60]
	Cu_2MnSn	0	247.0	234.1	[60]
Fe^{57}	Fe	0	46.65	339	[24]
		300	45.44	—	[29, 61, 62]
	Co	4.2	44.8	329	[63]
Co^{59}	Fe	0	289.2	286.3	[64–67]
	Co, fcc	0	217.2	215.0	[23]
	Co, hcp	0	228	226	[10, 68, 69]
	Ni	295	111.6	110.5	[64, 66, 70]
	Co-Pd	4.2	215.6	213.4	[71]
Ni^{61}	Fe	77	89.1	235	[73]
	Co	77	71.7	189	[70, 74]
	Ni	0	28.35	74.8	[75, 76]
		295	26.02	69	[65]
Cu^{63}	Fe	273	240.0	212.7	[68, 72]
	Co	282	117.9	157.5	[68, 72]
	Ni	290	53.27	47.2	[77, 78]
	Cu_2MnAl	0	242.0	229.3	[60]
	Cu_2MnIn	0	226.0	214.6	[60]
Cu^{65}	Fe	273	257.1	212.7	[72]
	Co	282	190.7	157.5	[72]
	Ni	290	—	47.2	[77]

TABLE II (continued)

Nucleus	Host	Temp. (°K)	Fequency (Mc/sec)	Field (| koe |)	Ref.
Cu^{65}	Cu_2MnAl	0	268.0	229.3	[60]
	Cu_2MnIn	0	241.8	214.6	[60]
As^{75}	MnAs	77	206.57	283.2	[79]
Sn^{117}	MnSn	77	196.9	124.8	[54]
	$Mn_{1.5}Sn$	77	322	200	[80]
Sn^{119}	MnSn	77	199.3	125	[54]
	$Mn_{1.5}Sn$	77	324	200	[80]
Sb^{121}	MnSb	273	358.2	352.5	[59]
	Mn_2Sb	196	230.7	228	[54]
Sb^{121}	MnNiSb	297	281.28	276.7	[81]
Sb^{123}	MnSb	273	194.80	352.5	[59]
	Mn_2Sb	196	125.7	228	[54]
	MnNiSb	297	152.70	276.7	[81]
Gd^{155}	GdN	2.2	44.204	~370	[82]
Gd^{157}	GdN	2.2	58.700	~370	[82]
Tb^{159}	Tb	77	3047	4000±1000	[83]

tion line shape. Increasing the rf power increases the dispersive character and broadens the line [76, 84].

a. **Sign of coupling.** It has been experimentally established by several techniques that the direction of the hyperfine field in the iron group metals is *opposite* to that of the electronic magnetization. The negative hyperfine field in cobalt metal was first established by nuclear resonance studies of single-domain particles [19]. The sense of the Co^{59} nuclear precession signal confirmed the negative hyperfine field [16]. A negative hyperfine field for Fe^{57} and Ni^{61} was first observed through Mössbauer studies which showed a reduction in the hyperfine splitting with the application of an external field [85, 86].

b. **Single-domain particles.** Nuclear resonance has been observed from single-domain fcc cobalt particles in the 100 Å to 150 Å diameter size range [19, 14]. The cobalt particles were dispersed throughout a porous alumina support, thus minimizing the particle-particle magnetic interaction. The Co^{59} resonance is observed at 218 Mc/sec, compared with the multidomain wall resonance at 213 Mc/sec. This zero-field shift arises from the single-particle demagnetizing field, which for spherical cobalt particles is approximately $-4\pi M/3 \simeq -5.6$ koe. The single-domain resonance is much weaker for the same particle losses, as expected from the analysis of Section II. In an applied field the single-domain resonance frequency decreases linearly at the rate of 1(Mc/sec)/koe for fields above the particle anisotropy field. Both the sign of the zero-field shift and the direction of frequency shift with applied field demonstrate that the hyperfine field of Co^{59} is negative.

Superparamagnetic behavior has been observed for cobalt particles in the 80–100 Å diameter size range [14]. The thermal fluctuations in alignment of magnetization become the principal nuclear relaxation mechanisms at a critical particle diameter and temperature, in accord with the Néel theory of critical particle diameter [87].

c. Crystal phases and defect structures. The effect of crystal phases on the hyperfine field is strikingly seen in the nuclear resonance studies of fcc and hcp cobalt. Near 0 °K the hexagonal resonance at 228 Mc/sec is 10.8 Mc/sec higher than the cubic resonance at 217.2 Mc/sec. At room temperature the difference has decreased to 7.9 Mc/sec [69, 88]. However, above the transition temperature of hcp cobalt (723 °K) the difference between hcp and fcc resonance is equal to 3.5 Mc/sec and is independent of temperature up to the maximum temperature measured, 923 °K [68].

F. Keffer [89] has suggested that the splitting may be associated with the deviation of the c/a ratio from the ideal value in hcp lattices and its approach toward the ideal value at elevated temperatures. Below 723 °K the temperature dependence of the splitting and the deviation of the c/a ratio appear to be in reasonable agreement. The constancy of the splitting above 732 °K suggests that there may be a systematic shift associated with the ideal hcp structure.

In addition to the two resonance lines for hcp and fcc cobalt discussed above, five more lines are observed in some cobalt samples [69, 90–92]. These lines have the dispersion derivative shape of the fcc resonance at high power levels [69] rather than the absorption derivative shape associated with the nuclei of the hcp phase, where the enhancement factor is evidently much smaller. An analysis of the types of stacking faults found in fcc structures, the change in density at these faults, and the calculated resonance frequencies for nuclei from such sites gives the following assignments at 27.4 °C [91]: slip along the $\langle 112 \rangle$ direction in the closed-packed plane, BCABCACABCAB ..., 215.5 and 218.4 Mc/sec; misstacked close-packed plane ABCABACABCA ..., 215.5 and 217.0 Mc/sec; and finally a twinning stacking fault, ABCABCBACBA ..., 215.5 and 217.0 Mc/sec. No adequate explanation for the resonance at 223.0 Mc/sec has been advanced.

The effects of strain on the Fe^{57} and Ni^{61} resonance have been studied by comparing the resonance in cold-rolled and annealed specimens.

A very weak resonance is observed for cold-rolled iron with a line width of several hundred kc/sec, compared with the 60 kc/sec line width of well-annealed specimens [24]. The resonance in nickel samples disappeared with cold working [76]. The role that volume imperfections play in the resonance line width is further seen in experiments on iron whiskers. Iron whiskers, which are substantially free of imperfections, yield a line width which is only one quarter of that obtained from micron-size particles [24].

d. *Temperature and pressure dependence.* The change of nuclear resonance frequency with temperature offers a sensitive way to obtain the temperature dependence of the magnetization. It was recognized by Robert and Winter [29] that such studies must be referred to constant volume to satisfy the relation

$$\nu = A\sigma, \qquad (3.1)$$

where A is the hyperfine coupling constant relating the nuclear resonance to the magnetization and σ is the specific magnetization, corrected for thermal expansion.

Table III summarizes the pressure studies on the nuclear resonance

TABLE III

Pressure Studies in Pure Metals

Nucleus	Host	$(\partial\nu/\partial P)_{20°C}$ (kc/kbar)	$(d\nu/dV)_{20°C}$	Max. pressure (kbar)	Ref.
Fe^{57}	bcc Fe	-7.67 ± 0.22	$+0.277\, \nu/V$	60	[93]
Co^{59}	fcc Co	128.0 ± 0.2	$-1.2\, \nu/V$	7.8	[94]
Co^{59}	hcp Co	134.3	—	10	[95]
Ni^{61}	bcc Ni	$+0.024$	—	—	[12]

frequencies of the iron group metals. For iron, after correcting ν and σ to constant volume, A is found to depend on temperature according to the following relation:

$$A = A_0(1 - 0.77 \times 10^{-7}\, T^2), \qquad (3.2)$$

where A_0 is the coupling constant at 0 °K [12]. For Ni^{61} A does not show an explicit temperature dependence larger than 0.5% (one-third that of iron as expressed by Eq. (3.2) [76]). Similarly, no temperature dependence was found for A in hcp Co^{59} up to 398 °K [95]. Thus, for nickel and hcp cobalt, the temperature dependence of the saturation magnetization can be directly found from the temperature dependence of the nuclear resonance at constant volume.

The temperature dependence of A for iron has been qualitatively interpreted as a d-band excitation effect [12]. As the temperature is increased, d electrons are excited to higher levels, resulting in a change in the average coupling constant.

A curious effect has been observed in the nuclear resonance of Co^{59} above 200 °C [10]. The resonance line due to the fcc phase splits into two lines at 235 °C; the resonance line due to the hcp phase shows a similar splitting above 260 °C. The fractional splitting as a function of temperature is the same for both phases. An attempt to interpret the splitting in terms of preferred orientations of spins within the domain wall [10] has not been successful.

4. Experiments in Alloys and Intermetallic Compounds

Nuclear resonance has been studied in a wide range of ferromagnetic alloys. Studies to date have used multidomain particles and the resonant nuclei have been in domain walls. It will be useful to study single-domain alloy particles where the magnetization direction is uniform for all nuclei so that the problem of nuclear spin anisotropy does not complicate the interpretation of the resonance line character. For purposes of dicussion, the alloy studies can be divided into three categories, based on whether the nuclei observed are (1) the host in a dilute alloy, (2) the solute in a dilute alloy, or (3) one of the components of a concentrated (greater than 4%) alloy.

a. Alloys. The effect of adding small amounts of a transition metal to an iron group host metal is to produce satellite lines in the wings of the central resonance line. These satellite lines have been interpreted both as to position and intensity by the following picture [92]. The added solute atom will have two effects on its neighboring host atoms. First, the magnetic moment of the solute atom will appear as an added dipolar field at the neighboring host nuclei; and, second, a pseudodipolar field will result from distortion of the neighboring host atom d shells by the unshielded charge on the solute atom. Analysis of the Co^{59} nuclear resonance spectra from dilute solutions of iron in cobalt [66] and nickel in cobalt [66, 74] in terms of the above dipolar picture has been extended to interpret the second-order satellite lines as due to two neighboring Fe atoms [96].

An alternate possibility is that the satellite lines are associated with neighbor shells around the impurity site [66].

The values of the hyperfine fields at the solute nuclei for dilute iron group alloys have been determined by nuclear resonance studies and are summarized in Table IV [73].

TABLE IV
HYPERFINE FIELDS AT SOLUTE NUCLEI IN DILUTE ALLOYS

Solute	Nuclear Moment	Values (koe) at 0° K			Ref.
		Iron	Cobalt, fcc	Nickel	
Fe^{57}	0.0905	−340	−330	−280	[63, 65]
Co^{59}	4.583	−286	−215	−120	[64, 65, 70]
Ni^{61}	0.749	−235	−189	−75	[73, 74, 97]

The hyperfine field has also been measured for Cu^{63} nuclei in iron, −213 koe [65]; in nickel, −47.2 koe [77]; and in cobalt, −157 koe [72]. In an attempt to interpret these results, three perturbations of the internal field at the solute atom nucleus have been considered [72, 98]: (1) $4s$ electron polarization by host $3d$ electrons, (2) covalent-type transfer of $3d$ electrons between host and solute atoms affecting the $4s$ electron polarization and the core polarization of the solute atom, (3) core s electron polarization by $3d$ electrons of neighboring host atoms.

The nuclear resonance of Cu^{63} and Cu^{65} has been studied in ferromagnetic alloys in an attempt to determine the dominant contribution to the hyperfine field. The results [72] given in Table II show that the hyperfine fields at Cu nuclei in iron and in cobalt are comparable in magnitude with the hyperfine fields in pure iron and cobalt. Further, the ratio of the hyperfine fields at copper nuclei in iron and cobalt is nearly equal to the ratio of saturation magnetization. These results point to conduction electron polarization as the dominant source of the hyperfine field; however, far more work is necessary, including a determination of the sign of the hyperfine field.

In concentrated alloys (greater than 4%), the local region around the solution atom may no longer be distinguishable and the host magnetization should be appreciably altered. An interesting example of this effect has been found in cobalt-iron alloys [64, 67]. Magnetic resonance studies of Co^{59} nuclei in the range of 1% to 8% cobalt show a 1.7% shift in hyperfine field at the Co^{59} nuclei. The hyperfine field at the host Fe^{57} nuclei increases by 4.7% over the same cobalt concentration range.

b. **Temperature and pressure dependence.** Measurement of the temperature and pressure dependence of the nuclear resonance in alloys provides information on the host-solute atomic interaction as seen by hyperfine field changes. As in the case of pure metals, one must correct the data to constant volume (this section, 3d), in order to compare temperature changes of hyperfine fields. Pressure studies have been made for only a few alloy systems. These include 1% cobalt in iron [72] and 1% copper in iron [68]. The hyperfine field of Co^{59} increases with pressure; $d \ln H_i/dP = 1.6 \times 10^{-7}/(kg/cm^2)$; whereas the hyperfine field for Cu^{63} decreases; $d \ln H_i/dP = -3.0 \times 10^{-7}/(kg/cm^2)$ [72]. The pressure dependence of the hyperfine field at Co^{59} nuclei in iron is opposite to that of iron, where $d \ln H_i/dP = -1.6 \times 10^{-7}/(kg/cm^2)$ [99].

When the temperature dependence of the hyperfine fields of Co^{59} in iron and Fe^{57} in iron are converted to constant volume [12, 72] and the reduced hyperfine fields are compared as a function of reduced temperature, one finds that the hyperfine field at the Co^{59} nucleus decreases more rapidly with increasing temperature than does the field at the Fe^{57} nucleus. The reverse effect has been noted for cobalt in nickel [70], uncorrected to constant volume.

c. **Intermetallic compounds.** As summarized in Table II, the principal intermetallic magnetic materials whose nuclear resonance has been studied are ferromagnetic manganese compounds [54]. Table V is a

TABLE V

HYPERFINE MAGNETIC FIELD AT NONMAGNETIC NUCLEI
IN FERROMAGNETIC MANGANESE COMPOUNDS AT 0° K [54]

Nonmagnetic nucleus	Compound	H_i (koe)	μ_A[Mn]	H_i/μ_A
B^{11}	MnB	25	1.94	13
P^{31}	MnP	44	1.29	34
As^{75}	MnAs	286	3.40	84
$Sn^{117,119}$	Mn_2Sn	128	1.2	106
Sb^{120}	MnSb	382	3.53	108
$Sb^{121,123}$	Mn_2Sb	253	1.74	145
Bi^{209}	MnBi	>585	3.95	>148

summary of the hyperfine field at the nonmagnetic nuclei for the ferromagnetic manganese compounds together with the magnetic moment in Bohr magnetons per manganese atom.

The line spectrum for Mn^{55} in Mn_5Ge_3 [56, 58] is closely analogous to that described for the Cr^{63} nucleus in $CrBr_3$ [42]. The Mn_5Ge_3 spectrum contains a broad intense line at 216.6 Mc/sec. There are, in addition, sharp, weak lines extending from 196.4 to 218.7 Mc/sec. The sharp lines presumably arise from quadrupole splitting within the domain volume. The broad intense line is evidently the central transition enhanced by wall displacement.

The observation of Gd^{155} and Gd^{157} nuclear resonance in ferromagnetic GdN [82] is of special interest since GdN appears to be a nearly cubic intermetallic compound with the rocksalt structure. The resonance for Gd^{155} and Gd^{157} was found at 44.2 Mc/sec and 58.7 Mc/sec, respectively, at 2.2 °K. No quadrupole interaction was observed, indicating nearly cubic symmetry.

IV. Nuclear Relaxation

One can distinguish two relaxation mechanisms of the excited nuclear spin system. First, there are the interactions between the nuclear spins themselves. These interactions may produce a decoherence of the transverse precessing component of spin. The characteristic time for such a process is identified as T_2', but in a system where the nuclei are nonuniformly excited as when they lie in domain walls, these same interactions may produce a spatial diffusion of energy by means of the nuclear spin-spin interaction [100]. In addition, if the interaction between nuclear spins is of sufficiently long range, resonant nuclei can distribute their energy to nuclei off resonance by a frequency diffusion process [101].

Second, the energy which is absorbed from the radio-frequency field [102] must ultimately find its way out of the nuclear system and into other kinds of excitations. The characteristic time for such a process is called T_1. In nonmagnetic insulators the principal energy reservoir is phonon excitation. In the nonmagnetic metals, it is the conduction electrons [103]. In magnetic materials we may expect that some of the energy will go to spin wave excitations. In addition to permitting energy transfer, these processes will also produce a transverse decoherence of spin. The transverse time which includes all decoherent processes is called T_2.

The relaxation times T_1 and T_2 have been obtained from the observation of the absorption line under adiabatic and nonadiabatic rapid passage [23, 104]. Although these measurements may be made with conventional absorption equipment, their interpretation is quite complex. One may more simply and directly interpret the results of spin-echo studies [16, 53]. The two nuclear relaxation processes will be discussed separately. The experimental results are summarized in Table VI.

TABLE VI

NUCLEAR RELAXATION IN MAGNETIC MATERIALS

Nucleus	Host	Half-width	T_2' (msec)	T_2 (msec)	T_1 (msec)	Temp. (°K)	Ref.
Mn^{55}	$MnFe_2O_4$	1.7 Mc/sec	—	—	1.5	1.65	[35]
Mn^{55}	Mn_5Ge_3	6g site	center of quadrupole multiplet		0.47	1.6	[58]
Fe^{57}	$Y_3Fe_5O_{12}$	Tetrahedral	—	3.8	24	4	[53]
		Octahedral	—	1.4	12.2	4	[53]
Fe^{57}	Fe^{57}	15 kc/sec	12	—	$2500/T$	298	[16]
Co^{59}	Co^{59}	800 kc/sec	0.025	—	$80/T$	295	[16]
Ni^{61}	Ni^{61}	100 kc/sec	0.27	—	$115/T$	295	[16]

1. Spin-Spin Interactions

The effect of decoherent spin processes is to reduce the intensity of the spin echo. In order to study spin-spin processes exclusively, one must have some way of separating out other interactions. This generally means going to very low temperatures to minimize decoherence by phonons. However, in cobalt and in isotopically enriched samples of iron and nickel, it has been possible to detect spin-spin relaxation at elevated temperatures [25].

One surprising feature of the relaxation is that it indicates an unusually strong nuclear spin-spin coupling. Suhl [105] and Nakamura [106] have independently proposed a coupling mechanism which is specific to magnetic materials. They point out that through the hyperfine interaction nuclei will virtually excite electronic spin waves. These spin waves may be reabsorbed by other nuclei, resulting in a coupling between the nuclei. The Suhl-Nakamura coupling has been extended [107, 108] to include nuclear spin waves. Because of the long range of the Suhl-Nakamura interaction, well-defined nuclear spin waves can exist at 1 °K where the nuclear polarization is less than 1% [109]. A high concentration of nuclear spins and a low anisotropy field are required.

Since the relevant spin wave energies, however, are large compared with the hyperfine interaction and the range of interaction is long, one might expect to be able to formulate a classical theory of the coupling. Such a theory has been developed by Simanek [110] by considering the local magnetic distortion produced by the nuclear spins.

A second surprising feature of the spin-spin relaxation is that it is nearly exponential, whereas the Suhl-Nakamura coupling does not lead to exponential relaxation. The exponential relaxation can be shown to result from the static broadening of the resonance line and the extremely long range of the nuclear interaction [16]. Winter [45] has pointed out that, since the relaxation experiments have been done on nuclei in walls, one should treat wall processes. He finds that the relaxation within a wall should be faster by a factor which is the ratio of the bulk anisotropy to the apparent wall anisotropy. With this modification the theory seems to be in general agreement with experiment.

a. Magnetic insulators. The strong nuclear spin-spin coupling, predicted by de Gennes *et al.* [108], has been observed in $MnFe_2O_4$ [111]. The Mn^{55} nuclear resonance frequency is "pulled" to lower values at low temperatures ($< 4\ °K$) by an effective field arising from a collective nuclear spin effect (the uniform mode of the nuclear spin wave spectrum [112]).

The nuclear relaxation of Fe^{57} in yttrium-iron garnet containing 0.5% rare earth impurity has been studied from $4\ °K$ to $100\ °K$ [113]. Below $20\ °K$ only spin-spin interactions are important for nuclear relaxation processes, as indicated by increase in T_1 below $20\ °K$, according to $T_2 = 2T_1$. The purely transverse character of the longitudinal spin-lattice relaxation can be interpreted by a Suhl-Nakamura coupling of the rare earth ions to the Fe^{57} nuclei [114]. The temperature minimum for T_1 at $20\ °K$ corresponds to the rare earth ion relaxation rate. The observed decrease of T_2 above $20\ °K$ is interpreted in terms of isotropic spin wave scattering, since $1/T_2$ varies as T^2 and becomes independent of an external field [113].

The behavior of the spin-spin relaxation of Fe^{57} in α-Fe_2O_3 has not been explained [41], largely owing to the lack of low-temperature information, since ferromagnetism disappears below $270\ °K$, the Morin transition [115]. Recently [116] it has been observed that the addition of small amounts of gallium depresses the Morin transition so that low-temperature data may be obtained.

b. Metals and alloys. Spin-spin relaxation in iron group metals furnishes confirmation of the Suhl-Nakamura picture of weak spin-spin coupling that extends over very long distances. The spin-spin time T_2 has been found to be inversely proportional to the concentration of the active isotope [16]; however, the relaxation has a pure exponential decay. A local static inhomogeneity of 100 oe in the resonance field would account for the observed line width and the discrepancy with theory [117]. Such a field inhomogeneity on the scale of the long-range interaction distance (remember that the excited nuclei lie in and near domain walls) would partially decouple the interacting nuclei, giving an exponential relaxation [16].

2. Spin-Lattice Interaction

The spin-lattice interaction produces a relaxation of the longitudinal component of nuclear magnetization as well as contributing to the transverse relaxation. Although the measurements of longitudinal relaxation give an exponential decay for long times, the short time decay is very rapid and nonexponential. The lattice interaction also contributes to T_2 and produces a nonexponential relaxation with qualitatively the same character as the longitudinal relaxation.

a. *Magnetic insulators.* The spin-lattice relaxation times for Fe^{57} in yttrium-iron garnets containing 0.5% earth ions [114] and for Mn^{55} in manganese ferrite with Fe^{2+} impurities present [118] have been interpreted by similar pictures. The minimum T_1 occurs when the temperature of the system is such that the impurity relaxation rate is comparable with the Fe^{57} nuclear frequency in the case of YIG and with the Mn^{55} frequency in the case of $MnFe_2O_4$ [119]. The purely transverse coupling of the longitudinal relaxation seen in the Fe^{57} nuclear resonance of YIG at 4 °K [30] is explained by the above picture.

The spin-lattice relaxation time of Fe^{57} in α-Fe_2O_3 [41] has been correlated with thermal domain-wall fluctuation. The behavior of the Fe^{57} relaxation in YIG and α-Fe_2O_3 at low power levels, where the nuclear resonance appears to originate in domain walls, shows similarities; both environments have a rapid, nonexponential decay and a similar dependence on power level and external field [30, 41]. The domain-wall fluctuation explanation agrees with experiment in both cases.

b. *Metals and alloys.* Korringa has shown that the main contribution to the nuclear relaxation rate in nonmagnetic metals is the interaction between nuclei and the conduction electrons and is proportional to the temperature [103]. The longitudinal relaxation in magnetic metals, however, is at least an order of magnitude faster than predicted by Korringa's theory, even in the long tail of decay.

Experimental investigations of the iron group metals show that the longitudinal relaxation time is highly sensitive to external fields. The longitudinal relaxation time of nuclei in domain walls is much shorter than in domains [25]. The faster relaxation in walls can be accounted for by a moderate amount of wall damping [45, 120]. Removing wall effects still leaves domain nuclei relaxing an order of magnitude faster in magnetic metals than in nonmagnetic metals. Since the relaxation rate for nuclei in domains appears to be linear in temperature over a wide temperature range, it is appealing to look to conduction electron processes as in the case of nonmagnetic metals.

Moriya [121] has treated the relaxation of nuclear spins in transition metals as it arises from coupling to the fluctuating orbital current of the d electrons, using a tight-binding approximation. He obtains a relaxation rate $1/T_1$ proportional to the temperature and in remarkably good agreement with experiment for Fe^{57}, Co^{59}, and Ni^{61}. According to this theory the rapid relaxation of nuclei in ferromagnetic metals is associated with the high density of states in the d band.

References

1. R. L. Mössbauer, Z. Physik **151**, 124 (1958).
2. H. Frauenfelder, "The Mössbauer Effect." W. A. Benjamin, New York, 1963.
3. M. A. Grace, C. E. Johnson, N. Kurti, R. G. Scurlock, and R. T. Taylor, Commun. Conf. Basses Temperatures, Paris, 1955 p. 263 (1955); Phil. Mag. [8] **4**, 948 (1959).
4. C. V. Heer and R. A. Erickson, Phys. Rev. **108**, 896 (1957); V. Arp, N. Kurti, and R. Petersen, Bull. Am. Phys. Soc. [2] **2**, 388 (1957).
5. A. C. Gossard and A. M. Portis, Phys. Rev. Letters **3**, 164 (1959).
6. W. Marshall, Phys. Rev. **110**, 1280 (1958).
7. A. Abragam, "The Principles of Nuclear Magnetism." Oxford Univ. Press, London and New York, 1961; G. E. Pake, Solid State Phys. **2**, 1 (1956).
8. P. Kusch, S. Millman, and I. I. Rabi, Phys. Rev. **57**, 765 (1940).
9. T. P. Das and E. L. Hahn, Solid State Phys. Suppl. 1 (1958).
10. R. C. La Force, L. E. Toth, and S. F. Ravitz, Phys. Chem. Solids **24**, 729 (1963).
11. R. J. Snodgrass and L. H. Bennett, Appl. Spectry. **17**, 53 (1963).
12. G. B. Benedek, "Interscience Tracts on Physics and Astronomy. No. 24. Magnetic Resonance at High Pressure." Wiley (Interscience), 1963.
13. R. L. Streever, Jr., Phys. Rev. Letters **10**, 232 (1963).
14. M. Rubinstein, Thesis, University of California, Berkeley, California, 1962.
15. E. L. Hahn, Phys. Rev. **80**, 580 (1950).
16. M. Weger, E. L. Hahn, and A. M. Portis, J. Appl. Phys. **32**, 124S (1961).
17. C. Kittel and J. K. Galt, in Solid State Phys. **3**, 437 (1956).
18. E. L. Boyd, Bull. Am. Phys. Soc. [2] **8**, 439 (1963).
19. R. H. Lindquist, A. C. Gossard, A. M. Portis, and M. Rubinstein, Bull. Am. Phys. Soc. [2] **5**, 491 (1960).
20. G. J. Perlow, in "Mössbauer Effect" (H. Frauenfelder and H. Lustig, eds.), p. 49. Univ. of Illinois Report, Urbana, Illinois, 1960.
21. W. Döring, Z. Naturforsch. **3A**, 373 (1948).
22. A. C. Gossard and A. M. Portis, Phys. Rev. Letters **3**, 164 (1959).
23. A. M. Portis and A. C. Gossard, J. Appl. Phys. **31**, 205S (1960).
24. J. I. Budnick, L. J. Bruner, R. J. Blume, and E. L. Boyd, J. Appl. Phys. **32**, 120S (1961).
25. M. Weger, Phys. Rev. **128**, 1505 (1962).
26. R. E. Watson and A. J. Freeman, Phys. Rev. **123**, 2027 (1961).
27. E. L. Boyd, V. L. Moruzzi, and J. S. Smart, J. Appl. Phys. **34**, 3049 (1963).
28. P. W. Anderson and A. M. Clogston, Bull. Am. Phys. Soc. [2] **6**, 124 (1961); A. M. Clogston and P. W. Anderson, Bull. Am. Phys. Soc. [2] **6**, 124 (1961).
29. C. Robert and J. M. Winter, Arch. Sci. (Geneva) **13**, 433 (1960).

30. C. Robert, *Compt. Rend.* **252**, 1442 (1961).
31. Le Dang Khoi and M. Buyle-Bodin, *Compt. Rend.* **253**, 2514 (1961).
32. S. Ogawa and S. Morimoto, *J. Phys. Soc. Japan* **17**, 654 (1962).
33. F. Boutron and C. Robert, *Compt. Rend.* **253**, 433 (1961).
34. A. Narath, *Phys. Rev. Letters* **7**, 410 (1961).
35. A. J. Heeger, S. K. Ghosh, and T. G. Blocker, III, *J. Appl. Phys.* **34**, 1034 (1963).
36. E. L. Boyd and J. C. Slonczewski, *J. Appl. Phys.* **33**, 1077S (1962).
37. E. J. W. Verwey and P. W. Haayman, *Physica* **8**, 979 (1941).
38. E. L. Boyd, *Phys. Rev.* **129**, 1961 (1963).
39. Le Dang Khoi and F. Bertaut, *Compt. Rend.* **254**, 1584 (1962).
40. D. H. Anderson, *Bull. Am. Phys. Soc.* [2] **7**, 537 (1962).
41. M. Matsuura, H. Yasuoka, A. Hirai, and T. Hashi, *J. Phys. Soc. Japan* **17**, 1147 (1962).
42. A. C. Gossard, V. Jaccarino, and J. P. Remeika, *J. Appl. Phys.* **33**, 1187 (1962).
43. A. C. Gossard, V. Jaccarino, and J. P. Remeika, *Phys. Rev. Letters* **7**, 122 (1961).
44. H. Suhl, *Bull. Am. Phys. Soc.* [2] **5**, 175 (1960).
45. J. M. Winter, *Phys. Rev.* **124**, 452 (1961).
46. H. Yasuoka, H. Abe, A. Hirai, and T. Hashi, *J. Phys. Soc. Japan* **18**, 593 (1963).
47. A. Narath, *Bull. Am. Phys. Soc.* [2] **7**, 481 (1962).
48. S. Ogawa, S. Morimoto, and Y. Kimura, *J. Phys. Soc. Japan* **17**, 1671 (1962).
49. Le Dang Khoi and M. Buyle-Bodin, *Compt. Rend.* **253**, 1783 (1961).
50. E. L. Boyd, J. I. Budnick, L. J. Bruner, and R. J. Blume, *J. Appl. Phys.* **33**, 2484 (1962).
51. H. Abe, M. Matsuura, H. Yasuoka, A. Hirai, T. Hashi, and T. Fukuyama, *J. Phys. Soc. Japan* **18**, 1400 (1963).
52. H. Yasuoka, A. Hirai, M. Matsuura, and T. Hashi, *J. Phys. Soc. Japan* **17**, 1071 (1962).
53. C. Robert, *Compt. Rend.* **251**, 2684 (1960).
54. E. Hirahara, unpublished data, 1963.
55. D. J. Lam, D. O. Van Ostenburg, M. V. Nevitt, H. D. Trapp, and D. W. Pracht *Phys. Rev.* **131**, 1428 (1963).
56. T. Hihara, Y. Koi, and A. Tsujimura, *J. Phys. Soc. Japan* **18**, 454 (1963).
57. Y. Koi and A. Tsujimura, *J. Phys. Soc. Japan* **18**, 1347 (1963).
58. R. F. Jackson, R. G. Scurlock, D. B. Utton, T. H. Wilmshurst, and M. Rubinstein, *Phys. Letters* **6**, 39 (1963).
59. A. Tsujimura, T. Hihara, and Y. Koi, *J. Phys. Soc. Japan* **17**, 1078 (1962).
60. K. Sugibuchi and K. Endo, *Phys. Chem. Solids* **25**, 1217 (1964).
61. C. Robert and J. M. Winter, *Compt. Rend.* **250**, 3831 (1960).
62. S. Ogawa and S. Morimoto, *J. Phys. Soc. Japan* **16**, 2065 (1961).
63. J. I. Budnick, R. C. La Force, and G. F. Day, *Proc. XI Colloq. AMPERE, Eindhoven,* 1962, 629-632, (1963).
64. R. C. La Force, S. F. Ravitz, and G. F. Day, *J. Phys. Soc. Japan* **17**, Suppl. B-I, 99 (1962).
65. Y. Koi, A. Tsujimura, T. Hihara, and T. Kushida, *J. Phys. Soc. Japan* **16**, 1040 (1961).
66. R. C. La Force, S. F. Ravitz, and G. F. Day, *Phys. Rev. Letters* **6**, 226 (1961).
67. E. Simanek and Z. Sroubek, *Czech. J. Phys.* **B12**, 202 (1962).
68. Y. Koi, A. Tsujimura, T. Hihara, and T. Kushida *J. Phys. Soc. Japan* **17**, Suppl. B-I, 96 (1962).
69. W. A. Hardy, *J. Appl. Phys.* **32**, Suppl. 3, 122S (1961).

70. L. H. Bennett and R. L. Streever, Jr., *J. Appl. Phys.* **33**, 1093 (1962).
71. S. Ehara and Y. Tumono, *J. Phys. Soc. Japan* **17**, 726 (1962).
72. T. Kushida, A. H. Silver, Y. Koi, and A. Tsujimura, *J. Appl. Phys.* **33**, 1079 (1962).
73. R. L. Streever, L. H. Bennett, R. C. La Force, and G. F. Day, *J. Appl. Phys.* **34**, 1050 (1963).
74. R. L. Streever, L. H. Bennett, R. C. La Force, and G. F. Day, *Phys. Rev.* **128**, 1632 (1962).
75. L. J. Bruner, J. I. Budnick, and R. J. Blume, *Phys. Rev.* **121**, 83 (1961).
76. R. L. Streever and L. H. Bennett, *Phys. Rev.* **131** (5), 2000 (1963).
77. K. Asayama, S. Kobayashi, and J. Itoh, *J. Phys. Soc. Japan* **18**, 458 (1963).
78. D. L. Weinberg and N. Bloembergen, *Phys. Chem. Solids* **15**, 240 (1960).
79. T. Hihara, A. Tsujimura, and Y. Koi, *J. Phys. Soc. Japan* **17**, 1320 (1962).
80. M. Asanuma, N. Sato, and R. Hoshino, paper at 18th Annual Meeting of Physical Society of Japan, unpublished (1963).
81. H. Suzuki and E. Hirahara, unpublished data, 1963.
82. E. L. Boyd and R. J. Gambino, *Phys. Rev. Letters* **12**, 20 (1964).
83. J. Hervé and P. Veillet, *Compt. Rend.* **252**, 99 (1961).
84. A. C. Gossard, thesis, University of California, Berkeley, California, 1960.
85. S. S. Hanna, J. Heberle, G. J. Perlow, R. S. Preston, and D. H. Vincent, *Phys. Rev. Letters* **4**, 513 (1960).
86. F. E. Obenshain and H. H. F. Wegener, *Phys. Rev.* **121**, 1344 (1961); private communication, 1961.
87. L. Néel, *Compt. Rend.* **228**, 664 (1949).
88. Y. Koi, A. Tsujimura, and T. Kushida, *J. Phys. Soc. Japan* **15**, 2100 (1960).
89. F. Keffer, private communication, 1960.
90. R. Street, D. S. Rodbell, and W. L. Roth, *Phys. Rev.* **121**, 84 (1961).
91. L. E. Toth and S. F. Ravitz, *Phys. Chem. Solids* **24**, 1203 (1963).
92. A. M. Portis and J. Kanamori, *J. Phys. Soc. Japan* **17**, 587 (1962).
93. J. D. Litster and G. B. Benedek, *J. Appl. Phys.* **34**, 688 (1963).
94. R. V. Jones and I. P. Kaminow, *Bull. Am. Phys. Soc.* [2] **5**, 175 (1960).
95. D. H. Anderson and G. A. Samara, *Bull. Am. Phys. Soc.* [2] **9**, 24 (1964).
96. G. F. Day, Ph.D. thesis, University of California, Berkeley, California, 1964.
97. P. R. Locher and S. Geschwind, *Phys. Rev. Letters* **11**, 333 (1963).
98. G. K. Wertheim, *J. Appl. Phys.* **32**, 110S (1961).
99. G. B. Benedek and J. Armstrong, *J. Appl. Phys.* **32**, 106S (1961).
100. N. Bloembergen, *Physica* **15**, 386 (1949).
101. A. M. Portis, *Phys. Rev.* **104**, 584 (1956); P. W. Anderson, *ibid.* **109**, 1492 (1958).
102. Y. Obata, *J. Phys. Soc. Japan* **18**, 1020 (1963).
103. J. Korringa, *Physica* **16**, 601 (1950).
104. D. L. Cowan and L. W. Anderson, *Phys. Rev.* **135**, A1046 (1964).
105. H. Suhl, *Phys. Rev.* **109**, 606 (1958); *J. Phys. Radium* **20**, 333 (1959).
106. T. Nakamura, *Progr. Theoret. Phys.* (*Kyoto*) **20**, 542 (1958).
107. P. G. de Gennes, *J. Phys. Radium* **23**, 510 (1962).
108. P. G. de Gennes, P. A. Pincus, F. Hartmann-Boutron, and J. M. Winter, *Phys. Rev.* **129**, 1105 (1963).
109. P. A. Pincus, *Phys. Rev.* **131**, 1530 (1963).
110. E. Simanek, *Czech. J. Phys.* **B11**, 711 (1961).
111. A. J. Heeger and T. Houston, *Bull. Am. Phys. Soc.* [2] **8**, 213 (1963).
112. P. G. de Gennes, F. Hartmann-Boutron, and P. A. Pincus, *Compt. Rend.* **254**, 1264 (1962).

113. C. Robert and J. M. Winter, *Compt. Rend.* **253**, 2925 (1961).
114. P. G. de Gennes and F. Hartmann-Boutron, *Compt. Rend.* **253**, 2922 (1961).
115. F. J. Morin, *Phys. Rev.* **78**, 819 (1950).
116. A. H. Morrish, G. B. Johnston, and N. A. Curry, *Phys. Letters* **7**, 177 (1963).
117. A. M. Portis, *J. Phys. Soc. Japan* **17**, Suppl. B-I, 81 (1962).
118. A. J. Heeger, T. G. Blocker, III, and S. K. Ghosh, *Phys. Rev.* **134**, A399 (1964).
119. F. Hartmann-Boutron, *J. Appl. Phys.* **35**, 889 (1964).
120. L. L. Buishvili and N. P. Giorgadze, *Zh. Eksperim. i Teor. Fiz.* **42**, 499 (1962).
121. T. Moriya, *J. Phys. Soc. Japan* **19**, 681 (1964).

7. Theory of Magnetism in the Rare Earth Metals

R. J. Elliott

Clarendon Laboratory,
Oxford, England

  I. Introduction 385
    1. Background 385
    2. Structures 387
    3. Conduction Electrons 388
    4. $4f$ Electrons 391
    5. Survey of Experimental Results 392
 II. Phenomenological Theory 397
    1. Molecular Fields 397
    2. Spatial Dependence of Exchange 399
    3. Low-Temperature Ordering 400
    4. Spin Waves 402
III. The Crystal Field 403
    1. The Heavy Rare Earths 403
    2. The Light Rare Earths 405
    3. Gadolinium 406
IV. Theory of Exchange Interactions 406
    1. The Rudermann–Kittel Interaction 406
    2. Superzone Boundaries 409
    3. Effect of Boundaries on R-K Interaction 411
    4. Effect of Scattering on R-K Interaction 414
    5. Other Exchange Mechanisms 416
    6. Electrical Resistivity 418
 V. Conclusion 421
    References 421

I. Introduction

1. Background

The fourteen elements of the rare earth group are characterized by the presence of $4f$ electrons in their normal electronic configurations.

The number varies from element to element and depends on the valency. In most chemical compounds the rare earths appear as tripositive ions, in which condition the number increases from $4f^1$ in Ce to $4f^{14}$ in Lu along the series. Two other elements are normally considered as rare earths because of their similar chemical properties; La, just before Ce in the period, and Y, which comes in the same position as La in the preceding (fifth) period. In fact Y is more like Lu in its properties than La.

The $4f$ electrons are closely bound inside the outer closed shells on the atoms. They therefore play a small role in the chemical bonding, and to a first approximation behave as they would in a free ion, giving it a resultant total angular momentum combined from the spin and orbital motion. This carries with it a magnetic moment, and all rare earth compounds have interesting magnetic properties. These vary considerably from substance to substance because in a solid the $4f$ electrons are subject to forces from their environment. These forces are of two main types. First, the array of charges in the crystal produces an electric field at any one ion called the crystalline electric field, which causes a Stark splitting of the free ion energy levels. This has been extensively investigated on the essentially isolated ions of paramagnetic salts. By modifying the electronic orbits it has an important influence on the magnetic properties. In dense materials then are also forces which directly couple the $4f$ electrons of different ions. These may take various forms but the most important are of the exchange type, giving rise to cooperative magnetic properties.

It will be shown that this basic picture of ions with $4f^n$ configurations acted on by the crystal field and coupled by exchange interactions is a satisfactory one in the metals, and can explain in a relatively simple way their complex properties.

Although some pioneering work was done before the war, the detailed investigation of the properties of the rare earth metals has been mainly carried out during the past 10 years. It owes much to the work of Professor F. H. Spedding and his collaborators at Ames, Iowa, in separating relatively large quantities of pure material, growing crystals, and subjecting them to systematic examination. Besides this work, probably the greatest single contribution to the understanding of the magnetic properties of these materials has come from the neutron diffraction investigations of the Oak Ridge group. But now that good single-crystal material is more easily available, many workers are actively engaged in the field.

2. Structures

Most of the metals conform to the simple picture outlined above of tripositive ions, embedded in a sea of conduction electrons formed from the three outer $5d$ and $6s$ valence electrons. However, even the valence properties differ a little, and the crystal structure of the elements in the second half of the series differs slightly from that common in the first half. Furthermore, two elements, Eu and Yb, have quite different properties. These elements come immediately before the middle and the end of the series and, in the tripositive form, would have configurations $4f^6$ and $4f^{13}$. In these circumstances they evidently prefer to gain the extra correlation energy of a half-filled or completed shell and take a divalent form with $4f^7$ and $4f^{14}$ configurations. This is borne out by their properties. They have a different crystal structure and much larger atomic volume than the other rare earths [1]. Ytterbium has only a small paramagnetism as is expected from a closed shell [2]. Europium shows large moments as expected from $4f^7$, and appears to have complicated magnetic ordering [3].

Since this article is mainly concerned with correlating the properties of different elements within a common theory, there will be little further reference to these anomalous materials.

For rather similar reasons Ce shows rather different anomalous properties. At the beginning of the series it is known from atomic properties that the $4f$ and $5d$ electrons have similar energies. The La^{2+} ion has a single $5d$ not a $4f$ electron [4] and Ce may be found in a fourvalent state. This apparently occurs in Ce metal at low temperatures or at high pressures when a dense fcc phase is stable with small paramagnetism and apparently four conduction electrons [5]. However, above 100 °K the stable form has one $4f$ electron and a crystal like that of its neighbors La and Pr.

Apart from this the elements at the beginning of the series are all similar. Lanthanum, Ce, Pr, and Nd have been reported with various crystal structures but apparently the usual form is a modified hcp structure [1]. The packing is such that it repeats after four layers along the hexagonal axis, i.e., ABACABAC... as opposed to the standard twolayer form ABAB... (see Fig. 1). The fcc structure along a (111) direction may also be written in the form ABCABC. Thus in these rare earth structures alternate layers B and C have hexagonal nearestneighbor environments while the others, A, have a cubic arrangement of near neighbors. Partly because of this complex structure the properties of these elements have so far not been investigated in great detail either experimentally or theoretically and it will only be possible to discuss them qualitatively.

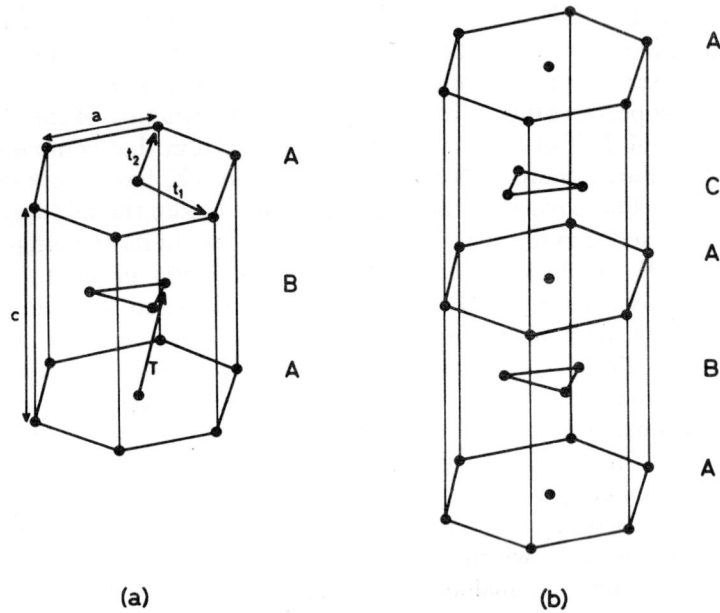

FIG. 1. Crystal structures common in the rare earth metals. (a) Hexagonal close packed; (b) double hexagonal structure.

The next element, Pm, has no stable isotope and so no information is available on the metallic phase. Samarium with $4f^5$ has an even more complex crystal structure [1] which repeats after nine hexagonal layers. It has been little investigated and will not be discussed in detail.

The rest of the elements, Gd–Lu, with the exceptions already cited, all have similar crystal structures, being hcp with c/a ratios 1.57–1.59, a little different from the ideal value of 1.63. They form an interesting series and have been extensively investigated. Their magnetic properties, while complex, show a certain regularity which may be traced to various properties of the exchange and crystal fields. Yttrium is very similar to these materials, and may be conveniently alloyed with them.

3. Conduction Electrons

Very little is known of the detailed band structure of the three conduction electrons in these materials. The only extensive calculation is that of Altmann *et al.* [6] on Y, and it shows a complex structure. Some aspects of the Fermi surface are reflected in the magnetic properties and some information will be deduced in this way. However, detailed understanding will have to await the production of pure material suitable for cyclotron resonance, de Haas van Alphen effect, and similar studies.

In the absence of such information theoretical work has normally made the crude assumption of nearly free electrons. The effect of lattice symmetry may to some extent be included by considering the Brillouin zone structure. This has been done by Harrison [7] for the hcp lattice applicable to the latter half of the series. The primitive translations are t_1, t_2, t_3. The first two are of magnitude a at 120° to each other in the basal plane, and the third has magnitude c along the hexagonal axis perpendicular to this plane. There are two atoms per cell, one at the origin and one at $T = \frac{1}{3}(t_1 + 2t_2) + \frac{1}{2}t_3$. The vectors of the reciprocal lattice also form a hexagonal lattice; τ_1, τ_2 have magnitude $4\pi/a\sqrt{3}$ and are at 120° to each other in the basal plane, τ_3 is of magnitude $2\pi/c$ and is perpendicular to this plane. The first Brillouin zone is therefore a hexagonal prism of side $4\pi/3a$ and height $2\pi/c$. The free electron Fermi surface is a sphere with radius

$$k_f = 2\sqrt{3}\pi/[2\pi a^2 c]^{1/3} \tag{1.1}$$

This stretches out over five zones (cf. Fig. 2).

The zone planes which perpendicularly bisect the reciprocal lattice vectors τ represent Bragg scattering of the electrons off the atomic lattice. The energy bands and hence the Fermi surface are discontinuous across these planes. Some idea of their effect is obtained by reflecting the Fermi sphere in these planes, back into the first zone [7]. The strength of these reflections may be estimated by considering the so-called form factor $F(\tau)$, defined as

$$F(\tau) = \tfrac{1}{2}(1 + \exp i\tau \cdot T) \tag{1.2}$$

For, if the two atoms in the cell produce the same potential, the interference in the scattering through τ is given by (1.2). Yosida and Watabe [8] have calculated the form factors of importance in this problem. It will be noted that $F(001) = 0$ and for this reason it is common to use a Brillouin zone of double extension along the c-axis, as is done in references [7] and [8], although this is not satisfactory in the presence of strong spin-orbit coupling [9]. In any case the Fermi surface produced in the first zone by this method is quite different from that calculated in reference [6] and is probably not a very good approximation. It does,

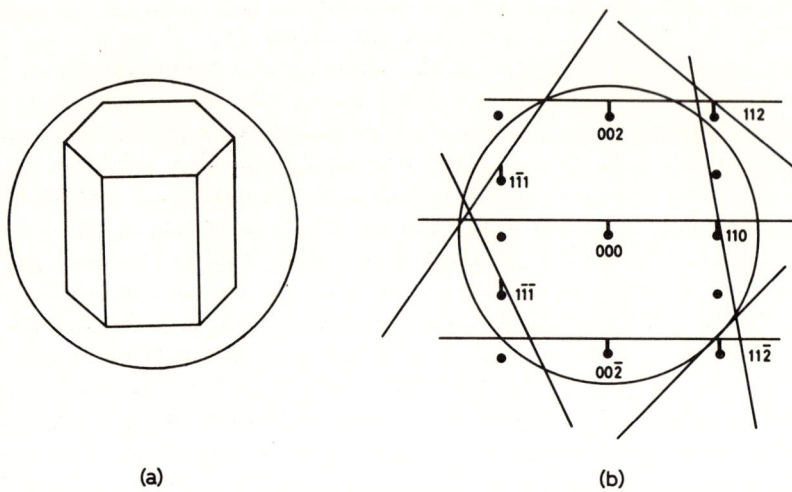

FIG. 2. (a) Free electron Fermi surface with first (double) Brillouin zone in hcp metals. (b) Position of zone boundaries in extended scheme relative to Fermi sphere. Points shown are half-way to reciprocal lattice points.

however, allow the calculation of properties which are not sensitive to the detailed form.

The double hexagonal structure found in the light rare earths may be considered in similar terms. The hexagonal cell has now a t_3 of magnitude $2c$, and four atoms per cell at

$$\mathbf{T}_1 = 0 \quad \mathbf{T}_2 = \tfrac{1}{3}(\mathbf{t}_1 + 2\mathbf{t}_2) + \tfrac{1}{4}t_3 \quad \mathbf{T}_3 = \tfrac{1}{2}\mathbf{t}_3 \quad \mathbf{T}_4 = -\mathbf{T}_2 \quad (1.3)$$

The first Brillouin zone is a hexagonal prism half the height (π/c) of that in the hcp structure. The stabilization of this structure suggests that the Fermi surface might lie close to these new boundaries in these metals. In any case the low $5d$ electron energy at this end of the series suggests that the band structure may be quite different from that at the other end.

In the one electron band picture the f electrons must occupy very narrow bands. For an atom with the $4f^n$ configuration n of these would lie well below the Fermi level and $(14 - n)$ well above. Because of the extremely narrow width the electron correlations are very important and will give rise to atomic configurations. In Ce the first f band must lie above the Fermi level in the dense low-temperature phase, and below it in the normal phase. Rocher [10] has argued that the transition indicates that it lies very close, and that this is confirmed by the high density of states reflected in the high low-T specific heat and in the anomalously large electrical resistivity. In Yb he shows by similar considerations that the fourteenth band must be only just below the Fermi level and that it may be raised to it by the application of high pressure. There appears to be no analogous effect in Eu.

Some indication of the position of the 4f band relative to the Fermi level may be obtained from optical properties. The absorption coefficient of Dy and Ho [11] rises sharply about 3 ev, indicating that a full 4f level is this far below the Fermi level, or that an empty one is the same distance above.

4. 4f Electrons

In the normal rare earths the n electrons in the 4f bands have strong intra-atomic coulomb forces which give them a value of total angular momentum L and total spin S in the ground state prescribed by Russell–Saunders coupling and Hund's Rule. The spin-orbit coupling combines these antiparallel in the first half of the series an parallel in the second half to give a total angular momentum J. The next excited J multiplet is usually higher in energy by more than 0.1 ev so that it plays no role in thermal properties. For this reason it is always sufficient to regard the ions as being confined to the lowest J multiplets.

Samarium and Eu are exceptions to this because of the low J arising from near cancellation of L and S [12]. However, the first excited multiplet in Sm^{3+} is about 0.12 ev and gives only a small correction. In Eu^{3+} it is much lower, but Eu metal does not exist in this state. Thus these corrections are unimportant in the metals.

The paramagnetic susceptibility of the metals above room temperature is found to be like that expected for an assembly of free ions

$$\chi = N\beta^2\lambda^2 J(J+1)/3kT \tag{1.4}$$

where β is the Bohr magneton and λ is Lande's factor obtained by projecting the magnetic moment $\mathbf{L} + 2\mathbf{S}$ on $\mathbf{J}$ in the lowest Russell–Sanders multiplet

$$\lambda = (\mathbf{L} + 2\mathbf{S}) \cdot \mathbf{J}/\mathbf{J}^2 \tag{1.5}$$

This indicates that the crystal field and the interionic forces give rise to energies somewhat lower than that corresponding to room temperature. This is borne out by the values of the temperatures of magnetic transitions and the anisotropy energies.

The values of L, S, J, λ, and other constants of importance in calculating the properties of ions in their ground multiplets are given in Table I [13].

TABLE I

Useful Constants Defining the Properties of the Ground States of Tripositive Rare Earth Ions

	$4f^n$	L	S	J	λ	$(\lambda-1)^2 J(J+1)$	$2\alpha J^2$	$8\beta J^4$	$16\gamma J^6$
Ce	1	3	1/2	5/2	6/7	0.18	−0.71	1.99	0
Pr	2	5	1	4	4/5	0.80	−0.62	−1.50	3.9
Nd	3	6	3/2	9/2	8/11	1.84	−0.26	−0.95	−5.1
Sm	5	5	5/2	5/2	2/7	4.46	0.52	0.78	0
Eu	6	3	3	0	0	0	0	0	0
Gd	7	0	7/2	7/2	2	15.75	0	0	0
Tb	8	3	3	6	3/2	10.5	−0.73	−1.27	−0.84
Dy	9	5	5/2	15/2	4/3	7.08	−0.71	−4.48	2.9
Ho	10	6	2	8	5/4	4.5	−0.28	−2.16	−4.7
Er	11	6	3/2	15/2	6/5	2.55	0.29	3.36	5.9
Tm	12	5	1	6	7/6	1.17	0.73	1.71	−4.2
Yb	13	3	1/2	7/2	8/7	0.32	0.77	−2.10	4.4

5. Survey of Experimental Results

The magnetic properties of the rare earth metals have now been investigated by many different techniques. Before expounding the present state of the theory it seems essential to provide a brief survey of the most important results which relate to the magnetic order. These will be amplified later in the discussion of the theory, together with an account of other effects. The information is summarized in Table II.

a. Neutron diffraction. The most important single technique in the study of rare earth metal magnetism has been neutron diffraction, which has allowed a determination of the complex orderings. In the second half of the series wave like variations of the magnetic moment from site to site are common. These are divided into two main types by the anisotropy, which gives a large difference between moments pointing along the hexagonal axis and those in the basal plane.

Along the axis the moment variation is a longitudinal wave (LW). The moment on an atom at $\mathbf{R}_n$ is

$$\mu_n^z = \lambda\beta J M' \sin(\mathbf{q} \cdot \mathbf{R}_n + \delta) \quad \text{where} \quad JM' = \langle J_z \rangle \quad (1.6)$$

Since $\mathbf{q} \parallel c$ this gives all atoms in any plane perpendicular to the c-axis the same moment. This is true in all the structures in this half of the series.

TABLE II
SUMMARY OF EXPERIMENTAL EVIDENCE ON RARE EARTH METALS[a]

	θ	$\theta_\parallel$	$\theta_\perp$	T_1	Order	T_2	Order	μ	N	M	c	ρ
Ce	−38	—	—	12.5	Complex	—	—	—	[14]	[15]	[16]	[17]
Nd	−15	−17	−11	7	Complex	19	Complex	—	[18]	[19]	[16]	[17]
Sm	−60	—	—	15	?	—	—	—	—	[15]	[20]	[21]
Gd	317	—	—	293	Ferro	—	—	7.55	—	[22]	[23]	[24]
Tb	236	195	239	228	Helix	220	Ferro	9.3	[25]	[26]	[27]	[26]
Dy	154	121	169	179	Helix	85	Ferro	10.0	[29]	[30]	[31]	[32]
Ho	85	—	88	132	Helix	?	—	9.5	[33]	[34]	[35]	[34]
Er	42	62	32	85	LW	50	Complex	9.0	[36]	[37]	[38]	[37]
						20	Cone					
Tm	20	—	—	56	LW	?	Antiphase	6.8	[39]	[40]	[41]	—

[a] Values of Curie–Weiss θ, transition temperatures, types of order, and saturation moment μ (in Bohr magnetons). The last four columns give references to experimental data from neutron diffraction, magnetic measurements, specific heat, and electrical resistivity, respectively. This list is not exhaustive; usually the most recent reference is given or that with the most detailed information.

For moments in the plane the structure is a helix

$$\mu_n^x = \lambda \beta JM \cos(\mathbf{q} \cdot \mathbf{R}_n)$$
$$\mu_n^y = \lambda \beta JM \sin(\mathbf{q} \cdot \mathbf{R}_n) \quad \text{where} \quad JM = \langle J_\perp \rangle \tag{1.7}$$

and the moment direction varies from layer to layer although the magnitude is always the same.

More complicated structures may occur, with a wave variation in all three components, such as the cycloidal structure described by Yoshimori [42] in his classic paper on spin structures. If the z-component orders ferromagnetically while the x- and y-components remain as in Eq. (1.7) the resulting structure is conical, the moments pointing along the generators of a cone. Furthermore, harmonics with variation like $n\mathbf{q}$ may occur, distorting the simple waves. Thulium shows this effect. Just below the Néel point it has a simple LW with $q = 4\pi/7c$ so that it repeats every seven layers. At lower T squaring begins and it becomes an antiphase domain structure with four layers pointing one way, then three the other. Examples of these structures are shown in Fig. 3.

In the first half of the series the magnetic order is even more complicated, partly because there are two types of atom. Lanthanum and Pr*

---

* Recent measurements indicate some antiferromagnetic ordering [43a].

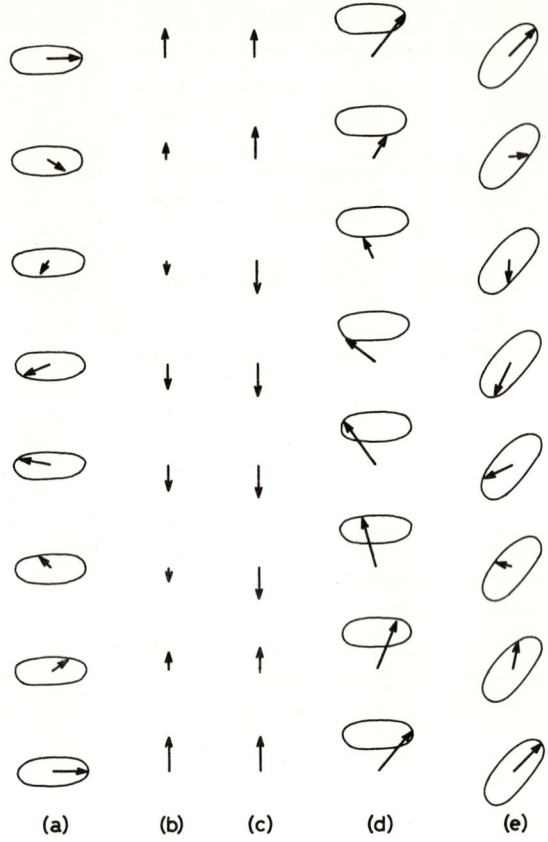

FIG. 3. Types of ordering. (a) Helix (e.g., Ho); (b) LW (Tm at high T); (c) antiphase (Tm at low T); (d) cone (Er at low T); (e) oblique helix (Er at intermediate T?).

are not ordered, Ce [14] and Nd [18] are, although the results have not been fully interpreted. The alloys of the heavy rare earths with Y and among themselves have also been investigated [25, 43]. They show remarkable uniformity in their properties (cf. Fig. 4).

Neutron diffraction also allows determination of the magnitude of the ordered moments and its variation with temperature although it is not easy to obtain these with high accuracy. From form-factor determinations it is in principle also possible to measure the spatial distribution of the moment, although this has not been done yet in the metals.

b. **Bulk magnetic measurements.** Measurements of the magnetic moments of the material as a function of T and of applied field **H** also lead to much important information. At high temperatures the susceptibility can be crudely described by a Curie–Weiss law

$$\chi = N\beta^2\lambda^2 J(J+1)/3k(T-\theta) \qquad (1.8)$$

The Weiss constant θ gives a rough value of the exchange energy favoring ferromagnetism. In fact, in these anisotropic materials

$$\theta = \tfrac{1}{3}(\theta_{\|} + 2\theta_{\perp}) \qquad (1.9)$$

measures this while $\theta_{\|} - \theta_{\perp}$ gives a rough measure of the anisotropy.

Once the system has ordered, a large enough field can change the ordering pattern. For example, a field in the basal plane will distort a helical structure and eventually change it into a ferromagnet. The value of the critical field at which this occurs is related to the energy difference between the ferromagnetic and helical states. This method also allows determination of the magnitude of the ordered moments.

The variation of the magnetic properties of alloys among the rare earths and with Y is remarkably uniform [28, 44]. Values of transition temperatures and θ are shown in Fig. 4 and 5.

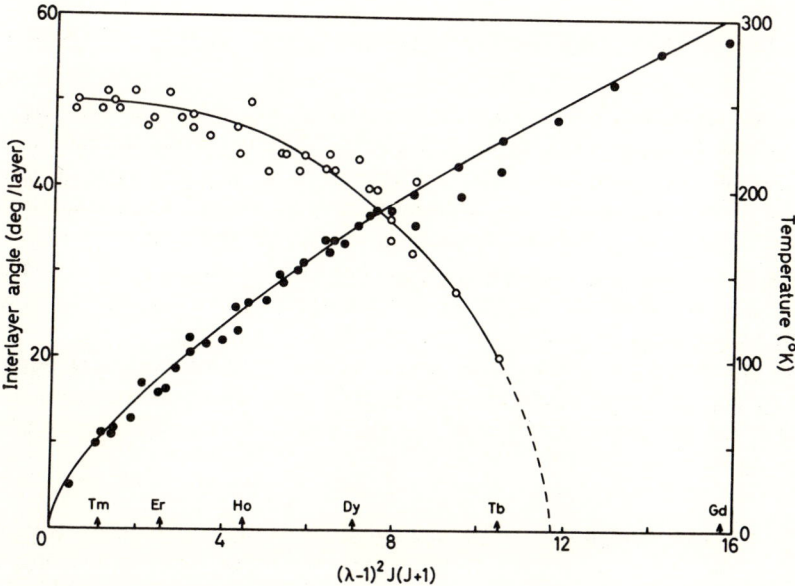

FIG. 4. Transition temperatures and interlayer turn angle $\tfrac{1}{2}qc$ in the heavy rare earths and their alloys with yttrium. (After reference [43].)

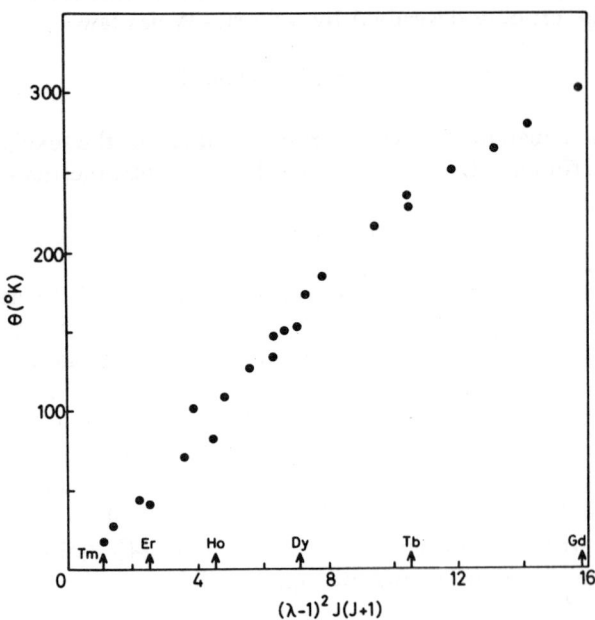

Fig. 5. Curie–Weiss constants for heavy rare earths and alloys with yttrium.

c. **Specific heat.** Sharp specific heat anomalies occur at the magnetic transition temperatures, and allow precise determination of their values. In general the transition from paramagnetic to ordered state is second order, while the anti-ferro transition is first order. This transition is accompanied by a relative lattice distortion of the order of 1% [45]. Since this transition may also be induced by a magnetic field this distortion sometimes gives a magneto-strictive effect several orders of magnitude larger than that observed in iron and nickel [46].

In the first part of the series, where crystal field effects are larger than the exchange, some remnants of Schottky anomalies may be seen. In Pr, which has no order, the whole specific heat may be interpreted in this way [47].

d. **Electron transport.** The electrical resistance and other transport properties show anomalies at the magnetic transitions, indicating that there must be a strong interaction between the conduction and localized electrons. In fact it will be shown that this interaction is at the basis of all the magnetic and electrical properties.

II. Phenomenological Theory

1. Molecular Fields

The essential requirement for the stabilization of magnetic orderings of the type described by Eqs. (1.6) and (1.7) is a long-range exchange interaction whose Fourier transform has a maximum at $\mathbf{q}$. Leaving aside for the moment the origin of such an exchange interaction we assume a simple Heisenberg form where atoms at $\mathbf{R}_n$, $\mathbf{R}_m$ are coupled by

$$-2\mathscr{J}(\mathbf{R}_n - \mathbf{R}_m)\mathbf{S}_n \cdot \mathbf{S}_m \qquad (2.1)$$

Since this exchange energy is generally found to be much smaller than the splitting of the J multiplets by the spin-orbit coupling, $\mathbf{S}$ for each atom must be projected on to the total momentum $\mathbf{J}$, and throughout

$$\mathbf{S} = (\lambda - 1)\mathbf{J} \qquad (2.2)$$

where the Lande factor λ in Eq. (1.5) is used.

For the LW order described by Eq. (1.6) the energy due to the exchange is

$$-\tfrac{1}{2}N\mathscr{J}(\mathbf{q})M'^2(\lambda - 1)^2 J^2 \qquad (2.3)$$

where

$$\mathscr{J}(\mathbf{q}) = \sum_{\mathbf{R}} \mathscr{J}(\mathbf{R}) \cos(\mathbf{q} \cdot \mathbf{R}) \qquad (2.4)$$

while for the helical order described by Eq. (1.7) it is

$$-N\mathscr{J}(\mathbf{q})M^2(\lambda - 1)^2 J^2 \qquad (2.5)$$

Here N is the number of atoms in the crystal. The helical order is thus stabilized at that $\mathbf{q}$ which makes $\mathscr{J}(\mathbf{q})$ a maximum. If $\mathscr{J}(\mathbf{R})$ is only large for nearest neighbors this $\mathbf{q}$ is at 0, if $\mathscr{J}(\mathbf{R})$ is positive, and gives ferromagnetism, while if $\mathscr{J}(\mathbf{R})$ is negative the $\mathbf{q}$ is half-way to a reciprocal lattice vector and gives simple antiferromagnetism. For $\mathscr{J}(\mathbf{R})$ with a long range and oscillatory behavior, $\mathbf{q}$ may take any value.

For the axial LW order the energy (2.3) is only half as great, because some layers have very little order when $\sin(\mathbf{q} \cdot \mathbf{R}_n + \delta) \simeq 0$. For $\mathbf{q} = 0$, all layers are ordered and this leads to $-N\mathscr{J}(0)(\lambda - 1)^2 J^2 M'^2$ without the factor $\tfrac{1}{2}$. Thus the energy favors ferromagnetism unless $\mathscr{J}(0)$ is much smaller than $\mathscr{J}(\mathbf{q})$. At low T the ordering is determined entirely by energy considerations. This is borne out by the fact that the LW phase does not persist to low T in Er and Tm, while the helical phase does in Ho. At high temperatures the extra entropy in the disordered planes of the LW structure helps to decrease the free energy $F = U - TS$ and so stabilize this structure. A molecular field or Bragg–Williams approximation gives the same transition temperature.

$$kT_N = \tfrac{2}{3}\mathscr{J}(\mathbf{q})(\lambda - 1)^2 J(J+1) \tag{2.6}$$

for both structures [48, 49, 50].

The relative stability of these two structures is determined by the anisotropy energy. The simplest form of this is a second-order spherical harmonic

$$K_2(3J_z^2 - J(J+1)) \tag{2.7}$$

If K_2 is positive the plane helical structure is favored, for K_2 negative, the LW. In general there will also be anisotropylike higher-order harmonics, which may tend to favor intermediate directions for the spin, as found in the cone structure of Er. These will be discussed later.

The anisotropy also affects the ordering temperature. If $|K_2| \gg \mathscr{J}(q)$ and K_2 is negative, the only states which are important have $\mathscr{J}_z = \pm J$ and an Ising model results with

$$kT_c = 2\mathscr{J}(q)(\lambda - 1)^2 J^2 \tag{2.8}$$

When K_2 is positive the important atomic states have $J_z = 0$ for integral J and $J_z = \pm \tfrac{1}{2}$ for half-integral. In the former case the ordering may be completely suppressed by the anisotropy; for the latter

$$kT_N = \tfrac{1}{2}\mathscr{J}(\mathbf{q})(\lambda - 1)^2 (J + \tfrac{1}{2})^2 \tag{2.9}$$

is greatly reduced. Bleaney [51] has considered the important effects of the anisotropy in reducing the ordering temperatures and ordered moments in the light elements, particularly Pr [47]. He regards the field as predominantly cubic in symmetry, but the argument is essentially the same. A modification of the molecular field theory gives the transition when

$$2\mathscr{J}(\mathbf{q})\left(1 - \frac{1}{\lambda}\right)^2 \chi_0/N\beta^2 = 1 \tag{2.10}$$

where χ_0 is the susceptibility of an assembly of N independent ions including crystal field effects. For K_2 comparable with $\mathscr{J}(\mathbf{q})$ the calculation is straightforward but very tedious, and has not been performed. For K_2 small,

$$kT_N = \tfrac{2}{3}(\lambda - 1)^2 J(J+1)\mathscr{J}(\mathbf{q}) - \tfrac{4}{5}K_2(J - \tfrac{1}{2})(J + \tfrac{3}{2}) \tag{2.11}$$

for the LW structure. Thus this temperature is increased by a negative K_2. For the helical structure

$$kT_N = \tfrac{2}{3}(\lambda - 1)^2 \mathscr{J}(\mathbf{q})J(J+1) + \tfrac{2}{5}K_2(J-\tfrac{1}{2})(J+\tfrac{3}{2}) \qquad (2.12)$$

is increased by a positive K_2.

The paramagnetic susceptibility may be derived in the same approximation in terms of a Curie–Weiss constant, as defined in Eq. (1.8). In fact, using definition (1.9), we find

$$k\theta = \tfrac{2}{3}(\lambda - 1)^2 \mathscr{J}(0)J(J+1) \qquad (2.13)$$

and

$$(\theta_\| - \theta_\perp) = -6K_2(J-\tfrac{1}{2})(J+\tfrac{3}{2})/5 \qquad (2.14)$$

Several authors have applied formulas of this type to an interpretation of the experimental data [51a]. $\mathscr{J}(0)/k$ proves to be remarkably constant (approximately equal to 36 °K) along the series Gd–Tm and in the alloys, and $\mathscr{J}(\mathbf{q})$ is almost as constant and slightly larger. K_2 fluctuates in sign and in magnitude, but in a systematic way. Both these variations will be interpreted in the light of specific mechanisms later.

2. Spatial Dependence of Exchange

Some indication of the range of $\mathscr{J}(\mathbf{R})$ may be obtained by analysis of the data with a particular model, such as one first used by Enz [52], with an effective exchange interaction A between layers of atoms perpendicular to the c-axis, extended as far as second-neighbor layers. The Hamiltonian is now written

$$-\sum_i \sum_{n=0,\pm 1,\pm 2} 2A_n \mathbf{S}_i \cdot \mathbf{S}_{i+n} \qquad (2.15)$$

where $\mathbf{S}_i$ is the average spin of an atom in the ith layer as defined by Eq. (2.2). Then

$$\mathscr{J}(q) = (A_0 + 2A_1 \cos \tfrac{1}{2}qc + 2A_2 \cos qc) \qquad (2.16)$$

for q along the z-axis. The maximum of $\mathscr{J}(q)$ is given by

$$\cos \tfrac{1}{2}qc = -A_1/4A_2 \qquad (2.17)$$

Using $\theta_\|$, $\theta_\perp$, T_N in conjunction is sufficient to obtain the three A's and the anisotropy constant K_2. For Dy, where there is complete experimental information, such an analysis [53] gives in degrees K,

$A_0/k = -24$, $A_1/k = 44$, $A_2/k = -15$, $K_2/k = 0.6$. $\mathscr{J}(\mathbf{R})$ must be rapidly oscillating and long range to produce such values.

Experimentally, it is found that q varies quite strongly with T (cf. Fig. 6). Phenomenologically this may be represented as a variation of the

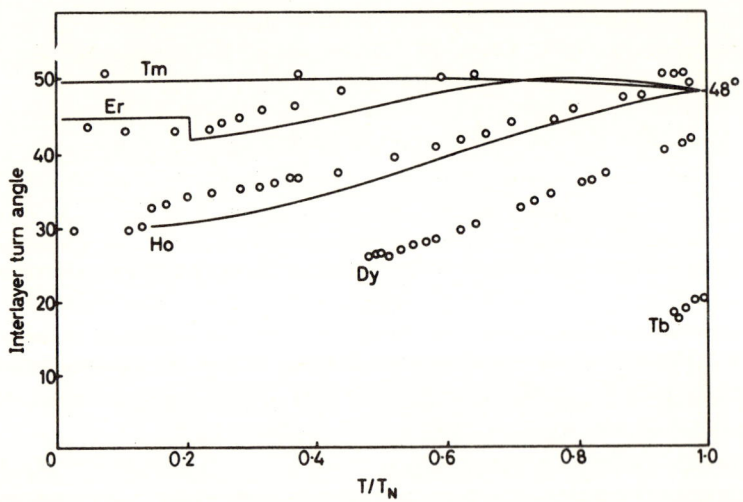

Fig. 6. Thermal variation of q. Points represent experimental results, curves theoretical values. (After reference [76].)

energy with powers of the ordered moment $\langle \mathbf{J}_i \rangle$ higher than the second, and indicates a more complicated exchange interaction than that so far assumed. Hysteresis effects shows that q depends on M rather than T. The results are adequately described by letting A_2 depend on M

$$A_2 \to A_2 - CM^2 \qquad (2.18)$$

For Dy this leads to a value of $C/k = 3\ {}^\circ\text{K}$.

3. Low-Temperature Ordering

Most of the heavy rare earth metals show a transition at a Curie temperature below the Néel temperature to a magnetic phase with a ferromagnetic component. The origin of these effects varies with the form of the high-T ordering. In Tb and Dy, which have a helical phase, the transition is to a simple ferromagnet with moments in the hexagonal plane. Holmium does not apparently show such a transition in the pure state but it is easily induced by an applied magnetic field [34], as in the other materials. This transition is apparently due to an anisotropy field reflecting the hexagonal symmetry. The simplest form of term in the phenomenological Hamiltonian would be

$$K_6{}^6[(J_x + iJ_y)^6 + (J_x - iJ_y)^6] \tag{2.19}$$

Since the transition takes place at a temperature where $\langle J_\perp \rangle \sim J$ and the entropy is low we may consider the stability in terms of the energy rather than the free energy. The change occurs when

$$2K_6{}^6 \langle J_\perp \rangle^6 = 2(\lambda - 1)^2 [\mathscr{J}(q) - \mathscr{J}(0)] \langle J_\perp \rangle^2$$
$$= 2A_1 \langle J_\perp \rangle^2 (\lambda - 1)^2 (1 - \cos \tfrac{1}{2} qc)^2 / \cos \tfrac{1}{2} qc \tag{2.20}$$

For Dy this gives $K_6{}^6 J^6 / k \sim 1\ °K$. The observed values of q become smaller along the series Ho, Dy, Tb, and $q = 0$ in Gd. From Eq. (2.20), this would indicate a rise in the Curie temperature as is observed. It is also possible if $\mathscr{J}(n\pi/3c) > \mathscr{J}(0)$ (for any n) to stabilize a structure which retains a helical component and keeps all moments in the easy hexagonal directions. A more complicated structure which reveals this tendency seems to occur in Ho for $T < 35\ °K$ [33].

Near the Néel point the magnitude of the anisotropy energy drops very rapidly because of the dependence on $\langle J_\perp \rangle^6$. If the transition is then induced by an applied field in the plane it occurs when

$$\lambda \beta H \langle J_\perp \rangle = 2A_1 \langle J_\perp \rangle^2 (\lambda - 1)^2 (1 - \cos \tfrac{1}{2} qc)^2 / \cos \tfrac{1}{2} qc \tag{2.21}$$

Nagamiya et al. [54] have shown that this transition is a little more complicated. At H_c given by Eq. (2.21) the ferromagnetic alignment is incomplete, and the moment structure has the form of a fan. As H increases further the fan closes up until at $H_t = 2H_c$ full ferromagnetic ordering is achieved.

Measurements of H_c such as those of Legvold et al. [30] allow from Eq. (2.21) an alternative method of obtaining A_1 [49]. It leads to results similar to those derived from θ and T_N.

If both the hexagonal anisotropy field and the applied field are important Nagamiya et al. [54] find that the fan phase is partly or wholly suppressed if H is parallel to the easy axis of the anisotropy and $H_t < 2H_c$. If H is along a hard axis, however, the range of the fan phase is increased and $H_t > 2H_c$.

In those metals which first display a LW order the low-T transitions occur for somewhat different reasons. In Er for $50 < T < 80\ °K$ there is a LW structure in agreement with the molecular field theory outlined above if $K_2 < 0$. At the lower temperature the transverse components order with the same wave-like variation as the longitudinal component. Below 20 °K it has a cone structure. The ordering of the transverse components is to be expected from Eq. (2.10) at

$$kT = \mathscr{J}(q)(\lambda - 1)^2[J(J+1) - \langle J_z^2 \rangle] \qquad (2.22)$$

The ordering of J_z causes the last term to vary rapidly with T, and the equation may not always have a solution. Kaplan [48] has considered this problem in some detail using a somewhat more complicated form of the anisotropy energy and finds a satisfactory agreement with experiment. The same problem has also been treated by Yosida and Miwa [50], who showed that a cycloidal spin structure might occur in certain circumstances. There is a greater tendency to order the component with higher anisotropy energy in the LW case, where there are planes of atoms with small moment and high entropy, than in the helical structure.

In Tm the anisotropy energy is apparently always sufficient to keep the moment parallel to the hexagonal axis. In this case the Ising model, as used by Elliott [49], should be an adequate approximation. In this theory some of the moments reach saturation at $T = \frac{1}{2}T_N$ so the structure must change. The transition to an antiphase domain structure has been considered by Wedgwood [55]; he finds it should occur smoothly at a higher temperature than $\frac{1}{2}T_N$. This is in agreement with the fact that the specific heat [41] shows only one sharp anomaly at T_N.

4. Spin Waves

The low-lying excitations of the magnetic structures with helical, conical, and ferromagnetic arrangements may be described by the method of spin waves. For the complicated structures the theory is very tedious, and has been pushed through for model Hamiltonians of varying complexity by a number of authors [56]. The most complete treatment is that of Cooper et al. [57], which is concerned with the explanation of magnetic resonance experiments. Spin wave theory has also been used to interpret low-temperature specific heat [53] and resistivity data [58], and to discuss the phase transitions [50].

The very large axial and hexagonal anisotropies lead to a spin wave spectrum quite different from that which occurs in normal ferro- and antiferromagnets. For the planar ferromagnet such as occurs in Tb and Dy the spin wave energies are

$$\hbar\omega(\mathbf{k}) = 2J\{[(\lambda-1)^2 2(\mathscr{J}(\mathbf{k}) - \mathscr{J}(0)) - \tfrac{1}{3}K_2 + 3K_6^6 J^4][(\lambda-1)^2$$
$$\times 2(\mathscr{J}(\mathbf{k}) - \mathscr{J}(0)) + 18K_6^6 J^4]\}^{1/2} \qquad (2.23)$$

where we have used a Hamiltonian containing Eqs. (2.1), (2.7), and (2.19), i.e., the exchange, a simple axial anisotropy, and the hexagonal anisotropy. This shows a minimum at $\mathbf{k} = \mathbf{q}$ but this is shallow since $\mathscr{J}(\mathbf{q}) - \mathscr{J}(0)$ is small. There is a gap of approximately $J^3(6K_2K_6^6)^{1/2}$ in the spectrum, which for Dy is estimated to be equivalent to about 22 °K using the parameters derived above, while experimentally [59] it appears to be about 30 °K.

In the helical phase, if the hexagonal anisotropy is neglected,

$$\hbar\omega(\mathbf{k}) = 4J(\lambda - 1)^2 \{[\mathscr{J}(\mathbf{q}) - \mathscr{J}(\mathbf{k}) + K_2/6(\lambda - 1)^2]$$
$$\times [\mathscr{J}(q) - \tfrac{1}{2}\mathscr{J}(\mathbf{q} + \mathbf{k}) - \tfrac{1}{2}\mathscr{J}(\mathbf{q} - \mathbf{k})]\}^{1/2} \quad (2.24)$$

where $\hbar\omega(\mathbf{k}) \to 0$ as $\mathbf{k} \to 0$ but is large, $\sim(K_2kT_N)^{1/2}$, when $\mathbf{k} \sim \mathbf{q}$. In a conical structure the expression is more complex but again $\omega(\mathbf{k}) \to 0$ as $\mathbf{k} \to 0$. In this case the frequencies of the spin waves with $\pm\mathbf{k}$ are different, as might be expected since the ferromagnetic component defines a unique direction in the crystal. Such materials will be optically active [60]. A magnetic field, as we have seen, can cause large modifications of these structures. Restricting attention to the helical case, it is found that at small H the changes in ω are proportional to H^2 with a small coefficient. This has been studied in some detail for $\mathbf{k} = 0$ and for combinations of $\mathbf{k} = \pm\mathbf{q}$ which are found to occur, since these are the modes which might be observed in a resonance experiment. There is an abrupt change in ω at H_c, and the $\mathbf{q}$ spin waves have zero frequency at $H = 2H_c$ where the fan structure collapses. So far attempts to find these resonances have not been successful.

For the LW structures, there is so much disorder that spin waves do not form a satisfactory description of the excited states [57].

III. The Crystal Field

1. The Heavy Rare Earths

The array of charges around any ion produces an electric field of appropriate symmetry. The theory of such fields has been extensively developed in ionic crystals. For hexagonal symmetry the potential energy of a single electron at (r, θ, ϕ) may be expanded in spherical harmonics [61] as

$$V = V_2^0 r^2 Y_2^0(\theta,\phi) + V_4^0 r^4 Y_4^0(\theta,\phi) + V_6^0 r^6 Y_6^0(\theta,\phi)$$
$$+ V_6^6 r^6 [Y_6^6(\theta,\phi) + Y_6^{-6}(\theta,\phi)] \quad (3.1)$$

Within a particular J manifold this may be written in terms of operator equivalents of spherical harmonics [62] as

$$V = V_2^0 \langle r^2 \rangle \alpha Y_2^0(\mathbf{J}) + V_4^0 \langle r^4 \rangle \beta Y_4^0(\mathbf{J}) + V_6^0 \langle r^6 \rangle \gamma Y_6^0(\mathbf{J})$$
$$+ V_6^6 \langle r^6 \rangle \gamma [Y_6^6(\mathbf{J}) + Y_6^{-6}(\mathbf{J})] \quad (3.2)$$

The α, β, γ are constants which have been evaluated [13] and are given in Table II. The $\langle r^n \rangle$ are the mean values of r^n over the $4f$ electron distribution, and may be computed [63]. However, it appears that polarization of the outer shells may cause very large changes in the effective values. The V_n^m depend on the particular model of the surrounding charges. This too is difficult to find because of uncertainties in this model and it is usual to regard $V_n^m \langle r^n \rangle$ as parameters. Some indication of relative magnitude may be obtained by placing tripositive charges on the nearest-neighbor sites and assuming the more distant ones to be screened by the conduction electrons. This gives

$$V_2^0 = -300 \text{ cm}^{-1}/\text{Å}^2 \quad V_4^0 = -60 \text{ cm}^{-1}/\text{Å}^4 \quad V_6^0 = +15 \text{ cm}^{-1}/\text{Å}^6$$
$$V_6^6 = -45 \text{ cm}^{-1}/\text{Å}^6 \quad (3.3)$$

The first term in Eq. (3.2) is exactly of the form (2.7) and the last term of the form (2.19) required in the phenomenological theory. The other terms complicate the axial anisotropy and may help to stabilize the cone type of structure as observed in Er. Assuming that the electric field remains the same in the various elements the different anisotropy must arise from variation in α, β, and γ. In the dominant term

$$\sum_i (3z_i^2 - r_i^2) = \langle r^2 \rangle \alpha [3J_z^2 - J(J+1)] \quad (3.4)$$

The left-hand side measures the quadrupole moment of the electron charge cloud, summed over all electrons. For the field given by Eq. (3.3) the low-energy state has its electrons concentrated in the hexagonal plane, and a negative quadrupole moment. This will correspond to a small J_z if α is negative and a large one if α is positive. Thus the sign of α determines the form of the anisotropy. In Table III, the values from Table II and the signs of Eq. (3.3) have been used to show the tendency of each term to produce alignment parallel, perpendicular, or at an angle to the hexagonal axis [49, 50]. Also shown is the easy axis in the hexagonal plane from the hexagonal anisotropy term, relative to a t_1 direction.

TABLE III

FORM OF CRYSTAL FIELD ANISOTROPY IN HEAVY RARE EARTHS

	Tb	Dy	Ho	Er	Tm
V_2^0	⊥	⊥	⊥	∥	∥
V_4^0	∥	<	<	∥	∥
V_6^0	∥	<	∥	<	∥
V_6^6	30	0	30	0	30

[a] The symbols ∥, ⊥, and < give tendency of crystal field to orient moments parallel, perpendicular, or at an angle to the c-axis. For V_6^6 angles of easy direction relative to t_1 are given.

The results show good agreement with experiment. From the dominant Y_2^0 term, Tb, Dy, and Ho are expected to have moments in the plane, while those of Er and Tm should be along the axis. Erbium does have some tendency to leave the axis, but Tm does not. The easy hexagonal directions are also in agreement with experiment [30, 34].

In order to calculate the magnetic anisotropy energy it is necessary to treat the exchange interaction at the same time as the crystal field. This has not been done, although estimates [64], based on the extrapolation of results obtained in the limit of small anisotropy, indicate a much stronger temperature dependence in the higher-order terms. This accounts for the fact that the anisotropy of the high-T phases is entirely determined by the Y_2^0 term but the others become more important at low T.

2. The Light Rare Earths

The crystal fields are relatively stronger in the light rare earths because the $\langle r^n \rangle$ are much bigger. In the usual crystal structure half the atoms have a hexagonal environment, but since the c/a ratio is that of perfect packing V_2^0 is zero in Eq. (3.1). The other half have a cubic environment, and the potential may be written

$$V = V_4^4 \langle r^4 \rangle \{Y_4^0 + 20\sqrt{2}\, Y_4^3\} + V_6^6 \langle r^6 \rangle \{Y_6^0 - (35/\sqrt{8})Y_6^3 + (77/8)Y_6^6\}$$
(3.5)

The energy levels in this field have been computed for all the rare earths [65], and are used by Bleaney [47] to investigate the properties of Pr. This has a singlet ground state, and no cooperative transition. He is able to find a field which accounts for the susceptibility and specific heat, although it is not proved to be unique.

In Ce and Nd, where the exchange effects are more important, no theory has yet been carried out to include this and the two crystal fields [65]. It is likely to be complicated and awaits more experimental information. Trammell [66] has made a theory of the magnetic anisotropy in the simpler but related system of the cubic rare earth nitrides which are antiferromagnetic.

3. Gadolinium

It is convenient at this point to discuss the special properties of gadolinium. It shows no wavelike phases and is a simple ferromagnet. However, it will be shown that the exchange interaction is similar to that in the rest of the heavy rare earths. What is different about it is the very low anisotropy energy, which is much closer to that found in the iron group than to that in the other rare earths. This is because it is an S-state ion; α, β, and γ are zero and there should be no crystal field splitting. It is known, however, from paramagnetic resonance on isolated ions [67] that a small potential persists in the form of (3.2)

$$V = B_2^0 Y_2^0(S) + B_4^0 Y_4^0(S) + B_6^0 Y_6^0(S) + B_6^6 [Y_6^6(S) + Y_6^{-6}(S)] \quad (3.6)$$

through high-order perturbations with the spin-orbit coupling and crystal potential breaking down the Russell-Saunders coupling. Usually $B_2^0 > B_4^0 > B_6^0$.

Gadolinium behaves like a classical Heisenberg ferromagnet with small anisotropy. The spin wave energies

$$\hbar\omega(\mathbf{k}) \sim 2S[\mathscr{J}(\mathbf{k}) - \mathscr{J}(0)] \quad (3.7)$$

are quadratic in k at small k. The specific heat [27] and magnetization change [26] are proportional to $T^{3/2}$ over a wide range, assisted by some fortuitous cancellation of higher terms [67a]. Ferromagnetic resonance [68] of the usual type has been observed. The magnetic anisotropy energy has a complicated temperature dependence [69] which remains if it is analyzed in terms of harmonics. However, the dependence of anisotropy caused by Eq. (3.6) is known to be complicated [70].

IV. Theory of Exchange Interactions

1. The Rudermann-Kittel Interaction

The analysis of the experimental data clearly shows that the exchange interaction must be long range and oscillatory in real space. It also appears to be closely proportional to $(\lambda - 1)^2 J(J + 1)$ as is expected of a simple Heisenberg interaction between spins. Moreover, throughout the alloy systems among the heavy rare earths and with yttrium all the properties vary smoothly with this quantity.

The indirect exchange interaction between localized $4f$ electrons via the conduction electrons has these essential properties. The exchange between a single $4f$ electron on the atom at $\mathbf{R}_n$ with wave function $\phi(\mathbf{r} - \mathbf{R}_n)$ may cause a scattering of a conduction electron between states $\mathbf{k}$ and $\mathbf{k}'$. The matrix element of the interaction

$$\iint u^*(\mathbf{r}_2, \mathbf{k}') \exp(-i\mathbf{k}' \cdot \mathbf{r}_2) \phi^*(\mathbf{r}_1 - \mathbf{R}_n) \frac{e^2}{r_{12}} u(\mathbf{r}_1, \mathbf{k})$$
$$\times \exp(i\mathbf{k} \cdot \mathbf{r}_1) \phi(\mathbf{r}_2 - \mathbf{R}_n) \, d\mathbf{r}_1 \, d\mathbf{r}_2 \quad (4.1)$$

where $u(\mathbf{r}, \mathbf{k}) \exp(i\mathbf{k} \cdot \mathbf{r})$ is a normalized Bloch function, has its main contribution from the u's on the atom at $\mathbf{R}_n$. The matrix element may then be written

$$-\frac{2}{N} V(\mathbf{k}, \mathbf{k}') \mathbf{s} \cdot \mathbf{S}_n \exp[i(\mathbf{k} - \mathbf{k}') \cdot \mathbf{R}_n] \quad (4.2)$$

where V will be similar to the atomic $6s$–$4f$, and $5d$–$4f$ exchange integrals. It will probably not depend very strongly on $\mathbf{k}$ and $\mathbf{k}'$. Here $\mathbf{s}$ is the conduction electron spin, and $\mathbf{S}_n$ the spin of the $4f$ electron. V is expected to be positive by Hund's rule. For an atom with total spin $\mathbf{S}_n$ projected onto $\mathbf{J}$ by Eq. (2.2), a similar formula holds with an appropriate average over V. In what follows V will be regarded as a parameter, and for convenience will be taken independent of $\mathbf{k}$ and $\mathbf{k}'$.

The effect of exchange of this type was first examined by Rudermann and Kittel [71] for the coupling of nuclei, and will be called by us the R-K interaction. Kasuya [72] and de Gennes [73] proposed that it was the important interaction in the rare earths, and Yosida [74] used it for electron spin coupling in other systems. Liu [74a] has examined the coupling V in more detail for the rare earth case. The spin on atom n may be regarded as setting up a spin polarization in the conduction electrons. Because the Fermi distribution restricts the wave vector of the electrons which carry the polarization, this tends to have an oscillatory component. The polarization couples to the spins of other atoms to produce an effective exchange with a long range.

This coupling may be calculated by perturbation theory. Using (4.2) for the interaction, we may write the second-order change in the energy of the system involving $\mathbf{S}_n$ and $\mathbf{S}_m$:

$$-\frac{2V^2}{N^2} \sum_{\mathbf{k}<k_F} \sum_{\mathbf{k}'} \frac{\exp[i\mathbf{q}(\mathbf{R}_n - \mathbf{R}_m)] \mathbf{S}_n \cdot \mathbf{S}_m}{\epsilon(\mathbf{k}') - \epsilon(\mathbf{k})} \quad (4.3)$$

where $\mathbf{q} = \mathbf{k} - \mathbf{k}'$. The first sum over $\mathbf{k}$ is over all occupied states inside the Fermi surface. The second sum is over all states $\mathbf{k}'$ without restriction. $\epsilon(\mathbf{k})$ is the energy of the conduction electron band states. The coefficient of $\mathbf{S}_n \cdot \mathbf{S}_m$ may be regarded as an effective exchange interaction $-2\mathscr{J}(\mathbf{R}_n - \mathbf{R}_m)$. It is clear that the first step in the calculation of (4.3) is to form one sum with $\mathbf{q}$ fixed which leads directly to

$$\mathscr{J}(\mathbf{q}) = \frac{V^2}{N} \sum_{k<k_F} \frac{1}{\epsilon(\mathbf{k}+\mathbf{q}) - \epsilon(\mathbf{k})} \tag{4.4}$$

$$= \frac{3ZV^2}{16\epsilon_F} \phi(\mathbf{q}) \tag{4.5}$$

where Z is the number of conduction electrons, ϵ_F the Fermi energy, and

$$\phi(\mathbf{q}) = 1 + \frac{1-x^2}{2x} \log \left| \frac{1+x}{1-x} \right| \tag{4.6}$$

in the free electron model with

$$x = q/2k_F \tag{4.7}$$

In a general band structure $\phi(\mathbf{q})$ will be different.

The perturbation theory clearly breaks down as $\mathbf{q} \to 0$. This case has $\mathbf{k} = \mathbf{k}'$ and should be excluded and treated by first-order perturbation theory. Van Vleck [75] and Yosida [74] have given more complete arguments to show that it should in fact be included. Performing the sum over $\mathbf{k}$ with this term gives

$$\mathscr{J}(\mathbf{R}) = \frac{3ZV^2\pi}{2\epsilon_F} \Phi(2k_F R) \tag{4.8}$$

where

$$\Phi(y) = (y \cos y - \sin y)/y^4 \tag{4.9}$$

If V depends on $\mathbf{k} - \mathbf{k}'$ only $V(\mathbf{q})$ should appear in Eqs. (4.4) and (4.5), and the sum which gives Eq. (4.8) is more complicated. However, the general long-range and oscillatory form of Eq. (4.8) is not affected.

There is an additional term in the energy giving an effective coupling $\mathscr{J}(0)$, a self-energy for every spin. From Eq. (4.9) with $y = 0$ this diverges, although it will be convergent if one includes the requirement that $V(\mathbf{q}) \to 0$ as $\mathbf{q} \to \infty$. Since, however, this energy does not affect the form of the magnetic ordering, it may be neglected in these considerations.

Although the R-K exchange has a long-range oscillatory form, the free electron result (4.5) gives a maximum of $\mathscr{J}(\mathbf{q})$ at $\mathbf{q} = 0$. A more realistic band structure may, however, produce a maximum for $\mathbf{q} \neq 0$, as was first demonstrated by Yosida and Watabe [8]. They included the symmetry of the hcp structure, while leaving the conduction electrons in the free electron model. In performing the sum of (4.3) to obtain (4.4), a contribution occurs for all $\mathbf{q} + \boldsymbol{\tau}$ where $\boldsymbol{\tau}$ is a reciprocal lattice vector. The interference of the exchange fields from the two atoms (with relative distance $\mathbf{T}$) in the unit cell means that V must be replaced by

$$\tfrac{1}{2}V[1 + \exp{(i\boldsymbol{\tau} \cdot \mathbf{T})}] = VF(\boldsymbol{\tau}) \qquad (4.10)$$

where $F(\boldsymbol{\tau})$ is the form factor (1.2). Assuming V remains independent of $\mathbf{q}$ and $\epsilon(\mathbf{k})$ is still given by the free electron value, Eq. (4.5) becomes

$$\mathscr{J}(\mathbf{q}) = \frac{3ZV^2}{16\epsilon_F} \sum_{\tau} |F(\boldsymbol{\tau})|^2 \phi(\boldsymbol{\tau} + \mathbf{q}) \qquad (4.11)$$

This sum was performed by Yosida and Watabe [8]. They found that $\mathscr{J}(\mathbf{q})$ had a maximum for $\mathbf{q}//z$ and $\tfrac{1}{2}qc = 48°$, remarkably close to the value of $\tfrac{1}{2}qc = 51°$ observed in Tm, Er, and Ho near T_N. The essential requirement for a maximum in $\mathscr{J}(\mathbf{q})$ away from $\mathbf{q} = 0$ is the existence of a $\boldsymbol{\tau}$ such that $|\boldsymbol{\tau} \pm \mathbf{q}|/2k_F \sim 1$, when ϕ is a rapidly varying function. This suggests that the observed values of $\mathbf{q}$ are closely associated with the actual form of the Fermi surface. This will now be investigated from a slightly different point of view.

2. Superzone Boundaries

If a coupling of the form (4.2) exists between the localized $4f$ and the conduction electrons, any ordering of the local magnetic moments will produce an effect on the conduction electrons. For an ordering like (1.6) or (1.7) a new exchange field acts on the conduction electrons which has a translational symmetry different from that of the perfect lattice. Strictly speaking, if the repeating distance of the order is not an integral number of lattice spacings there is no new zone which reflects the complete new symmetry. In a general band structure, confined to the first Brillouin zone, such a field couples electron states with wave vector $\mathbf{k} \pm \mathbf{q} + \boldsymbol{\tau}$ to those with $\mathbf{k}$. The coupling is strongest for those states $\mathbf{k}$ where

$$\epsilon(\mathbf{k} \pm \mathbf{q} + \boldsymbol{\tau}) = \epsilon(\mathbf{k}) \qquad (4.12)$$

A discontinuity appears in the new energy ϵ' which includes the exchange field along the surfaces in $\mathbf{k}$ space which satisfy Eq. (4.12).

These surfaces may be of two types, those imposed by symmetry which will occur for any band structure, and those which depend on the actual detailed form of $\epsilon(\mathbf{k})$. For $\mathbf{q}||z$ in the hcp structure, if we use the fact that $\epsilon(k_x, k_y, k_z) = \epsilon(k_x, k_y, -k_z)$, the symmetry surfaces are planes perpendicular to the z-axis and cutting it at $\pm \tfrac{1}{2}q$ and at $\tfrac{1}{2}q$ inside the zone faces. (The latter perpendicularly bisect $\pm (\tau - \mathbf{q})$ where $\tau = (0, 0, 2)$.) These are the only new zone boundaries which must occur by symmetry.

In the free electron picture it is more convenient to think in the extended zone scheme where the Fermi surface is a sphere. The surfaces of energy discontinuity are now planes which perpendicularly bisect all the $\tau + \mathbf{q}$ (cf. Fig. 2). In this special case there are then many more planes than those imposed by symmetry and these are all at some angle to the z-axis. If we use the interaction (4.2) with constant V as discussed in the preceding section, the magnitude of the energy gap in LW ordering (1.6) is

$$\Delta = \tfrac{1}{2}V(\lambda - 1)JM'F(\tau) \tag{4.13}$$

in bands of both spins, along the planes at $\pm \tfrac{1}{2}(\tau \pm \mathbf{q})$. For helical ordering the gap is

$$\Delta = V(\lambda - 1) JMF(\tau) \tag{4.14}$$

in one spin band along the plane at $\tfrac{1}{2}(\tau + \mathbf{q})$ and $\tfrac{1}{2}(-\tau + \mathbf{q})$ and in the other band along the planes at $\tfrac{1}{2}(\tau - \mathbf{q})$ and $-\tfrac{1}{2}(\tau + \mathbf{q})$. For a particular

$$\tau + \mathbf{q} = 2l \tag{4.15}$$

say, the energy in the bands caused by one boundary alone is

$$2\epsilon'(\mathbf{k}) = \epsilon(\mathbf{K}) + \epsilon(\mathbf{k} + 2l) \pm \{[\epsilon(\mathbf{k}) - \epsilon(\mathbf{k} + 2l)]^2$$
$$+ \tfrac{1}{4}V^2(\lambda - 1)^2 J^2 M_{\pm}^2 | F(\tau) |^2\}^{1/2} \tag{4.16}$$

where

$$M_{\pm}^2 = M'^2 + 2M^2 \pm 2M(M^2 + M'^2)^{1/2} \tag{4.17}$$

if M and M' are both non-zero. If $M = 0$, $M_{\pm} = M'$ and gives discontinuities in agreement with Eq. (4.13). If $M' = 0$, $M_{\pm} = 2M$ or 0 in agreement with Eq. (4.14).

The existence of these energy gaps has important consequences in the theory of many properties of the system. In the next few sections their effect on the detailed form of the R-K interaction, and on the electrical resistivity, will be described.

3. Effect of Boundaries on R-K Interaction

In particular the effective exchange interaction may be calculated from this point of view. In the presence of an order in the localized electrons, energy gaps appear in the conduction bands and the total energy of the system is changed. This energy will vary with the position of the gaps, i.e., with **q**. For small M it will be proportional to M^2 and to $\mathscr{J}(\mathbf{q})$; hence minimizing the energy with respect to **q** will predict the ordering. This method has the advantage that it is readily extended to larger M, to give what might be regarded as modification of the R-K interaction, and possibly predict a variation of q with M. It was noted in the phenomenological theory that such a variation of the effective exchange with M was required (cf. (2.18)).

This method has been used by Elliott and Wedgwood [76] to calculate the variation of **q** with T. Kaplan [77] has independently suggested that it might be an important mechanism, without detailed calculation. The essential improvement over the normal R-K interaction is the use of the energies (4.17) instead of the values computed in second-order perturbation theory. For the boundaries associated with a single τ with gaps $\varDelta$ it is convenient to define the total energy relative to that when $q = 0$:

$$E(\mathbf{q}, \varDelta) - E(0, \varDelta) = -\frac{3Z\varDelta^2}{8E_f} G(\mathbf{q}, \varDelta) \qquad (4.18)$$

where

$$G(\mathbf{q}, \varDelta) = \sum_{\tau} \tfrac{1}{2} |F(\tau)|^2 \left[\phi\left(\frac{|\tau + \mathbf{q}|}{2k_f}, \varDelta\right) + \phi\left(\frac{|\tau - \mathbf{q}|}{2k_f}, \varDelta\right) - 2\phi\left(\frac{|\tau|}{2k_f}, \varDelta\right) \right] \qquad (4.19)$$

and ϕ is the generalization of the R-K function (4.6). Assuming the various boundaries do not interfere, the total energy is a sum over τ in (4.19) as in (4.11).

In the computation of (4.18) it is essential to include the variation of the Fermi energy which occurs because of the new form of the energy bands. The Fermi surface also becomes very different. There are three cases which are illustrated in Fig. 7:

(a) The new boundary cuts the Fermi surface into two parts;
(b) the Fermi surface goes out to meet the boundary but does not have a second piece beyond;
(c) the new boundary does not cut the Fermi surface but merely causes a distortion.

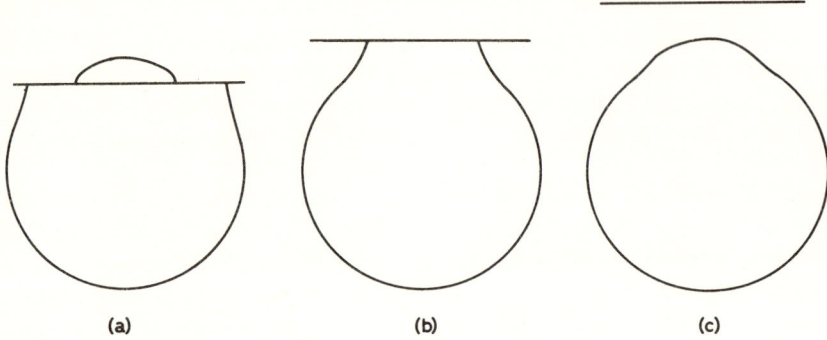

FIG. 7. Effect of one superzone boundary on spherical Fermi surface. These cases correspond to those discussed in Section IV, 3.

The largest effects arise for case (b) when the new boundary is very close to the old Fermi surface. The energy is in fact always minimized for such a value of q, although in a general case it is necessary to balance all the contributions from the different τ. The results of a computation for the free electron Fermi surface are shown in Fig. 8 (after reference [76]). It shows that as Δ/E_f becomes sizable and the distortion of the Fermi surface large, the actual value q_m of q which minimizes the energy varies as Δ increases; q_m eventually becomes smaller, and finally shows a sudden transition to ferromagnetism. This reproduces the essential property of the real rare earths. For a value of $V = 0.5$ ev and the free electron model a reasonable agreement is found for the variation of q with T in Tm, Er, and Ho (Fig. 6). For Dy and Tb, however, the situation is less satisfactory. In these materials the value of q observed at T_N where $\Delta \to 0$ is not given by the simple R-K value as it should be. This effect was first explained by de Gennes [78] as discussed below. The theory also explains the variation of q with concentration in alloys at low T. For Y-Tb alloys the theoretical variation is plotted in Fig. 9, assuming the same value of V and a Δ directly proportional to C, the concentration of Tb. For high concentrations the ferromagnetic phase appears at low T. Although this theory shows such a tendency toward ferromagnetism, the phase appears at values of C lower than the extrapolation of the curve, because of the assistance of the anisotropy energy.

Although the free electron theory has a remarkable quantitative success it will be seen later that the above explanation cannot be correct in detail. The actual Fermi surface must be very different from the free electron sphere. However, the appearance of new boundaries and the consequent distortion of the surface must be the explanation of the variation of q

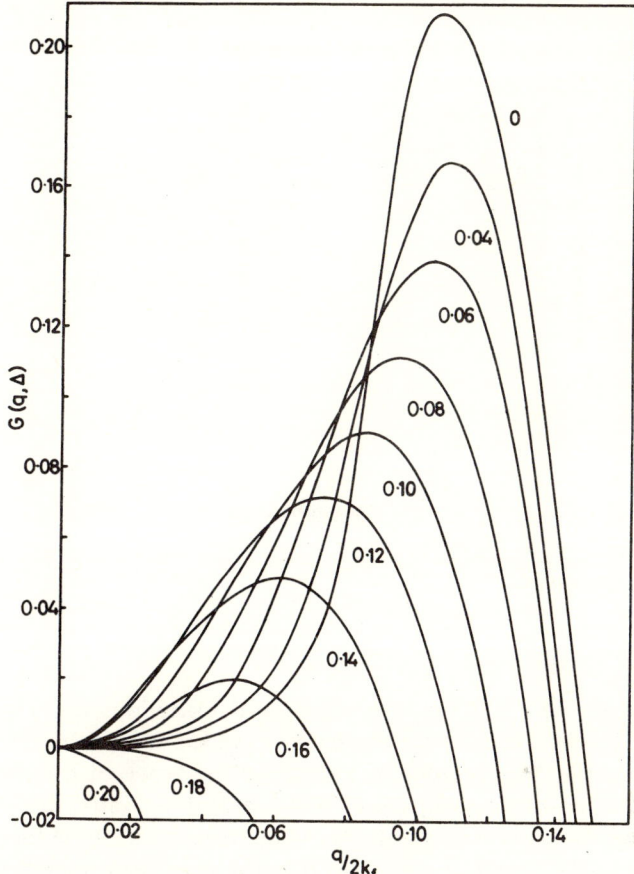

FIG. 8. Variation of the difference between the total energy in a helical structure and a ferromagnetic structure with q and the magnitude of the energy gaps. Numbers indicate values of Δ/ϵ_f. (After reference [76].)

with T. Other mechanisms for this variation such as the thermal expansion of the lattice [50] and the thermal redistribution of electrons at the Fermi surface have been shown to be too small. Other types of exchange interaction, to be discussed later, are also not adequate.

The relatively large distortion of the Fermi surface required here is reflected in other properties. It causes a moment to be induced in the conduction electrons parallel to the local moments. This may be seen most easily in the ferromagnetic state. In Gd the saturation moment [22] appears to have half a Bohr magneton from this effect. A value of $V \sim 0.3$ ev will produce this effect in the free electron model [74a].

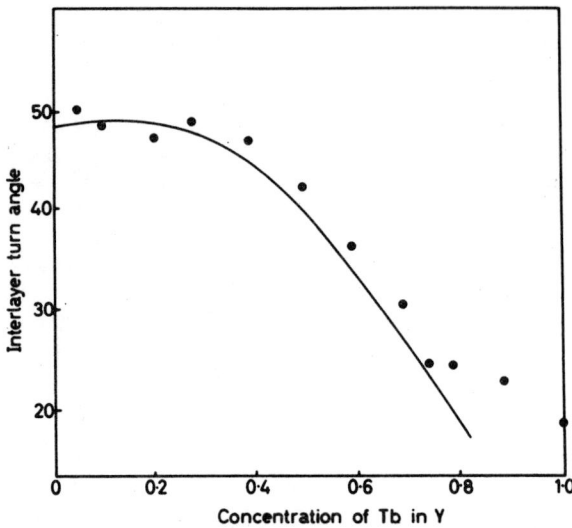

FIG. 9. Values of interlayer turn angle $\tfrac{1}{2}qc$ at 4°K in Tb-Y alloys. Full-curve theory. (After reference [76].)

4. Effect of Scattering on R-K Interaction

The theory of the R-K interaction and its modification given above assume that the conduction electrons have an infinite mean free path. De Gennes [78] has pointed out that the existence of a finite free path will affect this interaction, for it will certainly modify the polarization distribution set up in the conduction electrons by a single local moment. The dominant scattering mechanism in the rare earth metals arises from the disordered moment system through the interaction (4.2). In the approximation where V is independent of $\mathbf{k}$, the relaxation time of a conduction electron is

$$\frac{1}{\tau} = \frac{8\pi}{\hbar} V^2 \langle \mathbf{S}^2 \rangle \nu(\epsilon) \qquad (4.20)$$

where the $\nu(\epsilon)$ is the density of states. At the Fermi surface in the free electron model

$$\nu(\epsilon) = \frac{1}{2\pi^2} \left(\frac{2m}{\hbar^2}\right)^{3/2} \epsilon^{1/2} \qquad (4.21)$$

per unit volume. This relaxation time enters into the theory of the electrical resistance (Section IV, 6).

It is found that the effect of this scattering is to introduce an exponential decay into the effective exchange (4.8) crudely in the form $e^{-R/\lambda}$ where λ is the mean free path. However, partly because of the energy dependence of τ, the oscillatory part of Φ is also changed and so is the **q** which maximizes $\mathscr{J}(\mathbf{q})$. De Gennes used a Green's function method to calculate the general susceptibility of the free electron gas; the algebra is rather complicated and will not be reproduced here. Miwa [79] has extended the calculation to include the effects of the superzone boundaries and gaps using a simpler Green's function method.

In a band with energies $\epsilon(\mathbf{k})$, the Green's function defined by

$$G_0(\epsilon) = \sum_{\mathbf{k}} \frac{1}{\epsilon - \epsilon(\mathbf{k})} \qquad (4.22)$$

has an imaginary part for $\epsilon = \epsilon + i\delta$ which is simply proportional to $\nu(\epsilon)$. In the presence of scattering the G must be generalized to

$$G(\epsilon) = \sum_{\mathbf{k}} \frac{1}{\epsilon - \epsilon(\mathbf{k}) + \gamma(\mathbf{k}, \epsilon)} \qquad (4.23)$$

where γ may in general have a real and an imaginary part. To a first approximation

$$\gamma(\mathbf{k}, \epsilon) = i\hbar/\tau(\epsilon) \qquad (4.24)$$

as given by Eq. (4.20) and is independent of **k**. This assumption may be improved by a self-consistent procedure in which the $\nu(\epsilon)$ in τ is replaced by the new $\nu(\epsilon)$ calculated from Eq. (4.23). This new density of states leads to a new Fermi energy. If the effect of the gaps is now included, $\epsilon(\mathbf{k})$ must be replaced by $\epsilon'(\mathbf{k})$ in Eq. (4.16), and τ reduced because of the ordering; approximately replacing $\langle S^2 \rangle$ in (4.20) by $\langle S^2 \rangle - \langle S \rangle^2$ where the last term is related to the mean ordered moment. In fact τ now has a marked **k** dependence because of the detailed form of the energy bands.

From the new density of states the total energy is calculated and minimized with respect to **q** for a given M. In the limit of small M the energy is quadratic in M^2, but a modified R-K interaction results. The effective ϕ of Eq. (4.5) is found to be changed by scattering in a manner similar to that produced by the gaps. In fact de Gennes found that as the scattering increased, the value of **q** which maximized $\mathscr{J}(\mathbf{q})$ decreased in the free electron model. This accounts for the decrease in q at the Néel temperature along the series from Tm to Gd and for the absence of an antiferromagnetic phase in Gd (cf. Fig. 10). It also gives good

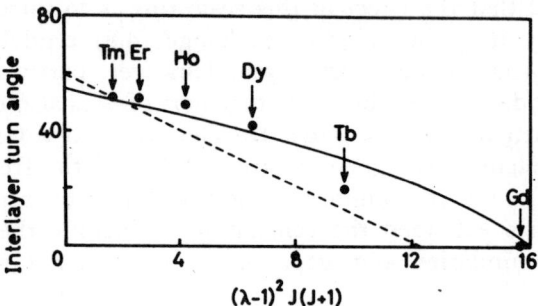

Fig. 10. Variation of q at T_N due to scattering. Two curves represent two theoretical calculations. (After reference [80].)

agreement with the observed variation of q in yttrium alloys [80] where for a concentration C_i of rare earth with spin S_i, $1/\tau$ is proportional to $C_i \langle S_i^2 \rangle$. As the temperature is lowered the scattering decreases but the new boundaries appear, to give two competing processes for the variation of **q**. This is found still to decrease with T but less rapidly than in the theory of Section IV, 3.

Thus it appears that the existence of wave structures and their variation with T are adequately accounted for entirely on the basis of the R-K interaction provided this is refined to include the effect of the new zone structure and the scattering on the Fermi surface.

5. Other Exchange Mechanisms

Several other interionic coupling mechanisms have been proposed in connection with the rare earth metals, many of them in an attempt to explain the **q** variation discussed above. Most of them have been shown to be small on theoretical grounds; none appear to be essential to account for empirical results. They will accordingly be discussed only briefly.

a. *Direct exchange.* Even in the transition group, the most recent computations [81] indicate that direct Heisenberg exchange due to overlap is very small. In the rare earths where the $4f$ electrons are more closely bound inside the $5s$ and $5p$ shells it is likely to be much smaller still. Such exchange as there is will not be simply written in the form $\mathbf{J}_i \cdot \mathbf{J}_j$ because the orbitals corresponding to different J_z states are directionally dependent through their orbital contribution and so have very different overlap. This overlap may be expanded in spherical harmonics of the electron position, and then related by operator equivalents [62] to complicated expressions in the operators $\mathbf{J}_i$ and $\mathbf{J}_j$.

This method has been used by Levy [82] to consider the exchange between a rare earth ion and the surrounding ferric ions in garnets, and even in this case leads to a complicated coupling. The observed simple systematics of the exchange in the heavy rare earth metals (cf. Figs. 4 and 5) would therefore seem to confirm that direct exchange is unimportant.

b. Quadrupole-quadrupole interaction. Because of the directional properties of the electronic wave functions, different J_z states have different quadrupole moments [cf. Eq. (3.4)]. The electrostatic forces between these quadrupoles will tend to line them up in a particular pattern. Such an interaction is thought to be important in solid hydrogen [83] and Bleaney [84] has suggested that it may be important in some rare earth materials. It is not an exchange interaction and only tends to align the moments (i.e., distinguish between $\pm J_z$) in higher order [85]. It has been proposed, however, that in association with normal exchange it would introduce terms in the energy quartic in M and so have the form required in Section II, 2 to explain the variation of q [49]. Although the actual value of the effective ion quadrupole is in doubt because of the induced polarization in the closed shells, it appears that the effect is too small to be observable.

c. Indirect exchange. In an insulating solid, the main exchange passes through the intervening nonmagnetic ions which have closed shells in the ground state. The most general treatment of this effect is given by Anderson [86]. If b is the width of the $4f$ band and U the energy required to transfer a $4f$ electron from one ion to another, there is an antiferromagnetic exchange of order b^2/U. The $4f$ band width is believed to be very small, although it may be increased by admixtures of the $5s$ and $5p$ electron wave functions. In general, however, the effect is expected to be small.

In a metal there is the possibility of a further mechanism in which a $4f$ electron is promoted into the conduction band, or alternatively a conduction electron is promoted into an empty $4f$ band. This mechanism has been considered by de Gennes [87]. He finds an effective exchange interaction of magnitude C^2K/Q^2 where K is the exchange interaction between two $4f$ electrons on the same atom, spectroscopically of order 1 ev. C is the Coulomb interaction between the conduction electron and a $4f$ electron [V in Eq. (4.2) is the exchange part of this interaction]. Q is the promotion energy of the electron between configurations; from optical data this may be as low as 3 ev. [11]. This energy depends on the **k** of the conduction state, but contains a constant term as well, and the variation is smaller than in (4.3). This leads to a sharper **R** dependence so that this interaction is likely to be short range, with an exponential fall-off. The numerical estimates are necessarily crude, but it could lead to an appreciable correction to Eq. (4.8) at the nearest-neighbor distance. Kondo [88] has applied a similar model to Gd, and concludes that it could explain certain properties which are difficult to interpret on the usual model.

d. *Orbital moment—conduction electron interaction.* In the theory of the R-K interaction the anisotropy of the local electron charge cloud was neglected. Expansion of this in spherical harmonics leads to a more complicated effective exchange than (4.3). The simplest correction term has the form

$$\sum_{i,j} T_{ij}[3J_{iz}^2 - J(J+1)]\mathbf{J}_i \cdot \mathbf{J}_j \tag{4.25}$$

This expansion, proposed by Liu [74a], was carried out by Kaplan and Lyons [89]. It is clear that it has precisely the form required of the phenomenological Hamiltonian (2.18) to give the variation of $\mathbf{q}$ with M. It proves, however, to be too small to have any appreciable effect.

e. *Spin density waves.* Overhauser [90] has put forward a theory of spin density waves which appears to account for the magnetic properties of materials like Cr. The theory predicts wavelike order in the conduction electrons because of the stabilization of a disturbance similar to that discussed in Section IV, 3, with new zone boundaries and a changed Fermi surface. In this form the change is stabilized by the interelectronic exchange interactions within the band, which are treated in a self-consistent manner, and the outside influence from the localized moments is not required. However, when local moments are present as in the rare earth case, these can help in the stabilization exactly as in Section IV, 3 [91]. Another way of including Overhauser's effect would be to include the exchange interaction between pairs of conduction electrons in the energy $\epsilon(\mathbf{k})$ in Eqs. (4.4), (4.16), and (4.22). These are of course many-body effects and difficult to handle.

Within the Hartree-Fock approximation he finds the equivalent of $\phi(\mathbf{q})$ in (4.6) to have a divergence at $q = 2k_F$. If this maximum persists it will give a stabilization of wave-like ordering without the interference of several τ effects as required in (4.19). No detailed calculations have yet been made for the rare earth case.

6. Electrical Resistivity

The appearance of the new superzone boundaries described in Section IV, 2 has important effects on other properties of the metals, particularly those associated with the conduction electrons. Perhaps the most important is the electrical resistivity, which shows marked anomalies at the magnetic transitions. In a simplified theory which allows the definition of a relaxation time for the conduction electrons the conductivity tensor may be written

$$\sigma_{ij} = \frac{e^2 \tau}{4\pi^2 \hbar} \int v_i \, dS_j \qquad (4.26)$$

where v_i is the group velocity in direction i of electrons at the Fermi surface. The integral is taken over the Fermi surface. The interaction (4.2) with the localized electrons enters into τ as in Eq. (4.20) [92, 93]. This may be seen in dilute alloys where the spin disorder scattering produces a small resistance. For small admixtures of Gd–Er in Lu, Mackintosh and Smidt [94] found a change in resistance proportional to concentration and to $\langle S^2 \rangle = (\lambda - 1)^2 J(J+1)$ with $V \sim 0.5$ ev. In the concentrated rare earths, this spin disorder scattering is the dominant effect in the paramagnetic region. As the spins order the scattering is reduced [93].

The total scattering may be roughly represented as a function of temperature by

$$\frac{1}{\tau} = \alpha + g(T) + \gamma(1 - M^2 - \tfrac{1}{2}M'^2). \qquad (4.27)$$

The first term arises from impurities, the second from lattice vibrations, and $g(T)$ will be a Gruneisen function and like βT at high T. The last term is the spin disorder scattering; γ is given by Eq. (4.20). This predicts a rapid fall in the resistance below a magnetic ordering transition as is observed in Fe, Ni, and Gd. There may in addition be a small peak at T_c because of critical scattering [93].

In an antiferromagnet, however, the existence of the new zone boundaries causes a strong reduction in the integral in Eq. (4.26) as Fermi surface is lost [95–97]. If the new boundary cuts the Fermi surface into two parts this reduction is proportional to $\Gamma(M'^2 + M^2)^{1/2}$ [96] so that

$$\rho = \frac{\alpha + \beta T + \gamma(1 - M^2 - \tfrac{1}{2}M'^2)}{1 - \Gamma(M^2 + M'^2)^{1/2}} \qquad (4.28)$$

where

$$\Gamma = \frac{3\pi V S}{4\epsilon_f} \sum_i \frac{l_i}{k_f} \qquad (4.29)$$

is a constant. This formula is compared with experiment in Fig. 11, and reproduces the essential form of the anomalies. There is a steep increase

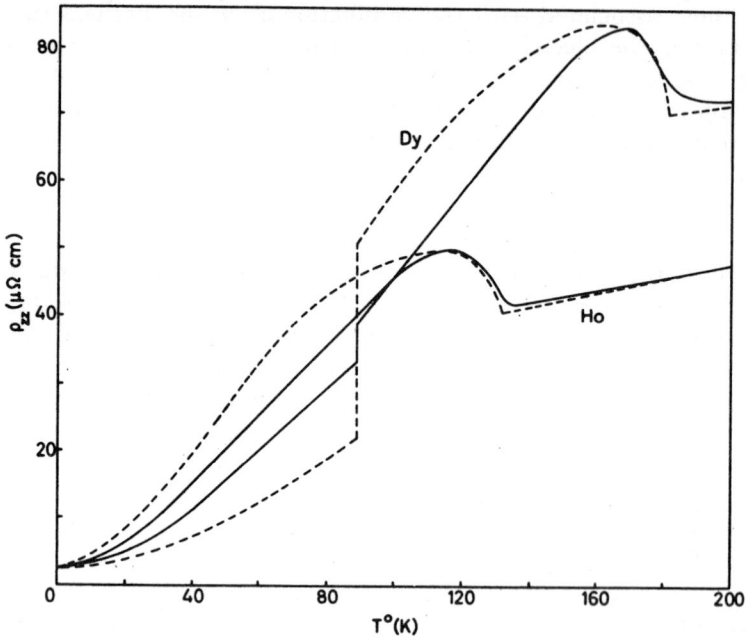

FIG. 11. Resistivity of Dy and Ho. Experimental results, full curves (after references [32, 34]); theoretical results using Eq. (4.28), dotted curves (after reference [96]).

in ρ below T_N as M increases rapidly and a sharp decrease at T_c when the zone boundaries disappear. The detailed fit is not, however, very satisfactory. But it was shown in Section IV, 3 that one boundary must fit onto the Fermi surface without overlap. This leads to a Γ which varies with q and with the precise form of this fit. The variation of q predicted by that theory leads to a variation of Γ of the required form [76].

The free electron theory does not, however, give a completely satisfactory account of the resistivity. Taking $V = 0.5$ ev as required for the fit of q with T leads to a $\Gamma > 1$. Moreover, the effect of the planes (111) and (112) is to give anomalies in the resistance in the c-plane about half as great as those along the c-axis. In this direction, however, no anomalies are observed experimentally requiring $\Gamma_x = 0$. This suggests that the only important boundary is perpendicular to z. Such boundaries must exist by symmetry, while the others need not in a general band structure. If only the (002) boundary is considered, since $F(\tau) = 1$ there, a value of $V = 0.3$ ev would fit the variation of q and give a more reasonable value of Γ. This suggests that the real Fermi surface has in fact a large portion just inside the hexagonal faces of the reduced zone.

In the first half of the series the resistivity anomalies are much less marked [17]. One other mechanism which may play a role there is the scattering of the conduction electrons from the f electrons in different states of the crystal field [98].

The effect of the gaps should also show up in the high-frequency conductivity, i.e., in the optical absorption. It is to be expected that the absorption will increase at a frequency corresponding to $\hbar\omega = \Delta$ when excitations become possible across the gap. Miwa [97] has estimated this effect in some detail. It should prove a useful way of directly determining Δ [97a].

Other transport properties [99] also show anomalies at the magnetic transitions which arise from the same effect but have not been interpreted in detail. The Hall effect is also extremely anomalous [100]. Kondo [101] has interpreted these results in terms of the scattering off the localized electrons which through the spin orbit coupling is not symmetrical about the current direction and hence helps to build up a Hall voltage.

V. Conclusion

The rare earth metals display a wealth of interesting and curious magnetic properties. In this article it has only been possible to review the most important and basic results. Subjects which seemed to the author peripheral, although important for other reasons, such as the hyperfine fields and the superconductivity of La and its alloys, have been excluded. The central theme has been the properties of the heavy rare earths, with their exotic magnetic order and powerful anisotropy. It has been shown that on the basis of a refined Rudermann–Kittel interaction and a crystal field these properties are satisfactorily accounted for. The elements in the first half of the series are less well understood. Preliminary results indicate that the complex crystal structure and the relatively strong crystal field make the results more complicated. The main need is for more experimental work and its interpretation.

It is a pleasure to acknowledge my gratitude to Dr. H. Miwa for comments on the manuscript and discussion of a wide range of problems in this field.

References

1. K. A. Gschneidner, "The Rare Earths" (F. H. Spedding and A. H. Daane, eds.), Chapt. 14. Wiley, New York, 1961.
2. J. M. Lock, *Proc. Phys. Soc.* (*London*) **B70**, 476 (1957).

3. C. E. Olsen, N. G. Nereson, and G. P. Arnold, *J. Appl. Phys.* **33**, 1135 (1962).
4. W. Hayes and J. W. Twidell, *Proc. Phys. Soc. (London)* **82**, 330 (1963).
5. K. A. Gschneidner, R. O. Elliott, and R. R. McDonald, *Phys. Chem. Solids* **23**, 555 (1962).
6. S. Altmann, C. J. Bradley, A. P. Cracknell, J. E. Jeacocke, and J. S. Rousseau, to be published.
7. W. A. Harrison, *Phys. Rev.* **118**, 1190 (1960).
8. K. Yosida and A. Watabe, *Progr. Theoret. Phys. (Kyoto)* **28**, 361 (1962).
9. M. H. Cohen and L. M. Falicov, *Phys. Rev.* **130**, 92 (1963).
10. Y. Rocher, *Advan. Phys.* **11**, 233 (1962).
11. C. Schuler and W. E. Muller, to be published.
12. J. H. Van Vleck, "Theory of Electric and Magnetic Susceptibilities." Oxford Univ. Press, London and New York, 1932.
13. R. J. Elliott and K. W. H. Stevens, *Proc. Roy. Soc.* **A218**, 553 (1953).
14. M. K. Wilkinson, R. H. Child, C. J. McHargue, W. C. Koehler, and E. O. Wollan, *Phys. Rev.* **122**, 1409 (1961).
15. J. M. Lock, *Proc. Phys. Soc. (London)* **B70**, 566 (1957).
16. D. H. Parkinson, F. E. Simon, and F. H. Spedding, *Proc. Roy. Soc.* **A207**, 137 (1951).
17. N. R. James, S. Legvold, and F. H. Spedding, *Phys. Rev.* **88**, 1092 (1952).
18. W. C. Koehler, E. O. Wollan, H. R. Child, and M. K. Wilkinson, Oak Ridge Natl. Lab. Rept. ORNL—2501, 41 (1958) unpublished.
19. D. R. Behrendt, S. Legvold, and F. H. Spedding, *Phys. Rev.* **106**, 723 (1957).
20. L. M. Roberts, *Proc. Phys. Soc. (London)* **B70**, 434 (1957).
21. J. K. Alstad, R. V. Colvin, S. Legvold, and F. H. Spedding, *Phys. Rev.* **121**, 1637 (1961).
22. H. E. Nigh, S. Legvold, and F. H. Spedding, *Phys. Rev.* **132**, 1092 (1963).
23. M. Griffel, S. Legvold, and F. H. Spedding, *Phys. Rev.* **120**, 741 (1961).
24. R. R. Birss and S. K. Dey, *Proc. Roy. Soc.* **A263**, 473 (1961).
25. W. C. Koehler, H. R. Child, E. O. Wollan, and J. W. Cable, *J. Appl. Phys.* **34**, 1335 (1963).
26. D. E. Hegland, S. Legvold, and F. H. Spedding, *Phys. Rev.* **131**, 158 (1963).
27. L. D. Jennings, R. M. Stanton, and F. H. Spedding, *J. Chem. Phys.* **32**, 620 (1960).
28. S. Weinstein, R. S. Craig, and W. E. Wallace, *J. Appl. Phys.* **34**, 1354 (1963).
29. M. K. Wilkinson, W. C. Koehler, E. O. Wollan, and J. W. Cable, *J. Appl. Phys.* Suppl. **32**, 48 (1961).
30. D. H. Behrendt, S. Legvold, and F. H. Spedding, *Phys. Rev.* **109**, 1544 (1958).
31. M. Griffel, R. E. Skochdopole, and F. H. Spedding, *J. Chem. Phys.* **25**, 75 (1956).
32. P. M. Hall, S. Legvold, and F. H. Spedding, *Phys. Rev.* **109**, 971 (1958).
33. W. C. Koehler, J. W. Cable, E. O. Wollan, and M. K. Wilkinson, *J. Phys. Soc. Japan* **17**, Suppl. B-III, 32 (1962).
34. D. L. Strandberg, S. Legvold, and F. H. Spedding, *Phys. Rev.* **127**, 2046 (1962).
35. B. C. Gerstein, M. Griffel, L. Jennings, R. Miller, R. Skochdopole, and F. H. Spedding, *J. Chem. Phys.* **27**, 394 (1957).
36. J. W. Cable, E. O. Wollan, W. C. Koehler, and M. K. Wilkinson, *J. Appl. Phys.* Suppl. **32**, 49 (1961).
37. R. W. Green, S. Legvold, and F. H. Spedding, *Phys. Rev.* **122**, 827 (1961).
38. R. E. Skochdopole, M. Griffel, and F. H. Spedding, *J. Chem. Phys.* **23**, 2258 (1955).
39. W. C. Koehler, J. W. Cable, E. O. Wollan, and M. K. Wilkinson, *Phys. Rev.* **126**, 1672 (1962).

40. D. D. Davis and R. M. Bozorth, *Phys. Rev.* **118**, 1543 (1960); W. E. Henry, *J. Appl. Phys.* Suppl. **31**, 323 (1960).
41. L. D. Jennings, E. Hill, and F. H. Spedding, *J. Chem. Phys.* **34**, 2082 (1961).
42. M. Yosimori, *J. Phys. Soc. Japan* **14**, 807 (1959).
43. W. C. Koehler, E. O. Wollan, and J. W. Cable, *Proc. 3rd Rare Earth Conf., Clearwater, Florida, 1963*, p. 199. Gordon and Breach, New York.
43a. J. W. Cable, R. M. Moon, W. C. Kochler, and E. O. Wollan, *Phys. Rev. Letters* **12**, 553 (1964).
44. W. C. Thoburn, S. Legvold, and F. H. Spedding, *Phys. Rev.* **110**, 1298 (1958).
45. F. J. Darnell, *Phys. Rev.* **130**, 1825; **132**, 1098 (1963).
46. S. Legvold, J. Alstad and J. Rhyne, *Phys. Rev. Letters* **10**, 509 (1963); F. J. Darnell, *Phys. Rev.* **132**, 128 (1963).
47. B. Bleaney, *Proc. Roy. Soc.* **A276**, 39 (1963); *Proc. 3rd Rare Earth Conf., Clearwater, Florida, 1963*, p. 417. Gordon and Breach, New York.
48. T. A. Kaplan, *Phys. Rev.* **124**, 329 (1961).
49. R. J. Elliott, *Phys. Rev.* **124**, 346 (1961).
50. K. Yosida and H. Miwa, *Progr. Theoret. Phys. (Kyoto)* **26**, 693 (1961).
51. B. Bleaney, *Proc. Roy. Soc.* **A276**, 19 (1964).
51a. R. Brout and H. Suhl, *Phys. Rev. Letters* **2**, 387 (1959).
52. U. Enz, *Physica* **26**, 698 (1960).
53. O. V. Lounasmaa and R. A. Guenther, *Phys. Rev.* **126**, 1357 (1962); O. V. Lounasmaa, *ibid.* **126**, 1352 (1962); **128**, 1136 (1962); **129**, 2461 (1963); **133A**, 502 (1964).
54. T. Nagamiya, K. Nagata, and Y. Kitano, *Progr. Theoret. Phys. (Kyoto)* **27**, 1253 (1962).
55. F. A. Wedgwood, Thesis, Oxford Univ., Oxford, England, 1963, unpublished.
56. K. Niira, *Phys. Rev.* **117**, 129)1960); see also refs. [42, 48, 50].
57. B. R. Cooper, R. J. Elliott, S. J. Nettel, and H. Suhl, *Phys. Rev.* **127**, 57 (1962); B. R. Cooper and R. J. Elliott, *ibid.* **131**, 1043 (1963).
58. A. R. Mackintosh, *Phys. Letters* **4**, 140 (1963); D. A. Goodings, *Phys. Rev.* **132**, 542 (1963).
59. B. R. Cooper, *Proc. Phys. Soc. (London)* **80**, 1225)1962).
60. H. Thomas, *Z. Angew. Phys.* **17**, 158 (1964).
61. R. J. Elliott and K. W. H. Stevens, *Proc. Roy. Soc.* **A219**, 387 (1953).
62. K. W. H. Stevens, *Proc. Phys. Soc. (London)* **A65**, 209 (1952).
63. R. E. Watson and A. J. Freeman, *Phys. Rev.* **131**, 250 (1963); **132**, 706 (1963).
64. S. H. Liu, D. R. Behrendt, S. Legvold, and R. H. Good, *Phys. Rev.* **116**, 1464 (1959).
65. K. Lea, M. J. M. Leask, and W. P. Wold, *Phys. Chem. Solids* **23**, 1381 (1962).
65a. T. Murao and T. Matsubara, *Progr. Theoret. Phys. (Kyoto)* **18**, 215 (1957).
66. G. Trammel, *Phys. Rev.* **131**, 932 (1963).
67. W. Low, *Solid State Phys.* Suppl. **2** (1960).
67a. D. A. Goodings, *Phys. Rev.* **128**, 1532 (1962).
68. A. F. Kip, *Rev. Mod. Phys.* **25**, 229 (1953).
69. G. D. Graham, *J. Phys. Soc. Japan* **17**, 1310 (1962); W. D. Corner, W. C. Roe, and K. N. R. Taylor, *Proc. Phys. Soc. (London)* **80**, 927 (1962).
70. E. R. Callen and H. B. Callen, *Phys. Chem. Solids* **16**, 310 (1960).
71. M. A. Rudermann and C. Kittel, *Phys. Rev.* **96**, 99 (1954).
72. T. Kasuya, *Progr. Theoret. Phys. (Kyoto)* **16**, 45 (1956).
73. P. G. de Gennes, *Compt. Rend.* **247**, 1836 (1958).
74. K. Yosida, *Phys. Rev.* **106**, 893 (1957).
74a. S. H. Liu, *Phys. Rev.* **121**, 451 (1960); **123**, 470 (1961).
75. J. H. Van Vleck, *Rev. Mod. Phys.* **34**, 681 (1962).

76. R. J. Elliott and F. A. Wedgwood, *Proc. Phys. Soc. (London)* **84**, 63 (1964).
77. T. A. Kaplan, *J. Appl. Phys.* **34**, 1339 (1963).
78. P. G. de Gennes, *J. Phys. Radium* **23**, 630 (1962).
79. H. Miwa, *Proc. Phys. Soc. (London)* to be published (1965).
80. P. G. de Gennes and D. Saint-James, *Solid State Commun.* **1**, 62 (1963).
81. R. Stuart and W. Marshall, *Phys. Rev.* **120**, 353 (1960); A. J. Freeman, R. K. Nesbet, and R. E. Watson, *ibid.* **125**, 1978 (1962).
82. P. M. Levy, Thesis. Harvard Univ., Cambridge, Massachusetts, 1963; *Phys. Rev.* **135A**, 155 (1964).
83. K. Tomita, *Proc. Phys. Soc. (London)* **A68**, 214 (1955).
84. B. Bleaney, *Proc. Phys. Soc. (London)* **77**, 113 (1961).
85. R. Finkelstein and A. Mencher, *J. Chem. Phys.* **21**, 437 (1953).
86. P. W. Anderson, *Phys. Rev.* **115**, 2 (1959).
87. P. G. de Gennes, *J. Phys. Radium* **23**, 510 (1962).
88. J. Kondo, *Progr. Theoret. Phys. (Kyoto)* **28**, 846 (1962).
89. T. A. Kaplan and D. H. Lyons, *Phys. Rev.* **128**, 2072 (1962).
90. A. W. Overhauser, *Phys. Rev.* **128**, 1437 (1962).
91. A. W. Overhauser, *J. Appl. Phys.* **34**, 1019 (1963).
92. T. Kasuya, *Progr. Theoret. Phys. (Kyoto)* **16**, 58 (1956).
93. P. G. de Gennes and J. Friedel, *Phys. Chem. Solids* **4**, 71 (1958).
94. A. R. Mackintosh and F. A. Smidt, *Phys. Letters* **2**, 107 (1962).
95. A. R. Mackintosh, *Phys. Rev. Letters* **9**, 90 (1962).
96. R. J. Elliott and F. A. Wedgwood, *Proc. Phys. Soc. (London)* **81**, 846 (1963).
97. H. Miwa, *Progr. Theoret. Phys. (Kyoto)* **28**, 208 (1962).
97a. C. C. Schüler, *Phys. Rev. Letters* **12**, 84 (1964).
98. R. J. Elliott, *Phys. Rev.* **94**, 564 (1954).
99. R. V. Colvin and S. Arajs, *Phys. Rev.* **133**, A 1076 (1964).
100. G. S. Anderson, S. Legvold, and F. H. Spedding, *Phys. Rev.* **111**, 1257 (1958).
101. J. Kondo, *Progr. Theoret. Phys. (Kyoto)* **27**, 772 (1962).

Author Index

Numbers in parentheses are reference numbers and indicate that an author's work is referred to although his name is not cited in the text. Numbers in italic show the page on which the complete reference is listed.

Abe, H., 365(46, 51), 368(46), *381*
Abragam, A., 169, 172, 175, 184(3), 208(169), 217, 220, 222(3), 240(3, 22), 247, 248, 260(22, 355), 261, 262, 264, 270(22), *292, 293, 297, 298, 302, 303,* 308(1), 312(1), *353,* 358(7), *380*
Adams, W. H., 237(280), *301*
Alexander, E., 111, 162, *164*
Alexander, S., *165*
Alfimenko, V. P., 205(159), *297*
Alff, C., 203(106), 204(106), 210(106), *295*
Alikhanov, A. I., 205(158), *297*
Allen, L. C., 177, *293*
Alstad, J. K., 393(21), 396(46), *422, 423*
Altick, P. L., 230(261a), *300*
Altmann, S., 389, *422*
Ambler, E., 185, *294, 299*
Amelinckx, S., 205(158), *297*
Amos, A. T., 233(275), *301*
Anderson, D. H., 233(275), *301,* 365(40), 367(40), 372(95), *381, 382*
Anderson, G. S., 421(100), *424*
Anderson, L. W., 376(104), *382*
Anderson, P. W., 271(395), 275(399, 400), 276, 283(395), *304,* 364(28), 376(101), *380,* 417, *424*
Angew, Z., 403(60), *423*
Antonoff, M., 84, 88, *102*
Arajs, S., 421(99), *424*
Armstrong, J., 375(99), *382*
Arnold, G. P., 387(3), *422*
Aronowitt, R., 169(4), *292*
Arp, V., 188, 218(60), *294,* 358(4), *380*
Arrot, A., 37(34), *41*
Arrott, A., 203(142), *296*

Asanuma, M., 370(80), *382*
Asayama, K., 369(77), 374(77), *382*
Avivi, P., 204(117), *295*
Axel, P., 204(129), *296*

Baqus, P. S., 288, *305*
Baker, G. A., 6, 18(6b), 19, 29(6b), *41*
Bailey, C. A., 190, *294*
Baker, J. B., 268(385), *303*
Baker, J. M., 267(375), 269(375), 272(375), 275(403), 281(414), *303, 304*
Barnes, R. G., 209(193, 194, 196), 211(193, 194), 273(196), *298*
Barrett, P., 202(105), 283(424), *295, 305*
Barrett, P. H., 204(156), 214(156), *296*
Bassani, F., 228(256c), *300*
Bauminger, R., 207(168), 202(108), 203(107, 108), 204(107, 116, 117, 118), 209(168), 210(107, 108, 168), 273 (168), *295, 296, 297*
Bazely, N. W., 223(244), *299*
Beards, G. B., 202(110), 203(110), *295*
Beck, P. A., 189, *294*
Becker, R. L., 213(200), 264(200), *297*
Behrendt, D. R., 393(19, 30), 401(30), 405(30, 64), *422, 423*
Belov, N. V., 111, 113(14), 114, 118, 122, 129, 155, 161, 164, *165*
Benczer-Koller, N., 204(150), *296*
Benedek, G. B., 214(205, 206), 238(292), 274(292), *298,* 308(2), 322(28), 331(28), 332, 342, 343, *354,* 359(12), 372(12, 93), 373(12), 375(12, 99), *301, 380, 382*
Bennett, L. H., 359(11), 369(70, 73, 74, 76), 370(76), 373(74), 374(70, 73, 74), 375(70), *380, 382*

425

Benson, R. E., 233(275), *301*
Benzie, R. J., 188(57), *294*
Berggren, K. F., 241(329), 245, *302*
Berlin, T. H., 3(2), *41*, 62(18), *102*
Berman, A., 285(424e), *305*
Bersohn, R., 177, *293*
Berstein, S., 222(238), *299*
Bertaut, F., 365(39), 367(39), *381*
Bertaut, E. F., 162, *165*
Berthier, G., 239(303), *302*
Bessis, N., 181, 182, 245, 247(337), 253, 254(41, 42), *293*, *302*
Bethe, H. A., 46, *102*, 289(439), *305*
Bettran-Lopez, V., 248(339), *302*
Beun, J. A., 218(214), *298*
Bienenstock, A., 113(25), *165*
Birss, R. R., 163, 164, *165*, 393(24), *427*
Blandin, A., 213(202, 203), *298*
Bleaney, B., 184(50), 185, 186, 187 (53, 56), 188, 191(53), 218(215), 260(357), 267 (373a, 376), 268(81, 316), 269(389), 281(414), *294*, *298*, *303*, *304*, 396(47), 398(47), 405, 417, *423*, *424*
Blinder, S. M., 240(316), *302*
Blin-Stoyle, 171(10), 268(10), *292*
Bloch, F., 83, *102*
Blocker, III, T. G., 365(35), 367(35), 377(35), *381*, *383*
Bloembergen, N., 369(78), 376(100), *382*
Bloom, M., 333(39), 337(39), *354*
Blum, N., 206(163), 207(163, 166) 282(163), *297*
Blumberg, W. E., 267(374), 269(374), 272(374), 281(415), *303*, *304*
Blume, M., 250, 274(398), *302*, *304*
Blume, R. J., 362(24), 365(50), 369(24, 75), 372(24), *380*, *381*, *382*
Bockmann, K., 205(159), *297*
Bolger, B., 313(7), *354*
Boorse, H. A., 285(424e), *305*
Borg, R. J., 205(158), *297*
Both, R., 205(158), *297*
Boutron, F., 208(169), *297*, 366(33), *381*
Boyd, E. L., 204, 272(397), *296*, *304*, 360(18), 362(24), 364(27), 365(36, 50), 367 (36, 38), 368(18), 369(24), 370(82), 372(24), 376(82), *380*, *381*, *382*
Boyd, R. G., 268, *303*

Boyle, A. J. F., 204(131, 154), 205(158), 211(154), *296*, *297*
Boys, S. F., 231(267), *300*
Bozorth, R. M., 393(40), *423*
Bradley, C. J., 389(6), *422*
Bragg, W., 49(16), *102*
Breit, G., 169(5),*292*
Brice, D. K., 261(361), *303*
Brigman, G. H., 257(348, 349, 351), *303*
Brillouin, L., 232, 258, *300*
Brix, P., 214, *298*
Brout, R., 3(4), 22(4), 29(4), 34(4), *41*, 46(7), 47(14), 51(7), 62(7), 63(7), 64(7), 71, 77(14), 80(14), 81(14), 83(7, 14), 85(32, 36), 87(32), 93, 94(7, 39), 96, 98(48), *102*, *103*, 399(51a), *423*
Brown, M. R., 322(22, 23), *354*
Brown, P. J., 203(121), *296*
Brown, T. H., 233(275), *301*
Bruner, L. J., 362(24), 365(50), 369(24, 75), 372(24), *380*, *381*, *382*
Bryukhanov, V. A., 205(158), 209(177, 192), 211(192), *297*, *298*
Buchanan, D. N. E., 203(119, 123), 208(119), 210(123), 285(429), 286(429), *296*, *305*
Bucharev, V. A., 205(158), *297*
Budnick, J. I., 286(429c), *305*, 362(24), 365(50), 369(24, 63, 75), 372(24), 374(63), *380*, *381*, *382*
Buishvili, L. L., 379(120), *383*
Bundury, D. St., 204(131, 154), 205(158), 211(154), *296*, *297*
Burford, R. J., 331(36), *354*
Burgiel, J. C., 218, *354*
Burke, E. A., 241(333), 246(333), 257(350), *302*, *303*
Burley, D. M., 29, 34(24b), *41*, *42*
Burns, G., 177, 179(35), 208(171, 174), 209(186, 197), 210(186), 257(35), *293*, *297*, *298*
Buryn, A., 221(235), *299*
Buyle-Bodin, M., 365(49), 366(31), *381*
Cable, J. W., 393(25, 29, 33, 36, 39, 43a), 394(25, 43), 395(43), 401(33), *422*, *423*
Cacho, C. F. M., 218(222), *299*
Caldow, G. L., 223(244b), *299*
Callaway, J., 238, *301*
Collen, E. R. 406(70), *423*

AUTHOR INDEX 427

Callen, H. B., 46, 51, 62, *102*, 406(70), *423*
Carr, Jr., W. J., 271, *304*
Carruthers, P. A., 85, *103*
Casimir, H. B. G., 172, *293*
Caspari, M. E., 212, 220, *298*, *299*
Chen, C. W., 203(147), *296*
Cheng, C. H., 189(63, 64, 65), *294*
Child, R. H., 393(14,18, 25), 394(14, 18, 25), *422*
Clogston, A. M., 238(294), 261, 271, 274(294), 275(417), 276, 278(412), 280 (412), 281(417, 420), 282(421), 283(421, 423), 284, *301*, *303*, *304*, *305*, 364(28), *380*
Cochran, D. R. F., 202(104, 105), 204(128), 205(104, 158) 283(424), *295*, *296*, *297*, *305*
Cohen, M. H., 172, 178, 219(16), 240(314), 254(314), 272(314), 280, 281, *293*, *302*, 389(9), *422*
Cohen, R. L., 204(157), 209(178, 196), 211 (178), 273(157, 196), *296*, *297*, *298*
Cohen, S. G., 202(108), 203(107, 108), 204(107, 116, 117, 118), 205(159), 210(107, 108), 220, *295*, *296*, *297*, *299*
Coleman, A. J., 224(247), 242(247), *300*
Coles, B. A., 322(22, 23), *354*
Collins, M. F., 286(430), *305*
Colvin, R. V., 393(21), 421(99), *422*, *424*
Compton, D. M. J., 201(100), *295*
Condon, E. U., 228(255), *300*
Cooke, A. H., 37(32), *42*, 188(57), *294*
Coolidge, A. S., 245, *302*
Cooper, B. R., 402, 403(57, 59), *423*
Coopersmith, M., 94(39), *103*
Corenzwit, E., 96(42), *103*, 282(421), 283 (421), 284(421), *305*
Corliss, L. M., 162(36), *165*
Corner, W. D., 406(69), *423*
Costa, N. L., 209(188), *298*
Coulson, C. A., 223(244b), 288, *299*, *305*
Cowan, D. L., 376(104), *382*
Cox, D. E., 202(109, 144), 203(109), 210(109, 144), *295*, *296*
Cracknell, A. P., 389(6), *422*
Craig, P., 285(424c), *305*
Craig, P. P., 189(67), 192(67), 202(105), 204(112, 128), 205(158), 207(165), 283 (424), 284(165), *294*, *295*, *296*, *297*, *305*

Craig, R. S., 395(28), *422*
Crangle, J., 278, *304*
Cranshaw, T. E., 202(103), 203(149), 210(149), 285(103, 425), *295*, *296*, *305*
Crawford, M. F., 223(244a), *299*
Culvahouse, J. W., 261, *303*
Curry, N. A., 378(116), *383*

Dabbs, J. W. T., 222(238), *299*
Dalgarno, A., 176(24), 177, 179(29), *293*
Dalton, N. W., 5(33), 15(33), 22(33), 34(33), *42*
Daniel, E., 213(202, 203), *298*
Daniels, J. M., 218(213, 215, 219), *298*
Danon, J., 209(188), *298*
Darnell, F. J., 396(45, 46), *423*
Das, T. P., 172, 177, 179(32), 238(292), 240(315), 274(292), *293*, *301*, *302*, 308(3), 309(3), *353*, 359(9), *380*
Dash, J.G., 189(67), 192, 204(112, 128), *294*, *295*, *296*
Davidson, E. R., 246(335, 336), 258, *302*
Davis, D. D., 393(40), *423*
Day, G. F., 369(63, 64, 66, 73, 74), 373(66, 74, 96), 374(63, 64, 73, 74), *381*, *382*
Dearman, H. H., 233(275), *301*
De Benedetti, S., 200(99), 208(99), 209(99, 179), 210(99, 179), 211, *295*, *297*
Debrunner, P., 218(224), *299*
deCoster, M., 205(158), *297*
de Gennes, P. G., 377(107, 108), 378(112, 114), *382*, *383*, 407, 412, 414, 416(80), 417, 419(93), *423*, *424*
de Groot, S. R., 183(45), 184(45), 215(45), 216(45), *293*
de Klerk, D., 222, *299*
Döring, W., 361, *380*
Delyagin, N. N., 205(158), 209(177, 192), 211(192), *297*, *298*
Dempesy, C. W., 222(72), *294*
De Pasquali, G., 194(89a), *295*
Deutsch, M., 221, *299*
Devons, S., 215(210), *298*
de Wette, F. W., 193, *294*
Dey, S. K., 393(24), *422*
Diaz, J., 204(153), *296*
Dienes, P., 5(6a), *41*
Dimmock, J. O., 109, 162, *164*, *165*, 285(424d), *305*

Dirac, P. A. M., 224(249), *300*
Domb, C., 3(3), 4, 5(3, 33), 6(3), 8(3), 10, 11(3, 13), 12(14), 13, 14, 15(3, 24a, 33), 16, 17, 18, 19, 20, 21, 22(33), 23, 24(15), 25(27a), 26(18a), 27(18), 29, 32(3), 34(24a, 33), 35, 37(27a), 38, *41*, *42* 45(4), 46(5), 59(4), *102*, 330(34), *354*
Donnay, G., 161(35), 162,*165*
Donnay, J. D. H., 161(35), *165*
Donnay, J. P. H., 162(36), *165*
Doyle, W. M., 268(382), *303*
Drever, R., 214(205), *298*
Dreyfus, B., 189(68, 69), 192, *294*
Dysony F. J., 47, 72, *102*
Dzialoshinskii, I. E., 163, *165*

Eaton, D. R., 233(275), *301*
Ebina, V., 267, *303*
Edmunds, D., 188(61), *294*
Edwards, C., 204(131, 154), 205(158), 211(154), *296*, *297*
Ehara, S., 369(71), *382*
Eicher, H., 205(159), 209(196), 273(196), *297*, *298*
Eisele, J. A., 330(33), *354*
Eisinger, J., 267(374), 269(374), 272(374), 281(415), *303*, *304*
Elinn, P. A., 203(147), *296*
Elliott, N., 162(36), *165*
Elliott, R. J., 94, 95, 96, *102*, *103*, 219, 267(227), 281(414), *299*, *303*, *304*, 392(13), 400(76), 401(49), 402(57), 403 (57, 61), 404(13, 49), 411, 412(76), 413(76), 414(76), 417(49), 417(96), 420 (76, 96), 421(98), *422*, *423*, *424*
Elliott, R. O., 387(5), *422*
Endo, K., 369(60), 370(60), *381*
Englert, F., 46, 47(14), 51, 67, 72, 77(14), 80(14), 81(14), 83(14), 84, 85(36), 86(23), 87, 88, 90, 98(48), *102*, *103*
Enz, U., 399, *423*
Epstein, L. M., 203(148), 204(148), 210(148), *296*
Erickson, R. A., 188, 218(58, 59), *294*, 358(4), *380*
Essam, J. W., 29(19a, 19b), *41*
Ewald, P. P., 113, *165*
Falicov, L. M., 389(9), *422*
Fermi, E., 169, 175, 240(20), *292*, *293*

Finkelstein, R., 417(85), *424*
Fisher, M. E., 4, 5, 17, 29(19a), 30, 31(22), 32, 34(24c), 37(35), 38, *41*, *42*, 331(36), *354*
Flinn, P. A., 205(158), 285(426), *297*, *305*
Fock, V., 224, *300*
Folen, V. J., 202(110), 203(110), *295*
Foley, H. M., 175(18, 18a,b), 176(18, 18a,b), 177, 178, 180(18a,b), 193(18, 18a,b), 240(18, 18a,b), *293*
Forrer, R., 35, *42*
Fournet, G., 6, *41*
Fox, D. W., 223(244), *299*
Fraenkel, G. K., 233(275), *301*
Frankel, R. B., 204(155), 214(155), *296*
Frankel, S., 217(212), 220(232, 212), *298*, *299*
Frauenfelder, H., 194(89a), 195(88), 197 (88), 201(88), 202(104), 205(104), 215 (208), 216(208), 222(208), *295*, *298*, 358(2), *380*
Freeman, A. J., 172, 178, 179(33), 180(37), 181(39, 40), 193(40), 206(164), 207(164, 166, 168), 208(39), 209(40, 166, 168), 210(168), 228(253b), 238(295, 299), 239(12, 299, 310, 312), 242, 247(337), 248, 249(12), 250(12, 340), 251(12, 310, 341), 252(341), 254(33, 39, 40, 310, 341, 343, 345), 255(346), 256(39), 257(40), 263(12), 264(12), 267(253b), 268(40), 269(40, 253b, 346), 270(12), 271(299, 312), 272(12), 273(168, 346, 398), 274(295, 299), 275(341), 279(412c), 281(343), 283(164), 285(424d), 287 (431), 289(431), *292*, *293*, *297*, *300*, *301*, *302*, *303*, *304*, *305*, 313(13), 316(13), *354*, 363, 366, *380*, 404(63), 416(81), *423*, *424*
Friedberg, S. A., 37, *42*
Friedel, J., 213, *298*, 419(93), *424*
Friedman, E. A., 203(122), 285(427), *296*, *305*
Froman, A., 223(244b), *299*
Fry, D. J. I., 281(415), *304*
Fukuda, N., 234(276), 237(276), *301*
Fukuyama, T., 365(51), *381*

Galperin, F., 206(162), 214(162), *297*
Galt, J. K., 360(17), *380*
Gambino, R. J., 272(397), *304*, 370(82), 376(82), *382*

Gammel, J., 24(18b)
Gastebois, J., 209(181, 184), 210(181), *297*
Gerber, W. D., 203(142), *296*
Gerstein, B. C., 191, 393(35), *294, 422*
Geschwind, S., 204, 250(240a), 264, 265, 272, 281(415), *296, 302, 303, 304,* 374(97), *382*
Ghosh, S. K., 365(35), 367(35), 377(35), 379(118), *381, 383*
Giles, J. C., 218(219), *298*
Gilleo, M. A., 217(212), 220(212), *298*
Giorgadze, N. P., 379(120), *383*
Giovannini, B., 279(412a), *304*
Glassgold, A. E., 230(261a), *300*
Gol'Danskii, V. I., 205(158), *297*
Goldfarb, L. J. B., 215(210), *298*
Golding, G., 218(223), *299*
Goldstone, J. 230, *300*
Good, R. H., 405(64), *423*
Goodings, D. A., 239(311), 240(313, 314), 241(313), 248(313), 254(314), 271(313), 272(314), 278(311), 280(311, 314), 281 (314), *302*, 402(58), 406(67a), *423*
Goodman, B. B., 189(68, 69), 192(69), *294*
Gordon, J. E., 222(72), *294*
Gordon, J. P., 238(294), 274(294), *301*
Gorodinskii, G. M., 205(158), 209(191), 211(191), *297, 298*
Gorter, C. J., 37(31), *42*, 184, 186(48), 206(48), 218(214), *294, 298*
Gossard, A. C., 218(217), 268(383), 278, 281(413, 419), *298, 303, 304, 305,* 358(5), 360(19), 362(5, 23), 365(42), 368(42, 43), 369(23), 370(19, 84), 375(42), 376(23), *380, 381, 382*
Grace, M. A., 218(213, 215, 222), *298, 299,*
Grace, M. A., 358(3), *380*
Graham, G. D., 406(69), *423*
Grant, I. P., 289(441), *305*
Grant, R. W., 204(151), 209(189), 211(189), *296, 298*
Gray, J. D., 257(349, 351), *303*
Green, R. W., 393(37), *422*
Griffel, M., 393(23, 31, 35, 38), *422*
Griffith, J. S., 260(358), 263(358), *303*
Grodzins, L., 206(163), 207(163, 166, 168), 209(168), 210(168), 221(235), 273(168), 282(163), *297, 299*

Gschneider, K. A., 387(1, 5), 388(1), *421, 422*
Guccione, R., 106, 109(2), 113(2, 23, 24), 116(2), 119(2), 123(2), 127(2, 23), 130(2), 152(2), 157(2), *164, 165*
Guenther, R. A., 191(75), 192, *294,* 399(53), 402(53), *423*
Gutousky, H.S., 233(275), *301*
Gutswiller, M., 84, *103*

Haas, W. P. A., 313(11, 12), 318(12), 327(11, 12), 328(12), 338(11, 12), *354*
Haayman, P. W., 367(37), *381*
Hahn, E. L., 172, *293*, 308(3), 309(3), *353*, 359(9, 15, 16), 361(16), 363(16), 370(16), 376(16), 377(16), 378(16), *380*
Haken, H., 71, *102*
Halban, H., 218(215), *298*
Halford, D., 268(381), *303*
Hall, G. G., 223(244b), 233(275), *299, 301*
Hall, H. E., 205(158), *297*
Hall, P. M., 393(32), 420(32), *422*
Hall, T. P., 238(299a), 274(299a), *301*
Ham, F. S., 261, *303*
Hanna, S. S., 194(89a), 201(101, 102), 202(101, 102, 139), 204(115, 120, 132, 153), 206(101, 102), 211(115), 270, *295, 296*, 370(85), *382*
Hardeman, G. E. G., 313(5-12), 314, 318(12), 327(5-12), 328(12), 329, 338 (5-12), *354*
Hardy, W. A., 369(69), 371(69), *381*
Harriman, J. E., 243(325), 253, *302*
Harris, J. K., 204(150), *296*
Harrison, W. A., 389, *422*
Hartmann-Boutron, F., 377(108), 378(108, 112, 114), 379(119), *382, 383*
Hartree, D. R., 224, 225(251), 226, 228, 287(251), 288, 289, *300*
Harvey, J. S. M., 182(43), *293*
Hashi, T., 365(41, 46, 51, 52), 367(41), 368(46), 378(41), 379(41), *381*
Hass, M., 202(110), 203(110), *295*
Hastings, J. M., 162(36), *165*
Hauser, U., 209(178), 211(178), *297*
Hay, H. J., 194(89b), *295*
Hayes, W., 238(299a), 267(373a), 274(299a), *301, 303*, 387(4), *422*
Heap, B. R., 94(40), 95(40), 96, *103*

Heberle, J., 194(89a), 201(101, 102), 202(101, 102, 139), 204(153), 206(101, 102), 270(101, 102), *295*, *296*, 370(85), *382*
Heeger, A. J., 365(35), 367(35), 377(35), 378(11), 379(118), *381*, *382*, *383*
Heer, C. V., 188, 218(58, 59), *294*, 358(4), *380*
Heer, E., 215(209), 216(209), 218, *298*
Heesch, H., 111, 113, 114(13), *164*, *165*
Hegland, D. E., 393(26), 406(26), *422*
Heine, V., 239(307), 240(314), 244(326), 253, 254(314), 272(314), 278(311), 280 (311, 314), 281(314), *302*
Heitler, W., 195, *295*
Helfand, E., 46, *102*
Heller, P., 322(28), 331(28), 332, 342, 343, *354*
Heltemes, E. C., 190, *294*
Henry, W. E., 393(40), *423*
Herber, R. H., 209(182), 210(182), *297*
Herman, F., 228(259), *300*
Hermann, C., 111, 127, *164*, *165*
Herrmann, K., 111, *164*
Hervé, J., 370(83), *382*
Hihara, T., 369(56, 59, 65), 370(59, 79), 374(65), 375(56), *381*, *382*
Hijikata, K., 233(275), *301*
Hiley, B. J., 6, 20, *41*
Hill, E., 393(41), 402(41), *423*
Hill, R. W., 186, *294*
Hirahara, E., 209(187), *297*, 369(54), 370 (54,81), 375(54), *381*, *382*
Hirai, A., 365(41, 46, 51, 52), 367(41), 368(46), 378(41), 379(41), *381*
Hoijtink, G. J., 233(275), *301*
Holstein, T., 75, *102*
Honma, A., 327(31), *354*
Horwitz, G., 46, 47(14), 51, 62, 67, 77(14), 80(14), 81(14), 83(14), 85, *102*, *103*
Horowitz, J., 175(22), 240(22), 248(22), 260(22), 261(22), 262(22), 264(22), 270 (22), *293*
Hoshino, R., 370(80), *382*
Houston, T., 378(111), *382*
Hudson, R. P., 218(222), *299*
Hüfner, S., 209(196), 214(204a), 273(196), *298*
Hughes, V. W., 248(339), *307*

Huiskamp, W. J., 218(218), 219, *298*, *299*
Hurrell, J. P., 275(403), *304*
Hurst, R. P., 257(349, 351), *303*
Huzinaga, S., 289, *305*
Hylleraas, E. A., 230, 257(347), *300*, *303*

Indenborn, V. L., 113, *165*
Ingalls, R. I., 179, 181, 200(99), 208(99, 172), 209(99, 179), 210(99, 179), 211(99), 255, *293*, *295*, *297*
Ishikawa, Y., 202(140, 141), 203(135, 141), 210(135, 140, 141), 285(141), *296*
Ito, A., 202(140, 141), 203(135, 141), 209(187), 210(135, 140, 141), 285(141), *296*, *297*
Itoh, J., 369(77), 374(77), *382*

Jaccarino, V., 171(10), 200(98), 209(190), 211(98, 190), 212(98), 237(287), 238 (290, 294) 268(10, 383), 274(290, 294), 275(417), 278(412), 280(412), 281(413, 417, 419, 420), 285(429), 286(98, 429), 287, *292*, *295*, *298*, *301*, *303*, *304*, *305*
Jaccarino, V., 317, 318(16b), 322(16, 26, 27), 324, 325(27), 326, 327(27), 332(27), 335, 343, 347, *354*, *355*, 365(41), 368(42, 43), 375(42), *381*
Jackson, J. A., 275(402), *304*
Jackson, R. F., 369(58), 375(58), 377(58), *381*
James, H. M., 245, *302*
James, N. R., 393(17), 421(17), *422*
Jeacocke, J. E., 389(6), *422*
Jefferts, K., 318, 347, *354*
Jeffries, C. D., 185, *294*
Jelinek, F. J., 191(82), *294*
Jennings, L. D., 190(81a), *294*, 393(27, 35, 41), 402(41), 406(27), *422*, *423*
Jørgensen, C. K., 269(386), *304*
Johnson, C. E., 202(103), 203(149), 208(170), 209(180), 210(149, 170, 180, 218(215, 222), 285(103, 425), *295*, *296*, *297*, *298*, *299*, *305*, 358(3), *380*
Johnson, F. M., 322(25), *354*
Johnston, G. B., 378(116), *383*
Jones, E. D., 318, 347, *354*
Jones, L. L., 246(336), 258, *302*
Jones, R. V., 372(94), *282*

Josey, A. D., 233(275), *301*
Joyce, G. S., 5(33), 15(33), 22(33), 34(33), *42*
Judd, B. R., 231, 267(377), 268(377), *300, 303*

Kac, M., 3(2), *41*, 45, 62(18), *102*
Kahn, B., 98, *103*
Kalvius, M., 205(159), 209(196), 273(196), *297, 298*
Kaminow, I. P., 372(94), *382*
Kanamori, J., 261, *303*
Kanamori, J., 371(92), 372(94), 373(92), *382*
Kaneko, S., 177, *293*
Kankeleit, E., 209(193, 194, 196), 211(193, 194), 273(196), *298*
Kaplan, M., 204(129, 151), *296*
Kaplan, N., 220(234), *299*
Kaplan, T. A., 402, 411, 418, *423, 424*
Karplus, R., 169(5), *292*
Kasuya, T., 237, 259, 270, 275, *301*, 407, 419(92), *423, 424*
Kaufman, B., 17, *41*
Kazus, O., 203(135), 210(135), *296*
Keffer, F., 74, *102*
Keffer, F., 238(291), 274(291), *301*, 320(19), 322(19), 323, 330(33), 348(19), *354*
Keffer, F., 371, *382*
Keller, D. A., 204(151), 209(189), 211(189), *298*
Keller, W. E., 204(128), *296*
Kelley, H. P., 230(262b), 233, *300*
Kelly, W. H., 202(110), 203(110), *295*
Kerler, W., 209(176, 183), 211(176, 183), *297*
Khrapou, V. V., 205(158), *297*
Khubchandani, P. G., 179, *293*
Khutsishuili, G. R., 218(216), *298*
Kienle, P., 205(159), 209(196), 214(204a), 273(196), *297, 298*
Kimball, C., 203(142), *296*
Kimura, Y., 365(48), *381*
Kip, A. F., 406(68), *423*
Kishida, T., 369(72), 374(72), 375(72), *382*
Kisliuk, P., 250(240a), *302*
Kistner, D. C., 209(190), 211(190), *298*
Kistner, O. C., 200(96) 203(96), 204(114), 208, 210(96), 211(114), *295*

Kitano, Y., 401(54), *423*
Kittel, C., 96(43), *103*
Kittel, C., 237, 259, 270, 275, *301*, 360(17), *380*, 407, *423*
Klein, A., 169(5), *292*
Klein, M., 96, *103*
Klein, M. P., 250(240a), *302*
Knight, W. D., 172, *292*
Knipper, A. C., 218(222), *299*
Know, R. S., 228(256c), *300*
Knox, K., 238(293), 274(293), *301*
Kobayashi, S., 369(71), 374(77), *382*
Kocher, C. W., 203(171), *296*
Koehler, W. C., 393(14, 18, 25, 29, 33, 36, 39), 394(14, 18, 25, 43), 395(43), 401(33), *422, 423*
Kogan, A. V., 219, *299*
Kohn, W., 234(276), 237(276), *301*
Koi, Y., 369(56, 57, 59, 65, 68, 72), 370(59, 79), 371(68, 88), 374(65, 72), 375(56, 68, 72), *381, 382*
Koide, S., 276, 277(404a), *304*
Kolos, W., 258, *303*
Komura, S., 203(136), *296*
Kondo, J., 263(367), 278, *303, 304*, 418, 421, *424*
Kopfermann, H., 170(6), 171, *292*
Korringa, J., 376(103), 379(103), *382*
Korytko, L. A., 205(158), *297*
Koster, G. F., 109, *164*, 175, 240(21), *293*
Kouvel, J. S., 206(161), 214(161), *297*
Krizhanskii, L. M., 205(158), 209(191), 211(191), *297, 298*
Kroll, N. M., 169(5), *292*
Kruglov, E. M., 209(191), 211(191), *298*
Kubo, R., 84, 88, *102*, 281(416), *305*, 320(17), 330(17), 333(42), 340(17), 348(17), *354*
Kündig, W., 218(224), *299*
Kulkov, V. D., 219(230), *299*
Kurti, N., 184, 188(60), 189, 190, 218(215, 60), *294, 298*
Kurti, N., 358(3, 4), *380*
Kusch, P., 248, *302*
Kusch, P., 359(8), *380*
Kushida, T., 238(292), 274(292), *301*, 369(65), 372(88), 374(65), *381, 382*
Lacaze, A., 189(69), 192(69), *294*
La Force, R. C., 359(10), 369(10, 63, 64, 66, 73, 74), 373(10, 66, 74), 374(63, 64, 73, 74), *380, 381, 382*

Lam, D. J., 369(55), *381*
Lam, D. M., 281(418), *305*
Lamb, W. E., 196, *295*
Lambe, J., 250(240a), *302*
Landaw, L. D., 84, *103*
Landau, L. L., 112, *165*
Lanq, G., 200(99), 208(99), 209(99), 210(99), 211(99), *295*
Lanq, L. G., 209(179), 210(179), *297*
Larin, S., 206(162), 214(162), *297*
Larson, A. C., 268(380), *303*
Lasheen, M. A., 37(31), *42*
Laurance, N., 250(240a), *302*
Lazenby, R., 37(32), *42*
Lea, K., 405(65), 406(65), *423*
Leask, M. J. M., 269(387), *304*, 405(65), 406(65), *423*
Le Blanc, M. A. R., 218(219), *298*
Le Corre, V., 162, *165*
Le Danq Khoi, 365(39, 49), 366(31), 367(39), *381*
Lee, Jr., L. L., 204(120), *296*
Lee, T. D., 66, *102*
Lefebvre, R., 233(275), 239(305), 245, 290, *301*, *302*, *305*
Lefebvre-Brion, H., 181(41, 42), 182(41, 42), 233(275), 245(320), 247(337), 253 (41, 42, 320), 254(41, 42), *293*, *301*, *302*
Legvold, S., 278(409b), *340*, 393(17, 19, 21, 22, 23, 26, 30, 32, 34, 36), 395(44), 396(46), 400(34), 401(30), 405(30, 34, 64), 406(26), 413(22), 420(32, 34), 421(17, 100), *422*, *423*, *424*
Lemons, J. F., 275(402), *304*
Levy, P. M., 417, *424*
Lewis, W. B., 275(402), *304*
Lidiard, A. B., 37(30), *42*
Lifschitz, E. M., 112, *165*
Lin, C. C., 233(275), *301*
Linderberg, J., 229(260), *300*
Lindgren, I., 231, 267(377, 378), 268(377, 378), *300*, *303*
Lindquist, R. H., 360(19), 370(19), *380*
Lines, M. E., 325(29), 326, *354*
Litster, J. D., 372(93), *382*
Littlejohn, C., 194(89a), 201(101), 206(101), 202(101), 270(101), *295*
Littlejohn Herzenberg, C., 204(120), *296*
Liu, S. H., 405(64), 407, 413(74a), 418, *423*

Llewelyn, P. M., 281(415), *304*
Lock, J. M., 387(2), 393(15), *421*
Locher, P. R., 204, 264, 265(157b), 272, 296, 374(97), *382*
Löwdin, P. O., 223(243b), 224(246), 225, 227, 239(246, 304), 242(246), 243(246), 244(327, 328), 246(246), 258(327, 328), *299*, *300*, *302*
Lomont, J. S., 109, 118, *164*
Loudon, R., 74, *102*
Lounasamaa, O. V., 189(65), 190, 191, 192, 193(77, 79, 80), 222(78), *294*, 399(53), 402(53), *423*
Lovejoy, J. F., 219(225), *299*
Low, F., 169(5), *292*
Low, G., 285(424b), *305*
Low, G. E., 286(430), *305*
Low, G. G., 322(21), 325(21), *354*
Low, W., 183, 260(44), 261(360), 262, 269(388), *293*, *303 304*, 406(67), *423*
Ludwig, G. W., 261(363), *303*
Lyons, D. H., 418, *424*
Lyubimov, V. A., 205(158), *297*

McClure, D. S., 266, *303*
McConnell, H. M., 233(275), *301*
McDonald, R. R., 387(5), *422*
McHargue, C. J., 393(14), 394(14), *422*
Mackay, A. L., 113, *165*
McKim, F. R., 37(32), *42*
Mackintosh, A. R., 402(58), 419(95), *423*
McLachlan, A. D., 233(275), *301*
Mc Weeny, R., 224, 242(247a), *300*
Madsen, P. E., 285(425), *305*
Makarov, E. F., 205(158), *297*
Margulies, S., 194(89a), *295*
Marinov, A., 202(108), 203(107, 108), 204(107, 116, 117, 118), 210(107, 108), *295*, *296*
Marrus, R., 268(382), *303*
Marshak, H., 222(240, 241), 253, *299*
Marshall, W., 24(18b), *41*, 96, *102*, *103*
Marshall, W., 187, 208(170), 209(180), 210(170, 180), 216(55), 238(296), 239 (296, 300, 306), 245, 269(387), 270, 274(296), *294*, *297*, *301*, *302*, *304*, 358(6), *380*, 416(81), *423*
Martis, J. B., 233, 241(273), 246, *301*
Matsen, F. A., 257(348, 349, 350), *303*

Matsubara, T., *423*
Matsuura, M., 365(41, 51, 52), 367(41), 378(41), 379(41), *381*
Matthias, B. J., 237(287), *301*
Matthias, B. T., 96(42), *103*, 205(104), 202(104), 282(421), 283(421), 284(421), *295, 305*
Matthiess, L. F., 254(344), 275(403a), *302, 304*
Matumura, O., 262, *303*
Mayers, D. F., 268(379), *303*
Mercher, A., 417(85), *424*
Metzger, F. R., 196(92), *295*
Meyer-Schützmeister, L., 204(115, 120, 132, 152), 211(115), *295, 296*
Midtal, J., 230(263), *300*
Miedema, A. R., 25(27a), 37(27a), *42*
Miller, R., 393(35), *422*
Millman, S., 359(8), *380*
Mitchell, A. H., 237(286), *301*, 333(40), 337(40), *354*
Miwa, H., 402, 404(50), 413(50), 415, 419(97), 421, *423, 424*
Mock, J. B., 277(405), *304*
Mössbauer, R. L., 194, 196, 197(86, 87), 205(159), 209(178, 193, 194, 196), 211(178, 193, 194), 273(196), *294, 297, 298*, 358, *380*
Montroll, E., 45, *102*
Moon, R. M., 278, 279, *304*
Moore, T. W., 278, *304*
Morgan, D. J., 94(40), 95(40), *103*
Morgan, L., 24(18b), *41*
Morimoto, S., 365(32, 48), 366(32), 367(32), 369(62), *381*
Morin, F. J., 378(115), *383*
Moriya, T., 333(37, 38, 40), 335, 337(37, 38), 338, 339(37, 41), 344(41), *354*, 380(121), *383*
Morrish, A. H., 378(116), *383*
Moruzzi, V. L., 364(27), *380*
Moser, C. M., 181(41, 42), 182(41, 42), 233(275), 245(320), 247(337), 253(41, 42, 320), 254(41, 42), *293, 301, 302*
Mozer, B., 285(428), *305*
Mukherjee, A., 240(315), *302*
Mukherji, A., 238(292), 274(292), *301*
Muller, W. E., 391(11), 417(11), *422*
Mulligan, J. F., 257(350), *303*
Murao, T., *423*

Nagamiya, T., 320(17), 330(17), 340(17), 348(17), *354*, 401, *423*
Nagata, K., 401(54), *423*
Nagle, D. E., 202(104, 105), 204(112, 128), 205(158), 207(165, 167), 283(165, 424), 284(165), *295, 296, 297, 305*
Nakamura, T., 344(44), 347, *355*, 377, *382*
Naomoto, Y., 203(143), 210(143), *296*
Narath, A., 318, *354*, 365(34,47), *381*
Nathans, R., 203(147), 285(428), *296, 305*
Néel, L., 35, 37, 42, 371(87), *382*
Nereson, N. G., 387(3), *422*
Neronova, N. N., 111, 113(14), 114(14), 118(14), 122(22), 129(22), 155, 161(35), 164, *165*
Nesbet, R. K., 231(264, 265), 232, 233, 235, 236(264, 265), 239(264, 265, 302, 309), 240, 241(309), 242(322, 323), 245, 246, 247(337), 254, 255, 269(264, 265), 287(309), 288, 289(309), 290, *300, 302, 305*, 416(81), *423*
Nethercott, A. H., 322(25), *354*
Nettel, S. J., 234(276), 237(276), *301*, 402(57), 403(57), *423*
Neuwirth, W., 209(176, 183), 211(176, 183), *297*
Nevitt, M. V., 281(418), *305*, 369(55), *381*
Newell, G., 45, *102*
Nicholson, D., 203(122), *296*
Nicholson, W. J., 209(186), 210(186), 285(427), *297, 305*
Niggli, A., 113, *165*
Nigh, H. E., 278(409b), *304*, 393(21), 413(22), *422*
Niira, K., 402, *423*
Nikitin, L. P., 219(230), *299*
Noakes, J. E., 37(34), *42*
Norem, P. C., 209(185, 195), 211(185, 195), *297, 298*
Novey, T. B., 215(209), 216(209), 218, *298*
Nowick, I., 205(159), 214(204), *297, 298*

Obata, Y., 281(416), *305*, 376(102), *382*
Obenshain, F. E., 203(146), 204(113), 213(200), 264 (200), *295, 296, 298*, 370(86), *382*
Ofer, S., 202(108), 203(107, 108), 205(159), 204(107, 116, 117, 118), 210(107, 108), 214(204), 220(234), *295, 296, 297, 298, 299*

Ogawa, S., 365(32, 48), 366(32), 367(32), 369(62), *381*
Oguchi, T., 75, *102*, 238(291), 274(291), *301*, 325(30), *354*
Ohlmann, R. C., 264(368), *303*, 327(32), *354*
Olsen, C. E., 202(105), 283(424), *295*, *305*, *387*(3), *422*
Ono, K., 202(140, 141), 203(141), 209(187), 210(140, 141), 285(141), *296*, *297*
Onsager, L., 3, 17, 31(1), *41*, *45*, *102*
Opechowski, W., 8(9), *41*, 106, 109(2), 113(2), 116(2), 119(2), 123(2), 127(2), 130(2), 152(2), 157(2), *164*
Orbach, R., 269(387), *304*
Ornstein, M. M., 46, 59, *102*
Ostanevich, Yu. M., 205(159), *297*
O'Sullivan, W., 238(291), 274(291), *301*
O'Sullivan, W. J., 318, 327(15), 328(15), 329, *354*
Overhauser, A. W., 96, *103*, 234, 237, 279(412b), 286, *301*, *304*, *305*, 418(91), *424*
Owen, J., 37(32), *42*, 322(22, 23), *354*

Pappalardo, R., 269(386), *304*
Park, D., 31, *41*
Parkinson, D. H., 393(16), *422*
Parks, R. D., 192, *294*
Paskin, A., 237(286), *301*
Pauli, W., 168, *292*
Pauthenet, R., 220(233), *299*
Pearce, R. R., 35, *42*
Peacock, R. H., 194(89a), *295*
Pekeris, C. L., 223(243a), 229(243, 243a), 230(243), 232(243, 243a), *299*
Perlow, G. J., 194(89a), 201(101, 102), 202(101, 102), 206(101, 102), 208(170), 209(180), 210(170, 180), 270(101, 102), *295*, *297*, 361(20), 370(85), *380*, *382*
Peter, M., 237(287), 238(294), 274(294), 276, 277(404a), 279(412a), 282(421), 283(421), 284(421, 424a), *301*, *304*, *305*
Petersen, R., 358(4), *380*, 188(60, 61), 218(60), *294*
Phillips, W. D., 233(275), 285(424b), *301*, *305*
Pichanick, F. M., 268(384), 269, *303*
Pincus, P. A., 352(47), *355*, 377(108, 109), 378(108, 112), *382*
Pines, D., 230(261), 289(261), *300*
Poindexter, J. M., 209(193, 194, 196), 211(193, 194), 273(196), *298*
Pollack, F., 169(5), *292*
Pollak, H., 205(158), *297*
Pople, J. A., 239(302), 290, *302*, *305*
Poppema, O. J., 218(214), *298*
Portis, A. M., 218(217), *298*, 358(5), 359(16), 360(19), 361(16), 362(5, 23), 363(16), 369(23), 370(16, 19), 371(92), 373(92), 376(16, 23, 101), 377(16), 378(16, 117), *380*, *382*, *383*
Postma, H., 219, 222(240, 241), *299*
Poulis, N. J., 313(4–12), 314, 318, 327 (4–12), 328(12), 329, 338(5–12), *353*, *354*
Pound, R. V., 172, 184(51), 194(89, 89a), 214(205), 217, 220, *293*, *294*, *295*, *298*
Powell, M. J. D., 245, 269(387), 287(432), *302*, *304*, *305*
Pracht, D. W., 281(418), *305*, 369(55), *381*
Pratt, Jr., G. W., 228(257, 258), 239(258, 301), 254(257), *300*, *301*
Preston, R. S., 194(89a), 201(101, 102), 202(101, 102, 139), 204(115, 132), 206(101, 102), 211(115), 270(101, 102), *295*, *296*
Preston, R. S., 370(85), *382*
Primakoff, H., 75, *102*
Pryce, M. H. L., 169, 175(22), 184(3), 240(3, 22), 247, 248(22), 260(22, 354, 355), 261(22), 262(22), 264(22), 270(22), 281(415), *292*, *293*, *303*, *304*

Quidort, J., 209(184), *297*
Quitmann, D., 214(204a), *298*

Rabi, I. I., 359(8), *380*
Radford, H. E., 248, *302*
Rasmussen, J. O., 219(226), *299*
Ravitz, S. F., 359(10), 369(10, 64, 66), 371(91), 373(10, 66), 374(64), *380*, *381*, *382*
Ray, D., 220(232), *299*
Rebka, G. A., 194(89, 89a), *295*
Redei, L., 223(244), *299*
Reif, F., 172, 178, 219(16), *293*
Reinov, M., 219(230), *299*
Remeika, J. P., 250(240a), *302*, 365(41), 368(42, 43), 375(42), *381*

Reno, R. W., 204(153), *296*
Reynolds, C. A., 222(240, 241), *299*
Rhyne, J., 396(46), *423*
Riedel, E. P., 162, *165*
Ridley, E. C., 228(253a), 268, *300*
Ridout, M. S., 202(103), 203(149), 210 (149), 285(103, 425), *295, 296, 305*
Ritter, G. J., 204, 268(157c), *296*
Roach, P. R., 190, 191(76), *294*
Robert, C., 206(160), 214(160), *297,* 365(53), 366(29, 30, 33, 53), 369(29, 61), 372(29), 376(53), 377(53), 378(113), 379(30), *381, 383*
Roberts, L. D., 201, 203(146), 204(130, 133, 134), 213, 222(238), 264(200), *296, 298, 299*
Roberts, L. M., 393(20), *422*
Robinson, F. N. H., 218(213, 215), *298*
Robinson, J. E., 228, *300*
Robinson, W. A., 318(15), 327(15), 328(15), 329(15), *354*
Robinson, W. K., 37, *42*
Rodbell, D. S., 278, *304,* 371(90), *382*
Rocher, Y., 390, *422*
Roe, W. C., 406(69), *423*
Roothaan, C. C. J., 258, 288, 289, *303, 305*
Rose, M. E., 184, 206(49), 222(236), *294, 299*
Roth, W. L., 371(90), *382*
Rothberg, G. W., 204(150), *296*
Rousseau, J. S., 389(6), *422*
Rubens, R. S., 269(388), *304*
Rubinstein, M., 359(14), 360(14, 19), 369(58), 370(14), 371(14), 375(58), 377 (58), *380, 381*
Ruby, S. L., 202(109, 127, 144), 205(158), 203(109), 210(109, 144), 285(426), *295, 296, 297, 305*
Rudermann, M., 407, *423*
Ruderman, M. A., 96(43), *103,* 237, 259, 270, 275, *301,* 407, *423*
Rukov, T., 205(159), *297*
Rushbrooke, G. S., 15, 23, 36, *41,* 94(40), 95(40), *103*

Sachs, L. M., 239(308), 240, 241, 243, *302*
Safrata, R.S., 189, 190, *294*
Sailor, V. L., 222(240, 241), *299*
Saint-James, S., 416(80), *423*

Sakamoto, M., 233(275), *301*
Salpeter, E. E., 169(5), 289(439), *292, 305*
Samara, G. A., 372(95), *382*
Samoilov, B. N., 218(218), 219, *298, 299*
Sandars, P. G. H., 268(384), 269(384), *303*
Sato, N., 370(80), *382*
Sawada, K., 234(276), 237(276), *301*
Scharenberg, P. P., 218(223), *299*
Schawlow, A. L., 318(16b), *354*
Schermer, R., 222(241), *299*
Scherrer, P., 218(224), *299*
Schmidtke, H. H., 269(386), *304*
Schoen, A., 201(100), *295*
Schooley, J. F., 219(226), *299*
Schreiner, W. N., 202(110), 203(110), *295*
Schreiffer, J. R., 228(256c), 279(412a), *300, 304*
Schuler, C., 391(11), 417(11), *422*
Schüler, C. C., 421(97a), *424*
Schwartz, C., 170, 223(245), *292, 300*
Scurlock, R. G., 218(215, 222), *298, 299,* 358(3), 369(58), 375(68), 377(58),*380, 381*
Segal, E., 202(108), 203(108), 210(108), *295*
Segnan, R., 285(424c, 428), *305*
Segrè, E., 175, 240(20), *293*
Sessler, A. M., 230(262b), *300*
Shaltiel, D., 277(405), 284(424a), *304, 305*
Shapiro, F. L., 205(159), *297*
Sharma, R. R., 179(32), *293*
Sherwood, R. C., 277(405), 282(421), 283(421), 284(421), *304, 305*
Shikazono, N., 203(136, 143), 210(143), *296*
Shirane, G., 202(109, 127, 144), 203(109 147), 205(158), 210(109, 144), *295, 296, 297*
Shirley, D. A., 204(129, 151, 155, 156), 209(189), 211(189), 214(155, 156), 219 (225, 226), *296, 298, 299*
Shishkov, A., 206(162), 214(162), *297*
Shore, F. J., 222(240, 241), *299*
Shortley, G. H., 228(255), 290, *300, 305*
Shpinel, V. S., 205(158), 209(177, 192), 211(192), *297, 298*
Shull, C. G., 278, *304*
Shull, H., 229(260), 244(327, 328), 258 (327, 328), *300, 302*
Shulman, R. G., 180(36), 238(290, 293, 297, 298), 263, 274(290, 293, 298), 275(401), *293, 301, 304,* 318, 322(16), 335, 343, 347, *354, 355*

Siegert, A., 46, *102*
Silver, A. H., 369(72), 374(72), 375(72), *382*
Simanek, E., 369(67), 374(67), 377, *381, 382*
Simmons, W. W., 318(15), 327(15), 328(15), 329(15), *354*
Simon, F. E., 184, 218(215), *294, 298,* 393(16), *422*
Sinanoğlu, O., 232, *300*
Singvi, K. S., 197, *295*
Sjölander, A., 197, *295*
Skillman, S., 228(259), *300*
Sklyarvskii, V. V., 218(218), *298*
Sklyarevsky, V. V., 219(229), *299*
Skochdopole, R. E., 393(31, 35, 38), *422*
Slater, J. C., 170(9), 226, 228(254), 231(254), 237(252), 239(281), 254, *292, 300, 301*
Slichter, C. P., 308(2), 333(2), *353*
Slonczewski, J. C., 204, *296,* 365(36), 367(36), *381*
Smart, J. S., 364(27), *380*
Smidt, F. A., 419, *423*
Smirnova, T. S., 111, 113(14), 114(14), 118(14), 122(22), 129(22), 164, *165*
Smit, J., 264(369), *303*
Snodgrass, R. J., 359(11), *380*
Sött, I., 220, *299*
Sokolov, I. A., 219(229), *299*
Soller, T., 222(72), *294*
Solomon, I., 200(97), 202(145), 210(145), 211(145), *295, 296*
Spedding, F. H., 190(81a), 191(82), 278(409b), *294, 304,* 393(16, 17, 19, 21, 22, 23, 26, 27, 30, 31, 32, 34, 35, 37,38, 41), 395(44), 400(34), 401(30), 402(41), 405(30, 34), 406(26, 27), 413(22), 420(32, 34), 421(17, 100), *422, 423, 424*
Sperce, R. D., 130, 162, *165*
Sroubek, Z., 369(67), 374(67), *381*
Stanck, F. W., 205(159), *297*
Stanford, C. P., 222(238), *299*
Stanton, R. M., 190, *294,* 393(27), 406(27), *422*
Stearns, M. B., 203(138), 279(412b), 285(138), 286(429b), *296, 304, 305*
Steenland, M. J., 218(214), 222, *293, 298, 299*
Steffen, R. M., 215(207), *298*
Stelmach, M. F., 219(230), *299*

Stepanov, E., 218(218), 219(229), *298, 299*
Stephenson, T. E., 222(238), *299*
Stern, F., 271(392), 278(392), *304*
Stern, H., 84, 85, 87, *103*
Sternheimer, R. M., 175(23), 176(18, a, b, 19), 177, 178, 180, 193(18, a, b), 208(173), 209(198), 228, 240(18, a, b, 19, 23), 255, 257(198), *293, 298, 300*
Stevens, K. W. H., 219, 260(356, 357), 263(356), 267(227), 269(356), *299, 303,* 392(13), 403(61), 404(13, 62), 416(62), *422, 423*
Stevenson, A. F., 223(244a), *299*
Stevenson, R. W. H., 238(299a), 274(299a), *301,* 322(23), *354*
Steyert, W. A., 207(165, 167), 283(165), 284(165), *297*
Stinchcombe, R. L., 47, 77(14), 80(14), 81(14), 83(14), *102*
Stolovy, A., 222(239, 242), 272(242), *299*
Stoner, E. C., 83, *102*
Stout, J. W., 322(24), *354*
Strandberg, D. L., 393(34), 400(34), 404(34), 420(34), *422*
Strathdee, J., 233(275), *301*
Street, R., 371(90), *382*
Streever, Jr., R. L., 272, 275(396), *304,* 359(13), 369(70, 73, 74, 76), 370(76), 373(74), 374(70, 73, 74), 375(70), *380, 382*
Strelhov, A. V., 205(159), *297*
Stroke, H. H., 171, 268(10), *292*
Stuart, R., 238(296), 239(296), 274(296), 416(81), *301, 423*
Sucksmith, W., 35, *42*
Sugano, S., 180(36), 238(297, 298), 239(298), 263, 274(298), *293, 301*
Sugibuchi, K., 369(60), 370(60), *381*
Suhl, H., 96(42), *103,* 237(287), *301,* 344(43), 347, 355, 368(44), 377, *381, 382,* 399(51a), 402(57), 403(57), *423*
Sunier, J., 218(224), *299*
Sunyar, A. W., 200(96), 203(96), 204(114), 208, 210(96), 211(114), *295*
Suzdaler, I. P., 205(158), *297*
Suzuki, H., 370(81), *382*
Swan, J. B., 204(114), 211(114), *295*
Swenson, C. A., 190, *294*
Swirles, B., 289, *305*

Sykes, M. F., 5(7), 11(13), 12(14), 14, 16, 18, 19, 21, 22, 23, 24(15), 26(18a), 27(18), 29(19b), 30, 31(22), 35, 38, *41*
Szasz, L., 233, *301*

Takei, W. J., 202(127, 144), 210(144), *295, 296*
Taub, H., 248, 275(402), *302, 304*
Tavger, B. A., 113, 114(21), 116, 153, *165*
Taylor, K. N. R., 406(69), *423*
Taylor, R. D., 189(67), 192(67), 202(104, 105), 204(112, 128), 205(104), 207(165, 167), 283(165, 424), 284(165), *294, 295, 296, 297*
Taylor, R. T., 218(215, 222), *298, 299,* 358(3), *380*
Temmer, G.M., 218(222), *299*
Temple, G., 223(244a), *299*
Ter Haar, D., 224, 242(247b), *300*
Teruya, S., 203(143), 210(143), *296*
Thoburn, W. C., 395(44), *423*
Thomas, H., 403(60), *423*
Thompson, E., 234, *301*
Thompson, J. O., 201, 204(130, 133, 134), 213, *296*
Thouless, D. J., 237(278, 279), *301*
Title, R. S., 262, *303*
Tinkham, M., 238(289), 263(289), 264(368), 274(289), *301, 303,* 327(32), *354*
Tolhoek, H. A., 183(45), 184(45), 215(45), 216(45), 218(218), *293, 298, 299*
Tomita, K., 333(42), *354,* 417(83), *424*
Toth, L. E., 359(10), 369(10), 371(91), 373(10), *380, 382*
Townsend, J., 233(275), *301*
Trammel, G., 406, *423*
Trapp. C., 322(24), *354*
Trapp, H. D., 281(418), *305,* 369(55), *381*
Trees, R. E., 170, *292*
Trolliet, G., 189(68, 69), 192(69), *294*
Tsujimura, A., 369(56, 57, 59, 65, 68, 72), 370(59, 79), 371(68, 88), 372(88), 374(65, 72), 375(56, 68, 72), *381, 382*
Tsuya, N., 267, *303*
Tumono, Y., 369(71), *382*
Twidell, J. W., 387(4), *422*
Tyablikov, S. V., 71, 72, *102*
Tycko, D., 175(18), 176(18), 177(18), 180(18), 193(18), 240(18), *293*

Unruh, U. P., 261(361), *303*
Utton, D. B., 369(58), 375(58), 377(58), *381*

van den Broek, J., 37(31), *41*
van der Lugt, W., 162, *165,* 313(10–12), 318(12), 327(10–12), 328(12), 338(10–12), *354*
Van der Waerden, B. L., *41*
Van Kranendonk, J., 320(18), 333(39), 337(39), 348(18), *354*
Van Ostenburg, D. O., 281(418), *305,* 369(55), *381*
Van Vleck, J. H., 267(372), 268(372), 281, *303,* 320(18), 345(45), 348(18), *354, 355,* 391(12), 408, *422, 423*
Van Wierengen, J. S., 262, 263, *303*
Veillet, P., 370(83), *382*
Verwey, E. J. W., 367(37), *381*
Vincent, D. H., 194(89a), 201(101, 102), 202(101, 102), 204(132), 206(101, 102), 270(101, 102), *295, 296,* 370(85), *382*
Vincow, G., 233(275), *301*
Violet, C. E., 205(158), *297*
Visscher, W. M., 197, 204(112), *295*

Waber, J. T., 268(380), *303*
Wagner, F. E., 205(159), *297*
Wakefield, A. J., 11(12), *41*
Walker, L. R., 200(98), 209(190), 211(190), 212, 238(294), 274(294), 286(98), *295, 298, 301,* 320(20), 322(26, 27), 324, 325(27), 326, 327(27), 331(27), 348(20), *354*
Wallace, W. E., 203(148), 204(148), 210(148), *296,* 395(28), *422*
Watabe, A., 389(8), 409, *422*
Watkins, G. D., 261(363), *303*
Watson, R. E., 172, 178, 179(33), 180(37), 181(39, 40), 193(40), 208(39), 209(40), 212, 214(199), 228(253b), 232, 238(295, 299), 239(12, 299, 309, 310, 312), 240, 241(309), 242, 247(337), 248, 249(12), 250(12, 340), 251(12, 310, 341), 252(341), 254(33, 39, 40, 310, 341, 343, 345), 253(346), 256(39), 257(40), 263(12), 264(12), 267(253b), 268(40), 269(40, 253b, 346), 270(12), 271(299, 312), 272(12), 273(346, 398), 274(295, 299), 275(341), 277, 279(412c), 281(343),

287(309, 431), 289(309, 431), *292*, 293, *298*, *300*, 301, 302, *303*, *304*, *305*, 313(13), 316(13), *354*, 363, 366, *380*, 404(63), 416(81), *423*, *424*
Weber, L., 111, *165*
Wedgwood, F. A., 400(76), 402, 411, 412(76), 413(76), 414(76), 419(96), 420(76, 96), *423*, *424*
Weger, M. 359(16), 361(16), 363(16, 25), 370(16), 376(16), 377(16, 25), 378(16), 379(25), *380*
Wegener, H. H. F., 203(146), 204(113), *295*, *296*, 370(86), *382*
Wei, C. T., 189(63, 64), *294*
Weil, L., 189(68), *294*
Weinberg, D. L., 369(78), *382*
Weinstein, A., 223(244a), *299*
Weinstein, S., 395(28), *422*
Weiss, A. W., 231(266), 232, 233, 240(266), 241(273), 246(266), *300*, *301*
Weiss, P., 35, *42*
Weissman, S. I., 233(275), *301*
Wernick, J. H., 202(125), 204(124), 210(125), 237(287), 268(383), 277(405), 281(419), 284(424a), 285(429), 286(429), *296*, *301*, *303*, *304*, *305*
Wertheim, G. K., 194(88a), 200(98), 202 (125, 126, 137), 208(111, 119), 209(182, 185), 203(106, 110, 119, 123, 137), 204(106, 120, 137), 210(106, 111, 119, 123, 137, 182), 211(98, 185, 195), 212(98), 285(429), 286(98, 429), 287, *295*, *296*, *297*, *298*, *305*, 374(98), *382*
Weyl, H., 112, *165*
Wheeler, R. G., 109, *164*
Wickman, H. H., 204(155), 214(155), *296*
Wiedemann, W., 205(159), 209(196), 273(196), *297*, *298*
Wigner, E., 109, *164*
Wijn, H. P. J., 264(369), *303*
Wikner, E. G., 177, 179(35), 209(197), 257(35), *293*, *298*
Wilkens, J., 238(299a), 274(299a), *301*
Wilkinson, M. K., 393(14, 18, 29, 33, 36, 39), 394(14, 18), 401(33), *422*
Williams, E., 49(16), *102*
Williams, F. I. B., 267(375), 268(385), 269(375), 272(375), *303*
Williams, H. J., 277(405), 282(421), 283(421), 284(421, 424a), *304*, *305*

Wilmshurst, T. H., 369(58), 375(58), 377(58), *381*
Wilson, R. H., 206(161), 214(161), *297*
Wilson, S. S., 286(429a), *305*
Winter, J., 352(47), *355*, 368(45), 378, 379(45), *381*
Winter, J. M., 206(160), 214(160), *297*, 366(29), 369(29, 61), 372(29), 377(108), 378(108, 113), *380*, *382*, *383*
Wohlfarth, E. P., 85, *103*, 213(201), *298*
Wold, W. P., 405(65), 406(65), *423*
Wolf, W. P., 37(32), *42*, 269(387), *304*
Wolff, P. A., 84, 85, 86, *103*, 283(422), *305*
Wollan, E. O., 393(14, 18, 25, 29, 33, 36, 39), 394(14, 18, 25, 43), 395(43), 401(33), *422*, *423*
Wood, D. L., 250(240a), *302*
Wood, D. W., 5(33), 15(24a, 33), 22(33), 34(24a, 33), *41*, *42*
Wood, G. T., 217(212), 220(232, 212), *298*, *299*
Wood, J. H., 228(258), 239(258), 271(390), *300*, *304*
Wood, J. W., 271(391), 278(390, 391), *304*
Wood, P. J., 15, 23, 36, *41*
Wood, R. F., 241(329), 245, *302*
Woodbury, H. H., 261(363), *303*
Woodgate, G. K., 268(384), 269(384), *303*
Wu-Kuang, Yen, 205(159), *297*
Wyluda, B. J., 275(401), *304*

Xavier, R. M., 209(188), *298*

Yafet, Y., 281(413, 420), *304*, *305*
Yamada, Y., 278, *304*
Yamashita, J., 238(291), 274(291), *301*
Yamazaki, M., 233(275), *301*
Yang, C. N., 66, *102*, 331(35), *354*
Yasuoka, H., 365(41, 46, 51, 52), 367(41), 368(46), 378(41), 379(41), *381*
Yosida, K., 237, 259, 270, 275, 277, *301*, 320(17), 330(17), 340(17), 348(17), *354*, 389(8), 402, 404(50), 407, 408, 409, 413(50), *422*, *423*
Yosimori, M., 393, *423*
Yvon, J., 6, *41*

Zaitsez, V. M., 113, 114(21), *165*
Zamorzaev, A. M., 111(15, 16), 112, 113, 117, 119(16), 127, 130, *165*
Zavoisky, E. K., 260(353), *303*
Zehler, V., 8, *41*
Zemansky, M. W., 285(424e), *305*
Zener, C., 237, 276, *301*
Zernike, F., 46, 59, *102*
Zmora, H., 220(234), *299*
Zvenglinskii, B., 209(177), *297*

Subject Index

A

Antiferromagnetism, 29–34
  perpendicular susceptibility, 33–34
  superexchange model, 32–33
  theoretical developments required, 34
Antiferromagnets
  nuclear resonance in, 307–355
  nuclear spin relaxation, 333–347
Antishielding factors, Sternheimer, 175–180

B

Band
  model of a ferromagnet, 45–47
  theory of ferromagnetism, 83–90
Boundaries, superzone, of rare earth metals, 409–410
  effect on electrical resistance, 418–421
  effect on Rudermann-Kittel interaction, 411–414

C

Cell, magnetic unit, 120–121
Cluster methods, 51–62
Cobalt, low-temperature specific heat, 188–189
Configuration interaction method, 231–232
Correlation in atoms, 229–233
Critical behavior in magnetic systems, statistical mechanics of, 1–42
Crystal(s)
  electric field
    gadolinium, 406
    heavy rare earths, 403–405
    light rare earths, 405–406
  single, whose space group is F, 133–150
Curie
  point, precursor phenomena associated with, 85–88
  temperature of a ferromagnet, 1

D

Domain-wall displacement, 362–363

E

Electron transport of rare earths, 396
Energy
  critical values, 24–27
  gap, large, effect on $M(T)$, 327
Entropy, critical values, 24–27
Exchange
  direct, 416–417
  indirect, 417–418
  interactions, theory of, 406–421
  polarization, 237–239
    for iron series ions, 248–251
    for rare earth ions, 251
    for some iron series atoms, 247–248
    for some neutral atoms, 247
  spatial dependence of rare earth metals, 399–400
Expansion(s)
  high-temperature, 7–15
    general methods, 7–8
    Heisenberg model, 15
    Ising model, 8–15
  linked-cluster, of Heisenberg model, 78 ff

F

Fe^{57}, nuclear properties in presence of electric and magnetic fields, 198–200
Fermi exchange hole, 227–228
  and the Hartree-Fock method, 224–228
Ferromagnetism theoretical developements required, 34
Ferromagnets
  band theory, 83–90
  statistical mechanics of, 43–103
Field(s)
  hyperfine

conduction electron field, 364
core field, 363–364
direction in iron group metals, 370
effect of crystal phases, 371
effect of defect structures, 371–372
with external magnetic field, 206–207
in ferromagnetic metals, 270–274
in iron series salts, 259–266
local field, 363
in magnetic materials, 363–376
in metals, 280–282
in nonmagnetic ions in magnetic materials, 274–280
at nuclei in magnetic materials, 201–207
in rare earth salts, 266–270
variation with temperature, 205–206
internal, origins and magnitudes, 312–319
magnetic solid, 312–318
nonmagnetic solid, 312
molecular, of rare earth metals, 397–399
theory, molecular, 330–331
F^{19} nuclear magnetic resonance in MnF_2, 324–325, 326, 332, 343–344

G

Gamma-rays
angular correlation, 215–221
anisotropy emission, 216
recoiless emission and absorption, 194–214
Groups
magnetic, 106–113
construction, 114–133
noncrystallographic, 116–117
point, 114–122
Shubnikov, 112
space, 127–130

H

Hamiltonian, nuclear, 312–319
Hartree-Fock method, 223–259
extended functions for closed-shell systems, 257–258
and the Fermi exchange hole, 224–228
interelectronic correlation effects, 236
restrictions associated with, 233–237
spin-, or exchange-, polarized, 239–254

Heisenberg model, 22–24, 67–83
effect of lattice structure, 24
ferromagnet, 44–45
graphical analysis, 78–83
linked-cluster expansion, 78 ff
spin waves, 67–77
Hyperfine
effects and nuclear properties, 197–201
field at nuclei in magnetic materials, 201–207
interactions, 167–182
and aspherical distortions, 254–257
interpretation of measured, 259–287
isotropic, expressed in terms of spin wave operators, 351–362
Hylleraas method, 230–231

I

Integration, functional, 46
Interactions
electric quadrupole, 207–211
electrostatic between electrons and nuclei, 172–175
exchange, 406–421
hyperfine, 167–182
due to closed-shell distortions in open-shell ions, 180–182
experimental methods, 182–222
multielectron contributions to, 175–182
one-electron theory, 169–175
indirect nuclear spin-spin, 344–347
magnetic, between atomic nucleus and its electrons, 169–172
orbital moment-conduction electron, 418
quadrupole-quadrupole, 417
Ising model, two dimensional, 2 ff, 15–22, 47–67, 331
closed-form approximation, 6–7
cluster methods, 51–62
ferromagnet, 44
high temperature expansions, 8–15
partition function, 8, 13
ring approximation, 51–67
series expansion, 5–6
the Weiss field, 47–51
Isomer shift, 200–201, 211–215
in dilute ferromagnetic gold alloys, 213–214
of Fe^{57} in iron, 214

K

Knight shift, 280–282

L

Lattices, magnetic
  Bravais class, 118–119
  construction of, 114–122

M

Magnetic disturbance, spatial extent, due to impurity in alloy, 286–287
Magnetization
  of antiferromagnet at low temperatures, 319–327
    interacting spin waves, 325–327
    noninteracting spin waves, 320–325
  spontaneous, 29
  temperature dependence, 372–373, 375
  sublattice in antiferromagnet, 319–333
Moment, quadrupole, 173
Mössbauer effect, 194–215
  experimental results, 201–215
  with external magnetic field, 206–207, 282–285
  and hyperfine fields in Fe with dilute impurities, 285–287
  method of measurement, 197

N

Neutron
  diffraction
    measurement of rare earth metals, 392–394
    studies of Shull and Yamada, 278–280
  polarized, interaction with polarized nuclei, 221–222

O

Onsager's exact solution, 2
Ordering, low temperature, in rare earth metals, 400–402
Orientation, nuclear, 183–185
  contribution to low-temperature specific heats, 186–187
  methods for achieving, 184–185

P

Padé approximant, 18–20
Partition function of Ising model
  with general s, 13
  for $s = \frac{1}{2}$, 8
Point groups, magnetic, construction of, 114–122
Polarizabilities, quadrupole, 175–180

R

Radermann-Kittel interaction, 406–409
  effect of scattering, 414–416
  effect of superzone boundaries, 411–414
Rare earth metals
  band structures, 388–391
  crystal structures, 387–388
  4f electrons, 391–392
  low-temperature specific heat, 189–193
  theory of magnetism, 385–424
    phenomenological, 397–403
Relaxation, nuclear, in magnetic materials, 376–380
Resistance, electrical, effect of superzone boundaries, 418–421
Resonance excitation, radiofrequency, 358–363
  apparatus, 359–360
  intensity of absorption, 360–362
Resonance, nuclear
  in antiferromagnets, 307–355
    basic relations, 308–309
  criteria for observing in magnetic materials, 309–312
  in ferromagnetic materials, 357–383
    alloys and intermetallics, 373–376
    insulators, magnetic, 364–368
    metals, 368–373
  line widths, 333–347
  magnetic of F^{19} in MnF_2, 324–325, 326, 332, 343–344
  pressure dependence of frequency, 372–373, 375
  in single-domain particles, 370–371

S

Specific heat
  anomaly for ferromagnets, 187
  cobalt low-temperature, 188–189

critical values, 27–28
low temperature, contribution of nuclear spin orientation, 186–187
measurement, 37
nuclear, 186–194
rare earth, 189–193, 396

Spin
  arrangements
    invariant, 133–163
    in perfect magnetic crystals, 105–106
  -lattice interaction in ferromagnetic materials, 379–380
  relaxation, nuclear, in antiferromagnets, 333–347
    fluctuating local field, 333–344
  -spin interaction
    in ferromagnetic materials, 377–378
    indirect nuclear, 344–347
  systems, random, 91–96
  wave(s)
    description of rare earth metals, 402–403
    Heisenberg model, 67–77
    -phonon interactions, 352–353
    on a two-sublattice antiferromagnet, 348–351

Statistical mechanics
  of critical behavior in magnetic systems, 1–42
  of ferromagnets, 43–103
Sternheimer antishielding factors, 175–180
Stoner theory, elementary, 84–85
Susceptibility
  of an antiferromagnet, 37–38
  of a ferromagnet, 35–37
    above the Curie point, 15–24
  of Ising model ($s = \frac{1}{2}$), 30–32
  of three-dimensional Heisenberg model, 23–24
  zero field, 10, 12, 14
Symmetry, magnetic, 105–165

T

Temperature
  Curie, of a ferromagnet, 1
  dependence
    of magnetization, 372–373, 375
    of sublattice magnetization of $CuCl_2 \cdot 2H_2O$, 327–330

W

Waves, spin
  density, 418
  Heisenberg model, 67–77
Weiss field, 47–51